Barron's Review Course Series

Let's Review:

Biology—
The Living Environment

Sixth Edition

Gregory Scott Hunter, M.A.
has served New York State public education as

Superintendent of Schools
Chatham Central School District
Chatham, New York

Superintendent of Schools
Mexico Academy and Central School District
Mexico, New York

School Business Administrator and
Biology and General Science Teacher
Schodack Central School District
Castleton-On-Hudson, New York

Consultant
New York State Education Department
Bureaus of Science Education, Educational Testing,
and Curriculum Development
Albany, New York

All inquiries should be addressed to:
Barron's Educational Series, Inc.
250 Wireless Boulevard
Hauppauge, New York 11788
www.barronseduc.com

ISBN: 978-1-4380-0216-3

ISSN 1945–1768

10%
POST-CONSUMER WASTE
Paper contains a minimum of 10% post-consumer waste (PCW). Paper used in this book was derived from certified, sustainable forestlands.

PRINTED IN THE UNITED STATES OF AMERICA

9 8 7 6 5 4 3 2

Contents

Preface

TO THE STUDENT

For Which Course Can This Book Be Used?

This book is designed to be used as a review text for the New York State Regents course in The Living Environment. The material presented illustrates and complements the Core Curriculum for this course. Don't let the words "Core Curriculum" scare you. The Core Curriculum simply lists and describes what topics you must learn to complete this course. Because this book is comprehensive, it can be used to supplement any college-preparatory course in biology taught anywhere in the United States.

What Special Features Does This Book Have?

The topics in this book parallel those of the New York State Core Curriculum for The Living Environment. The Core Curriculum is reproduced in its entirety within the text of this book. All Key Ideas, Performance Indicators, and Major Understandings for Standards 1 and 4 of the Core Curriculum appear within the text. In addition, this edition contains information about the laboratory experiences required of all New York State biology students.

Each section contains the material considered to be most relevant to a Major Understanding. However, the contents of this book do not represent all the information you will be required to learn for this course. Requirements will vary among teachers and school districts. This book illustrates and explains the concepts contained in the Major Understandings. Students are expected to do more than simply memorize facts. You are also required to demonstrate understanding of scientific concepts. Your teacher may supplement material in the basic curriculum due to interest expressed by a class or a particular student. Feel free to ask your teacher about the areas of biology that interest you.

Practice question sets that mirror the style and content of the Regents examination appear in this book. The multiple-choice questions usually require that you use your knowledge of biology to make judgments and to select an appropriate choice from among a number of possible responses. Free-response questions ask you to construct single-word, complete-sentence, or essay answers that are understandable as well as scientifically accurate. Graphical analysis questions challenge you to organize, represent in graphic form, and draw inferences from experimental data.

Answers Explained for the practice multiple-choice questions provide an extensive analysis of each question, not simply an answer key. The explanations give you information about why choices are correct or incorrect. Sample free-response and essay answers guide you toward the standard required to perform successfully on a Regents examination in biology. Graphs and charts provide guidance for construction of responses that will meet new Regents standards.

An extensive glossary provides an alphabetical listing of terms commonly used in biology courses together with simple, easy-to-understand definitions. Many of these terms will appear in Regents examination questions or in your reading assignments.

Finally, two full-length Regents examinations are included. They will give you an opportunity to test your knowledge of biology and to practice your question-answering skills before taking the Regents examination.

Who Should Use This Book?

Any student enrolled in the New York State course on The Living Environment will find this book to be a valuable supplement to the regular class textbook. In addition, this book can be used to prepare for tests throughout the school year and for the year-end Regents examination on The Living Environment. It will help you do your best on the Regents examination by increasing your self-confidence in your test-taking abilities.

Students in any secondary-level biology course anywhere in the United States can use this book in much the same way. Its clear, concise style helps to clarify concepts. The absence of extraneous material enables you to concentrate on the learning standards without being bogged down in unnecessary details.

NEW YORK STATE LEARNING STANDARDS

Graduation Requirements

Several graduation or commencement standards are required of students in New York State public schools regarding their performance in math, science, and technology. The Living Environment Core curriculum addresses two of these standards as follows:

Standard 1: Students will use mathematical analysis, scientific inquiry, and engineering design, as appropriate, to pose questions, seek answers, and develop solutions.

Standard 4: Students will understand and apply scientific concepts, principles, and theories pertaining to the physical setting and living environment and recognize the historical development of ideas in science.

The Core Curriculum for The Living Environment was devised from these two commencement standards. The Core Curriculum is not a detailed outline, or syllabus, for biology. It does not prescribe what must be taught and learned in any particular classroom. Instead, it defines the skills and major understandings that you must master to achieve graduation standards for life science.

Key Ideas, Performance Indicators, and Major Understandings

Each topic within the graduation standards is broken down into a number of Key Ideas. Key Ideas are broad, general statements about what you as a student will be expected to know. Within Standard 1, three Key Ideas are concerned with laboratory investigation and data analysis. Within Standard 4, seven Key Ideas present a set of concepts that are central to the science of biology. The unifying principles develop your understanding of the essential characteristics of living things and how these characteristics allow them to be successful in diverse habitats.

Each Key Idea contains Performance Indicators. The Performance Indicators spell out what skills you should be able to demonstrate when you have mastered the Key Idea. Performance Indicators help you understand what is expected of you.

Performance Indicators are further subdivided into Major Understandings. The Major Understandings contain specific concepts that you must master in order to demonstrate the skills required by a Performance Indicator. The Major Understandings contain the material that will appear on your Regents exam.

Laboratory Component

A meaningful laboratory experience is essential to the success of any science course. You're expected to develop a good sense of how scientific inquiry is carried out by a professional scientist and how these same techniques can assist in the full understanding of scientific concepts. The Regents requires 1,200 minutes of laboratory experience coupled with satisfactory written reports of your findings.

The Living Environment Regents: Format and Scoring

The format of the Living Environment Regents is as follows:

Part A: 35 multiple-choice questions that test your knowledge of specific factual information. All questions in Part A must be answered.

Part B: Variable number of questions representing a mixture of multiple-choice and free-response items. Questions may be based on your direct knowledge of biology, interpretation of experimental data, analysis of readings in science, and the ability to deal with representations of biological phenomena. All questions must be answered in Part B.

Part C: Variable number of free-response questions. Questions may be based on your direct knowledge of biology, interpretation of experimental data, analysis of readings in science, and ability to deal with representations of biological phenomena. All questions must be answered in Part C.

Part D: Laboratory component of the Regents examination in which four required laboratory experiences are tested. See pages xvii and xviii for additional information on the Part D requirement. All questions must be answered in Part D.

HOW TO STUDY

You've spent the school year learning many different facts and concepts — far more than you could hope to remember from one review session. Your teacher has drilled you on these facts and concepts. You've done homework, taken quizzes and tests, and reviewed the material at intervals throughout the year. Now it's time to put it all together. The Regents exam may be only a few weeks away. If you and your teacher have planned properly, you will have finished the course material about a month before the exam. You now have to make efficient use of the days and weeks ahead to review all that you have learned and demonstrate your mastery of the material on the Regents.

It may seem like an impossible task, but you shouldn't be discouraged. You've actually retained much more of the year's material than you realize. The review process should be one that helps you to recall the many facts and concepts you have stored away in your memory. This book will help you to review this material as efficiently as possible.

You also have to get yourself in the right frame of mind. Being nervous and stressed during the review process won't help you one bit. The best way to avoid being stressed during an exam is to be well rested, prepared, and confident. This book is here to help you get prepared and build your confidence. So, now is the time to get started on the road to a successful exam experience. To begin, carefully read and follow the following steps:

1. *Get started!* Start your review early; don't wait until the last minute. Allow at least two weeks to prepare for the Regents exam. Set aside an hour or two a day over the next two weeks for your review. Less than an hour a day is not enough time for you to concentrate on the

material in a meaningful way; more than two hours a day will give diminishing returns on your investment of time.

2. *Get lost!* Find a quiet, comfortable place to study. You should seat yourself at a well-lit work surface free of clutter and in a room without distractions of any kind. (You may enjoy watching TV or listening to music curled up in a nice, soft chair, but these and other distractions should be avoided when studying.)

3. *Get prepared!* Make sure you have the tools you need to work, including this book, a pen and pencil, and some scratch paper for taking notes and doing calculations. Keep your class notebook at hand for looking up information between test-taking sessions. Have this book available for quick and efficient review of important concepts.

4. *Get focused!* Concentrate on the study material in the question sets in this book. Think about the questions that you review. Read carefully and thoughtfully. Try to make sense out of the questions. Think carefully about the answers that you choose. See the section "Using This Book for Study" for additional information about question-answering techniques.

5. *Get help!* Use available resources, including a dictionary and the glossary of this book, to look up the meanings of unfamiliar words in the practice questions. Remember that these same terms can appear on the Regents exam, so take the opportunity to learn them now.

6. *Go get 'em!* Remember—study requires time and effort. Your investment in study now will pay off when you take the Regents exam.

Using This Book for Study

This book will be an invaluable tool for you if used properly. Answer all of the questions in the question sets and practice exams even though the exam directions may allow you to choose which questions to answer. The more you study and practice, the better your chances of increasing your knowledge about biology and of obtaining a high grade on the Regents exam. To maximize your chances, use this book as follows.

1. Answer all of the questions in each question set. Check and correct your responses by using Answers Explained and Wrong Choices Explained at the back of the book. Follow this procedure for each question set. Study your areas of weakness. Then redo the questions you missed on the first round. Make sure you now understand fully what the question is asking and what makes one answer choice the correct answer.

2. When you've completed the questions in all the question sets, go on to the examination section. Take the exam under test conditions. Plan to answer all test questions; allow yourself no choices at this stage of the study process.

3. Test conditions should include the following:

   - Be well rested—get a good night's sleep before attempting any exam
   - Find a quiet, comfortable room in which to work
   - Allow no distractions of any kind
   - Have a pen, pencil, and some scratch paper handy
   - Set an alarm clock or watch for the three-hour exam limit

4. This step is very important. Take a deep breath, close your eyes for a moment, and RELAX! Tell yourself that you know this material. You have lots of time to take the Regents exam; use it to your advantage by reducing your stress level. Forget about your plans for later—your first priority right now is to do your best on the Regents exam, whether it is a practice exam in this book or the real thing.

5. Read all test directions carefully. Be sure you know how many questions you must answer to complete each part of the exam. If test questions relate to a reading passage, diagram, chart, or graph, be sure you fully understand the supplemental information before you attempt to answer the questions that relate to it.

6. When answering the multiple-choice questions on the Regents exam, TAKE YOUR TIME! Be sure to read the stem of the question very carefully; read it over several times. These questions are painstakingly written by the test preparers, and every word is chosen to convey a specific meaning. If you read the questions *carelessly*, you may answer a question that was never asked. Read each of the four choices carefully, using a pencil to mark the test booklet next to the answer you feel is correct.

7. Remember that three of the choices are incorrect; these incorrect choices are called distracters because they are selected by the test preparers to seem like plausible answers to poorly prepared or careless students. Make sure that you think clearly, using everything you have learned about biology since the beginning of the year, to eliminate each of the distracters as correct answers. This elimination process is just as important to your success on the The Living Environment Regents exam as knowing the correct answer. If more than one answer seems to be correct, reread the question to find the words that will help you to distinguish between the correct answer and the distracters. When you have made your best judgment about the correct answer, circle its number in pencil in your test booklet.

8. Free-response questions appear in a number of different forms. Students may be asked to select a term from a list, write the term on the answer sheet, and define the term. Students may be asked to describe some biological phenomenon or state a true fact in biology using a complete sentence. Students may be asked to read a value from a diagram of a measuring instrument and write that value in a blank on the answer sheet. Exercise care when answering this type of question—follow directions precisely. A complete sentence must contain a noun and a verb, must be punctuated, and must be written in an understandable way in addition to accurately answering the scientific part of the question. Values must be written clearly and accurately. They must include a unit of measure, if appropriate. Failure to follow the directions for a question may result in a loss of credit for that question.

9. The essay or paragraph question is a special type of free-response question. Typically, essay or paragraph questions provide an opportunity to earn multiple credits for answering the question correctly. As in the free-response questions described above, you must follow the directions for a question if you hope to earn the maximum number of credits for the question. Typically, the question will outline exactly what must be included in your essay to gain full credit. Follow these directions step by step, double-checking to be certain that all question components are addressed in your answer. In addition, your essay or paragraph should follow the rules of good grammar and good communication so that it is readable and understandable. Of course, it should contain correct information and answer all parts of the question asked.

10. Graphs and charts are used in a special type of question that requires the student to organize and represent data in graphic format. Typically, for such questions you are expected to place unorganized data into ascending order in a data chart or table. You may also be asked to plot unorganized data on a graph grid, connect the plotted points, and label the graph axes appropriately. Finally, questions regarding data trends and extrapolated projections may be asked, requiring you to analyze the data in the graph and draw inferences from it. As with all examination questions, always follow all directions for the question. Credit can be granted only for correctly followed directions and accurate interpretation of the data.

11. When you've completed the exam, relax for a moment. Check your time; have you used the entire three hours? Probably not. Resist the urge to quit. Go back to the beginning of the exam and, in the time remaining, **retake the exam in its entirety**. Try to ignore the penciled notations you made the first time through the exam. If you come up with a different answer the second time through, *stop* and read over the question with extreme care before deciding which is the correct

response. Once you've decided on the correct response, finalize the answer in ink in the answer booklet.

12. Score the exam using the answer key at the end of the exam. Review the Answers Explained section of *Barron's Regents Exams and Answers—The Living Environment* for each question to aid your understanding of the exam and material. Remember that understanding why an answer is incorrect is just as important as understanding why an answer is correct.

13. Finally, focus your between-exam study on your areas of weakness in order to improve your performance on the next practice exam. Complete all the practice exams using the preceding techniques.

Test-Taking Techniques

The following is a summary of test-taking techniques you should follow when taking the actual exam.

1. Complete your study and review at least one day before taking this examination. Last minute cramming may actually hurt, rather than help, your performance on the exam.

2. Be well rested the day of the exam. Get a good night's sleep before taking any examination.

3. Bring two pens, two pencils, and an eraser to the exam. If your school requires it, bring some form of identification with you, as well. Wear a watch, or sit where you see a clock. Before entering the room, remember that you will remain for the entire three-hour examination period.

4. Be familiar with the format of the examination. You must answer all questions in Part A (multiple-choice questions), all questions in Part B (mixed-format questions), all questions in Part C (free-response questions), and all questions in Part D (mixed format questions).

5. Before beginning the exam, take a deep breath, close your eyes for a moment, and RELAX. Use this technique any time you feel yourself tensing up during the exam.

6. Read all exam directions carefully. Be sure you fully understand supplemental information (reading passages, charts, diagrams, graphs) before you attempt to answer the questions that relate to it.

7. When answering questions on the Regents exam, TAKE YOUR TIME. Be sure to read the stem of multiple-choice questions very carefully. Read each of the four answer choices carefully, as well.

With a pencil, make a mark in the test booklet next to the answer you feel is correct. If you're temporarily stumped by a question, put a check mark next to it and go on to the next question. Come back to the question later, when your mind is clear.

8. Remember that three of the multiple-choice answers are incorrect (known as distracters). If more than one answer seems to be correct, reread the question to find the words that will help you distinguish between the correct answer and the distracters. When you have made your best judgment about the correct answer, circle its number in pencil on your answer sheet.

9. Be certain to follow directions when completing free-response questions. Use complete sentences whenever they're required.

10. Complete charts and graphs according to instructions. If you're instructed to circle some part of an answer, be sure to circle it. If the directions call for you to connect points or list items in ascending order, then be certain to do so.

11. Answer essay or paragraph questions according to instructions. Be sure to include correct information that addresses every required point. Doing so will help to ensure that you receive full credit for your answers.

12. When you have completed the exam, relax for a moment. Go back to the beginning and, in the time remaining, **retake the exam in its entirety**. Once you have decided on the correct response, finalize the answer by marking an X in *ink* through the penciled circle on the answer sheet for multiple-choice questions. Be sure to finalize all of the answers for all parts of the exam.

13. Be certain to sign the declaration on your answer sheet. Unless this declaration is signed, your exam cannot be scored.

TIPS FOR TEACHERS

For teachers and administrators in New York State public schools, this book will provide an excellent source of primary instructional material for locally developed curricula for The Living Environment. The examples given and the factual material presented represent a reasonable body of knowledge to complement the commencement standards for life science. In addition, it will provide a ready source of review material to prepare students for the New York State Regents examination on The Living Environment.

All teachers will be able to use this book with their students as a companion to their regular textbooks and will find that students will gain considerable self-confidence and facility in test taking through its consistent use. Some school systems may wish to use this book as the primary text for their courses in college-preparatory biology; others may wish to employ it as the review text for test preparation.

Local Curricula

Teachers and administrators need to develop local curricula that complement the Core Curriculum, promulgated by the State in 1999. It is up to the teacher/administrator to decide what examples and factual knowledge will best illustrate the concepts presented in the Core Curriculum, what concepts need to be reinforced and enhanced, what experiences will add measurably to the students' understanding of science, and what examples of local interest should be included. The teacher will immediately recognize the need to go beyond this level in the classroom, with examples, specific content, and laboratory experiences that complement and illuminate these Major Understandings. It is on this level that the locally developed curriculum is essential. Each school system is challenged to develop an articulated K-12 curriculum in mathematics, science, and technology that will position students to achieve a passing standard at the elementary and intermediate levels, such that success is maximized at the commencement level.

The addition of factual content must be accomplished without contradicting the central philosophy of the learning standards. If local curricula merely revert to the fact-filled syllabi of the past, then little will have been accomplished in the standards movement other than to add yet another layer of content and requirements on the heads of students. A balance must be struck between the desire to build students' ability to think and analyze and the desire to add to the content they are expected to master.

ESSENTIAL QUESTIONS

A key component of a thoughtfully developed local curriculum is the development of Essential Questions. Essential Questions state, in interrogative form, the key concepts we expect our students to master. Posed correctly by the teacher, they provide a pathway for students to follow as they pursue their study of the subject matter, in this case, the subject matter of biology. Essential Questions also help students answer the questions "Why do I have to know this material? What will I be able to do with what I learn? How will knowing this help me to achieve my life goals? How will having this skill contribute to my growth as a person and a citizen?"

Teachers should be skilled in the art of developing Essential Questions, which serve as a guide not only to students but also to teachers. Essential Questions should be used by teachers in the preparation of class activities and of student performance assessments.

A feature of this book is the inclusion of Essential Questions at key points throughout the text. These Essential Questions organize and frame the concepts to be understood and the skills to be mastered through the student's study of biology. Likewise, they focus the concepts to better enable the teacher to develop class activities and assessments that truly add to the student's experience in measurable ways. They are meant to stimulate the teacher to develop his or her own Essential Questions so as to make the local curriculum come alive and provide an opportunity for student engagement in the learning process.

Included below is a listing of Essential Questions included in this book as samples to be considered by the teacher in delivering the Living Environment local curriculum. These questions are meant to be samples only, to help stimulate the teachers' thinking as they develop their own Essential Questions to assist in student learning.

Essential Questions:

Standard 1:

- Why is science important?
- How is good science done?

Key Idea 1.1

- How do models help scientists understand natural phenomena?

Performance Indicators:

1.1.1 How does expressing an idea in visual or mathematical terms assist in our understanding?

1.1.2 What can we learn from studying the work of others?

1.1.3 How can we tell the difference between good information and bad information?

1.1.4 How can data be effectively represented in a manner that assists in our understanding of science?

Key Idea 1.2

- What are the steps in the scientific method?
- How do these steps ensure that science is valid?

Performance Indicators:

1.2.1 How do questions and observations help scientists to conduct good experiments?

1.2.2 How do scientists use the work of other scientists to help them in their own work?

1.2.3 What is a hypothesis and how does it help the scientist to focus an experiment?

1.2.4 What techniques and tools should be selected when designing an experiment?

Key Idea 1.3

- What are the different ways that scientists analyze scientific data?

Performance Indicators:

1.3.1 How can tables, charts, and graphs of data help to illustrate the findings of a scientific inquiry?

1.3.2 How do scientists measure the accuracy of their own experimental results?

1.3.3 How does a scientist draw inferences and conclusions based on data?

1.3.4 How does one experiment lead to another once results are known?

1.3.5 What are the appropriate ways to present the findings of a scientific investigation?

Standard 4:

- What are the principles that guide scientists in their study of the natural world?

- How is modern science built on the work of scientists of the past?

Key Idea 4.1

- How are living things similar to and different from other living things and nonliving things?

Performance Indicators

4.1.1 How does the great variety of living species impact environmental stability?

4.1.2 How are humans adapted to survive in their environment?

4.1.3 How are unicellular life forms able to carry out the life processes needed to sustain life?

Key Idea 4.2

- How does genetic inheritance help ensure the continuity of life?

Performance Indicators

4.2.1 How does DNA control the inheritance of genetic traits in living things?

4.2.2 What effect will genetic engineering have on the future of genetic inheritance?

What cautions should be considered in the use of genetic engineering?

Key Idea 4.3

- What has caused the wide variety of living things that have inhabited Earth over the past 3.5 billion years?

Performance Indicator

4.3.1 How do new species arise from previously existing species?

Key Idea 4.4

- What mechanisms are at work to create and perpetuate life on Earth?

Performance Indicator

4.4.1 How do living things produce more living things of the same kind?

Key Idea 4.5

- How do living things create a balanced steady state that maintains the living condition?

Performance Indicators

4.5.1 What biochemical reactions are common in living cells and promote the living condition?

4.5.2 How does the body defend itself against disease?

What happens when a living organism gets sick?

4.5.3 How do the activities that sustain the life of a cell contribute to the survival of a multicellular organism?

Key Idea 4.6

- How do species of different kinds depend on each other and contribute to each other's survival?

- How do conditions in the nonliving environment contribute to the survival of plant/animal communities?

Performance Indicators

4.6.1 What forces of nature manage the balance of species populations?

4.6.2 Why is it important to protect endangered/threatened species and habitats?

4.6.3 How do Earth's environments affect, and how are they affected by, the living community?

Key Idea 4.7

- What is the proper role of the human species in the ecological community?

- What responsibility does the human species have to protect the natural ecosystem?

- What trade-offs must be considered when making economic and political decisions that affect the environment?

Performance Indicators

4.7.1 How do humans positively and negatively affect the ecosystem?

4.7.2 How has human technology and the growth of the human population affected the quality of life for other species?

4.7.3 How can humans take positive action to reverse the degradation of the natural environment caused by human technology?

Lab Needs

A positive benefit of the reduction of factual detail in the Core Curriculum is that it should allow more in-depth treatment of laboratory investigations to be planned and carried out than was possible under the previous syllabus. Laboratory experiences should be designed to address Standard 1 (inquiry techniques) but should also take into account Standards 2 (information systems), 6 (interconnectedness of content), and 7 (problem-solving approaches). They should also address the laboratory skills listed in Appendix A of the Core Curriculum and of this book.

It is important to recognize that this and other assessment tools based on the learning standards may undergo transition in the first few years of implementation. Teachers and students should remain alert to the possibility of changes in the format and scoring of the Regents examination.

New York State is a participant in a multi-state effort to develop "Common Core Standards" in biology. These standards are scheduled to be implemented in the near future. When implemented, these standards will require some changes in the way that biology is taught and tested.

UNIT ONE

Scientific Inquiry

STANDARD 1

Students will use mathematical analysis, scientific inquiry, and engineering design, as appropriate, to pose questions, seek answers, and develop solutions.

Essential Questions:
- *Why is science important?*
- *How is good science done?*

Good science is a logical, yet creative, process that applies human intelligence to discovering how the world and the universe work. It depends as much on the published reports of past scientific work as on the ongoing experimental method used to create new knowledge. Scientists continually question the meaning of scientific discoveries and develop new theories based on combining old knowledge with new knowledge. In this way, there is a continual turnover of knowledge that serves the best interests of all people everywhere.

Discovering Truths

Understanding scientific inquiry and discovery is important because the truths that emerge from this discovery can and should be used in everyday social and ethical decision making. These decisions often involve issues of personal and community health, economic considerations, and technological claims. To fail to develop basic scientific literacy is to place oneself and one's community at great potential risk since proposed changes may carry significant negative consequences to any or all of these items.

I. EXPLANATIONS OF NATURAL PHENOMENA

> ## KEY IDEA 1—PURPOSE OF SCIENTIFIC INQUIRY
> The central purpose of scientific inquiry is to develop explanations of natural phenomena in a continuing and creative process.

Essential Question:

- *How do models help scientists understand natural phenomena?*

Scientific inquiry is used as a tool to aid scientists in the study of the natural world. Scientists ask questions, pose hypotheses, design experiments to test these hypotheses, collect and organize data, analyze collected data, draw inferences, and develop conclusions. Major scientific principles are developed from the combined knowledge that results from such experimentation. Scientific inquiry is a creative process that has been ongoing since humans have been on the Earth and will continue long into the future. We are only beginning to unlock the many secrets of the natural world.

Following Procedures

Scientific inquiry depends on an objective application of certain procedures. Some of these include modeling, research, reconciliation of opposing viewpoints, and presentation of explanations. The following sections will discuss each of these processes.

Modeling

Essential Question:

- *How does expressing an idea in visual or mathematical terms assist in our understanding?*

Performance Indicator 1.1.1 *The student should be able to elaborate on basic scientific and personal explanations of natural phenomena and develop extended visual models and mathematical formulations to represent one's thinking.*

Modeling is a way of putting complex scientific data into understandable terms. These models may be mathematical models, schematic models, or physical models. Models frequently use easy references to help you understand complex principles and relationships.

Scientific explanations are built by combining evidence that can be observed with what people already know about the world. Scientific explanations of natural phenomena come from many different sources, including scientific and personal observations. The modern scientist builds on the work of many earlier scientists who asked questions, developed hypotheses, designed experiments, made observations, analyzed experimental results, and drew inferences and conclusions. These scientists also wrote about their experiments and results, creating vast amounts of scientific literature. Scientists often compare their results with the results of other scientists performing the same or similar experiments.

In many cases, these scientists developed models to better understand and to help others understand their thinking. A biochemist, for example, may develop a model of a biochemical pathway in the form of a series of formulas that represent different steps in that pathway. An ecologist may model a complex set of nutritional relationships in the form of a food web or pyramid. A geneticist may use three-dimensional models to illustrate how the shape of a molecule of DNA helps to predict how it will carry out its genetic control. A cytologist may find that a physical model of a cell helps to explain the movement of materials through the cell membrane. In each of these examples, a model was developed to help explain an abstract scientific concept. During your study of biology, you will be expected to be able to develop models to illustrate what you learn.

Decision Making

An important way to enlarge one's scientific knowledge is to use it to make sound decisions. Science provides knowledge, but values are also essential to making effective and ethical decisions about the application of scientific knowledge. You need to understand scientific knowledge and its implications for social action. Technological advance can impact the lives of individuals, communities, and civilizations. Therefore, the decision must be made whether or not an advance should be introduced, controlled, or set aside.

As a responsible individual, you have an obligation to use your knowledge of scientific principles to make such judgments and act accordingly. You have an ethical and moral responsibility to exercise sound judgment based on a solid understanding of scientific fact.

Research

Essential Question:

• *What can we learn from studying the work of others?*

Performance Indicator 1.1.2 *The student should be able to hone ideas through reasoning, library research, and discussion with others, including experts. Humans sharpen their intellect by reasoning, reading the ideas of* *others, and discussing topics with people whose ideas may differ from their own.*

By starting with the knowledge base of past generations, we develop ever greater understanding of the world around us. We constantly seek to understand natural phenomena in greater and greater detail. This is possible only by applying the scientific method of inquiry.

Inquiry involves asking questions and locating, interpreting, and processing information from a variety of sources. Scientists do not rely solely on the results of their own experiments when developing scientific principles. To broaden their knowledge, they depend heavily on a large body of **scientific literature** created from centuries of experimentation, thinking, and discussion. Each new generation of scientists adds to this body of knowledge and moves forward with the discovery of scientific truths.

Reconciliation of Opposing Viewpoints

Essential Question:

- *How can we tell the difference between good information and bad information.*

Performance Indicator 1.1.3   *The student should be able to work toward reconciling competing explanations and clarify points of agreement and disagreement.*

Inquiry involves making judgments about the reliability of the source and relevance of information. Not all material written about science is reliable or accurate. In some cases, such material may be misleading or biased and may unfairly represent the data in order to support a particular point of view. Good scientific writing involves complete explanation of the experimental method used, a clear presentation of the data, and an objective analysis of the data with respect to other, prior experimentation.

In order to be useful to science, the results of these reported experiments have to be **reproducible** by other scientists using the same experimental methods. Experiments that are not reproducible are considered unreliable and are usually not considered in this process. A good scientist also needs to be able to sort through scientific literature and discard material that does not appear to be relevant to the inquiry at hand. Relevant material should be reviewed and summarized as an introduction to the new inquiry, even though it may suggest conclusions that differ from the scientist's hypothesis.

Scientific explanations are accepted when they are consistent with experimental and observational evidence and when they lead to accurate predictions. A scientific principle, if it is to gain widespread acceptance, must enable scientists to make **accurate predictions** of how the natural world operates under a given set of conditions. The conclusions drawn from experimental results must accurately reflect an objective analysis of the data gathered from such experiments. These conclusions must predict the outcome of similar experimentation with reasonable accuracy in order to be considered valid. When they do not do so, conclusions are thrown into question until further investigation is conducted to determine their validity.

All **scientific explanations are tentative** and subject to change or improvement. Each new bit of evidence can create more questions than it answers. This increasingly leads to better understanding of how things work in the living world. Scientists are always at work to add to the body of knowledge we call science. This new knowledge often supports, but sometimes contradicts, previously held understandings. When contradictions occur, further scientific work is needed to determine where the truth lies.

Scientists do not always agree on the interpretation of scientific data. Divergent ideas and attitudes, as well as diametrically opposed conclusions and theories, also exist. Far from impeding the development of scientific thinking, these divergent concepts are used to stimulate new thinking and experimentation in a never-ending quest for the truths behind the theories. Ultimately, some theories are proven incorrect. However, the process of continual questioning and inquiry is responsible for moving science in a positive direction.

Presentation of Explanations

Essential Question:

- *How can data be effectively represented in a manner that assists in our understanding of science?*

Performance Indicator 1.1.4 *The student should be able to coordinate explanations at different levels of scale, points of focus, and degrees of complexity and specificity, and recognize the need for such alternative representations of the natural world.*

The information drawn from scientific investigation is discussed and analyzed from different points of view. By changing the way they look at the data, scientists begin to see patterns and trends not immediately apparent from a one-dimensional analysis.

When data collected from an experiment is represented graphically, the **scale** used to represent the dependent variable may be large or small. A large scale includes a wide range of possible data points. A small scale uses fewer possible data points. When a large scale is used, small variations in the data may not be easily seen. By decreasing the scale of the graph, the range of data represented is reduced so that small variations can be seen more easily. Analytical methods may involve decreasing the scale of the data to amplify small changes, or increasing it to examine general trends.

Looking at a broad range of data or focusing on specific components of the data may isolate the specific change being measured. The **focus area** may be the point at which the experimental variable has the most significant effect on the dependent variable. By focusing on this narrow range of data, the researcher can learn much about the changes that occur within this range. The scientist may then design a new experiment that breaks the experimental variable into smaller increments in order to determine more precisely how small changes in the experimental variable work to affect the dependent variable.

Changing the way data is viewed may involve **describing the data** in new ways to highlight a researcher's interpretation of the results of one or more scientific studies. Many concepts in the life sciences are extremely complex, with multiple variables operating at once to produce conditions that effect the changes noted by the researcher. It is often the job of the researcher to reduce this complexity for the reader by increasing the specificity of the study. By dealing with a **single variable** at a time, the researcher determines the effect of each variable independent of the others. When the isolated effect of each variable is understood, the reader can then comprehend how multiple environmental conditions affect the dependent variable in complex ways.

Well-accepted theories are the ones that are supported by different kinds of scientific investigations often involving the contributions of individuals from different disciplines. These varying methods of interpretation help to illuminate the data in new ways and to stimulate still more ideas about its meaning. Well-accepted theories are those **supported by investigations** involving different scientific disciplines (for example, biochemistry, cytology, genetics). As more and more evidence mounts supporting a theory, the more credibility it gains with the scientific community. Each discipline reports its findings independently of the others so that independent corroboration is achieved. In this way, an inference becomes a hypothesis, which then may develop into an accepted theory, and finally it becomes established scientific principle.

II. TESTING OF PROPOSED EXPLANATIONS

> ## KEY IDEA 2—METHODS OF SCIENTIFIC INQUIRY
> Beyond the use of reasoning and consensus, scientific inquiry involves the testing of proposed explanations involving the use of conventional techniques and procedures that usually require considerable ingenuity.

Essential Questions:

- *What are the steps in the scientific method?*
- *How do these steps help to ensure that science is valid?*

Scientific inquiry involves several well-defined procedural steps known collectively as the **scientific method**. These steps include several elements, the first five of which are discussed in the following section:

1. Pose a scientific question based on observation of the natural world.
2. Review available scientific literature relating to the scientific question.
3. Develop a hypothesis about the proposed answer to the scientific question.
4. Design an appropriate experimental procedure to test the hypothesis.
5. Collect and record data resulting from the experimental procedure.

By following this design, the researcher defines his/her goals prior to commencing an experiment. Doing so helps the researcher to focus on the problem at hand. The researcher can then conduct the experiment in a manner that will generate useful data. This data can then be analyzed to yield new knowledge about how the natural world operates.

Question and Observation

Essential Question:

- *How do questions and observations help scientists to conduct good experiments?*

Performance Indicator 1.2.1   *The student should be able to devise ways of making observations to test proposed explanations.*

Scientific inquiry, as its name implies, always begins with the fundamental step of asking a question. A nonscientist may be curious about something he/she observes in nature and ask a question something like "How do leaves turn color in the fall?" or "Why do fruit flies have red eyes?" While questions such as these provide a starting point for inquiry, they do not lend themselves to easy and understandable answers. They do not point to aspects of the observed phenomena that can be measured. They are focused more on reasons for the phenomena than on their observable and measurable characteristics.

By contrast, a **scientific question** is always asked in terms of observable qualities and measurable quantities. Scientific questions might ask "What pigments are found in the leaves of sugar maple trees?" or "What is the inheritance pattern of red eyes in fruit flies?" Questions such as these lead the researcher toward established laboratory methods. For instance, chromatography (separating pigments of differing molecular characteristics) and genetic crosses (studying the inheritance of genetic traits) provide clues to the broader questions posed above.

These questions, even though less general, can be broken down still further into even more specific, measurable questions. For example, "What pigment is most abundant in sugar maple leaves?" "How does temperature affect the production of chlorophyll in sugar maple leaves?" "What is the relative rate of destruction of the leaf pigments chlorophyll and anthocyanin under constant, intense white light?" and "At what temperature does chlorophyll function most effectively? Least effectively?" The researcher uses these, more focused, questions to help point the way toward research studies. These studies should produce data that lead to new knowledge about natural phenomena.

The simplest method for testing is direct **field observation** of the experimental subject. Such observations may be conducted informally but by taking careful notes. Later, more formal and **controlled observations** may be made as a part of an experimental procedure. Such observations are always made objectively, accurately, and in great detail.

Conventional **laboratory techniques** may take on various characteristics depending on what idea is being tested and what variables are being measured. For example, an investigation into the anatomic characteristics of an earthworm should be conducted using dissection tools. An experiment designed to discover the different pigments present in a maple leaf should employ techniques of chromatography. An inquiry into the genetics of the fruit fly should include the tools and techniques necessary to isolate, mate, and count the offspring of fruit flies displaying specific genetic characteristics.

Review of Literature

Essential Question:

- *How do scientists use the work of other scientists to help them in their own work?*

Performance Indicator 1.2.2   *The student should be able to refine research ideas through library investigations, including electronic information retrieval and reviews of the literature, and by peer feedback obtained from review and discussion.*

A vast body of **scientific information** is available on a wide range of scientific topics. This information is available in books, professional journal articles, research databases, and other reference sources. This literature may contain the findings of other scientists who have conducted similar studies in the past. It may provide good clues to the researcher as to the likely outcome of his/her investigation. Pertinent sources in the scientific literature are normally cited in the introductory material of a report on the experiment at hand.

Increasingly, these resources are available though networked computer databases. You should possess the skills needed to search the Internet to locate reliable research information and to **discriminate among sources** based on their authenticity. This may require focusing on the archived databases of universities and research laboratories while excluding data sources that do not carry the endorsement of such institutions. You should recognize that posting information on the Internet does not require the editorial scrutiny of competent scientists that is necessary to publish in a professional journal. For this reason, exercise care when using the Internet for conducting literature reviews since the data and analyses presented may be inaccurate, misleading, or biased.

Development of a **research plan** involves researching background information and understanding the major concepts in the area being investigated. Recommendations for methodologies, use of technologies, proper equipment, and safety precautions should also be included. Such sources provide information about the appropriate methods and techniques to be used in conducting research in a particular area of scientific inquiry. They also inform the reader about what equipment and **safety precautions** should be used in such investigations. Having this information available in the literature saves valuable time since these techniques and precautions do not have to be discovered by each research scientist conducting a study in a similar area of inquiry. Having this information available also assists the researcher in developing a formal hypothesis.

Finally, being able to review and summarize information gained from such sources and to discuss it with other researchers engaged in the same sorts of investigations is an important skill. **Peer feedback** can benefit all participants since the shared information and ideas provide everyone with a better understanding of the phenomena being studied. Collaborative sharing is also an important skill that you, as a student, should learn.

Hypothesis and Research Plan

Essential Question:

- *What is a hypothesis and how does it help the scientist to focus an experiment?*

Performance Indicator 1.2.3   *The student should be able to develop and present proposals including formal hypotheses to test explanations, that is, predict what should be observed under specific conditions if the explanation is true.*

Hypotheses are predictions based upon both research and observation. A **hypothesis** is an educated guess made by a researcher about the likely outcome of an investigation. It is a prediction of what should be observed in an experiment run under a specific set of conditions. Hypotheses are based upon both research and observation.

Hypotheses are widely used in science for determining what data to collect and as a guide for interpreting the data. The hypothesis serves as a guide. It assists the researcher in designing the experiment and interpreting the data derived from that experiment. For example, a hypothesis concerning the effect of temperature on bacterial growth might state, "*Escherichia coli* bacteria will grow best at human body temperature." This hypothesis suggests that a series of identical experimental setups be created that provide a nutrient medium to a single species of bacterium. It further suggests that these setups be placed under a range of temperatures including temperatures below, at, and above human body temperature. Finally, it suggests that bacterial growth be measured in some way that will give the researcher a clear, objective understanding of differential growth rates.

If the researcher has interpreted the literature appropriately, the hypothesis will often be correct, and the results of the experiment will be as predicted. However, the experimental results may not match the hypothesis. When this occurs, it is *not* viewed as a failure by the researcher but as a means to suggest more probing questions that get at the real relationships among the variables in the experiment.

Experimentation

Essential Question:

- *What tools and techniques should be selected when designing an experiment?*

Performance Indicator 1.2.4 *The student should be able to carry out a research plan for testing explanations, including selecting and developing techniques, acquiring and building apparatus, and recording observations as necessary.*

Carrying out a research plan requires a **procedure**, or set of steps. These steps are laid out within the plan and are specific to the experimental question being asked.

Development of a **research plan** for testing a hypothesis requires planning to avoid bias (for example, repeated trials, large sample size, and objective data-collection techniques). A research plan is developed in such a way as to provide objective data that will help to answer the experimental question without introducing bias into the experiment. **Bias** is the tendency to prejudice the results of an experiment with one's own assumptions about the likelihood of its outcome. While developing a hypothesis is an important step in scientific research, the researcher must not let that hypothesis color his/her objective interpretation of the data. To do otherwise would be to introduce bias into the experiment. To ensure objective results, researchers use techniques such as repeated trials, large sample size, adequate controls, and objective data collection techniques.

The **equipment** being used, the **method** being followed, and the means of **recording and analyzing data** must be appropriate to answering that question. For example, an experiment to determine the structural similarities of five species of worm would probably require dissection equipment and a dissecting microscope. It would involve studying and collecting data about the external and internal structures of a reasonable number of representatives of each species. An experiment on the effects of auxins on geranium stem growth, on the other hand, would require several geranium plants, various concentrations of auxins, a constant light source, and instruments to measure stem elongation and rate of bending. Each experiment has its own unique setup and design. Following is a description of some of the commonly used techniques and tools used by biologists when performing scientific investigations.

The Compound Light Microscope

The compound light microscope, developed hundreds of years ago by early scientists such as Anton von Leeuwenhoek and Robert Hooke, still represents a major tool of cell study used by biologists all over the world. Compound light microscopes of the type used in most school laboratories produce magnifications of 50× to 500× and are suitable for viewing whole cells and large organelles such as nuclei and chloroplasts. Since this tool is used commonly in the biology classroom, it is important that you understand its parts and workings thoroughly. The critical parts of the microscope include:

- **Eyepiece lens** (ocular) is the lens closest to the eye during study.
- **Objective lens** is the lens closest to the object of study.
- **Coarse-adjustment knob** is used to make large changes in the focus of the microscope; fine adjustment knob is used for small changes.
- **Light source** provides light, which must pass through the object in order to make it visible through the microscope.
- **Diaphragm** is used to make changes in the quality and amount of light passing through the object and entering the objective lens.

Compound Light Microscope

The compound light microscope provides two major advantages to the scientist studying the cell structure of living things:

- **Magnification** is the quality of the compound light microscope that makes the image of an object appear larger than the object itself. The magnification of a compound light microscope is determined by multiplying the power of the ocular by the power of the objective lens (for example, 10× ocular × 40× objective = 400× total magnification).
- **Resolution** is the quality of the compound light microscope that makes it possible to see two or more objects that are very close together as separate objects.

Use of the microscope is an important skill best learned in the laboratory. Practical laboratory experience is the best way to master the skills involved in preparing wet mount slides, focusing the microscope, and making observations of biological specimens. Students may wish to attempt to answer the practice questions below in conjunction with such practical laboratory experience. These questions are provided here and in the Laboratory Skills section to illustrate the kind of question that may be asked on the Regents examination. See the Laboratory Skills section for additional information.

Scientific Measurement

The researcher must also have a good understanding of the various units of measurement and means of data collection used in each type of research study. The **metric system** is used universally in scientific measurement. Knowledge of the relative sizes of various metric measures is important to accurate data collection. For example, the units of linear measure include the **meter** (m), the **centimeter** (cm—1/100 of a meter), **millimeter** (mm—1/1,000 of a meter), and **micrometer** (μm—1/1,000,000 of a meter). The units of volume include the **liter** (l) and **milliliter** (mL). The metric unit of temperature is the **degree** (°C). Metric units of mass include the **gram** (g), **kilogram** (kg), and **milligram** (mg).

1mm = 1,000 μm   Ruler Edge

Cell Size

Typical Low-Power Field

← 15 Micrometers →

Cell Measurement

The Micrometer

Understanding the use of the compound microscope and the units of measurement for microscopic objects is very important for conducting research in the biological sciences. The unit of linear measurement used to measure cells and their organelles is the **micrometer (μm)**. Micrometers are so small that 1,000 of them can fit inside 1 millimeter, and a million of them are required to equal 1 meter. Cells typically are found to have diameters of between 10 and 50 micrometers. If the diameter of the microscope field of view is known, it can be used to estimate the size of objects under it. The low-power field can be measured directly by placing a metric ruler under the low-power field of view. Most low-power fields have diameters of between 1,200 and 1,600 micrometers (1.2 and 1.6 millimeters). The diameter of a high-power field can be determined mathematically using a simple proportion as follows:

$$\frac{\text{Low-Power Magnification}}{\text{High-Power Magnification}} = \frac{\text{High-Power Field Diameter}}{\text{Low-Power Field Diameter}}$$

To illustrate the use of this proportion, assume you are using a microscope with a 100× low-power magnification and a 400× high-power magnification. Direct measurement shows that the low-power field diameter is 1,200 micrometers (1.2 mm). The diameter of the high-power field is determined by substituting the known values into the formula and solving algebraically for the unknown value as follows.

$$\frac{100\times}{400\times} = \frac{\text{High-Power Field Diameter}}{1,200 \ \mu\text{m}}$$

(400×) (High-Power Field Diameter) = (100×) (1,200 μm)
High-Power Field Diameter = 120,000/400 μm
High-Power Field Diameter = 300 μm

Indicators, Stains, Tools, and Safety

The use of indicators and stains is also important for working in this field, as is a basic knowledge of the structure of the cell. Indicators are used to help determine the chemical composition of materials used in the laboratory. Common indicators include **litmus** (acid/base indicator), **Benedict's solution** (simple sugar indicator), **Lugol's iodine** (starch indicator), and **bromthymol blue** (carbon dioxide indicator). Stains are used to increase the contrast of cell organelles so that they can be viewed under the microscope more easily. See the Laboratory Skills section for additional information.

Being able to use and accurately read a variety of different measuring tools, including metric rulers, thermometers, balances, and graduated cylinders, is important to good data collection. The techniques for using these measuring tools are best learned in the laboratory, where demonstration and practice will provide valuable teaching and learning opportunities. Students may wish to attempt to answer the practice questions that follow in conjunction with such practical laboratory experience. See the Laboratory Skills section for additional information.

Finally, a good understanding of **laboratory safety** is essential for conduction laboratory work in any area of study. Like the use of measuring tools, laboratory safety is something that is best learned in practice. A few simple safety rules should always be followed, however. A sampling is shown below. See the Laboratory Skills section for additional information.

- Always follow instructions given to you by your instructor and by your laboratory manual.
- Never use laboratory equipment that is damaged or with which you are unfamiliar.
- Never endanger yourself or others with dangerous horseplay in the laboratory.
- Never touch, taste, or smell objects, chemicals, or solutions unless you know what they are and know they are safe.
- Always wear protective eyewear when working with chemicals, hot liquids, or sharp tools.

QUESTION SET 1.1—EXPLANATIONS OF NATURAL PHENOMENA (ANSWERS EXPLAINED, P. 251)

1. A biologist plans to spend a year investigating the mating behavior of a certain species of frog. To make meaningful observations, the biologist should observe
   (1) a small number of frogs in their natural habitat
   (2) a large number of frogs in their natural habitat
   (3) several groups of frogs maintained in different temperatures in the laboratory
   (4) several groups of frogs maintained on different diets in the laboratory

2. In the diagram below, letter A represents the starting volume of liquid in a graduated cylinder. Letter B represents the volume after 8 milliliters of this liquid was removed.

   This information indicates that the scale on this graduated cylinder is in milliliter increments of
   (1) 1   (3) 8
   (2) 2   (4) 4

3. Which sentence represents a hypothesis?
   (1) Environmental conditions affect germination.
   (2) Boil 100 milliliters of water, let it cool, and then add 10 seeds to the water.
   (3) Is water depth in a lake related to available light in the water?
   (4) A lamp, two beakers, and elodea plants are selected for the investigation.

4. Zebra finches are small black-and-white birds that lay eggs about the size of a bean seed. Which unit of measurement is best for accurately measuring the length of these eggs?
   (1) millimeters   (3) feet
   (2) micrometers   (4) meters

5. Each division of the metric ruler shown in the diagram below equals 1 millimeter.

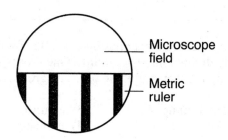

The diameter of the field of vision is approximately
   (1) 2,800 μm   (3) 4,400 μm
   (2) 3,700 μm   (4) 4,700 μm

6. How does a control setup in an experiment differ from the other setups in the same experiment?
   (1) It tests a different hypothesis.
   (2) It has more variables.
   (3) It differs in the one variable being tested.
   (4) It utilizes a different method of data collection.

7. When a test tube of water containing elodea (an aquatic plant) is placed near a bright light, the plant gives off gas bubbles. When the light is placed at different distances from the plant, the rate of bubbling is affected. The experimental variable in this demonstration is the
   (1) concentration of gas in the water
   (2) type of aquatic plant in the test tube
   (3) amount of water in the test tube
   (4) distance of the plant from the light

8. Which procedure would be part of a laboratory investigation designed to determine if a specific nutrient is present in a food?
   (1) Test a moist sample of the food with pH paper.
   (2) Add Lugol's iodine solution to a sample of the food.
   (3) Place a sample of the food into a test tube containing methylene blue.
   (4) Add bromthymol blue to a sample of the food.

9. What is the first step of a scientific investigation?
   (1) Perform the experiment.
   (2) Analyze the experimental data.
   (3) Formulate a hypothesis.
   (4) State the problem.

10. By using one or more complete sentences, state one safety procedure that a student should follow when dissecting a preserved frog.

   _____

11. A new drug for the treatment of asthma is tested on 100 people. The people are evenly divided into two groups. One group is given the drug and the other group is given a glucose pill. The group given the glucose pill serves as the
   (1) experimental group   (3) control
   (2) limiting factor   (4) indicator

12. A scientific study showed that the depth at which algae were found in a lake varied from day to day. On clear days, the algae were found as much as 6 meters below the surface of the water but were only 1 meter below the surface on cloudy days. Which hypothesis best explains these observations?
   (1) Light intensity affects the growth of algae.
   (2) Wind currents affect the growth of algae.
   (3) Nitrogen concentration affects the growth of algae.
   (4) Precipitation affects the growth of algae.

13. The diagram below represents a hydra as viewed with a compound light microscope.

If the hydra moves toward the right side of the slide preparation, which diagram best represents what will be observed through the microscope?

  (1)   (2)   (3)   (4)

14. The diagram below represents two cells next to a metric measuring device as seen under low power with a compound light microscope.

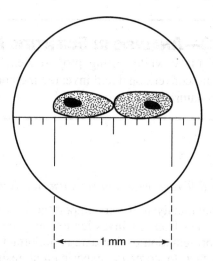

What is the length of a nucleus of one of these cells?
(1) 100 μm   (3) 1,000 μm
(2) 500 μm   (4) 1,500 μm

15. The diagram below represents measurements of two leaves.

Leaf A

Leaf B

The difference in length between leaves A and B is closest to
(1) 20 mm   (3) 0.65 m
(2) 20 cm   (4) 1.6 μm

III. ANALYSIS OF TESTING RESULTS

> **KEY IDEA 3—ANALYSIS IN SCIENTIFIC INQUIRY** The observations made while testing proposed explanations, when analyzed using conventional and invented methods, provide new insights into natural phenomena.

Essential Question:

- *What are the different ways that scientists analyze scientific data?*

Good, objective data analysis is very important to the scientific method of inquiry. The raw data from an investigation may be extensive, so proper organization and presentation of that data is important. Analysis of this data allows the researcher to compare experimental results against predicted results, to perform additional research, and finally to report confidently on the results of the research. The following section discusses the last five steps involved in data analysis.

6. Organize data in a manner that facilitates analysis.
7. Analyze data to identify trends and points of focus.
8. Summarize the data into a conclusion that answers the experimental question.
9. Propose and/or conduct additional research.
10. Report on the results of the investigation.

By following this design, the researcher is able to analyze effectively and report on the results of an experiment. This organization and analysis follows certain conventions, allowing researchers from different parts of the world to understand each other's work.

Organization of Data

Essential Question:

- *How can tables, charts, and graphs of data help to illustrate the findings of a scientific inquiry?*

Performance Indicator 1.3.1 *The student should be able to use various methods of representing and organizing observations (for example, diagrams, tables, charts, graphs, equations, matrices) and insightfully interpret the organized data.*

Diagrams of observational data can be important for certain kinds of investigations, such as those carried out with the help of the compound microscope. The researcher focuses on the components being studied and sketches his/her interpretations of what is seen. The researcher emphasizes the details pertaining to the study at hand. Diagrams must be accurately labeled to be useful in research. An important skill to be used with this method is estimating the sizes in micrometers (μm) of microscopic objects when seen in the low-power and high-power fields.

Charts or tables are appropriate for presenting many kinds of data. These grids have the experimental variable presented in ascending order on the left side of the chart or table and the dependent variable labeled at the top. The dependent variable values are placed next to the corresponding experimental variable data. A data matrix operates in much the same way. It organizes data by category for ease of comparison.

When data elements relate to each other in complex ways, representing the data as **mathematical equations** is sometimes preferred. Such equations are derived only after extensive field and laboratory study have supplied extensive data that supports the mathematical model.

Often, a **graphic representation** of the data will aid in its interpretation, especially where trends are clearly identifiable. Selecting the kind of graph that will best illustrate the data being analyzed is important. A line graph is useful when both experimental variable and dependent variable are numerical. A bar graph is used when the experimental variable cannot be expressed as a number but the dependent variable can be. Pie graphs are useful for depicting data that adds up to 100 percent of some value. Appropriately title the graph in each graph style. Label and scale the axes in line and bar graphs. Accurately plot data points in line and bar graphs or determine percentage sections in pie graphs. Connect points in line graphs, complete bars in bar graphs or complete sections in pie graphs.

From these depictions of the data, seeing broad trends that would not be obvious if the data were not organized in chart or graphic formats becomes possible. By studying the trends in these data elements, researchers are able to develop new and more accurate hypotheses. These hypotheses are then tested using the scientific method of inquiry. With further data to review, researchers can make generalizations, draw inferences and conclusions, and formulate explanations of natural phenomena. At this point, the research is considered to have added new knowledge to science.

Statistical Analysis of Data

Essential Question:

- *How do scientists measure the accuracy of the results of their own experiments?*

Performance Indicator 1.3.2 *The student should be able to apply statistical analysis techniques when appropriate to test if chance alone explains results.*

One of the simplest statistical methods used in science is calculating the **percentage of error**. In this method, the value obtained in the laboratory investigation is compared with the result expected through a mathematical formula. The difference between these values is divided by the expected value, and the result multiplied by 100 to obtain the percentage of error in the experiment. The researcher should always speculate on the probable causes of the percentage of error. These causes generally fall into three categories: (1) experimenter error (the researcher fails to follow proper procedure), (2) instrument error (the laboratory equipment introduces error by being inaccurate), or (3) calculation error (calculating the values used in the experiment is inaccurate or rounded inappropriately).

A slightly more complex method used to determine whether the data is valid within statistical norms is the **test of statistical significance**. The accuracy of any investigation depends on the size of the experimental sample being taken. Small samples often yield unreliable results, while large samples tend to yield more reliable results when probabilities are involved. The test of statistical significance compares what is expected according to probability against results obtained in the laboratory in the light of the size of the sample taken. Specific examples of tests of statistical significance are beyond the scope of this course of study, but may be implemented at the local level as necessary.

Summary and Conclusion

Essential Question:

- *How does a scientist draw inferences and conclusions based on data?*

Performance Indicator 1.3.3 *The student should be able to assess correspondence between the predicted result contained in the hypothesis and the actual result.*

The student should reach a **conclusion** as to whether the explanation on which the prediction was based is supported. After conducting a well-designed experiment with appropriate controls; having collected, organized, and objectively analyzed data; and determining the reliability of this data, the researcher is then ready to summarize his/her findings in the form of a conclusion. In its simplest form, the conclusion restates the experimental question and hypothesis. Then it objectively assesses the accuracy of that hypothesis in light of the data presented. If the hypothesis and the experimental results closely correspond, the conclusion should then go on to examine whether the results were produced in a manner that follows the logic used by the researcher in developing the hypothesis. In a well-researched investigation, both the hypothesis and the logic behind it should be reasonably accurate.

When a hypothesis is not supported by the experimental evidence, the conclusion should state that fact. The scientist should examine the literature review, experimental technique, and data analysis to try to detect flaws in the methodology used. Some fundamental omission in the literature review may have lead to the incorrect hypothesis. The techniques used may have been flawed, producing inaccurate data. The organization and analysis of the data may have been insufficient to permit trends to be appropriately identified. Far from being viewed as a failure, this situation is used by professional scientists to stimulate new questioning, additional research, and honed data analysis. Eventually, better results are obtained.

Additional Research

Essential Question:

- *How does one experiment lead to another once results are known?*

Performance Indicator 1.3.4 *The student should be able to revise the explanation and contemplate additional research based on the results of the test and through public discussion.*

Investigations in science do not always work out as predicted. Frequently, the results vary in some way from those anticipated in the hypothesis. When this happens, a competent researcher will use the opportunity to examine the experiment, make improvements, and conduct further research. Performing additional research is basic to the pursuit of scientific knowledge.

Hypotheses, even if found to be untrue, are valuable because they help to stimulate further investigation. As a part of the investigation, the researcher makes an objective judgment as to the accuracy of his/her hypothesis. If it is substantially correct, the hypothesis may add measurably to the information available to answer the question asked. Instead, it may take a partially answered question to new levels of understanding. If incorrect, the hypothesis leads the researcher to question the premise upon which the hypothesis was based, reexamine the literature for possible misinterpretation, and review the methods used to conduct the experiment. After performing this review, the researcher may well enter into a new round of experimentation to address any deficiencies found.

Claims should be questioned if the data are based on samples that are very small, biased, or inadequately controlled or if the conclusions are based on the faulty, incomplete, or misleading use of numbers. Claims should be questioned if the experimental design is not consistent with generally accepted experimental procedure. If samples are small or unrepresentative, the data collected may not fairly represent the population as a whole, leading to skewed results. If the data are biased or slanted to match the researcher's hypothesis, the objectivity of the entire investigation is called into question. Since science is based on objective interpretation of data, bias invalidates the inferences drawn from such experiments.

If the experimental method is flawed (uses inappropriate tools or techniques), the data will be meaningless and not able to answer the question. An experiment that lacks appropriate controls (allows multiple variables to operate) does not allow the researcher to determine which variable is at work in producing the results reported. If data analysis is not accurate and objective, but rather is faulty, incomplete, or misleading, then trends in the data may be difficult or impossible to identify and project beyond the limited scope of the experiment. Conclusions drawn from flawed data or flawed data analysis are invalid since the basis of all science is objective data.

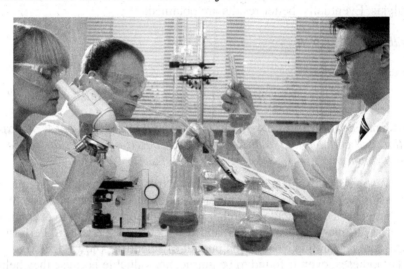

Claims should be questioned if fact and opinion are intermingled, if adequate evidence is not cited, or if the conclusions do not follow logically from the evidence given. Claims should be questioned if the data and its analysis are not objective in nature. If fact is mingled with opinion when reporting results, if adequate evidence is not cited, or if the conclusions do not follow logically from the evidence provided, be skeptical of the validity of the experiment. These flaws call into question the objectivity and completeness of the data collection and data analysis procedures. Good science is not based on opinion but on factual evidence. This evidence must be collected in sufficient quantity so it can provide a firm foundation for the conclusions drawn analyzing it. Conclusions must come from an objective analysis of the data collected and presented. Those that seem to ignore or make peripheral use of the data are immediately suspect and should be discarded.

Report of Experimental Results

Essential Question:

- *What are the appropriate ways to present the findings of a scientific investigation?*

Performance Indicator 1.3.5 *The student should be able to develop a written report for public scrutiny that describes the proposed explanation, including a literature review, the research carried out, its result, and suggestions for further research.*

When scientific investigation results in valid conclusions, the researcher may choose to summarize his/her findings in a written report. If the findings are of sufficient significance, the report may be published in a professional journal or other medium for review by other researchers and the general public.

Repeatability

One assumption of science is that other individuals could arrive at the same explanation if they had access to similar evidence. Scientists make the results of their investigations public; they should describe the investigations in ways that enable others to **repeat the investigations**. A major assumption of professional science is that other researchers, by following the explicit descriptions found in such a report, could arrive at the same results as those reached in the original study.

A **scientific report** should accurately and clearly summarize the literature examined when formulating the hypothesis, describe the methods used in designing and carrying out the investigation, present the data and its analysis in an organized fashion, state the conclusions drawn from this analysis, and suggest additional research to carry the investigation further. These items must be described in sufficient detail to allow others to duplicate the experiment in their own laboratories. If this is done and the same results are obtained, the additional research helps to validate the original investigation. If this is done but different results are obtained, doubt is cast on the original investigation and its conclusions.

Peer Review

Scientists use **peer review** to evaluate the results of scientific investigations and the explanations proposed by other scientists. They analyze the experimental procedures, examine the evidence, identify faulty reasoning, point out statements that go beyond the evidence, and suggest alternative explanations for the same observations. Peer review either lends support to, or helps to discredit, the reported investigation. It also guides the researcher in the investigation's design, execution, analysis, and reporting.

Peer review may be able to offer alternative explanations not presented by the researcher. Such peer review is meant to strengthen the quality of scientific inquiry and reporting. As this quality improves, so does the body of useful knowledge available to the public.

QUESTION SET 1.2—ANALYSIS OF TESTING RESULTS
(ANSWERS EXPLAINED, P. 255)

1–4 Base your answers to questions 1 through 4 on the information and data table below and on your knowledge of biology. The table shows the average systolic and diastolic blood pressure measured in millimeters of mercury (Hg) for humans between the ages of 2 and 14 years.

DATA TABLE

Age	Average Blood Pressure (mm of Hg)	
	Systolic	Diastolic
2	100	60
6	101	64
10	110	72
14	119	76

Directions (1–4): By using the information in the data table, construct a line graph on the grid provided, following the directions below.

1. Mark an appropriate scale on each labeled axis.

2. Plot the data for systolic blood pressure on your graph. Surround each point with a small triangle, and connect the points.   Example:

3. Plot the data for diastolic blood pressure on your graph. Surround each point with a small circle, and connect the points.   Example:

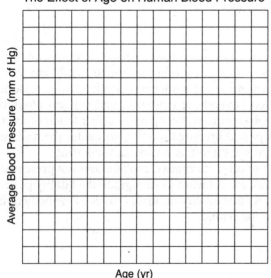

The Effect of Age on Human Blood Pressure

Average Blood Pressure (mm of Hg)

Age (yr)

Key

Systolic blood pressure
Diastolic blood pressure

4. By using one or more complete sentences, state one conclusion that compares systolic blood pressure with diastolic blood pressure in humans between the ages of 2 and 14 years.

_____

_____

5. The graph below shows the results of an experiment.

Effect of Temperature on Growth of Fish

At 16°C, what percentage of the brook trout reached maximum size?
(1) 30%   (3) 75%
(2) 55%   (4) 95%

6. An experiment is represented in the diagram below.

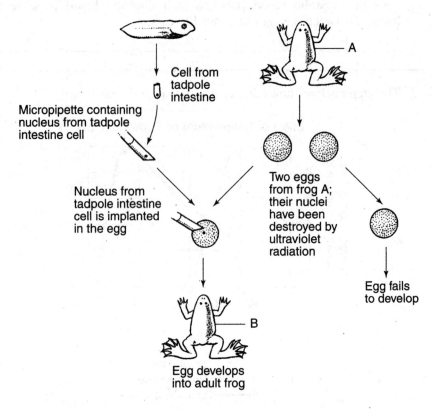

An inference that can be made from this experiment is that
(1) adult frog B will have the same genetic traits as the tadpole
(2) adult frog A can develop from only an egg and a sperm
(3) fertilization must occur in order for frog eggs to develop into adult frogs
(4) the nucleus of a body cell fails to function when transferred to other cell types

7. The charts below show the relationship of recommended weight to height in men and women age 25–29.

HEIGHT–WEIGHT CHARTS

MEN Age 25–29 Weight (lb)					WOMEN Age 25–29 Weight (lb)				
Height		Small Frame	Medium Frame	Large Frame	Height		Small Frame	Medium Frame	Large Frame
Feet	Inches				Feet	Inches			
5	2	128–134	131–141	138–150	4	10	102–111	109–121	118–131
5	3	130–136	133–143	140–153	4	11	103–113	111–123	120–134
5	4	132–138	135–145	142–156	5	0	104–115	113–126	122–137
5	5	134–140	137–148	144–160	5	1	106–118	115–129	125–140
5	6	136–142	139–151	146–164	5	2	108–121	118–132	128–143
5	7	138–145	142–154	149–168	5	3	111–124	121–135	131–147
5	8	140–148	145–157	152–172	5	4	114–127	124–138	134–151
5	9	142–151	148–160	155–176	5	5	117–130	127–141	137–155
5	10	144–154	151–163	158–180	5	6	120–133	130–144	140–159
5	11	146–157	154–166	161–184	5	7	123–136	133–147	143–163
6	0	149–160	157–170	164–188	5	8	126–139	136–150	146–167
6	1	152–164	160–174	168–192	5	9	129–142	139–153	149–170
6	2	155–168	164–178	172–197	5	10	132–145	142–156	152–173
6	3	158–172	167–182	176–202	5	11	135–148	145–159	155–176
6	4	162–176	171–187	181–207	6	0	138–151	148–162	158–179

The recommended weight for a 6'0" man with a small frame is closest to that of a
(1) 5'10" man with a medium frame
(2) 5'9" woman with a large frame
(3) 6'0" man with a medium frame
(4) 6'0" woman with a medium frame

8–11 Base your answers to questions 8 through 11 on the information below and on your knowledge of biology.

A group of biology students extracted the photosynthetic pigments from spinach leaves using the solvent acetone. A spectrophotometer was used to measure the percent absorption of six different wavelengths of light by the extracted pigments. The wavelengths of light were measured in units known as nanometers (nm). One nanometer is equal to one-billionth of a meter. The following data were collected:

yellow light (585 nm)—25.8% absorption
blue light (457 nm)—49.8% absorption
orange light (616 nm)—32.1% absorption
violet light (412 nm)—49.8% absorption
red light (674 nm)—41.0% absorption
green light (533 nm)—17.8% absorption

8. Complete all three columns in the data table below so that the wavelength of light either increases or decreases from the top to the bottom of the data table.

Color of Light	Wavelength of Light (nm)	Percent Absorption by Spinach Extract

Directions (9–10): By using the information in the data table, construct a line graph on the grid provided, following the directions below.

9. Mark an appropriate scale on the axis labeled "Percent Absorption."

10. Plot the data from the data table on your graph. Surround each point with a small circle, and connect the points.

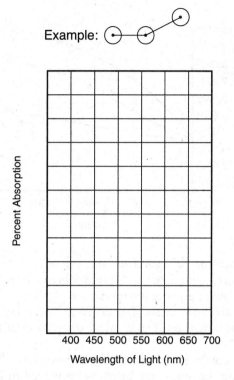

Example:

11. Which statement is a valid conclusion that can be drawn from the data obtained in this investigation?
    (1) Photosynthetic pigments in spinach plants absorb blue and violet light more efficiently than red light.
    (2) The data would be the same for all pigments in spinach plants.
    (3) Green and yellow light are not absorbed by spinach plants.
    (4) All plants are efficient at absorbing violet and red light.

12. The graph below represents the results of an investigation of the growth of three identical bacterial cultures incubated at different temperatures.

Which inference can be made from this graph?
(1) Temperature is unrelated to the reproductive rate of bacteria.
(2) Bacteria cannot grow at a temperature of 5°C.
(3) Life activities in bacteria slow down at high temperatures.
(4) Refrigeration will most likely slow the growth of these bacteria.

13. A study was conducted using two groups of ten plants of the same species. During the study, the plants were placed in identical environmental conditions. The plants in one group were given a growth solution every 3 days. The heights of the plants in both groups were recorded at the beginning of the study and at the end of a 3-week period. The data showed that the plants given the growth solution grew faster than those not given the solution.

When other researchers conduct this study to test the accuracy of the results, they should
(1) give growth solution to both groups
(2) make sure that the conditions are identical to those in the first study
(3) give an increased amount of light to both groups of plants
(4) double the amount of growth solution given to the first group

14. Worker bees acting as scouts are able to communicate the distance of a food supply from the hive by performing a waggle dance. The graph below shows the relationship between the distance of a food supply from the hive and the number of turns in the waggle dance every 15 seconds.

By using one or more complete sentences, state the relationship between the distance of the food supply from the hive and the number of turns the bee performs in the waggle dance every 15 seconds.

_____

15. Based on experimental results, a biologist in a laboratory reports a new discovery. If the experimental results are valid, biologists in other laboratories should be able to perform
    (1) an experiment with a different variable and obtain the same results
    (2) the same experiment and obtain different results
    (3) the same experiment and obtain the same results
    (4) an experiment under different conditions and obtain the same results

IV. LABORATORY REQUIREMENTS FOR PART D

New York State's biology students are required to complete four state-developed laboratory activities as part of their laboratory experience.

In addition to performing these laboratory activities and completing laboratory reports on your findings for each of them, you will be expected to answer questions concerning them on The Living Environment Regents examination, Part D. Following this section are practice questions and answers based on the first four of these required laboratory activities as known at the time of this publication.

Required Laboratory: The Beaks of Finches

This laboratory exercise explores the competitive advantages of beak sizes and shapes by means of a simulation activity. You are asked to examine a variety of common tools with which to pick up seeds of different sizes and to predict which tools will prove most and least effective in picking up a particular seed.

In the data collection portion of the experiment, you and your partner will use the tool assigned to you to pick up seeds over four equal-length time trials. You are asked to record on a data chart the number of seeds picked up during each trial.

In a variation of the data-collection process described above, you and your partner will join other lab teams to compare the effectiveness of your assigned tool against those of the other teams.

By completing this laboratory activity, you will be able to:

- Describe how structural differences in beak shape may affect the survival rate of members of a species;
- Analyze the role of competition for limited resources in the survival of members of a species with varying beak shapes;
- Comprehend how environmental conditions act to select favorable variations within a species;
- Draw inferences and conclusions concerning the accuracy of your hypothesis.

Required Laboratory: Relationships and Biodiversity

This laboratory exercise helps you to explore similarities and differences among different plant species. You will carry out observations of the gross and microscopic anatomical features, chemical characteristics, and genetic makeup of four simulated plant species.

Based on your observations of anatomical features, you will be expected to develop a hypothesis concerning the species that is most similar to a hypo-

34

thetical species, *Botana curus*. You and your lab partner will then conduct a number of tests, including chromatography, use of indicators, gel electrophoresis, and genetic sequencing, to test your hypothesis.

As a result of completing this activity you should be able to:

- Accurately predict the relative degree of relatedness of two or more plant species based on their anatomical features;
- Analyze the results of several laboratory tests to determine the biochemical relatedness of two or more plant species;
- Draw inferences and conclusions concerning the accuracy of your hypothesis.

Required Laboratory: Making Connections

In this laboratory activity, you and your lab partner will explore the reactions of the human body to fatigue. An important skill needed for this activity is the ability to accurately take another person's pulse. Once you have mastered this skill, you will be asked to use it as one of your primary data-collection techniques.

You are asked first to determine your average resting pulse rate. This data will become the "baseline" data for your experiment. You will then be expected to pool this data with that of the rest of your class and to represent class data graphically on a histogram.

In the next segment of the laboratory, you are asked to gather data about the effect of fatigue on muscle performance. Squeezing a clothespin repeatedly is the measurable activity.

Finally, you will be presented with hypothetical conflicting claims concerning the relationship between exercise and the rate at which a clothespin can be squeezed. You and your lab partner are asked to develop a hypothesis concerning which of these claims you believe to be true. You will then be expected to design a scientific experiment to test your hypothesis.

As a result of completing this experiment you will be able to:

- Accurately determine the pulse rate of another person;
- Collect, organize, and analyze class data on average pulse rate;
- Develop a hypothesis concerning the effect of exercise on muscle performance;
- Design and carry out an experiment to test your hypothesis;
- Draw inferences and conclusions concerning the accuracy of your hypothesis;
- Synthesize a report of your findings and present it to the class.

Required Laboratory: Diffusion Through a Membrane

This laboratory experience involves the creation of a simulated "cell" from dialysis tubing and some simple mixtures of common substances. The cell you create will behave in much the same way as a real cell. Like a real cell, it will contain a mixture of complex organic molecules suspended or dissolved in water. Like a cell membrane, the dialysis tubing will be a semipermeable membrane, allowing some substances to pass through and others not.

Once constructed, the cell will be subjected to two chemical tests to determine the nature of its semipermeability. These tests will involve the use of indicator solutions that indicate the presence of either glucose (blue-colored Benedict's solution) or starch (tan-colored Lugol's solution).

In the second part of the experiment, you will be asked to investigate the nature of a special kind of diffusion known as osmosis (the diffusion of water). You and your lab partner will place living cells in a solution containing dissolved salt. Your observations will help you to understand the nature of osmosis in living tissues.

By the time you complete this activity you should be able to:

- Demonstrate the use of chemical indicators to identify substances dissolved or suspended in water;
- Explain diffusion and osmosis as they relate to semipermeable membranes;
- Describe how the various chemical substances used in this experiment diffused in or out of the semipermeable cell constructed from dialysis tubing.

Required Laboratory: Adaptations for Reproductive Success in Flowering Plants

No information was available about the details of this laboratory activity at the time of this revision. Presumably, this laboratory will deal with the structure and function of the flower as a reproductive structure found in many plant species.

Required Laboratory: DNA Technology

No information was available about the details of this laboratory activity at the time of this revision. Presumably, this laboratory will deal with the various techniques used to analyze the structure and function of DNA.

Required Laboratory: Environmental Conditions and Seed Germination

No information was available about the details of this laboratory activity at the time of this revision. Presumably, this laboratory will deal with how environmental conditions such as temperature, moisture, and oxygen affect the germination of seeds of various types.

QUESTION SET 1.3—PART D REQUIRED LABORATORY EXPERIENCES (ANSWERS EXPLAINED, P. 259)

Base your answers to questions 1 through 3 on the diagram below which represents various finch species on the Galapagos Islands, your specific knowledge of the laboratory entitled "The Beaks of Finches," and your general knowledge of biology.

— from *Galapagos: A Natural History Guide*

1. An isolated island in the Galapagos supports four native finch species. These species are the large ground finch, the large tree finch, the small ground finch, and the small tree finch. A new species (the sharp-billed ground finch) is introduced to the island. Which of the native species will be most strongly affected by the introduction of the sharp-billed ground finch, and how will it be affected? In your answer, be sure to include:

• the native finch species affected [1]
• what will likely occur that will affect this native species [1]

_____

_____

2. A major environmental change occurs on this island that eliminates most of the plants that produce small seeds, leaving only plants that produce large seeds with thick, hard coverings. In terms of food gathering, the finch species having the greatest adaptive advantage under this changed set of conditions is most probably the
   (1) large ground finch   (3) large tree finch
   (2) small ground finch   (4) small tree finch

3. When one finch species tries to gain an advantage in obtaining the specific type of food it needs for survival to the disadvantage of other species in the same habitat, this represents
   (1) variation within a species   (3) interspecies competition
   (2) environmental change   (4) mutagenic agents

Base your answers to questions 4 through 6 on the partial genetic code chart and information below, your specific knowledge of the laboratory entitled "Relationships and Biodiversity," and your general knowledge of biology.

A plant species with the scientific name *S. hunta* is known to produce a protein substance with valuable medicinal qualities. Its genetic makeup is analyzed, and the gene for this protein is located at a specific gene locus on a certain chromosome. This gene is found to contain the following sequence of DNA codons:

GAA-CAA-TGA-CTT-GTA-GTA-GGG-CAA

Partial Genetic Code Chart

DNA Codons	mRNA Codons	Amino Acids
CTT	GAA	Glu
CTC	GAG	
GTA	CAU	His
GTG	CAC	
GAA	CUU	Leu
GAC	CUG	
GGA	CCU	Pro
GGG	CCC	
TGA	ACU	Thr
TGG	ACC	
CAA	GUU	Val
CAC	GUG	

4. The messenger RNA (mRNA) codon sequence that will result from this DNA sequence is
   (1) CUU-GUU-ACU-GAA-CAU-CAU-CCC-GUU
   (2) GAA-CAA-UGA-CUU-GUA-GUA-GGG-CAA
   (3) CTT-GTT-ACT-GAA-CAT-CAT-CCC-GTT
   (4) UGG-AGG-CUG-ACC-UCG-UCG-UUU-AGG

5. The amino acid sequence in a protein segment that will result from this DNA codon sequence during protein synthesis is
   (1) Leu-Glu-His-Pro-Val-Val-Pro-Val
   (2) Leu-Val-Thr-Leu-Glu-Glu-His-Pro
   (3) Leu-Pro-His-Val-Thr-Thr-Glu-Pro
   (4) Leu-Val-Thr-Glu-His-His-Pro-Val

6. Four new plants (species W, X, Y, and Z) are discovered that appear to be related to *S. hunta*. Upon analysis of their genetic makeup, the following DNA sequences were found at the same gene locus on the same chromosome as was found in *S. hunta*:
   Species W: GAC-CAC-TGA-CTT-CAA-GAA-TGA-CAA
   Species X: GAA-CAA-TGG-CTT-GTA-GTG-GGA-CAA
   Species Y: GAC-GGG-GTG-GTA-CAC-GTA-GGG-CAC
   Species Z: GAA-CAA-GAA-GGG-CTT-CTC-TGG-CAC

   Which species produces a protein segment most similar to that produced by *S. hunta*?
   (1) Species W   (3) Species Y
   (2) Species X   (4) Species Z

Base your answers to questions 7 through 9 on the experimental data below, your specific knowledge of the laboratory entitled "Making Connections," and your general knowledge of biology.

Data on Resting Pulse Rate (Beats per Minute)

Student (M/F)*	Average Pulse Rate (3 Trials)	Student (M/F)*	Average Pulse Rate (3 Trials)
Juan (M)	58	Jessica (F)	66
Sue (F)	81	Sean (M)	75
Tae Moon (M)	67	Mai Le (F)	86
Louise (F)	69	Cesar (M)	80
Sam (M)	72	Rita (F)	63
Erika (F)	60	Guiseppe (M)	71
Jose (M)	85	Meredith (F)	68
Tatiana (F)	65	Bill (M)	88
Brad (M)	91	Concetta (F)	74
Madelena (F)	76		
Greg (M)	62	AVERAGE	72.85

*M/F = male/female.

7. Represent the data in the chart in the form of a histogram (bar chart) in the grid below. Shade the histogram bars that result from your data analysis. [1]

Number of Students	10						
	9						
	8						
	7						
	6						
	5						
	4						
	3						
	2						
	1						
		<51	51–60	61–70	71–80	81–90	>90

Average Pulse Rate Range (beats per minute)

8. In a complete sentence, describe one pattern that is evident in the data that would help someone else to understand how pulse rate is distributed among these 20 students. [1]

_____

9. State a question that someone might ask about pulse rate that could be answered by further analyzing the data above or by collecting additional data about the same 20 students. [1]

_____

Base your answers to questions 10 through 12 on the experimental data below, your specific knowledge of the laboratory entitled "Diffusion Through a Membrane," and your general knowledge of biology.

Initial State   **Final State**

10. The following chart shows data that was collected by a student laboratory group as a result of performing the laboratory entitled "Diffusion Through a Membrane." It shows the results of chemical tests that were done on various fluids in the initial and final setups shown above.

Indicator Fluid Tested	Initial Setup		Final Setup	
	Lugol's Solution	Benedict's Solution	Lugol's Solution	Benedict's Solution
Beaker fluid	Tan	Blue	Tan	Red
Dialysis tube fluid	Blue-black	Red	Blue-black	Red

The difference between the fluids in the initial and final setups is that
(1) Lugol's solution has been converted to Benedict's solution
(2) starch has been converted to glucose by the indicator solutions
(3) starch has diffused from the fluid inside the dialysis tubing, across the membrane, and into the beaker fluid
(4) glucose has diffused from the fluid inside the dialysis tubing, across the membrane, and into the beaker fluid

11. A biologist observed a plant cell in a drop of water as shown in diagram A. The biologist then added a 10% salt solution to the slide and observed the cell as shown in diagram B.

The change in appearance of the cell resulted from
(1) more salt moving out of the cell than into the cell
(2) more salt moving into the cell than out of the cell
(3) more water moving into the cell than out of the cell
(4) more water moving out of the cell than into the cell

12. A farmer irrigating his corn fields irrigates field A with normal well water. In preparing to irrigate field B, he makes an error and loads a 10% saltwater solution (instead of well water) into his irrigation tanker. He then proceeds to irrigate field B with this 10% saltwater solution. Within 24 hours, he notes that all the corn plants in field A are standing straight and healthy, whereas all the corn plants in field B have wilted. Fortunately, a rainstorm the next day restores the corn plants in field B to a healthy condition.

Write a brief paragraph that describes the biological explanation for the farmer's observations of the corn plants in field B. In your answer, be sure to:

• define the terms *osmosis* and *equilibrium* [2]
• describe the *movement of water* into and out of the corn plants in field B during *each of the two days* of this observation [2]

V. LABORATORY SKILLS AND CURRENT EVENTS IN BIOLOGY

The New York State Regents Examination on The Living Environment tests the student's knowledge of both content and practical skills. In general, the skills required include basic laboratory technique, analysis of readings in science, written expression of biological concepts, and general awareness of current events in biology.

LABORATORY SKILLS

The Laboratory Checklist (see pages 47 and 48) mentions a set of specific laboratory skills that should be mastered in preparation for the year-end Regents examination. Students should be able to do the following.

1. Formulate a question or define a problem and develop a hypothesis to be tested in an investigation.
2. When given a laboratory problem, select suitable lab materials, safety equipment, and appropriate observation methods.
3. Distinguish between controls and variables in an experiment.
4. Identify parts of a light microscope and their functions, and focus in low and high power.
5. Determine the size of microscopic specimens in micrometers.
6. Prepare wet mounts of plant and animal cells, and apply staining techniques using iodine and methylene blue.
7. Identify cell parts under the compound microscope, such as the nucleus, cytoplasm, chloroplast, and cell wall.
8. Use and interpret indicators, such as pH paper, Benedict's (Fehling's) solution, iodine (Lugol's) solution, and bromthymol blue.
9. Use and read measurement instruments, such as metric rulers, Celsius thermometers, and graduated cylinders.
10. Dissect plant and animal specimens for the purpose of exposing major structures for suitable examination. Suggestions of specimens include seeds, flowers, earthworms, and grasshoppers.
11. Demonstrate safety skills involved in heating materials in test tubes or beakers, using chemicals, and handling dissection instruments.
12. Collect, organize, and graph data.
13. Make inferences and predictions based on data collected and observed.
14. Formulate generalizations or conclusions based on the investigation.

15. Assess the limitations and assumptions of the experiment.
16. Determine the accuracy and repeatability of the experimental data and observations.

These laboratory skills may be subdivided into three broad areas:

A. The methods by which a laboratory experiment is designed.
B. The techniques used in conducting a laboratory experiment.
C. The skills related to interpreting the results of a laboratory experiment.

A. Methods Used in Designing a Laboratory Experiment (Skills 1–3)

Skill 1 Students should be able to express an experimental problem as a statement or as an experimental **question** to be answered. Such questions should be written so as to indicate a quantity to be measured in a laboratory experiment. Students also should be able to state the expected results of such an experiment in the form of a **hypothesis**.

Skill 2 When presented with a hypothetical experiment to perform, students should be able to **select the group of tools** that would be most appropriate to use for conducting that experiment. For example, an experiment involving comparative anatomy would probably be carried out most effectively with dissecting instruments; one involving the chemical nature of an unknown food would be performed most effectively using chemical indicators; one involving plant growth might use a variety of measuring instruments.

Skill 3 When given an outline of an experiment, students should be able to identify the **variables**, which are the (changing) quantities being measured in the experiment. Certain variables are manipulated by the investigator (independent variables), whereas others vary as a result of experimental factors (dependent variables). Students should also be able to identify the aspects of the experiment, known as **controls**, designed to exclude possible interference by unwanted variables.

B. Techniques Used in Conducting Laboratory Experiments (Skills 4–11)

Skill 4 Students should have a basic familiarity with the **compound light microscope**—its parts, the function of each part, its use as a tool for measuring small objects, the procedure for determining its magnification, and the appearance of objects within its visual field. Students should be prepared to describe each of these aspects in sentence form.

Skill 5 Students should be able to express the sizes of microscopic objects in metric units. An understanding of the **micrometer (μm)**, a metric unit of linear measure, is required. Students should be able to convert measurements in micrometers to millimeters (or centimeters) and from larger units to micrometers (1 μm = 0.001 mm = 0.0001 cm = 0.000001 m).

Skill 6 Students should be able to describe, in sentence form, how to prepare a wet-mount slide for examination under the compound light microscope. A basic familiarity with the application of biological stains and the stains **iodine** and **methylene blue** is also required.

Skill 7 Students should be able to recognize and label the major **organelles** of typical plant, animal, and protozoan cells as shown in photomicrographs (photographs taken through a microscope). Students should also be prepared to describe the major functions of these organelles in sentence form as well as to judge their sizes in micrometers.

Skill 8 Students should be familiar with the use of **indicators** used to determine the chemical characteristics of food samples and solutions. Students should be prepared to describe, in sentence form, the use of each of the following indicators.

- **pH paper** is used to determine the relative acidity (pH) of a solution. A pH paper containing litmus will turn red in acid solution and blue in basic solution.
- **Bromthymol blue** turns yellow under acid conditions and remains blue under basic conditions. It may be used to detect the presence of carbon dioxide in solution since carbon dioxide forms a weak acid when dissolved in water.
- **Benedict's (Fehling's) solution** is used to detect the presence of simple sugars in food samples or solutions. Benedict's solution is blue when first prepared. When heated in the presence of simple sugar, it turns color, ranging from yellow to brick red.
- **Iodine (Lugol's) solution**, normally light tan, turns blue-black when applied to a food sample or solution containing starch.

Skill 9 Students should be able to determine **quantities and dimensions**, using a variety of metric measuring instruments. The dimensions of common objects should be determinable in millimeters and centimeters, using metric rulers and scales. Students should be able to read correctly the volume of liquid in a graduated cylinder by sighting the bottom of the meniscus. Students should be able to read the temperature indicated on a Celsius thermometer.

Skill 10 Students should be able to recognize, identify, and label the major **organs and organ systems** of common dissection specimens, including earthworms, grasshoppers, seeds, and flowers.

Skill 11 Students should be able to describe the proper methods of dealing with a variety of **laboratory situations**. This includes the correct means of handling chemicals so as not to cause harm to oneself or others. Also included is the safe use of dissecting tools, such as scalpels and other sharp instruments. Students should also be able to describe the approved techniques for heating liquids and for handling hot objects. Students should be prepared to describe these methods in sentence form.

C. Skills Used in Interpreting Experimental Results (Skills 12–16)

Skill 12 When given unorganized raw data, students should be able to **collect and organize** these data in chart form according to increasing values of the independent variable. Students should also be familiar with the proper techniques to use in representing data in graph form. Knowledge of the correct methods for constructing both bar charts and line graphs is required. Students should be able to label and increment graph axes properly, appropriately title graphs, and correctly plot graphic data points.

Skill 13 Students should be able to analyze the data that result from a laboratory experiment and to draw **inferences** (conclusions based on facts) that help to solve the experimental problem or answer the experimental question. Students should also be able to predict the outcome of experiments that broaden the range of the independent variable. In order to do this, students should understand how to interpret data organized in either chart or graph form. Students should be prepared to describe their inferences and predictions in sentence form.

Skill 14 When given an experiment whose data have been properly organized and analyzed, students should be able to develop generalizations (broad **conclusions**) concerning the effect of the test variable on the experimental question. It should then be possible to project these generalizations onto situations outside the laboratory where similar variables interact. Students should be prepared to express such generalizations in sentence form.

Skill 15 Students should recognize the **limitations** of their experimental methods in the context of professional science. This requires an understanding of the limits of accuracy of measuring equipment, the errors that may be introduced through incompletely developed laboratory skills, the ambiguities that result from inadequate experimental controls, and other limitations affecting experimental results.

Skill 16 Once a laboratory experiment is completed, students should be able to **determine the accuracy of the experimental results** through a review of the experimental methods. Where appropriate, calculating the percent error may assist in determining experimental accuracy. The experimental methods should allow for repeatability of the experiment to assist in the verification of experimental accuracy.

LABORATORY CHECKLIST

In addition to demonstrating performance indicators relating to scientific inquiry described in Standard 1, biology students need to develop proficiency in certain laboratory or technical skills in order to successfully conduct investigations in biological science. During the school year, students must develop the capacity to successfully perform each of the laboratory skills listed below. Proficiency in performing these laboratory skills may also be evaluated by items found on certain parts of the New York State Living Environment assessment, including the new Part D.

Follows safety rules in the laboratory

Selects and uses correct instruments

- Uses graduated cylinders to measure volume
- Uses metric ruler to measure length
- Uses thermometer to measure temperature
- Uses triple-beam or electronic balance to measure mass

Uses a compound microscope/stereoscope effectively to see specimens clearly, using different magnifications

- Identifies and compares parts of a variety of cells
- Compares relative sizes of cells and organelles
- Prepares wet-mount slides and uses appropriate staining techniques
- Uses other laboratory skills
- Designs and uses dichotomous keys to identify specimens
- Makes observations of biological processes
- Dissects plant and/or animal specimens to expose and identify internal structures
- Follows directions to correctly use and interpret chemical indicators
- Uses chromatography and/or electrophoresis to separate molecules
- Demonstrates ability to design, carry out, and report the results of simple biological experiments
- Designs and carries out a controlled, scientific experiment based on biological processes
- States an appropriate hypothesis
- Differentiates between independent and dependent variables
- Identifies the control group and/or controlled variables
- Collects, organizes, and analyzes data, using a computer and/or other laboratory equipment
- Organizes data through the use of data tables and graphs
- Analyzes results from observations/expressed data

- Formulates an appropriate conclusion or generalization from the results of an experiment
- Recognizes assumptions and limitations of the experiment

ANALYSIS OF READINGS IN SCIENCE

Part B and C of the New York State Regents examination may include items that require the analysis of short readings in science. These readings may deal with New York State Core Curriculum understandings or with concepts related to, but outside of, the curriculum. Students are expected to comprehend the meanings of technical terms that appear in the curriculum (see the glossary of this book). If other technical terms are used, they will be defined within the passage.

Students are expected to be able to read through the passage and answer questions based on it. Students should also be able to draw on their knowledge of Major Understandings to answer some questions about these reading passages.

WRITTEN EXPRESSION OF BIOLOGICAL CONCEPTS

Students in the Regents Biology—The Living Environment course are expected to be able to express themselves in complete sentences concerning biological principles. Although this requirement is not meant to be a test of the grammatical skills of the student, it is a test of the student's ability to express himself/herself clearly in scientific terms. Students should be prepared to answer Part B and C questions in sentence form.

CURRENT EVENTS IN BIOLOGY

Students should maintain an awareness of current events in biology that have reached statewide, national, or international prominence. Students will not be tested directly on current events but rather through use of reading comprehension and analysis of graphs and other data representations. Topics that may be selected for inclusion on Parts B and C include environmental situations (for example, acid precipitation, toxic waste disposal), advances in genetic research (for example, genetic engineering), aspects of biomedical research (for example, immunology, AIDS research), and others of a similar nature.

Following are typical Part B and C questions, grouped by the skills tested, that have appeared on actual Regents examinations in recent years. Students can find additional practice questions in the Regents examinations at the end of this book.

QUESTION SET 1.4—LABORATORY SKILLS QUESTION SET (SKILLS 1–14) (ANSWERS EXPLAINED, P. 264)

Skill 1

1. A student reported that a wilted stalk of celery became crisp when placed in a container of ice water. The student then suggested that water entered the stalk and made it crisp. This suggestion is considered to be
   (1) a control
   (2) a hypothesis
   (3) an observation
   (4) a variable

Skill 2

Base your answers to questions 2 through 4 on the four sets of laboratory materials listed below and on your knowledge of biology.

Set A
Light source
Colored filters
Beaker
Test tubes
Test tube stand

Set C
Scalpel
Forceps
Scissors
Pan with wax bottom
Pins
Stereomicroscope
Goggles

Set B
Droppers
Benedict's solution
Iodine
Test tubes
Starch solution
Sugar solution
Test tube holder
Test tube rack
Heat source
Goggles

Set D
Compound light microscope
Glass slides
Water
Forceps

2. Which set should a student select in order to test for the presence of a carbohydrate in food?

3. Which set should a student select to determine the location of the aortic arches in the earthworm?

4. Which set should a student use to observe chloroplasts in elodea (a green water plant)?

Skill 3

5. Some scientists have concluded that stressful situations cause a decrease in the normal operation of the immune system in human beings. In a recent study, people who were under severe stress were examined to measure how well their immune systems were functioning. These people showed poorer immune system function during times of severe stress than when they were under less stress.

   If the experimental group studied consisted of truck drivers who drove daily for 8 hours in very heavy traffic, a corresponding control group would most likely consist of truck drivers who drove
   (1) daily for 12 hours in very heavy traffic
   (2) every *third* day for 8 hours in very heavy traffic
   (3) every other day for 12 hours in very light traffic
   (4) daily for 8 hours in very light traffic

6. A student is studying the effect of temperature on the hydrolytic action of the enzyme gastric protease, which is contained in gastric fluid. An investigation is set up using five identical test tubes, each containing 40 milliliters of gastric fluid and 20 millimeters of glass tubing filled with cooked egg white. The five test tubes are each placed in a different temperature-controlled environment at 0°C, 10°C, 20°C, 30°C, and 40°C. After 48 hours, the amount of egg white hydrolyzed in each tube is measured. Which is a variable in this investigation?
   (1) gastric fluid
   (2) length of glass tubing
   (3) temperature
   (4) time

Skill 4

7. The diagram below represents a compound light microscope. Choose *one* of the numbered parts. *In a complete sentence,* name the part selected and describe its function.

8. To view cells under the high power of a compound microscope, a student places a slide of the cells on the stage and moves the stage clips over to secure the slide. She then moves the high-power objective into place and focuses on the slide with the coarse adjustment.

  Two steps in this procedure are incorrect. For this procedure to be correct, she should have focused under
  (1) low power using the coarse and fine adjustments, and then under high power using only the fine adjustment
  (2) high power first, and then under low power using only the fine adjustment
  (3) low power using the coarse and fine adjustments, and then under high power using the coarse and fine adjustments
  (4) low power using the fine adjustment, and then under high power using only the fine adjustment

Skill 5

Base your answers to questions 9 and 10 on the information following and on your knowledge of biology.

  A student was using a microscope with a 10× eyepiece and 10× and 40× objective lenses. He viewed the edge of a metric ruler under low power and observed the following field of vision.

9. What is the diameter, in micrometers, of the low-power field of vision?
  (1) 1   (3) 1,000
  (2) 2   (4) 2,000

10. The diameter of the high-power field of vision of the same microscope would be closest to
  (1) 0.05 mm   (3) 5 mm
  (2) 0.5 mm   (4) 500 mm

Skill 6

11. Which substance, when added to a wet mount containing starch grains, would react with the starch grains and make them more visible?
    (1) litmus solution   (3) distilled water
    (2) iodine solution   (4) bromthymol blue

Base your answers to questions 12 and 13 on the diagrams below and on your knowledge of biology. The diagrams show wet-mount microscope slides of fresh potato tissue.

Slide A   Slide B

12. The formation of air bubbles on slide *A* could have been prevented by
    (1) using a thicker piece of potato and less water
    (2) using a longer piece of potato and a coverslip with holes in it
    (3) holding the coverslip parallel to the slide and dropping it directly onto the potato
    (4) bringing one edge of the coverslip into contact with the water and lowering the opposite edge slowly

13. A drop of stain is put in contact with the left edge of the coverslip on slide *B*, and a piece of absorbent paper is placed in contact with the right edge of the coverslip. What is the purpose of this procedure?
    (1) It prevents the stain from getting on the ocular of the microscope.
    (2) It prevents the water on the slide from penetrating the potato tissue.
    (3) It allows the stain to penetrate the potato tissue without the removal of the coverslip.
    (4) It helps increase the osmotic pressure of the solution.

Skill 7

Base your answers to questions 14 and 15 on the drawing below, which shows a piece of tissue stained with iodine solution, as viewed with a microscope under high power.

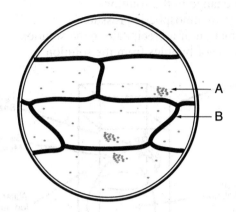

14. The tissue represented in the drawing is most likely made up of
    (1) onion epidermal cells
    (2) ciliated protists
    (3) cardiac muscle cells
    (4) blue-green algae

15. The organelle labeled *B* in the drawing is most likely a
    (1) mitochondrion   (3) lysosome
    (2) centriole   (4) cell wall

---

16. Diagram *A* represents the appearance of a wet mount of plant tissue as seen through a compound light microscope. Diagram *B* represents the appearance of the same field of view after the fine adjustment knob is turned. What is the best conclusion to be made from these observations?

A   B

    (1) The tissue is composed of more than one layer of cells.
    (2) The tissue is composed of multinucleated cells.
    (3) The cells are undergoing mitotic cell division.
    (4) The cells are undergoing photosynthesis.

Skill 8

17. A student was testing the composition of exhaled air by exhaling through a straw into a solution of bromthymol blue. The presence of carbon dioxide in the exhaled air would be indicated by
    (1) a color change in the solution
    (2) a change in atmospheric pressure
    (3) the formation of a precipitate in the solution
    (4) the release of bubbles from the solution

18. A student tested a sample of the fluid in the glass container for glucose 30 minutes after the apparatus had been set up. Which indicator should be used for this test?
    (1) iodine solution
    (2) bromthymol blue
    (3) Benedict's solution
    (4) pH paper

Skill 9

19. What is the total volume of water indicated in the graduated cylinder illustrated below?

    (1) 10 mL   (3) 12 mL
    (2) 11 mL   (4) 13 mL

20. The diagram below represents a segment of a metric ruler and part of an earthworm. What is the length of the part of the earthworm shown? You must include the correct units in your answer.

21. Which group of measurement units is correctly arranged in order of increasing size?
    (1) micrometer, millimeter, centimeter, meter
    (2) millimeter, micrometer, centimeter, meter
    (3) meter, micrometer, centimeter, millimeter
    (4) micrometer, centimeter, millimeter, meter

Skill 10
22. In earthworms and grasshoppers, which structure is ventral to the esophagus?
    (1) gizzard
    (2) brain
    (3) intestine
    (4) nerve cord

Base your answers to questions 23 and 24 on the illustration below of the flower of an amaryllis plant.

23. Name the circled part of the stamen.

24. Using a complete sentence, state a process carried out within the circled structure.

Skill 11

25. A student performing an experiment noticed that the beaker of water she was heating had a slight crack in the glass, but was not leaking. What should the student do?
    (1) Discontinue heating and attempt to seal the crack.
    (2) Discontinue heating and report the defect to the instructor.
    (3) Discontinue heating and immediately take the beaker to the instructor.
    (4) Continue heating as long as fluid does not seep from the crack.

26. Which would be the proper laboratory procedure to follow if some laboratory chemical splashed into a student's eyes?
    (1) Send someone to find the school nurse.
    (2) Rinse the eyes with water and do not tell the teacher because he or she might become upset.
    (3) Rinse the eyes with water; then notify the teacher and ask further advice.
    (4) Assume that the chemical is not harmful and no action is required.

27. While a student is heating a liquid in a test tube, the mouth of the tube should always be
    (1) corked with a rubber stopper
    (2) pointed toward the student
    (3) allowed to cool
    (4) aimed away from everybody

Skill 12

A student was investigating the relationship between different concentrations of substance X and the height of bean plants. He started with six groups, each of which contained the same number of bean plants with identical heights. Conditions were kept the same except that each group was watered with a different concentration of substance X for a period of 2 weeks. Then the concentration of substance X used in watering each group of plants and the average height for each group of plants were recorded by the student as follows:

Group A—6%, 32.3 cm   Group D—8%, 37.1 cm
Group B—0%, 28.7 cm   Group E—4%, 31.5 cm
Group C—2%, 29.4 cm   Group F—10%, 30.7 cm

For questions 28 and 29, organize the above data by filling in the Data Table on page 57, following the directions given in the questions.

28. Label column III with an appropriate heading. [*Include the proper unit of measurement.*]

29. Complete all three columns in the Data Table so that the concentrations of substance X are increasing from the top to the bottom of the Data Table.

DATA TABLE

I	II	III
Group	Concentration of Substance X (%)	

The Data Table shows the wolf and moose populations recorded at the end of June from 1970 to 1980 on an isolated island national park where no hunting by human beings is allowed. Before the arrival of wolves on the island (1965), the moose population had increased to more than 300 members. Wolves have been observed many times on this island hunting cooperatively to kill moose.

DATA TABLE

	Number of Members	
Year	Wolf Population	Moose Population
1970	10	90
1972	12	115
1974	20	145
1976	25	105
1978	18	95
1980	18	98

For questions 30 through 33, use the information in the Data Table to construct a line graph on the grid, following the directions given in the questions.

57

30. Mark an appropriate scale on the axis labeled "Number of Members of Each Population."

31. Mark an appropriate scale on the axis labeled "Year."

32. Plot the data for the wolf population on the graph. Surround each point with a small triangle and connect the points.

    Example

33. Plot the data for the moose population on the graph. Surround each point with a small circle and connect the points.

    Example

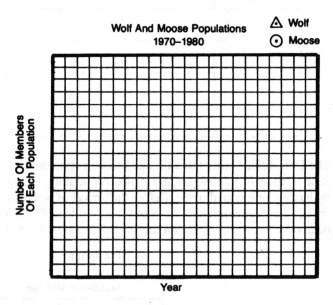

Skill 13

Base your answers to questions 34 through 36 on the following graph representing survival rates of fish species at various pH levels.

**The Effect Of pH On Survival Rates
Of Selected Species In Certain Adirondack Lakes**

Key:
A - pH Of A Certain Group of
    Adirondack Lakes, 1880

B - pH Of Rainfall, 1880

C - pH Of The Same Group Of
    Adirondack Lakes, 1980

D - pH Of Rainfall, 1980

— *National Geographic* (Adapted)

34. Which species can tolerate the highest level of acidity in its water environment?
    (1) mussels   (3) brown trout
    (2) smallmouth bass   (4) lake trout

35. In the years between 1880 and 1980, which species would most likely have been eliminated first because of the gradual acidification of Adirondack lakes?
    (1) mussels   (3) brown trout
    (2) smallmouth bass   (4) lake trout

36. What is the total change in the pH value of rainwater from 1880 to 1980?
    (1) 1.3   (3) 5.3
    (2) 1.7   (4) 9.7

Base your answers to questions 37 and 38 on the information and the graph given below, which shows the effect of sugar concentration on osmotic balance.

Four pieces of apple were cut so that all were the same mass and shape. The pieces were placed in four different concentrations of sugar water. After 24 hours, the pieces were removed and their masses determined. The graph below indicates the change in the mass of each piece.

37. What was the change in mass of the apple piece in the 10% sugar solution?
    (1) a decrease of 0.45 gram
    (2) an increase of 0.30 gram
    (3) a decrease of 0.30 gram
    (4) an increase of 0.10 gram

38. At approximately what sugar concentration should the pieces neither lose nor gain weight?
    (1) 6%   (3) 3%
    (2) 10%   (4) 20%

Skill 14

39. The graph below shows the average growth rate for 38 pairs of newborn rats. One of each pair was injected with anterior pituitary extract. The other member of each pair served as a control.

Average Growth Of 38 Untreated Rats (Control)

— — — — — Average Growth Of 38 Rats Injected With Anterior Pituitary Extract (Experimental)

Based on the graph, it can be correctly concluded that the pituitary extract
(1) is essential for life
(2) determines when a rat will be born
(3) affects the growth of rats
(4) affects the growth of all animals

40. Plant hormones are chemical regulators that stimulate or inhibit growth depending on their concentration and the type of tissue in which they are found.

Based on the information in the preceding graph, which is a correct conclusion about plant hormones?

(1) They stimulate maximum root growth and stem growth at the same concentration.
(2) They stimulate maximum stem growth at low concentrations.
(3) They most strongly inhibit root growth at low concentrations.
(4) They stimulate maximum root and stem growth at different concentrations.

41. The data below are based on laboratory studies of male *Drosophila,* showing the inherited bar-eye phenotype.

Culture Temperature (°C) During Development	15	20	25	30
Number of Compound Eye Sections	270	161	121	74

Which is the best conclusion to be drawn from an analysis of these data?
(1) The optimum temperature culturing *Drosophila* is 15°C
(2) *Drosophila* cultured at 45°C will show a proportionate increase in the number of compound eye sections.
(3) Temperature determines eye shape in *Drosophila.*
(4) As temperature increases from 15°C to 30°C, the number of compound eye sections in male *Drosophila* with bar-eyes decreases.

42. The diagram below represents the result of spinning a suspension of broken cells in an ultracentrifuge. Which is a correct conclusion?
(1) Ribosomes are more dense than mitochondria.
(2) Nuclei are more dense than mitochondria.
(3) Mitochondria and ribosomes are equal in density.
(4) The cell consists of only solid components.

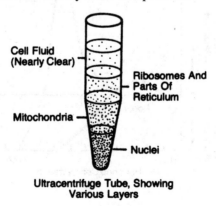

Cell Fluid (Nearly Clear)

Ribosomes And Parts Of Reticulum

Mitochondria

Nuclei

Ultracentrifuge Tube, Showing Various Layers

READING, WRITING, AND CURRENT EVENTS IN SCIENCE
QUESTION SET (ANSWERS EXPLAINED, P. 279)

Base your answers to questions 1 through 5 on the passage below.

Gene Splicing

Recent advances in cell technology and gene transplanting have allowed scientists to perform some interesting experiments. Some of these experiments have included splicing a human gene into the genetic material of bacteria. The altered bacteria express the added genetic material.

Bacteria reproduce rapidly under certain conditions. This means that bacteria with the gene for human insulin could multiply rapidly, resulting in a large bacterial population which could produce large quantities of human insulin.

The traditional source of insulin has been the pancreases of slaughtered animals. Continued use of this insulin can trigger allergic reactions in some humans. The new bacteria-produced insulin does not appear to produce these side effects.

The bacteria used for these experiments are *E. coli,* bacteria common to the digestive systems of many humans. Some scientists question these experiments and are concerned that the altered *E. coli* may accidentally get into water supplies.

For each statement below, write the number 1 if the statement is true according to the paragraph, the number 2 if the statement is false according to the paragraph, or the number 3 if not enough information is given in the paragraph.

1. Transplanting genetic material into bacteria is a simple task.
2. Under certain conditions bacteria reproduce at a rapid rate.
3. Continued use of insulin from other animals may cause harmful side effects in some people.
4. The bacteria used in these experiments are normally found only in the nerve tissue of humans.
5. Bacteria other than *E. coli* are unable to produce insulin.

Base your answers to questions 6 through 8 on the reading passage below. Write your answers in complete sentences.

Time Frame for Speciation

Evolution is the process of change through time. Theories of evolution attempt to explain the diversification of species existing today. The essentials of Darwin's theory of natural selection serve as a basis for our present understanding of the evolution of species. Recently, some scientists have suggested two possible explanations for the time frame in which the evolution of species occurs.

Gradualism proposes that evolutionary change is continuous and slow, occurring over many millions of years. New species evolve through the accumulation of many small changes. Gradualism is supported in the fossil record by the presence of transitional forms in some evolutionary pathways.

Punctuated equilibrium is another possible explanation for the diversity of species. This theory proposes that species exist unchanged for long geological periods of stability, typically several million years. Then, during geologically brief periods of time, significant changes occur and new species may evolve. Some scientists use the apparent lack of transitional forms in the fossil record in many evolutionary pathways to support punctuated equilibrium.

6. Identify one major difference between gradualism and punctuated equilibrium. [1]
7. According to the theory of gradualism, what may result from the accumulation of small variations? [1]
8. What fossil evidence indicates that evolutionary change may have occurred within a time frame known as gradualism? [1]

---

9. An organism contains many structures that enable it to survive in a particular environment. The human body has many such structures that have adaptive value. Several of these structures are listed below.
   - sweat gland
   - pancreas
   - liver
   - epiglottis
   - capillary
   - villus
   - kidney
   - platelet

Choose *three* (3) of the structures listed above. For *each* one chosen, write the name of the structure and then, using a complete sentence, describe one of its adaptive values to the human body. [3]

Base your answers to questions 10 through 14 on the information below and on your knowledge of biology.

Acid rain is a serious environmental problem in large areas of Canada and the northeastern United States, including New York State. It is partly created as rain "washes out" sulfur and nitrogen pollutants from the air. Acid rain alters the fundamental chemistry of sensitive freshwater environments and results in the death of many freshwater species. The principal sources of this pollution have been identified as smokestack gases released by coal-burning facilities located mainly in the midwestern United States.

"Unpolluted" rain normally has a pH of 5.6. Acid rain, however, has been measured at pH values as low as 1.5, which is more than 10,000 times more acidic than normal. Commonly, acid rain has a pH range of 3 to 5, which changes the acidity level of the freshwater environment into which it falls. The effect of the acid rain depends on the environment's ability to neutralize it. Evidence is accumulating, however, that many environments are adversely affected by the acid rain. As a result, the living things within lakes and streams that cannot tolerate the increasing acidity gradually die off.

There are many environmental problems that result from acid rain. Most of these problems center around the food web upon which all living things, including humans, depend. If freshwater plants, animals, and protists are destroyed by the acid conditions, then terrestrial predators and scavengers dependent on these organisms for food are forced to migrate or starve. These changes in a food web can eventually affect the human level of food consumption.

10. The accompanying scale shows the pH values of four common household substances. Acid rain has a pH closest to that of which of these substances?

(1) ammonia
(2) tap water

(3) baking soda
(4) vinegar

11. What is most likely the source of acid rain in New York State?
(1) far western United States
(2) midwestern United States
(3) far eastern Canada
(4) far western Europe

12. Which food chain includes organisms that would most immediately be affected by acid rain?
(1) grass → rabbit → fox → decay bacteria
(2) algae → aquatic insect → trout → otter
(3) shrub → mouse → snake → hawk
(4) tree → caterpillar → bird → lynx

13. Acid rain is generally considered a negative aspect of human involvement with the ecosystem. As such, it would most correctly be classified as a type of
    (1) biological control
    (2) conservation of resources
    (3) technological oversight
    (4) land-use management

14. A strain of fish that could survive under conditions of increased acidity could best be obtained by
    (1) binary fission
    (2) vegetative propagation
    (3) selective breeding
    (4) budding

UNIT TWO

Understanding and Applying Scientific Concepts

STANDARD 4

Students will understand and apply scientific concepts, principles, and theories pertaining to the physical setting and living environment and will recognize the historical development of ideas in science.

Essential Questions:

- *What are the principles that guide scientists in their study of the natural world?*
- *How is modern science built on the work of scientists of the past?*

I. SIMILARITY AND DIVERSITY AMONG LIVING THINGS—LIFE PROCESSES

KEY IDEA 1—APPLICATION OF SCIENTIFIC PRINCIPLES Living things are both similar to and different from each other and from nonliving things.

Essential Question:

- *How are living things similar to and different from other living things and nonliving things?*

Life Processes

Living things rely on many of the same processes to stay alive. Yet the ways that these processes and interactions are carried out are diverse. The **life processes** carried out by living things include nutrition, transport, respiration, regulation, reproduction, and (in animals) locomotion. These life

67

processes provide mechanisms for living things to obtain and process foods (nutrition), circulate essential materials around their bodies (transport), convert food energy to cell energy (respiration), coordinate life activities (regulation), produce more members of their kind (reproduction), and move from place to place in their environment (locomotion). Although almost all living things perform these same life functions, they are performed in many different ways in different kinds of living organisms. These different mechanisms are known as adaptations.

Nearly all living things are composed of cells. Cells are small living units that serve as the structural and functional building blocks of most living organisms. Cells carry on dynamic chemical activities that together constitute the cell's metabolism (**metabolic activity**). By carefully balancing this metabolic activity, cells are able to maintain a balanced internal stability known as **homeostasis**. Nonliving things differ from living things in that nonliving things lack the ability to perform these life processes.

The components of living systems must work together to maintain homeostatic balance. This is true whether the systems are represented by cells, multicelled organisms, or ecosystems. Each level of complexity brings a new degree of variation to the theme of homeostasis. On the cellular level, life functions are performed by **organelles**, or small organs, found within the cell. Multicellular organisms, depending on their complexity, are made up of specialized organ systems that perform different aspects necessary to carry out separate, but interrelated, life functions. On the ecosystem level, species populations of all kinds interact to help achieve homeostatic balance in the living environment.

Diversity and Ecosystem Stability

Essential Question:

- *How does the great variety of living·species impact environmental stability?*

Performance Indicator 4.1.1 *The student should be able to explain how diversity of populations within ecosystems relates to the stability of ecosystems.*

Recognizing the interrelatedness of all living things, including humans, is important. It is also important to understand the unique, yet interdependent, roles that these organisms play in maintaining the stability of life on Earth. It is the richness of **species diversity** that provides for this stability. When even one species in an ecosystem is eliminated, several other species are adversely affected, interrupting stability. See material under "Interdependence of Living Things" for a complete discussion of ecology (pages 200–224).

Life Functions in Humans

Human Physiology

Essential Question:

- *How are humans adapted to survive in their environment?*

Performance Indicator 4.1.2 *The student should be able to describe and explain the structures and functions of the human body at different organizational levels (for example, systems, tissues, cells, organelles).*

Humans, like all living things, must perform basic life functions in order to survive. These life functions include nutrition, transport, respiration, excretion, regulation, and locomotion. The bodies of human beings are organized to provide efficient and effective means of performing these life functions. Each life function in a human is carried out by an **organ system** that contains **organs** and **tissues** specialized for the tasks involved. In turn, these structures are composed of cells whose metabolic processes are specialized for assisting these life functions. In addition, the activities of each organ and organ system must be coordinated so as to provide an integrated, homeostatic balance that promotes the maintenance of life. When any part of this complex system breaks down or becomes inefficient, this balance may be disrupted. **Diseases** may result from such disturbances.

Important levels of organization for structure and function include **organelles**, **cells**, **tissues**, **organs**, **organ systems**, and whole **organisms**. Several specific adaptations are found in humans (and in many other living things) to carry out these life functions. The simplest of these adaptations is the **cell**, which is the unit of structure and function of all living things. Cells contain subcomponents, known as **organelles**, that are specialized to perform aspects of these life functions. Some important organelles and their respective functions are discussed on pages 106–107.

Above the cellular level of organization, the bodies of living things contain groupings of cells that are similar in structure and function. Such groupings of similar cells are known as **tissues**. For example, muscle tissues in humans are composed of spindle-shaped cells rich in contractile fibers. These cells are specialized to act together and assist in body movement. Because of their similar structure and function, muscle cells comprise a tissue. The bodies of complex organisms such as humans contain hundreds of distinct tissues.

An **organ** is a structure, composed of several different tissues, that plays a major role in the performance of a life function. Examples of organs in the human include the heart, the stomach, the liver, and the kidney. A series of organs that function together to assist in the performance of a life function is known collectively as an **organ system**. For example, the digestive system of the human is made up of the esophagus, stomach, small intestine, and large intestine, as well as several accessory organs. Each organ in the system has a specific role to play in the performance of the life function. Organ systems of humans are described in detail below.

Human Organ Systems

Humans are complex organisms. They require multiple systems for digestion, respiration, circulation, excretion, movement, coordination, and immunity. The systems interact to perform the life functions. Humans are complex organisms whose existence depends on the coordinated functioning of several integrated organ systems. Each organ system is specialized for the tasks involved in carrying out a separate and distinct life function. By utilizing these systems, humans are able to:

- take in and process complex foods to produce simpler subunits;
- move essential materials to all parts of the body;
- obtain respiratory gases and metabolize these foods to produce cellular energy;
- remove potentially harmful wastes from the body;
- coordinate all life functions to produce an efficient, integrated system;
- move from place to place in the environment; and
- provide immunity from disease.

Perhaps the most essential feature of this process is the coordination of the separate systems into an integrated whole, promoting the maintenance of life in each individual human being. Without this integration, complex functional living units would probably not exist on Earth.

Nutrition is the life function by which human beings obtain materials needed for energy, growth and repair, and other life functions. As part of this process, these materials are converted to a simplified form that can be used by the cell. Nutritional requirements of humans are similar to those of other animals. A diet that includes the proper balance of carbohydrate, protein, lipid, roughage, vitamins, and minerals is essential to the physical health of the human body. These requirements are known to vary, however, with the age, sex, and physical activity of the individual.

The digestive system enables the human to carry out the life function of nutrition. In general, the human digestive tract resembles that of simpler organisms whose body plan is that of a one-way tube within a tube. The outer tube is the body exterior; the inner tube is the digestive tract. Within the digestive tract, food materials are progressively converted into molecular end products. The components of this system are as follows.

- **Oral cavity**—The human oral cavity (mouth) is used to ingest food. The teeth and tongue help to manipulate and break the food down mechanically. This process increases the surface area of the food, thereby aiding the process of chemical digestion by enzymes in saliva. Saliva is produced by salivary glands and secreted into the mouth cavity. The enzymes in saliva are responsible for the partial digestion of complex carbohydrates, such as starch, into double sugars, such as maltose.
- **Esophagus**—The esophagus is a short tube that connects the oral cavity to the stomach. The swallowing action initiated at the back of the oral cavity continues in the esophagus as a wave of muscular contraction known as **peristalsis**. This peristaltic action moves the chewed food to the stomach for further digestive action.
- **Stomach**—The stomach is a muscular organ whose main function is to liquefy and further digest food materials. The lining of the stomach contains digestive glands that secrete the digestive enzymes and hydrochloric acid that make up gastric fluid. Stomach enzymes are specifically designed to digest proteins. Unlike most of the body's enzymes, gastric enzymes work best in an acidic condition. This acidic condition is provided by the hydrochloric acid secreted by the stomach's digestive glands.
- **Small intestine**—Liquefied and partially digested food enters the small intestine from the stomach. There, enzymes secreted by intestinal glands complete the digestive process. These enzymes include protease (to digest proteins), lipase (to digest fats and oils), and enzymes such as maltase and sucrase (to digest maltose and sucrose sugars). In addition to the digestive process carried on within the small intestine, the lining of the small intestine acts as the principal surface for the absorption of the molecular end products of digestion. To facilitate this absorption, the lining of the small intestine contains millions of microscopic projections known as **villi**. The villi contain microscopic blood vessels, known as **capillaries**, and extensions of the lymphatic system, known as **lacteals**, that receive the dissolved nutrients and conduct them throughout the body.
- **Large intestine**—The large intestine receives food materials that have passed through the entire digestive tract but have not been digested. These materials are normally in a liquid state when they pass from the small to the large intestine. The large intestine reabsorbs much of the water from the waste matter and condenses it into semisolid waste known as **feces**. The feces are stored in the lower end of the large intestine in an area known as the **rectum**. The feces pass out of the body via the anus by means of strong muscular (peristaltic) contractions. This action constitutes the process of egestion.

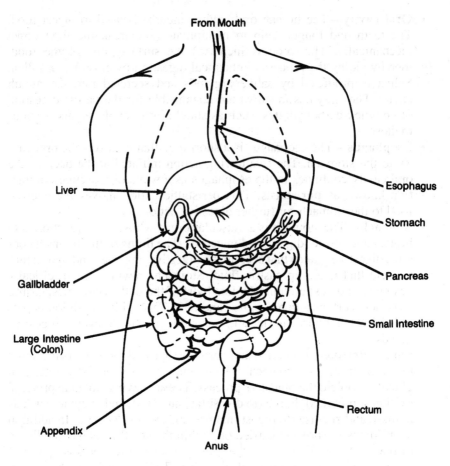

Digestive Tract

- **Accessory organs**—Accessory digestive organs include the salivary glands, liver, gallbladder, and pancreas. The term accessory refers to the fact that these organs are not a part of the food tube. The liver produces bile, which is important in the digestion of fats. Bile acts by breaking up (emulsifying) fat into small droplets. The gallbladder stores bile for release into the small intestine. The pancreas manufactures and stores several digestive enzymes important for the complete digestion of several kinds of complex molecules into simple, soluble end products. These end products include glucose, amino acids, fatty acids, and glycerol.

The underlying mechanism of digestion is the chemical process known as **hydrolysis**. This term literally means "splitting by adding water." In this reaction, large, insoluble molecules are converted to small, soluble molecules. To accomplish this, the cell uses enzymes to add the atoms that constitute water (hydrogen and oxygen) to the structure of the complex molecule. This reaction breaks the complex molecule at specific locations, producing smaller subunit molecules that can be absorbed by the cell.

This enzyme-catalyzed reaction may be accomplished with each of the major types of food and is illustrated by the following equation:

$$\text{maltose} + \text{water} \xrightarrow{\text{maltase}} 2 \text{ glucose}$$

$$C_{12}H_{22}O_{11} + H_2O \xrightarrow{\text{maltase}} 2\, C_6H_{12}O_6$$

In this reaction, the double sugar maltose is hydrolyzed to two molecules of the simple sugar glucose by the addition of a single molecule of water. The enzyme maltase catalyses this reaction. In similar reactions, starch molecules are hydrolyzed to double sugars, protein molecules are hydrolyzed to amino acid molecules, and fat molecules are hydrolyzed to fatty acid and glycerol molecules.

Transport is the life function by which human beings absorb and distribute the materials necessary to maintain life. The human circulatory system is specially adapted to move essential materials through the body to all cells. At the same time, waste materials resulting from cellular metabolism are carried to areas where they can be released to the environment away from living cells. **Immunity** from disease is also provided by transport mechanisms. Specialized tissues and organs assist in the transport function.

Transport media include blood and lymph. The blood is a fluid tissue suspended in liquid plasma. Plasma is made up of water containing dissolved salts, nutrients, gases, and molecular wastes. Also found in the plasma are hormones and a large variety of manufactured proteins such as antibodies, clotting proteins, and enzymes. The plasma also suspends the cellular fraction of the blood.

The cellular fraction of the blood contains several different cell types. **Red blood cells** are the most abundant cell type in the blood. These small (about 8 micrometers in diameter), dish-shaped cells lack nuclei and cannot reproduce. Red blood cells contain **hemoglobin**, a red oxygen-carrying pigment that makes the blood an efficient medium for transporting oxygen to all body tissues. **Phagocytes**, a type of white blood cell, engulf and destroy bacteria that enter the bloodstream though breaks in the skin surface. **Lymphocytes**, another type of white blood cell, produce **antibodies** specifically designed to recognize and attack particular types of proteins (**antigens**) that may enter the blood by various routes. **Platelets** are small, noncellular components of the blood that contain chemicals important to the clotting process. The body's immune reactions are discussed in detail in the section "Disease as a Failure of Homeostasis."

Intercellular fluid (ICF) and lymph are abundant in human tissues. All cells in the body are bathed in ICF, which is rich in salts and various components important to the homeostatic balance of the cell. ICF drains from the tissues within lymphatic vessels, where it is known as lymph.

The **circulatory system** includes a series of structures (vessels and organs) designed to move the transport fluids throughout the body efficiently. These transport vessels and organs include the following.

- **Arteries**—Arteries are relatively thick-walled blood vessels that contain cardiac muscle tissues. These muscles enable the artery to maintain blood flow via rhythmic contractions known as the **pulse**. Arteries are always involved with conducting the blood away from the heart toward the body's tissues.
- **Veins**—Veins are relatively thin-walled blood vessels that lack muscular tissues. Veins contain one-way valves that aid the forward movement of blood by preventing backflow within the vein. Veins are always involved with conducting the blood back toward the heart and away from the body tissues.
- **Capillaries**—Capillaries are microscopic blood vessels whose walls are only one cell in thickness. Capillaries branch from the ends of small arteries and carry blood rich in oxygen and nutrients to all tissues in the body. Dissolved materials are readily exchanged by diffusion between the blood and the body tissues through the thin walls of the capillaries.
- **Lymph vessels**—Lymph vessels form a branching series of microscopic vessels containing lymph. The lymph vessels carry lymph to and from the body tissues where it bathes the cells. To aid the movement of the lymph, the lymph vessels contain valves similar to those found in veins. Lymph vessels become enlarged and are gathered in masses known as lymph nodes at specific parts of the body. These nodes contain phagocytic white blood cells that attack and destroy bacteria in the lymph.
- **Heart**—The structure of the heart permits efficient movement of blood throughout the body within blood vessels. The heart is a muscular pump with four chambers. Two of these chambers, the **atria**, receive blood from veins leading from body organs. Two other chambers, the **ventricles**, have thick, muscular walls that contract to force blood out under great pressure through arteries to other organs.

Heart Circulation

The circulatory function can be traced from the entry of deoxygenated blood (blood whose hemoglobin has given up its oxygen) into the **right atrium** of the heart via the **vena cava**. The vena cava is a large vein that collects blood from smaller veins. This blood then passes through a one-way valve to the **right ventricle**. Strong muscular contractions in this ventricle force the blood out through the **pulmonary arteries** and into the lungs. A valve in the pulmonary artery prevents the backflow of blood. In the lungs, the blood passes through **capillaries**, where gas exchange occurs. This process adds oxygen to the blood and removes carbon dioxide from the blood. The oxygenated blood then returns via the **pulmonary veins** to the heart, entering through the **left atrium**. From the left atrium, the blood passes through another one-way valve on its way to the **left ventricle**. Contraction of the left ventricle sends the oxygenated blood out of the heart to body organs by way of the **aorta**. As in the pulmonary artery, a valve prevents the blood from flowing backward into the heart. The aorta branches into a series of smaller arteries. They eventually terminate in capillary networks within body tissues. In the capillaries, oxygen and nutrients are absorbed from the blood and carbon dioxide and metabolic wastes are absorbed into the blood. The capillaries then carry the blood to veins, which eventually lead back to the vena cava and into the heart.

The circulation of blood through the lungs is known as **pulmonary circulation**. The circulation of blood through the body organs is known as the **systemic circulation**. The movement of blood through the blood vessels serving the heart muscle is known as the **coronary circulation**.

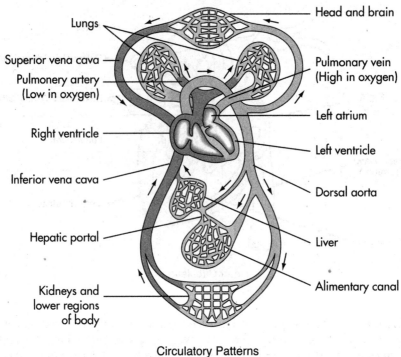

Lungs

Superior vena cava

Pulmonery artery
(Low in oxygen)

Right ventricle

Inferior vena cava

Hepatic portal

Kidneys and
lower regions
of body

Head and brain

Pulmonary vein
(High in oxygen)

Left atrium

Left ventricle

Dorsal aorta

Liver

Alimentary canal

Circulatory Patterns

Blood pressure from both the pumping action of the heart and the contractions of the muscular artery walls maintains the flow of blood through the arteries and capillaries. During blood pressure testing, the higher pressure (**systole**) is registered when the ventricles contract; the lower pressure (**diastole**) is registered when the ventricles relax.

QUESTION SET 2.1—HUMAN PHYSIOLOGY 1
(ANSWERS EXPLAINED, P. 282)

1. Some characteristics of digestive systems are listed below.

   A Food is moved along by peristalsis.
   B Food is moved along by involuntary muscles.
   C Accessory organs are present.

   Which characteristics best describe the human digestive system?
   (1) *A*, *B*, and *C*
   (2) *A* only
   (3) *B* only
   (4) *B* and *C* only

2. Which statement most accurately describes the human heart?
   (1) It has two atria and one ventricle, and it pumps blood directly into veins.
   (2) It has one atrium and one ventricle, and it is composed of cardiac muscle.
   (3) It has one atrium and two ventricles, and it is composed of visceral muscle.
   (4) It has two atria and two ventricles, and it pumps blood directly into arteries.

3. A malfunction of the lymph node would most likely interfere with the
   (1) release of carbon dioxide into the lymph
   (2) filtering of glucose from the lymph
   (3) release of oxygen into the lymph
   (4) filtering of bacteria from the lymph

4. A heart attack may be due to all of the following except
   (1) an increase in arterial blood pressure
   (2) oxygen deprivation of cardiac muscle
   (3) narrowing of the arteries transporting blood to the heart muscle
   (4) decreased consumption of complex carbohydrates

5–7 Base your answers to questions 5 through 7 on the diagram below of the human digestive system and on your knowledge of biology.

5. In which structure does the initial hydrolysis of carbohydrates occur?
   (1) A   (3) C
   (2) E   (4) D

6. From which structure are glucose and amino acids normally absorbed into the circulatory system?
   (1) F   (3) C
   (2) H   (4) E

7. In which structure does extracellular chemical digestion of protein begin?
   (1) G   (3) C
   (2) B   (4) E

---

8. A pulse can be detected most easily in
   (1) an artery   (3) a capillary
   (2) a vein   (4) a lacteal

9–10 Base your answers to questions 9 and 10 on the diagram below of the human heart and on your knowledge of biology.

9. Which structures are most closely associated with the transport of deoxygenated blood?
   (1) A, B, and C
   (2) B, F, and I
   (3) C, D, and E
   (4) D, H, and I

10. A structure that prevents the backflow of blood into an atrium is indicated by letter
    (1) G
    (2) B
    (3) C
    (4) H

11. Which transport structures have specialized regions for filtering out bacteria and dead cells?
    (1) arteries
    (2) capillaries
    (3) veins
    (4) lymph vessels

Respiration is the life function by which human beings convert the chemical energy stored in foods to a form that the cells can use more easily. The production of cellular energy is one of the most essential functions carried on by living things. Organs and tissues specially adapted to promote the absorption of oxygen are important parts of the human respiratory system. The components of this system are as follows.

- **Nasal cavity**—The nasal cavity is composed of a series of channels through which outside air is admitted to the body interior. The cavity is lined by ciliated cells capable of producing mucus. Hair is also present; it catches dust particles. Air passing through the nasal cavity is moistened, warmed, and filtered in the nasal cavity, making it more compatible with the environment of the lung.
- **Pharynx**—The pharynx (throat) is the area in the back of the oral cavity where the nasal cavity joins in. In the pharynx, a flap of tissue, the epiglottis, covers the open end of the trachea to prevent food from entering the respiratory tubes.
- **Trachea**—The trachea (windpipe) is a cartilage-ringed tube used to conduct air from the pharynx deeper into the respiratory system. The cartilage rings maintain the open condition of the trachea at all times. The trachea is lined with ciliated tissues that sweep dust particles up and out of the trachea so they can be swallowed or expelled.
- **Bronchi**—Two bronchi branch from the the end of the trachea and lead to the two lungs. Like the trachea, the bronchi are ringed with cartilage and lined with ciliated mucous membrane.
- **Bronchioles**—The bronchioles are highly branched tubules that subdivide from the ends of the bronchi and become progressively smaller as they pass deeper into the lungs. The bronchioles lack cartilage rings.
- **Alveoli**—The alveoli, tiny air sacs, are found at the ends of each of the bronchioles. The alveoli are lined with cells that constitute the actual respiratory surface of the lung. Hence, the alveoli are the functional unit of the lung. Capillaries surround the alveoli. They carry away the oxygen absorbed in the moist lining so it can be transported to body tissues.
- **Lungs**—The lungs are composed of the bronchi, bronchioles, alveoli, and their supporting tissues. The functional unit of the lung is the alveolus (alveoli, plural), whose function is described above.

Breathing is a mechanical process used to move air into the lungs as efficiently as possible. This process involves muscular movements of the **diaphragm** and rib cage, which raise and lower pressures within the chest cavity. As the diaphragm is lowered, pressure in the chest cavity decreases and air containing oxygen is forced into the lungs by normal atmospheric pressures. As the diaphragm is raised, pressure in the chest cavity increases and air containing carbon dioxide and water vapor is expelled.

Deoxygenated blood enters through capillaries *(A)* branching from the pulmonary artery. As the blood passes through the capillary network *(B)* surrounding the alveoli, oxygen diffuses into the blood and carbon dioxide diffuses out of the blood. Oxygenated blood then returns to the heart via capillaries *(C)* leading to the pulmonary veins.

Adaptations for Human Respiration

Gas exchange in the alveoli is accomplished by simple diffusion. Oxygen in the bloodstream combines with hemoglobin to form **oxyhemoglobin**. In a reversal of the process that forms oxyhemoglobin, oxygen is released into the cells, where it is used in the chemical process of **aerobic respiration**. At the cells, carbon dioxide formed as a waste product of cellular respiration diffuses into the blood and is carried in the plasma in the form of the **bicarbonate ion**. Carbon dioxide and water vapor are released from the blood into the alveoli for removal from the body.

At the cellular level, food molecules are broken down chemically into energy and certain waste products. When this process occurs in the absence of molecular oxygen, the process is known as **anaerobic respiration**. Both anaerobic and aerobic respiration are discussed in detail on pages 178–180.

Excretion is how human beings remove metabolic wastes from their cells and release them into the environment. The by-products of metabolic activity in humans may be harmful to the human body if they are not removed from where they are produced. Excretion is carried out by a set of specialized excretory tissues and organs, with the help of the transport system. The organs that assist the process of excretion are as follows.

- **Lungs**—The lungs are responsible for the excretion of carbon dioxide and water vapor, which result from the process of aerobic respiration. These materials diffuse out of the blood into the alveoli and are expelled in the breathing process when exhaling.
- **Liver**—The liver's role in excretion includes the recycling of worn-out red blood cells and the production of urea, which results from the breakdown of amino acids. The chemical breakdown of amino acids is known as **deamination**.

- **Sweat glands**—The sweat glands of the skin play a role in excretion by removing water, salts, and urea from the blood and excreting them as perspiration. Heat, a by-product of metabolism, is also removed from the tissues by the sweat glands. As water in the perspiration evaporates, it carries with it body heat that would otherwise overheat the cells.

- **Kidneys**—The kidneys in the urinary system are major excretory organs in human beings. They help regulate the chemical composition of the blood and thereby the chemical composition of the body tissues. Two arteries that branch off the aorta carry blood to the kidneys for filtering. These arteries quickly branch into many capillary networks, each known as a **glomerulus**. The glomerulus is nested inside a cup-shaped structure, **Bowman's capsule**, where many soluble blood components (including water, salts, urea, and soluble nutrients) are absorbed from the blood by diffusion. Bowman's capsule is part of a larger structure, known as the **nephron**, that is the functional unit of the kidney. In a looped portion of each nephron, active transport is used to reabsorb most of the soluble nutrient molecules, certain mineral ions, and some water. These reabsorbed components are returned to the blood flowing out of the kidney by way of veins to the vena cava. The concentrated mixture of waste materials remaining in the nephron, including water, salts, and urea, is known as urine.

- **Ureters**—The two ureters, which are small tubes, conduct urine from the kidneys through the lower abdomen to the urinary bladder.

- **Urinary bladder**—The urinary bladder collects urine from the ureters and stores it for periodic excretion from the body.

- **Urethra**—The urethra is a small tube that leads from the urinary bladder to the outside of the body. Urine is released to the environment through the urethra.

Nephron (Magnified)

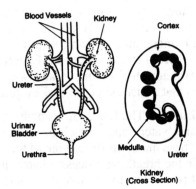

Urinary System

1. When humans exhale, air passes from the trachea directly into the
   (1) bronchioles   (3) bronchi
   (2) alveoli   (4) pharynx

2. Which human excretory structure aids in the maintenance of normal body temperature?
   (1) sweat gland   (3) liver
   (2) nephron   (4) urinary bladder

3. What is the principal function of structure X represented in the diagram below?

   (1) filtration of cellular wastes from the blood
   (2) transport of urine out of the body
   (3) storage of urine
   (4) secretion of hormones

4. The diagram below represents part of a capillary in a specific region of the human body.

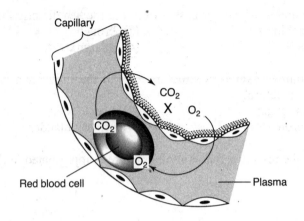

The region labeled X represents part of
(1) a glomerulus   (3) a villus
(2) an alveolus   (4) the liver

5. An individual running a marathon may experience periods of oxygen deprivation that can lead to
(1) anaerobic respiration in muscle cells, forming lactic acid
(2) aerobic respiration in muscle cells, generating glycogen
(3) anaerobic respiration in liver cells, producing glucose
(4) aerobic respiration in liver cells, synthesizing alcohol

6. Which structure is lined with a ciliated mucous membrane that warms, moistens, and filters air?
(1) pharynx   (3) epiglottis
(2) alveolus   (4) nasal cavity

7. In humans, the ureter transports urine from the
(1) blood to the kidney
(2) liver to the kidney
(3) kidney to the urinary bladder
(4) urinary bladder to outside the body

8. Which set of symptoms would most likely lead to a diagnosis of asthma?
(1) enlargement and degeneration of the alveoli
(2) constriction of the bronchial tubes and wheezing
(3) inflammation and swelling of the epiglottis
(4) constriction of the nasal cavity and watery eyes

9. Which structure shown in the diagram below contracts, causing a pressure change in the chest cavity during breathing?

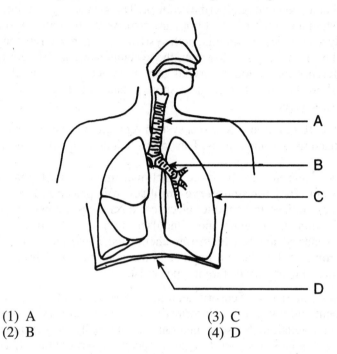

A

B

C

D

(1) A   (3) C
(2) B   (4) D

10. What is the function of the nephron?
   (1) It breaks down red blood cells to form nitrogenous waste.
   (2) It regulates the chemical composition of the blood.
   (3) It forms urea from the waste products of protein metabolism.
   (4) It absorbs digested food from the contents of the small intestine.

Regulation is the life function by which human beings control and coordinate other life functions to maintain existence. Regulation involves both nerve control and chemical (endocrine) control. The nervous system and the endocrine system are similar in that both are responsible for helping to regulate the body's activities. To do this, both systems use chemical messengers that affect other living tissues. However, the systems work at different rates and over different time periods. Nervous responses occur very rapidly and have a very short duration. Endocrine responses take longer to initiate, but their effect lasts longer.

Nervous System Regulation—Nervous system regulation is carried out by a set of specialized tissues and organs. The organs that assist the process are as follows:

- **Neurons**—Neurons are the basic functional units of the human nervous system. They transmit **nerve impulses** from place to place in the body. They initiate and conduct electrical discharges, known as nerve impulses, along their membranes. Impulses are transmitted from cell to cell by means of **neurotransmitter** chemicals. These are secreted by one neuron and stimulate an impulse in a second neuron. Three structurally and functionally different types exist.

  ➤ **Sensory neurons**—Sensory neurons receive stimuli from the environment and transmit this information to the central nervous system for interpretation. Sensory neurons are normally concentrated in organs specially designed to receive specific types of stimuli. These sensory receptor organs include the eyes, ears, nose, tongue, and skin.

  ➤ **Interneurons**—Interneurons exist primarily in central nervous system organs, although they may also be found in nerve centers in other parts of the body. The interneurons are responsible for interpreting sensory impulses brought to them by the sensory neurons. Interneurons also transmit commands to those motor neurons leading back to the body's effector organs (muscles or glands).

  ➤ **Motor neurons**—Motor neurons carry impulses from the command center in the central nervous system to effector organs (muscles or glands), where an appropriate response is initiated.

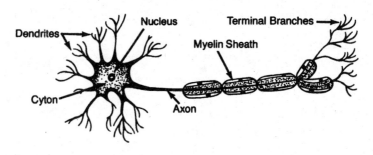

Vertebrate Neuron

- **Nerves**—Nerves are bundles of neurons that may contain a single type of neuron (sensory or motor nerves) or two separate types (mixed nerves). A fatty sheath protects the neurons in a nerve from coming into contact with each other and short-circuiting the impulses they carry. Nerves are specialized for conducting impulses over comparatively long distances at high rates of speed.
- **Brain**—The brain is a large organ composed of a mass of interneurons located within the cranial cavity in humans. The brain is one of the most highly specialized parts of the human body. It is responsible for regulating everything from the simplest to the most complex human activity. The brain is subdivided into three major regions, each responsible for a specific aspect of human behavior. The **cerebrum** is responsible for conscious thought, memory, sense interpretation, reasoning, and other voluntary activities. The **cerebellum** is responsible for coordinating muscular activities and helping the body maintain physical balance relative to its surroundings. The **medulla oblongata** is responsible for the regulation of automatic activities including heartbeat, breathing, blood pressure, and peristaltic activity.
- **Spinal cord**—The spinal cord is continuous with the brain. It extends downward from the base of the brain along the dorsal surface of the body. It is encased within the bony vertebral column, which protects it from mechanical damage. The major functions of the spinal cord include connecting the brain to the peripheral nerves and coordinating the reflex response. Reflexes are simple, inborn, involuntary patterns of behavior that permit immediate, unthinking responses to potentially dangerous situations.

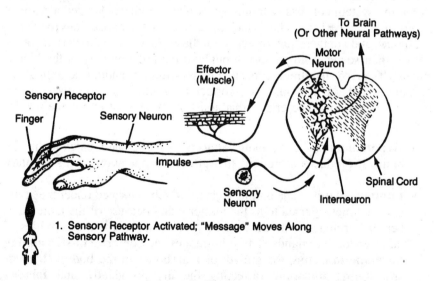

1. Sensory Receptor Activated; "Message" Moves Along Sensory Pathway.

Reflex Arc

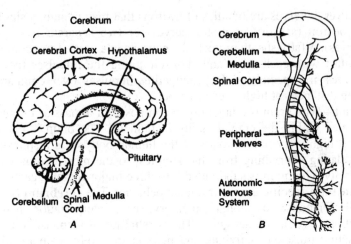

Central Nervous System

- **Peripheral nervous system**—The peripheral nervous system consists of all types of nerves that branch through the body from the central nervous system. The peripheral nervous system is organized into two divisions. The **somatic nervous system** includes the nerves that control the actions of the voluntary skeletal muscles. The **autonomic nervous system** consists of nerves regulating automatic functions such as the actions of glands and involuntary muscle.

Endocrine Regulation—Endocrine regulation is carried out by a set of specialized tissues and organs, with the help of the transport system. The endocrine system consists of a number of discrete glands located at various points in the body. These glands lack ducts to carry their secretions to the target tissues and are therefore known as ductless glands. The endocrine glands deliver their secreted **hormones** throughout the body by way of the bloodstream. **Negative feedback** controls many aspects of endocrine regulation. Negative feedback and other self-correcting mechanisms are discussed in further detail below. The organs that assist the process of endocrine regulation are as follows.

- **Hypothalamus**—The hypothalamus is a small gland located within the brain. It produces secretions that affect the operation of the pituitary gland.
- **Pituitary gland**—The pituitary gland, which is located under the brain, is sometimes referred to as the master gland because of the large number of hormones it produces. Many of them control the activities of other endocrine glands. These hormones include **growth-stimulating hormone** (affecting the growth of long bones in the body), **thyroid-stimulating hormone** (affecting the thyroid gland), and **follicle-stimulating hormone** (affecting the ovaries in females).

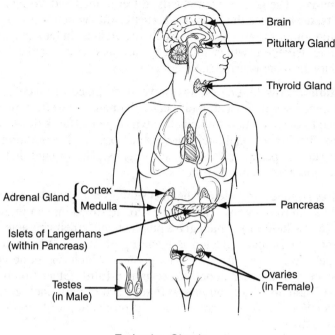

Endocrine Glands

- **Thyroid gland**—The thyroid gland is located in the neck region, surrounding the trachea. Its principal hormone is **thyroxin** (affecting the general metabolic rate of the body).
- **Parathyroid gland**—The parathyroid glands are embedded within the thyroid gland. The hormone produced by the parathyroids is **parathormone** (affecting the metabolism of calcium in the body).
- **Adrenal gland**—The adrenal glands, located on the kidneys within the abdomen, consist of two separate regions, the **cortex** (outer region) and the **medulla** (inner region). The adrenal cortex secretes steroid hormones that regulate water balance and blood pressure as well as the conversion of complex protein and fat molecules into simpler glucose. The adrenal medulla secretes the hormone **adrenaline**. It increases the rates of metabolism, heartbeat, and breathing and stimulates the conversion of complex glycogen molecules into simpler glucose.
- **Islets of Langerhans**—The islets of Langerhans, small groups of glandular tissue scattered throughout the pancreas, produce the hormones **insulin** and **glucagon**. These two hormones have opposite effects on the storage of sugar in the liver and muscles of the body. Insulin promotes the storage of excess blood sugar as glycogen, thus lowering blood sugar levels. Glucagon stimulates the conversion of glycogen back into glucose, thus raising blood sugar levels.

- **Gonads**—The gonads (sex glands) differ in men and women. In males, the **testes** produce the hormone **testosterone**, which is important in the promotion of male secondary sex characteristics. In females, the **ovaries** secrete **estrogen**, which influences various aspects of the female secondary sex characteristics.

Locomotion is how human beings move from place to place within their environment. As with most animal species, humans are well adapted to move from place to place in their immediate environment. The skeletal system and its muscle attachments make this movement possible. Locomotion is carried out by a set of specialized tissues and organs. The organs that assist the process of locomotion are as follows.

- **Bones**—Bones provide mechanical support and protection for the body and its internal organs. These bones are arranged into an internal skeleton (**endoskeleton**), which gives human beings their characteristic body shape. The human skeleton is made up of over 200 bones of varying shapes, sizes, and functions. The point at which one bone comes into contact with another bone is known as a **joint**. Other functions of the bones include protection of soft tissues and organs, anchorage sites for muscles, leverage for movement, and production of blood cells in the bone marrow.

Skeletal System

90

- **Cartilage**—Cartilage is a flexible, fibrous connective tissue that pads the joints between bones. Cartilage is found in many of the flexible parts of the body, including the outer ear and the nose. In embryos and very young children, cartilage is abundant, making up the major portion of the skeleton. In adults, it is much reduced, being found only at bone ends and in flexible portions of the body such as the trachea.

- **Muscles**—Muscles in the human being are of three types. **Visceral** and **cardiac muscles** are involuntary muscles controlled primarily by the autonomic nervous system. **Skeletal muscles** are voluntary and are those most directly involved in human locomotion. Skeletal muscles normally work in opposing pairs, each pulling a joint in a different direction. By coordinating these muscle pairs, the body is able to move skeletal levers to accomplish movement from place to place. Muscles that extend (open) joints are known as **extensors**, while muscles that flex (close) joints are known as **flexors**.

  Muscles require energy in order to contract. During heavy exercise, oxygen supplies may not be sufficient to release energy to the muscle cells by aerobic means. At such times, the muscle cell has the capacity to use **lactic acid fermentation** to supply additional energy. The buildup of lactic acid in the muscle tissues is known as **muscle fatigue**.

- **Tendons**—Tendons are responsible for attaching muscles to the bones of the skeleton. They are composed of tough, inelastic connective tissue.

- **Ligaments**—Ligaments attach bones to other bones at the skeletal joints and are composed of tough, elastic connective tissues that are able to stretch as the joints flex.

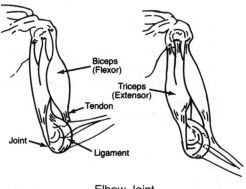

Elbow Joint

QUESTION SET 2.3—HUMAN PHYSIOLOGY 3
(ANSWERS EXPLAINED, P. 289)

1. During a race, the body temperature of a runner increases. The runner responds by perspiring, which lowers body temperature. This process is an example of
   (1) maintenance of homeostasis
   (2) an antigen-antibody reaction
   (3) an acquired characteristic
   (4) environmental factors affecting phenotype

2. Which substances are secreted at the endings of nerve cells?
   (1) antibodies   (3) neurotransmitters
   (2) antigens   (4) lipids

3. Which row in the chart below contains the words that best complete this statement?

   "The <u>I</u> glands produce <u>II</u>, which are transported by the <u>III</u> system."

Row	I	II	III
A	Digestive	Hormones	Circulatory
B	Endocrine	Enzymes	Lymphatic
C	Endocrine	Hormones	Circulatory
D	Digestive	Enzymes	Lymphatic

   (1) A   (3) C
   (2) B   (4) D

4. All living things carry out a variety of life functions such as coordination, excretion, digestion, circulation, and synthesis. Select *two* of the life functions listed. Define the two life functions you selected and explain how they interact to keep an organism alive. [4 points]

   _____

5. Which statement does *not* correctly describe a function of cartilage?
   (1) It anchors muscles to bones.
   (2) It provides flexibility in an embryo.
   (3) It makes up the outer ear.
   (4) It cushions bones at a joint.

6. A reflex arc is illustrated in the diagram below.

Muscle

X

Spinal cord

Structure X represents
(1) an effector
(2) a motor neuron
(3) an interneuron
(4) a receptor

7. One of the functions of the human endoskeleton is to
(1) transmit impulses
(2) produce blood cells
(3) produce lactic acid
(4) store nitrogenous wastes

8. A human skeleton is shown in the illustration below.

The elongation of structures A and B was stimulated by a hormone produced by the

(1) islets of Langerhans
(2) liver
(3) pituitary gland
(4) striated muscles

A

B

9. When a child runs to his/her mother after hearing a clap of thunder, the child is using
   (1) the central nervous system only
   (2) the peripheral nervous system only
   (3) both the central and the peripheral nervous systems
   (4) neither the central nor the peripheral nervous systems

10. The humerus, the bone in the upper arm of a human, is directly connected to other bones in the arm by
    (1) cartilage   (3) extensors
    (2) tendons   (4) ligaments

11–12 For each phrase in questions 11 and 12, select from the list below the answer best described by that phrase. Then record its number.

<u>Endocrine Glands</u>
(1) thyroid
(2) adrenal
(3) islets of Langerhans
(4) parathyroid

11. Secretes a hormone in times of emergency, accelerating metabolic activities

12. Requires a supply of iodine to synthesize its hormone

II. HOMEOSTASIS

The components of the human body, from organ systems to cell organelles, interact to maintain a balanced internal environment. To accomplish this successfully, organisms possess a diversity of control mechanisms that detect deviations and make corrective actions.

When all components of the human body are functioning at peak efficiency, each helps to establish a balanced set of conditions that supports the continuation of life. This balance, or steady state, is known as **homeostasis**. In this state, all systems work together to perform their specialized functions and to synthesize their specialized products. The resulting products are used to perform other life processes. The human body contains sensory tissues that can detect and measure the concentrations of many organic and inorganic compounds. When inappropriate levels of these materials are found in the body, a response is initiated that helps to correct the imbalance and reestablish homeostasis. Many different mechanisms, known as **feedback loops**, exist to assist in the maintenance of homeostasis. Examples of feedback loops include the following.

- As carbon dioxide builds up in the body tissues as a result of aerobic respiration, it diffuses into the blood for elimination from the body as respiratory waste. Elevated concentrations of dissolved carbon dioxide are detected by the central nervous system. Commands are sent to the breathing mechanisms that increase the breathing rate. The increased breathing rate allows dissolved carbon dioxide to exit the bloodstream at an accelerated pace and introduces atmospheric oxygen into the blood at the same time. As the central nervous system detects a reduced carbon dioxide concentration in the blood, commands are sent that slow the breathing rate to normal levels as homeostatic balance is reestablished once more.
- When human skin comes into contact with a hot surface, the body's tissues may be damaged as a result. A **reflex arc** consisting of a sensory neuron, interneuron, and motor neuron respond to this danger. The sensory neuron detects the stimulus and identifies it as heat. This information is transmitted to an interneuron in the spinal cord. A command response is then initiated that is sent along a motor neuron to muscles. The muscles are commanded to move the part of the body away from the hot surface. This reflex action is involuntary and very rapid. The quick response helps to protect the skin and underlying tissues from damage and maintain homeostatic balance.
- The mechanism that controls many aspects of endocrine regulation is **negative feedback**. It operates on the principle that the effects of a particular hormone may inhibit the further production of that hormone but may stimulate the production of another hormone. A good example of negative feedback is the insulin-glucagon feedback loop. Elevated

sugar levels in the blood stimulate the production of insulin (which functions to store glucose in the liver) while inhibiting the production of glucagon. As blood sugar levels drop as a result of insulin's action, glucagon production increases. As a result, glucose is released from the liver, while insulin production is inhibited.

- Sensors in the central nervous system are specialized to monitor the temperature of the blood. When this temperature exceeds 98.6°F (38°C), by a few tenths of a degree, the central nervous system sends out a series of commands that have various effects. Breathing deepens and quickens, allowing heat to escape from the body by means of exhaled air and water vapor. Sweat glands on the face and body open. Perspiration beads begin to appear on the skin surface, where they can evaporate and cool the skin further. By initiating these responses, the body avoids overheating and damaging sensitive tissues throughout. This protection is important in maintaining overall homeostatic balance.

If there is a disruption in any human system, there may be a corresponding imbalance in homeostasis. When the balance of the body's systems is disrupted, cells, tissues, and their associated organs are affected. This may lead to failure of one or more organs or systems within the body. When such a failure occurs, the homeostatic balance of the entire organism can be thrown off. This can lead to illness or death, depending on the severity of the imbalance.

For example, untreated high blood pressure can damage the nephrons, the functional units of the human kidney. If the high blood pressure condition persists for a long period of time, the tissues of the kidneys can become so badly damaged that the kidneys cease functioning altogether. As the kidneys stop functioning to filter the blood of urea and salts, the concentration of these wastes builds up to unsafe levels in the blood. As this unfiltered blood reaches other body tissues, it poisons them and causes them to slow or stop functioning. Eventually, the body's tissues become so badly poisoned with toxic metabolic wastes that they can no longer support life activity. At this point, organs and body systems fail, and the person may die if he/she is not treated. See the section "Disease as a Failure of Homeostasis" for more information about disease as a failure of homeostasis (see page 184).

III. CELL FUNCTION AND STRUCTURE

The organs and systems of the body help to provide all the cells with their basic needs. The cells of the body are of different kinds and are grouped in ways that enhance how they function together. Each of the body systems and its associated organs mentioned provides a different functional life activity. These systems and organs work together to support life. Life really begins with the biochemical activity of the individual cells that make up a multi-celled organism such as the human being. Each living cell in the body needs food, water, and oxygen to function appropriately These cells must also be cleaned constantly to free them of toxic metabolic wastes such as urea and carbon dioxide.

To illustrate, consider how the body's systems work together to provide for the needs of an individual cell living deep inside the body of a human being. The digestive system provides food for this cell by breaking down the complex structure of ingested foods into soluble components that can be absorbed by the cell. Oxygen needed by the cell for respiration is absorbed from the atmosphere by the respiratory system. The circulatory system absorbs these soluble food and oxygen molecules from their respective points of entry into the body. It moves them near the cell so the molecules can be absorbed through the cell membrane. The cell metabolizes these molecules, producing cell components and energy needed to support the cell's life activities. The cell also produces a variety of waste materials, including carbon dioxide, salts, and urea, that must be removed from the cell and from its immediate environment. Once again, the circulatory system assists by absorbing these waste molecules and moving them to organs that specialize in the release of wastes to the outside environment. Carbon dioxide is removed from the blood in the respiratory system, while salts and urea are filtered out of the blood by the excretory system. Meanwhile, all these systems are being coordinated by the regulatory system. As a result, a homeostatic balance is struck. All systems operate in a coordinated fashion to sustain life.

Tissues

Each cell in the body is a part of the complex fabric of the organism. Cells are specialized to perform specific functions in the body. Groups of such specialized cells are known as **tissues**. They are often found clustered together within organs where they carry on their specialized activities with great efficiency. Examples of human body tissues include:

- skin tissues that provide an impervious layer surrounding and protecting interior organs;
- bone tissues that secrete bone mineral and maintain the body's endoskeleton;
- blood tissues that transport oxygen and other components as well as protect the body against disease;
- muscle tissues that move skeletal levers during locomotion and specialize in the physical movement within certain body organs;
- nerve tissues that provide a communication link among all body systems; and
- endocrine tissues that produce a variety of chemical substances that control the functioning of other body tissues.

Many other specific tissues operate within the body. Each of the tissue types listed on page 97 can be subdivided into several highly specialized tissues that carry on very narrow roles in maintaining homeostatic balance. For example, the broad category of muscle tissue contains skeletal muscle that moves skeletal levers, visceral muscle that creates internal movements such as peristalsis in the intestines, and cardiac muscle that establishes and maintains rhythmic contractions of the heart and arteries. Blood tissues include red blood cells that carry oxygen, phagocytes that engulf bacteria, lymphocytes that produce antibodies against disease, platelets that specialize to produce clots, and a variety of other cell types, each with its own special function in the body. Nerve tissues are subdivided into sensory neurons that receive environmental stimuli, motor neurons that control muscles and glands, and interneurons that process sensory input and send motor commands.

Cells have particular structures that perform specific jobs. These structures perform the actual work of the cell. Just as systems are coordinated and work together, cell parts must also be coordinated and work together. The cell is extremely complex and is composed of smaller functional parts known as cell organelles. Many of these organelles were unknown to scientists until 20th-century scientific techniques and equipment (electron microscope, ultracentrifuge) made them visible to researchers. Each type of organelle has special work to do that promotes the survival of the cell. In a sense, these tiny structures are similar to organs and organ systems. They provide specialized functions coordinated to establish and maintain homeostasis within the cell. Organelles are specialized to carry on the life functions of nutrition, respiration, circulation, excretion, regulation, reproduction, and locomotion within their particular cell.

Cell Membrane

Each cell is covered by a membrane that performs a number of important functions for the cell. These include separation from its outside environment, controlling which molecules enter and leave the cell, and recognition of chemical signals. The processes of diffusion and active transport are important in the movement of materials in and out of cells. The outer membrane of the cell, which regulates the transport of materials into and out of the cell, is known as the **plasma (cell) membrane**. The plasma membrane also serves as the interface between the cell interior and its outside environment. Recent research points to the plasma membrane's function as a site of cell-cell recognition, which has important implications for the control of disease, including cancer.

The structure of this membrane has been the subject of study for many years. Several models have been developed to describe it. The **fluid-mosaic model** (developed by Singer and Nicholson in 1972) is the currently accepted model for this structure. This model shows the principal component of the plasma membrane to be a double layer of lipid molecules with many embedded proteins, known as receptors. This structure allows the passage of small, soluble molecules such as monosaccharides, amino acids, and dissolved gases

Fluid-Mosaic Model

while preventing the passage of larger molecules (for example, starches and proteins). Current research indicates that size may be only one factor affecting the passage of materials through the cell membrane. Chemical recognition factors are also important and are regulated by a "lock-and-key" mechanism in which chemical receptor sites attract, link with, and transport specific molecules across the membrane boundary to the cell interior.

Passive and Active Transport

All particles in the environment of the cell contain energy of motion (kinetic energy). As these particles collide with the plasma membrane, they may pass through by slipping between the lipid molecules making up the membrane. Because no cellular energy is expended in this absorption process, it is known as **passive transport**. **Diffusion** is a form of passive transport in which a dissolved substance passes through the plasma membrane from a region of higher relative concentration to a region of lower relative concentration of that substance. The diffusion of water molecules into or out of the cell (through the plasma membrane) is known as **osmosis**.

Some cells are capable of expending cellular energy in order to pump materials into (or out of) themselves from regions of lower relative concentration to regions of higher relative concentration. Because cellular energy must be expended to accomplish this process, it is known as **active transport**. The plasma membrane may be involved in active transport through the use of special carrier proteins embedded in the membrane's lipid layers. These carrier proteins take advantage of the specific chemical nature of desired components by attaching to them and carrying them through the membrane to the cytoplasm.

IV. CHEMISTRY OF LIVING ORGANISMS

Many organic and inorganic substances are necessary to permit life-sustaining chemical reactions to take place in the cell. Large organic food molecules such as proteins and starches must initially be broken down (digested to amino acids and simple sugars, respectively) in order to enter cells. Once nutrients enter a cell, the cell will use them as building blocks in the synthesis of compounds necessary for life. In addition, molecular oxygen, water, and other inorganic molecules provide materials needed to complete these reactions.

The same chemical elements that comprise the Earth's crust, water, and atmosphere also make up the bodies of living things. However, the proportion of elements in living matter is different from the proportion of elements in nonliving matter. Living things are made up of relatively large percentages of the elements **carbon** (chemical symbol **C**), **hydrogen** (chemical symbol **H**), **oxygen** (chemical symbol **O**), and **nitrogen** (chemical symbol **N**). Elements found in lower percentages include sulfur (chemical symbol S), phosphorus (chemical symbol P), magnesium (chemical symbol Mg), iodine (chemical symbol I), iron (chemical symbol Fe), calcium (chemical symbol Ca), sodium (chemical symbol Na), chlorine (chemical symbol Cl), and potassium (chemical symbol K).

Electron

1 p
Hydrogen

2 p
2 n
Helium

6 p
6 n
Carbon-12

6 p
8 n
Carbon-14

Atomic Structure

H Atom   H Atom
Hydrogen Gas

H Atom   O Atom   H Atom
Water

4 H Atoms
1 C Atom
Methane

Sharing of Electrons in Covalent Bonds

Compounds

Compounds are substances composed of two or more different elements bound together chemically. The bonds holding these compounds together may be ionic (involving the transfer of electrons between atoms) or covalent (involving the sharing of electrons between atoms). [*Note: The chemistry of bonding is not a required understanding for Biology—The Living Environment. For details on this topic, consult a chemistry textbook.*]

Two broad categories of compounds exist in nature: **inorganic** and **organic** compounds. All organic compounds contain the elements carbon and hydrogen. Inorganic compounds may contain any of Earth's elements but rarely contain carbon and hydrogen together. Living things are composed of both inorganic and organic compounds.

Inorganic Compounds
Inorganic compounds commonly found in living things include the following.

- **Water**—Living things consist of 60 to 98 percent water, which acts as a medium for transport and for chemical activities within the cell.
- **Salts**—Salts are chemicals composed of metallic and nonmetallic ions combined by ionic bonds. They are important for maintaining osmotic balance in the cell and for supplying ions necessary for many of the cell's chemical reactions.
- **Acids and bases**—These compounds are important for maintaining the proper balance of hydrogen ion concentration (acidity/alkalinity) in the cell.

Organic Compounds
Organic compounds commonly found in living things include the following.

- **Carbohydrates** are organic compounds and so contain the elements carbon and hydrogen. In addition, they contain the element oxygen. Carbohydrates generally contain hydrogen and oxygen in the ratio of 2:1. This ratio differs from that found in other organic compounds. Like many organic compounds, carbohydrates are actually formed around a skeleton of carbon atoms linked together in a chain. Some chains are quite short (six or fewer carbon atoms), while others are very long (thousands of carbon atoms).

    The simplest stable carbohydrate is the monosaccharide (**simple sugar**). Monosaccharides form around a chain of 5 or 6 carbon atoms. An example of a monosaccharide is the simple sugar **glucose**, whose chemical formula is $C_6H_{12}O_6$. Glucose is the basic chemical unit of nearly all the more complex carbohydrates.

    A more complex form of carbohydrate is the disaccharide (double sugar), composed of two chemically linked monosaccharide molecules. Disaccharides are commonly formed around a chain of 12 carbon atoms. The chemical process by which this linkage occurs is **dehydration synthesis**. The atoms making up water molecules are removed from the monosaccharide molecules and the two monosaccharides join to form a single disaccharide molecule. Examples of disaccharides are maltose and sucrose.

A complex form of carbohydrate is known as a polysaccharide (many sugar). It forms by chemically combining many monosaccharide molecules via dehydration synthesis. The chain of monosaccharide units forming a polysaccharide may be thousands of units long and may be highly branched. Starch (amylose), glycogen, and cellulose are examples of polysaccharides.

Monosaccharide
Structural Formula: Glucose

Disaccharide
Structural Formula: Maltose

- **Lipids** are composed of the elements carbon, hydrogen, and oxygen. Many of the more common lipids, known as triglycerides, are constructed of a unit of **glycerol** (a three-carbon alcohol) combined chemically via dehydration synthesis with three molecules of **fatty acid** (a hydrocarbon chain with an attached carboxyl group). Examples of lipids include animal fats, plant oils, and waxes.

Glycerol

Fat

Lipid Structural Formula: Triglyceride

- **Proteins** contain the elements carbon, hydrogen, and oxygen, as do carbohydrates and lipids. In addition, proteins contain a substantial proportion of nitrogen. Certain proteins may also contain sulfur. Proteins are polymers made up of repeating units of other, simpler compounds known as **amino acids**. Although there are more than 20 different types of amino acids, they all share a common general structure. All amino acids contain an amino group (–NH$_2$) and a carboxyl group (–COOH) attached to a central carbon atom. A third attached group (radical) varies and gives each type of amino acid its unique properties.

Two amino acid molecules may be joined together chemically by dehydration synthesis to form a dipeptide. The term **peptide** refers to the name of the actual chemical bond between carbon and nitrogen atoms (C–N) that joins the two amino acid units together. As more and more amino acids link together by dehydration synthesis, an amino acid chain, known as a **polypeptide**, is formed. Polypeptide chains form the basis of protein molecules. Because of the almost endless variations in which amino acids may be arranged in a polypeptide, proteins are found in thousands of different forms. The extreme variability of proteins is thought to be the chemical basis for the individual variations in living things. Examples of proteins are structural proteins and enzymes. See below for additional information about the mechanisms of protein synthesis.

R = variable (radical) group of atoms

Structural Formula: Amino Acid

Structural Formula: Dipeptide Formation

- **Nucleic acids** are extremely complex polymers that function to determine the characteristics of individual cells and thereby the characteristics of the entire organism of which they are a part. Principal nucleic acids include **deoxyribonucleic acid (DNA)** and **ribonucleic acid (RNA)**. Nucleic acids are polymers composed of a simpler chemical subunit known as a **nucleotide**. Thousands of these units comprise a single DNA molecule, making DNA one of the largest and most complex of all organic compounds. DNA exists in hundreds of thousands of different forms, depending on the precise arrangement of nucleotides in the molecule. Its variability and its role in protein synthesis are the keys to genetic variation in living things.

DNA nucleotides themselves are quite complex. They are composed of three separate subunits: phosphate group, deoxyribose, and one of four nitrogenous bases, including adenine (A), thymine (T), cytosine (C), and guanine (G). DNA's function in the cell is to code for the production of specific protein types in the cell. This function is discussed in more detail in the section "Genetic Material." RNA nucleotides are similar to those comprising DNA molecules, but substitute ribose sugar for deoxyribose and uracil (U) in place of thymine.

Chemical Reaction: Hydrolysis

When taken into the body as food, complex food molecules such as starches, proteins, fats, and nucleic acids must be broken down into their simpler chemical subunits before they can be used by the living cell. These large molecules cannot pass readily through the plasma membrane, but their smaller, soluble subunits can. The act of breaking down these complex molecules is known as chemical digestion and is accomplished by the chemical process of **hydrolysis**. Hydrolysis (splitting with water) involves replacing existing chemical bonds within a complex molecule with the atoms comprising a water molecule, thereby breaking the complex molecule apart at those points and producing simpler molecules. The following word equations will illustrate this process:

$$\text{starch} + N \text{ water} \longrightarrow N \text{ glucose}$$
$$\text{maltose} + \text{water} \longrightarrow 2 \text{ glucose}$$
$$\text{lipid} + 3 \text{ water} \longrightarrow \text{glycerol} + 3 \text{ fatty acids}$$
$$\text{protein} + N \text{ water} \longrightarrow N \text{ amino acids}$$

The end products of hydrolysis are as follows:

Complex Food Molecules	Molecular End Products
Carbohydrates (starches, double sugars)	Simple sugars
Lipids (fats, oils)	Fatty acids and glycerol
Proteins	Amino acids
Nucleic acids	Nucleotides

Hydrolysis

Cell Organelles.

Inside the cell, a variety of specialized structures, formed from many different molecules, carry out the transport of materials (cell membrane and cytoplasm), extraction of energy from nutrients (mitochondria), protein building (ribosomes), waste disposal (cell membrane), storage (vacuole), and information storage (nucleus). Living cells are extremely complex structural and functional units. The major cell organelles and their functions are described below.

- **Plasma (cell) membrane**—the outer membrane of the cell, which regulates the transport of materials into and out of the cell. The cell membrane serves as the interface between the cell and its environment, and ultimately controls the nature of the cell's internal environment. See pages 98–99 for additional information concerning the structure and function of the plasma membrane.
- **Cytoplasm**—a watery medium for the suspension of cell organelles and the circulation of soluble material throughout the cell. The cytoplasm also serves as the site for many of the cell's chemical reactions.
- **Nucleus**—a spherical organelle, usually located near the center of the cell, that contains the cell's genetic information in the form of chromosomes. The nucleus allows the free transfer of that genetic information during synthesis and reproduction.
- **Ribosome**—a small, dense organelle that serves as a site for the manufacture of protein molecules within the cell. The ribosome may be attached to the endoplasmic reticulum or may be floating free in the cytoplasm.
- **Mitochondrion**—a small organelle that contains the enzymes necessary to allow the cell to perform certain aspects of chemical respiration.
- **Vacuole**—a membrane-bound organelle containing water, enzymes, and other substances. The vacuole may serve to store food molecules, nonremovable wastes, or secretion products.
- **Nucleolus**—a small organelle, located within the nucleus, that functions in the cell's synthesis mechanism. It forms ribosomes involved in the manufacture of proteins.
- **Endoplasmic reticulum**—a series of intracellular membranes that functions in the cell's synthesis mechanism. It houses ribosomes, accepts manufactured proteins, and transports these proteins to the plasma membrane for incorporation into the membrane or for secretion to the cell exterior.
- **Golgi complex**—a series of membrane-bound organelles that functions in the cell's synthesis mechanism. It accepts manufactured proteins and transports these proteins to the plasma membrane for incorporation into the membrane or for secretion to the cell exterior.
- **Lysosome**—a specialized vacuole that aids nutrition by carrying digestive enzymes and by merging with food-containing vacuoles. The lysosome may also help to recycle aging or defective cells.

Plant Versus Animal Cells

- **Centriole**—a cylindrical structure, found primarily in animal cells, that apparently functions in the process of cell division.
- **Chloroplast**—a chlorophyll-containing structure, found primarily in plant and algae cells, in which the chemical reactions of photosynthesis occur.
- **Cell wall**—a structure, found primarily in plants, that provides mechanical support and protection for the cell.

Cellular Interaction

Receptor molecules play an important role in the interactions between cells. Two primary agents of cellular communication are hormones and chemicals produced by nerve cells. If nerve or hormone signals are blocked, cellular communication is disrupted and the organism's stability is affected.

Receptor molecules are specialized proteins embedded in the plasma membrane. These receptor proteins are able to capture specific types of molecules that come into contact with the membrane and move them to the cell interior. Once inside the cell, these captured molecules can affect the behavior of the cell if they have the capacity to do so. Two types of molecules that have this capacity are hormones and neurotransmitters.

- **Hormones** are biochemicals produced and secreted by endocrine glands. Each hormone exerts its effect on specific target tissues in the body. In order to be absorbed by the individual cells of a target tissue, these neurotransmitters must be captured by receptor molecules and

107

moved into the cell interior. Endocrine glands are located far from the tissues that they affect, and they have no tubes to connect them to these tissues. Since no ducts carry them, hormones are distributed to other parts of the body by means of the circulatory fluid.

- **Neurotransmitters** are biochemicals produced and secreted by the neuron's terminal branches. They carry the nerve impulse from one neuron to the next. An example of such a neurotransmitter is acetylcholine. In order to be absorbed by the dendrites of a subsequent neuron, these neurotransmitters must be captured by receptor molecules and moved into the cell interior.

Because these chemicals function in controlling the body's metabolic processes, they are essential in maintaining homeostatic balance. When an endocrine gland fails to produce its hormone in appropriate quantities, the life process controlled by it does not operate properly. This can lead to organ or system failure. It may even threaten the survival of the entire organism. For example, if the pancreas malfunctions and causes insulin production to cease, then blood sugar is not converted to glycogen. This will cause the sugar concentration to build in the blood until it reaches dangerous levels. Tissues throughout the body may be affected, but some, such as kidney tissues, may be destroyed and cease to function. If the kidneys fail, then the bloodstream has no way to cleanse itself, and all the body's tissues will be damaged or destroyed.

Unicellular Life Forms

Essential Question:

- *How are unicellular life forms able to carry out life processes needed to sustain life?*

Performance Indicator 4.1.3 *The student should be able to explain how a one-celled organism is able to function despite lacking the levels or organization present in more complex organisms.*

The organellar structures present in some single-celled organisms (known as **protists** and **monera**) act in a manner similar to the tissues and systems found in multicellular organisms. This enables single-celled organisms to perform all of the life processes needed to maintain homeostasis. While these protists and monera exist in nature in the millions, a few species are well-known to scientists. These include the **paramecium**, the **ameba**, and the euglena. The paramecium is a slipper-shaped protist that inhabits freshwater environments. At nearly 400 micrometers (400 millionths of a meter), one of the largest of the paramecia is *Paramecium caudatum*. Its organelles function efficiently to allow the organism to thrive in its environment.

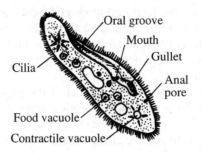

Paramecium

- Nutrition depends on the oral groove and food vacuoles. The oral grooves allow the paramecium to ingest food particles. The food vacuoles permit the digestion of these particles to a simple chemical form that can be used by the cell. Egestion is accomplished by means of the anal pore.
- Transport is accomplished by means of the plasma membrane and the cytoplasm, which permit the absorption and circulation of materials.
- Respiration takes place in the mitochondria, which contain the enzymes needed to release energy from food molecules.
- Excretion of metabolic wastes occurs directly though the plasma membrane. A specialized organelle, the contractile vacuole, expels excess water from the cell.
- Regulation is the function of the nucleus, which manufactures proteins that control the cell's metabolic activities.
- Locomotion is accomplished in paramecia by cilia that sweep in a coordinated motion to propel the cell through the watery environment.

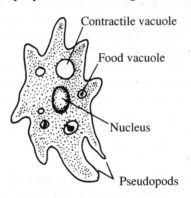

Ameba

The ameba, *Ameba proteus*, is also a relatively large freshwater protozoan but lacks a defined shape. The ameba is also successful in its environment because it contains structural adaptations on the organella level that allow it to carry on the basic life functions efficiently.

- Nutrition is aided by the pseudopods and food vacuoles. The pseudopods allow the ameba to ingest food particles. The food vacuoles permit the digestion of these particles to a simple chemical form that can be used by the cell. Egestion is accomplished by means of reverse pinocytosis at the cell membrane.
- Transport is accomplished by means of the plasma membrane and the cytoplasm, which permit the absorption and circulation of materials.
- Respiration takes place in the mitochondrion, which contains the enzymes needed to release energy from food molecules.
- Excretion of metabolic wastes occurs directly though the plasma membrane.
- Regulation is the function of the nucleus, which manufactures proteins that control the cell's metabolic activities.
- Locomotion is accomplished in ameba by the use of pseudopods, which flow forward on a solid surface to propel the cell through its environment.
- **Algae** is a term used to refer to a large number of unicellular and colonial organisms that resemble plants in their ability to carry on photosynthesis as their nutritional process. Some algae (kelp, seaweeds) form large aggregations that mimic true multicellular plants and contain specialized structures that resemble, in appearance and function, leaves, stems, and roots. Algae are found in virtually all aquatic and marine ecosystems where light is present and are responsible for providing a significant amount of Earth's atmospheric oxygen. Like other unicellular organisms, algae are capable of performing all of the basic functions necessary to sustain life.
- **Bacteria** are simple unicellular organisms that inhabit many of Earth's least-hospitable environments. These simple organisms generally lack organized nuclei but are still able to reproduce rapidly under favorable conditions. Bacteria fill many different environmental niches, including roles as decomposers, pathogens (disease organisms), and converters of atmospheric nitrogen into soil nutrients that help plants grow. Some bacteria are chemosynthesizers capable of deriving cell energy from the chemical bond energy of inorganic compounds. A few species of bacteria even thrive in deep-ocean thermal vents under conditions of pressure, toxicity, and temperature that would prove lethal to almost any other form of life.
- **Fungi** are organisms whose structures range from unicellular to loosely multicellular. Fungi occupy an essential niche in the natural environment, along with bacteria, as nature's decomposers and recyclers. In this role, they break down the complex bodies of other living things and recycle them for reuse in the environment. Fungi gain their nutrition by sending root-like structures (mycelia) into the living or dead organic tissues of plants and/or animals and secreting digestive enzymes that

break down the complex tissues into simple, absorbable nutrient molecules. Fungi include molds, yeasts, rusts/smuts, and mushrooms.

Multicellular Life Forms

More familiar to us due to their size and proximity to human experience are the many species of multicellular organisms that inhabit Earth. Multicellular organisms may contain as few as hundreds to as many as several millions of cells. These cells are typically specialized to perform unique functions in the organisms. Despite this specialization, each cell must be able to perform the essential functions necessary to sustain life. It is not the intention of this book to provide a comprehensive survey of all multicellular species, or even all major groupings of these species. Following is a brief overview of the major species groupings (phyla and selected classes), which may be expanded upon at the discretion of the classroom teacher.

Animal-Like Organisms
Animals, in general, are multicellular, heterotrophic ("other-feeding") organisms. Because their nutritional habit involves consuming the bodies of other organisms for food, animals are known collectively as **consumers**. Animals that consume plants are known as **herbivores**, while those that consume animals are known as **carnivores**. Two major subdivisions of the animal kingdom include invertebrates and vertebrates (chordates).

Invertebrates (non-chordates) are those animal-like organisms that lack a chord or spine as part of their structure. Many invertebrate species are soft-bodied, while others develop a rigid exoskeleton to protect inner organs and to provide levers for locomotion. Invertebrates include phyla such as porifora (e.g., sponges), coelenterates (e.g., coral, jellyfish, hydra), flat and round worms (e.g., planaria, nematodes), echinoderms (e.g., sea urchins, sand dollars, starfish), annelids (e.g., earthworms), and arthropods (e.g., grasshoppers, orb spiders), among several others.

Chordates include true chordates and vertebrates. True chordates (e.g., lancelet, amphioxus) are simple multicellular animals that resemble small worms except for the significant inclusion of a simple notochord located along their dorsal aspect. The notochord provides an anchor for the attachment of simple muscles and support of body shape. The earliest direct ancestors of vertebrates are thought by scientists to have resembled these simple chordates. **Vertebrates** are chordates whose dorsal supporting structure has evolved into a jointed adaptation (vertebral column) of cartilage or bone. The vertebral column provides multiple functions for these animals, including basic body support and articulation, neural protection, and muscle attachment. Examples of vertebrates include subgroupings (classes) of animals such as fish (e.g., trout, bass, shark), amphibians (e.g., grass frog, spotted salamander, red eft), reptiles (e.g., painted turtle, garter snake, anolis lizard), birds (e.g., robin, bluebird, red-tailed hawk), and mammals (e.g., gray whale, manatee, timber wolf, white-tail deer, human being).

The adaptations for survival of complex multicellular animals (vertebrate mammals) are typified by those of the human being. These adaptations are detailed in Unit 2, Chapter 1 (pages 67–94) of this book.

Plant-Like Organisms

Plants, in general, are multicellular, autotrophic ("self-feeding") organisms. Two major subdivisions of the plant kingdom include the non-vascular plants and vascular plants. Plants function in the ecosystem as **producer** organisms, a term that relates to their ability to produce simple sugars (glucose) through the process of **photosynthesis**. The sugars are produced in sufficient quantities to serve the nutritional needs of both plants and animals. Plants also produce oxygen gas as a by-product of photosynthesis, which is used by both plants and animals in the process of respiration.

Non-vascular plants, also known as **bryophytes**, are those lacking true roots, stems, or leaves and the vascular (tubular) tissues found in these structures. Examples of bryophytes include mosses and liverworts. These organisms are found typically in environments containing poor or rocky soil, high moisture, and low light conditions. They are low-lying plants that provide ground cover for bare soils and function in these environments to break down rock into smaller fragments that form the basis of soil. This function alters the environment and paves the way for subsequent communities of plants that need deeper soils to germinate, root, and grow. In this role, bryophytes inhabit their ecological niche as pioneer organisms.

Vascular plants, also known as **tracheophytes**, are those having true roots, stems, and leaves that contain vascular (tubular) tissues. These vascular tissues (**xylem** and **phloem**) function to transport water and dissolved minerals upward from the roots through the stems to the leaves (xylem) and transport dissolved sugars downward from the leaves through the stems to the roots (phloem). Examples of tracheophytes include cone-bearing plants (e.g., white pine, blue spruce, juniper) and flowering plants (e.g., wild geranium, shadbush, sugar maple).

All vascular plants contain three types of specialized structures: roots, stems, and leaves. Each of these structures performs a set of functions that help the plant survive in its environment.

- **Roots** penetrate and hold onto the soil, absorbing from it the water and dissolved minerals needed by the plant to support its metabolic activities. Roots function to support the plant, absorb needed materials, and store complex carbohydrates (starches) for use by the plant in its nutritional activities. Roots contain the vascular tissues xylem and phloem.

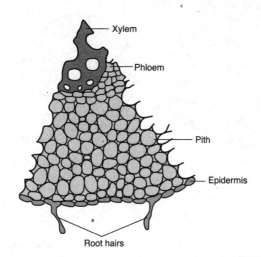

- **Stems** provide a branching structure that is attached to the roots and supports the leaves (or needles) and reproductive structures (flowers or cones) of the plant. The principal role of the stem is to provide for the transport of materials between the roots and the leaves. Stems contain the vascular tissues xylem and phloem.
- **Leaves** (or needles) are specialized to carry on the nutritional activities of the plant and are highly adapted to perform this function at maximum efficiency. Leaves of flowering plants are broad, maximizing the surface area available for the efficient absorption of light needed to energize the photosynthetic process. Leaves contain specialized tissues that further maximize the efficiency of photosynthesis. The **upper epidermis** contains the other leaf tissues and help to minimize moisture loss. The **palisade layer** contains the majority of the chloroplasts of the leaf and is the primary photosynthetic tissue of the leaf. The **spongy layer** of loosely packed cells allows for the free circulation of atmospheric

gases (carbon dioxide and oxygen) that are needed for, or produced by, the photosynthetic reactions. The **lower epidermis** helps to contain the other leaf tissues, but also contains specialized **stomates** (pores) that are bounded by specialized guard cells. **Guard cells** regulate the passage of atmospheric gases into (carbon dioxide) and out of (oxygen) the leaf through the stomates. Finally, the leaf contains **veins** that function to transport materials into (water and dissolved minerals) and out of (dissolved sugars) the leaf. Leaf veins contain the vascular tissues xylem and phloem.

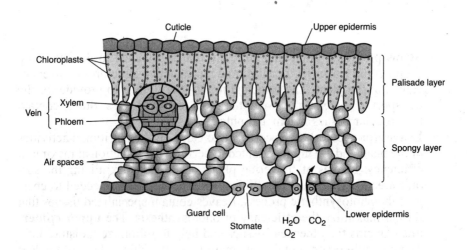

QUESTION SET 2.4—CELL ORGANELLES, BIOCHEMISTRY, UNICELLULAR LIFE FORMS
(ANSWERS EXPLAINED, P. 293)

1. Which structures could most likely be observed in cells in the low-power field of a compound light microscope?
   (1) cell walls and chloroplasts
   (2) ribosomes and endoplasmic reticula
   (3) lysosomes and genes
   (4) nucleotides and mitochondria

2. What would most likely happen if the ribosomes in a cell were not functioning?
   (1) The cell would undergo uncontrolled mitotic cell division.
   (2) The synthesis of enzymes would stop.
   (3) The cell would produce antibodies.
   (4) The rate of glucose transport in the cytoplasm would increase.

3. Which activity is illustrated in the diagram below?

   (1) a virus destroying a cell by extracellular digestion
   (2) a moss plant performing intercellular digestion
   (3) a protozoan ingesting food during heterotrophic nutrition
   (4) a lysosome egesting a food particle into the cytoplasm

4. Which activity is represented by the arrows in the diagram below?

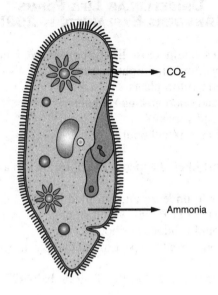

(1) anaerobic respiration
(2) autotrophic nutrition
(3) deamination of amino acids
(4) excretion of metabolic wastes

5. The diagram below represents a green alga.

Which process is most closely associated with structure A?
(1) excretion   (3) locomotion
(2) transport   (4) reproduction

6. Which formula represents an organic compound?
(1) $Mg(OH)_2$ (3) $C_{12}H_{22}O_{11}$
(2) $NaCl$ (4) NH_3

7. The diagram below shows the same type of molecules in area A and area B. With the passage of time, some molecules move from area A to area B.

The movement is the result of the process of
(1) phagocytosis   (3) diffusion
(2) pinocytosis   (4) cyclosis

8–9. Base your answers to questions 8 and 9 on the structural formula of a molecule below and on your knowledge of biology.

8. Which statement best describes this molecule?
   (1) It has the ability to control heredity.
   (2) It has the ability to control reactions.
   (3) It has a high energy content.
   (4) It is involved in photosynthesis.

9. Which formula represents an end product derived from the chemical digestion of this molecule?

(1) $O{=}C{=}O$ (3) $H{-}O{-}H$

(2)

           H
           |
   H—C—OH
           |
   H—C—OH
           |
   H—C—OH
           |
           H

(4)

        H
        |
  H—C—H
        |
        H

10–11. Base your answers to questions 10 and 11 on the diagram below and on your knowledge of biology.

Molecule A   Molecule B

10. In molecule B, what type of group is contained in box Y?
    (1) an amino group   (3) a carboxyl group
    (2) a variable group   (4) a peptide group

11. How many peptide bonds are present in molecule A?
    (1) 1   (3) 3
    (2) 2   (4) 4

---

12. A student observed a wet mount of some stained plant cells in the high power field of a compound light microscope. Diagram A represents the general appearance of these cells. The student then added several drops of a liquid to the wet mount and continued the observations. Diagram B represents the general appearance of the cells a few minutes after adding the liquid.

A   B

The liquid that the student added to the wet mount was most likely
(1) salt water   (3) pond water
(2) distilled water   (4) tap water

13. Which statement is a valid inference concerning structure X represented in the diagram below?

(1) Structure X contains guard cells that regulate glucose intake.
(2) Structure X carries out heterotrophic nutrition.
(3) Structure X produces gametes for asexual reproduction.
(4) Structure X tansports materials for metabolic activities.

V. GENETIC CONTINUITY

> **KEY IDEA 2—GENETIC CONTINUITY** Organisms inherit genetic information in a variety of ways that result in continuity of structure and function between parents and offspring.

Essential Question:

• *How does genetic inheritance help ensure continuity of life?*

A significant unifying principle of biology is the concept that all living organisms possess a set of instructions, in the form of genes, that determine the characteristics of the organisms. These instructions are unique to the species. They determine the general physical and biochemical traits of the group. They are also unique to the individual, coding for the specific set of characteristics that sets one individual organism apart from all others of that species. Students should be familiar with the

• mechanisms by which genetic traits are passed from generation to generation to maintain genetic continuity of the species.
• molecular basis of genetics that is used to maintain genetic continuity mechanism by which it changes through recombination, mutation, and genetic engineering.

You should be familiar with the mechanism by which DNA replicates its structure during normal cell operations and in the process of reproduction. You must also understand the roles of DNA and RNA in the coding of cell-specific proteins and understand the role of these structural and functional proteins (enzymes, hormones, and other substances) in the operation of the cell.

Humans have long used their understanding of heredity to selectively breed new plant and animal varieties as well as to maintain hybrid varieties. Students should appreciate the economic role that these activities have played over the past centuries as well as the scientific basis of such activities. A relatively new branch of science, genetic engineering, has taken on an increasingly important and visible place in the scientific community. You should understand and appreciate this concept and make informed judgments as to the ethical considerations surrounding such research.

Essential Question:

- *How does DNA control the inheritance of genetic traits in living things?*

Performance Indicator 4.2.1 *The student should be able to explain how the structure and replication of genetic material results in offspring that resemble their parents.*

The Genetic Code

Every organism requires a set of coded instructions for specifying its traits. For offspring to resemble their parents, information must be reliably transferred from one generation to the next. Heredity is the passage of these instructions from one generation to another. As scientists have continued their study of genetics, they have learned more and more of the details of the mechanisms of genetics. In one of the most significant discoveries in genetic science, **deoxyribonucleic acid (DNA)** was revealed to be the chemically active agent of the gene. As is now known, DNA replicates itself when chromosomes replicate in the early stages of cell division. DNA is passed from generation to generation during reproduction and acts as genetic factors. DNA interacts with the cell's chemical factory and produces the observable effects of the phenotype when genes are inherited by a cell or an organism. DNA regulates the production of enzymes in the cell and thereby enables the cell to perform the complex cellular chemical reactions necessary to sustain life. (See "DNA Code," page 128, for further information.)

DNA Structure

In order to understand the role of DNA in the maintenance of genetic continuity, you must first understand the currently accepted model of the DNA structure. DNA is a polymer made up of a repeating chemical unit known as the **nucleotide**. Thousands of these units are known to comprise a single DNA molecule, making DNA one of the largest of all organic compounds. DNA exists in hundreds of thousands, if not millions, of different forms, depending on the precise arrangement of nucleotides in the molecule. Its variability is the key to genetic variation in living things. DNA nucleotides themselves are quite complex, being composed of three separate subunits.

- Phosphate group—a chemical group made up of phosphorus and oxygen.
- Deoxyribose—a five-carbon sugar made up of carbon, oxygen, and hydrogen.
- Nitrogenous base—a chemical unit composed of carbon, oxygen, hydrogen, and nitrogen. Bases found in DNA are **adenine** (A), **thymine** (T), **cytosine** (C), and **guanine** (G). In chemical terms, the bases A and T are

complementary as are the bases C and G. In other words, A always links to T and C always links to G.

Phosphate

Deoxribose

Nitrogenous☐
Base

Adenine → Complements → Thymine

Guanine → Complements → Cytosine

Complementary Nucleotides (DNA)

The Watson-Crick model, developed by James Watson and Francis Crick, is an attempt to describe the physical and chemical structure of DNA in a way that would explain its known characteristics, including its ability to replicate. Watson and Crick's model was developed using the best experimental evidence available at the time and involves the following points.

- Nucleotide units are joined end to end, forming a long chain of alternating deoxyribose and phosphate units with nitrogenous bases sticking out on one side of the chain. The specific arrangement of nitrogenous bases on the chain makes up the genetic code.

Hydrogen Bond

Base

Phosphate

Sugar

DNA: Single Strand

DNA Model

Base

Hydrogen Bond

Phosphate

Sugar

DNA Helix

DNA: Double Strand

- A second chain, with complementary nitrogenous bases, is aligned with the first, forming a double-stranded, ladderlike molecule. In this formation, the repeating sugar and phosphate units form the uprights of the ladder, while the pairs of linked nitrogenous bases form the rungs connecting the two strands.
- The nitrogenous bases paired to form the ladder rungs are always adenine to thymine (A-T) and cytosine to guanine (C-G). The bases are held together by weak hydrogen bonds.
- Finally, the double strands of DNA twist around each other to form a "double helix," or "twisted ladder," shape.

The processes that occur during cell division provide a reliable mechanism for transferring this genetic code from one generation to the next. Of central importance in these processes is the process of **DNA replication**. Replication is the exact self-duplication of the genetic material in the early stages of cell division. Chromosome replication is actually a function of the replication of the DNA making up the chromosome strand. DNA replication is thought to occur as follows.

- The two strands of the DNA molecule separate by unzipping between pairs of nitrogenous bases.
- Unbound (free) nucleotides floating freely in the cytoplasm are attracted to and incorporated into the unzipped portion of the DNA molecule. In the building process that follows, complementary nucleotides are attracted to each other to ensure that the new strands produced are exact duplicates of the original strands.

- When the unzipping process is complete and all bonding sites are filled with free nucleotides, two identical DNA molecules result; these are free to separate into two chromosome strands.

DNA Replication: Phase I

DNA Replication: Phase II

The Role of Genes in Heredity

Hereditary information is contained in genes, located in the chromosomes of each cell. An inherited trait can be determined by one or by many genes, and a single gene can influence more than one trait. A human cell contains many thousands of different genes in its nucleus. Hereditary factors known as **genes** are thought to exist as discrete portions (known as loci) of chromosomes. The term "discrete" in this case refers to the concept that genes are always located at the same point (or locus) on a chromosome. It is believed that pairs of homologous chromosomes contain linear, matching arrangements of genes exerting parallel control over the same traits. Pairs of genes that exercise such parallel control over the same traits are known as **alleles**. The control mechanism used by genes in genetic inheritance is now thought to be chemical in nature and is described in greater detail on the next page.

124

Genetic characteristics can be extremely complex and may require the actions of several separate genes to be expressed in the adult organism. Examples of traits that require more than a single pair of genes to be expressed are human height and human hair color. Other traits may be controlled by a single pair of genes. Some of these traits may have extreme phenotypes, such as albinism (lack of skin pigment) in humans.

Mapping of the human genome is a project that has intrigued scientists for many years. As of this writing, it has been completed in draft form, with many years of research still necessary to confirm the locations and actions of all human genes. Among the pieces of data to be determined from this project is an exact count of the total number of separate genes found on human chromosomes. The number will certainly be in the thousands. Each of these genes codes for the production of a different, unique protein. Each protein has a separate and distinct function to perform in the cell.

Effect of the Environment

Genes are inherited, but their expression can be modified by interactions with the environment. Genetic traits are determined largely through the precise information found in the cell's gene structure. Although this information provides a basis for each individual organism's characteristics, it is not the only force at work in shaping the actual phenotype. Another major force in shaping the final phenotype is the environment in which the gene has to operate. A variety of factors in the environment can actually alter the effects of a particular gene. Some examples of this effect are as follows.

- Effect of light on chlorophyll production—Although most plants have the genetic ability to produce chlorophyll, they will do this only in the presence of light. Without light, these plants produce only a light-yellow pigment and therefore appear pale and sickly until they are exposed to sunlight. After a few days of exposure to sunlight, the chlorophyll production mechanism is enabled and green color returns.
- Effect of temperature on hair color in the Himalayan hare—In their native Arctic environment, Himalayan hares have white body hair with black hair on their extremities. However, when raised in warm climates, they are entirely white. In exploring this phenomenon, scientists shaved off some hair from the hare's back (normally white) and strapped an ice pack onto the bare skin. Under these experimental conditions, the hare's hair grew back black, indicating the role of environmental temperature on the production of hair color in this species.

Cell Division

In asexually reproducing organisms, all the genes come from a single parent. Asexually produced offspring are normally genetically identical to the parent. Asexual reproduction involves the production of a new organism of a species from a cell or cells of a single parent organism. This is made possible by a type of cell division known as **mitotic cell division**. In this type of cell division, the genetic material is duplicated, followed by the splitting of a single cell or group of cells. This results in the production of more cells with characteristics identical to those of the single parent organism.

The **Cell Theory** states that "all cells arise from preexisting cells by cell division." The type of cell division involved is known as mitotic cell division and involves two distinct stages.

- **Mitosis** is a precise duplication of the contents of the parent cell nucleus followed by an orderly separation of those contents into two new, identical nuclei. The events of mitosis are as follows.

  ➤ Replication (exact self-duplication) of each chromosome strand in the nucleus of the parent cell. This results in the doubling of each chromosome strand to form double-stranded chromosomes. Each of the two strands of a double-stranded chromosome is known as a **chromatid**. Chromatids are chemically identical to each other and carry identical genetic information. These chromatids are held together by a **centromere**.

  ➤ Disappearance of the nuclear membrane surrounding the chromosomes.

  ➤ Appearance of the **spindle apparatus**, a series of fibers that attach to the double-stranded chromosomes at the centromere during the early stages of mitosis.

  ➤ Replication of the centromere of each double-stranded chromosome. This is followed by separation of the two chromatids of the double-stranded chromosomes to form single-stranded chromosomes.

  ➤ Migration of the single-stranded chromosomes along the spindle apparatus toward opposite ends of the cell. The chromatids in each pair separate in this stage of mitosis, allowing a full set of single-stranded chromosomes to migrate to each pole.

  ➤ Reformation of the nuclear membrane around the chromosomes grouped at the ends of the cell. These two daughter nuclei are identical to each other and to the parent cell nucleus in terms of the number and type of chromosomes as well as in the particular genetic information found within these chromosomes. The significance of mitosis is the *exact duplication* of the parent nucleus.

- **Cytoplasmic division** is the separation of the two new nuclei into two new daughter cells as the cytoplasm of the parent cell divides. After the formation of the **daughter nuclei**, the cytoplasm of most (though not all) cells is divided into two roughly equal portions, each enclosing one of the daughter nuclei. The daughter cells will grow and eventually go through the cycle of mitotic cell division themselves. The mechanisms governing this cytoplasmic division are not fully understood but are known to occur differently in different types of organisms. For example, in plants and animals, mitosis occurs in much the same way as described on page 126. However, in animals, the centriole is known to function in the formation of the spindle apparatus. In plants, the spindle forms without the presence of a centriole. In animals, cytoplasmic division is accomplished by the formation of a constriction that separates the two daughter cells. In plants, the division of the cytoplasm occurs after the formation of a cell plate that separates the daughter nuclei.

Mitosis

Reproduction

A common element of asexual reproduction is the fact that offspring produced by this mechanism are genetically identical to the parent organism. See the section "Mechanisms of Reproduction" (pages 159–171) for additional information on asexual reproductive patterns.

In sexually reproducing organisms, the new individual receives half of the genetic information from its mother (via the egg) and half from its father (via the sperm). Sexually produced offspring often resemble, but are not identical to, either of their parents. Sexual reproduction involves the production of a new offspring organism from the fusion of two sex cells, known as **gametes**, each of which is contributed by a different parent organism. It is made possible by a special kind of cell division known as **meiotic cell division**. In this type of cell division, the genetic information is halved. It is restored to its full complement during fertilization.

Diploid and Haploid Explained

Each body cell of an organism contains a number of chromosomes characteristic of the species. This number is known as the **diploid** chromosome number. The term "diploid" refers to the fact that the chromosomes in the nucleus are found in pairs with similar structure. These chromosome pairs are known as homologous chromosomes; they carry genes for the same traits. The symbol **2n** is used to represent the diploid chromosome number.

In gametogenesis (formation of gametes), cells located in the sex organs (gonads) undergo a special type of cell division (meiotic cell division) that results in the formation of sperm cells (in males) or egg cells (in females). In this process, the number of chromosomes is reduced by half. This reduced number of chromosomes is known as the haploid chromosome number and is represented by the symbol **n**.

Because each parent contributes 50 percent of the genetic information to the offspring, the offspring resemble, but do not look exactly like, the parents. See the section "Mechanisms of Reproduction" (pages 159–171) for additional information concerning meiosis, fertilization, and sexual reproduction.

DNA Code

In nearly all organisms, the coded instructions for specifying the characteristics of the organism are carried in DNA, a large molecule formed from subunits arranged in a sequence with bases of four kinds (represented by A, G, C, and T). The chemical and structural properties of DNA are the basis for how the genetic information that underlies heredity is both encoded in genes (as a string of molecular bases) and replicated by means of a template. Recall that DNA is a complex organic molecule composed of thousands of repeating nucleotide molecules and that each free nucleotide carries with it one of four nitrogenous bases. The particular sequence of nitrogenous bases adenine, thymine, cytosine, and guanine (A, T, C, and G) comprising a strand of DNA provides the type of chemical code that is understood by the chemical mechanisms of the cell. The DNA code is used by these mechanisms to manufacture specific enzymes and other proteins through the process of protein synthesis. A DNA strand provides a **template** (pattern) for the formation of **messenger RNA** (mRNA). The DNA code is transcribed (read) by mRNA as the latter is synthesized in a pattern complementary to the DNA strand. The process by which the DNA code is transferred to mRNA code is known as **transcription**. In RNA molecules, uracil (U) nucleotides are substituted for DNA's thymine (T).

Each group of three nitrogenous bases, known as a triplet **codon**, provides the information necessary to code for the insertion of a single, specific amino acid into a building protein molecule. The particular sequence of triplet codons on DNA (and transcribed to mRNA) enables amino acids to be linked together in a specific sequence during protein synthesis.

Cells store and use coded information. The genetic information stored in DNA is used to direct the synthesis of the thousands of proteins that each cell requires. The cell contains many thousands of such codes in its chromosomes. Each strand of DNA in the chromosome has the potential to provide the complete chemical code for the manufacture of at least one complete protein. These proteins are highly specific. They result in the expression of some specific trait or portion of a trait in the living cell and, consequently, in the organism of which they are a part.

Protein Synthesis

The actual process of protein synthesis begins when an mRNA molecule migrates out of the nucleus, where it was formed, and attaches to a ribosome located on the surface of the endoplasmic reticulum. Recall from the previous section that mRNA carries the information, read from a genetic blueprint housed in a molecule of DNA, needed to encode the creation of a single protein strand. Once attached to the ribosome, the mRNA begins the process of protein synthesis by insertion of a "start" codon. The codon sequence of the mRNA provides the pattern upon which a new polypeptide (protein) strand will be built.

In this role, mRNA acts like a piece of manufacturing machinery in a factory, while the ribosome represents the factory within the machinery operates.

Floating free in the cytoplasm of the cell are molecules of **transfer RNA** (tRNA). Though composed of the nucleotides A, U, G, and C, tRNA differs in structure and function from mRNA. Each type of tRNA is specifically designed to link to one of 20 different types of amino acid at a chemically active site and carries an **anticodon** that is designed to link to a complimentary codon on the mRNA molecule. As protein synthesis continues to completion, tRNA molecules link in a specific order to mRNA on the ribosome. While so linked, the tRNA molecules facilitate the formation of chemical **peptide** bonds between adjacent amino acid molecules. The chemical linkage of many amino acids by means of peptide bonds results in a chain molecule known as a **polypeptide** or protein. The synthesis of an individual protein molecule ceases when the "stop" codon is reached in the mRNA sequence. The result of this process is the formation of a polypeptide chain of a specific chemical composition that is released into the cytoplasm to perform a particular biochemical function. The process by which the mRNA code is decoded by tRNA and transferred to the amino acid sequence in a polypeptide molecule is known as **translation**.

To continue the factory analogy, tRNA molecules act as factory workers who carry manufacturing components (amino acids) to the mRNA machinery and determine their exact placement in the protein product being manufactured.

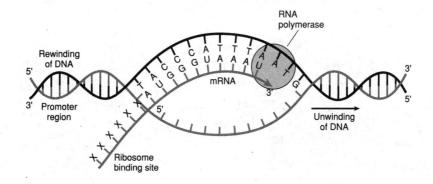

One-Gene–One-Polypeptide Hypothesis

Each gene in the cell's nucleus contains the coded information required to synthesize a single polypeptide chain. A particular gene is believed to operate throughout the life of the cell to produce its specific polypeptide and only that polypeptide. This concept is known as the **one-gene–one-polypeptide hypothesis**. The modern concept of the gene defines it as that sequence of nucleotides and codons in a molecule of DNA necessary to code for a complete polypeptide chain.

① Initiation: First, the small subunit of the ribosome binds the mRNA. Second, the first tRNA with the anticodon to the start codon binds to the mRNA. Then, the large subunit binds. The second codon is now available in the A site.

② Elongation: tRNAs carrying amino acids enter the A site. An enzyme in the ribosome catalyzes bond formation between the adjacent amino acids. The ribosome-mRNA complex shifts. The first tRNA exits from the E site. A new tRNA enters the A site, and the process repeats.

③ Termination: When a stop codon is present in the A site, the enzyme release factor enters the ribosome and releases the polypeptide chain. After that, the ribosomal subunits separate from the mRNA.

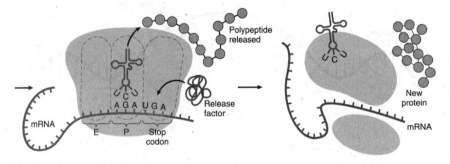

Gene Mutations

Genes are segments of DNA molecules. Any alteration of the DNA sequence is a mutation. Usually, an individual cell's altered gene will be passed on to every cell that develops from it. Gene mutations may be defined as being any changes in the nitrogenous base sequence of a molecule of DNA. When the base sequence of DNA is altered, the amino acid sequence of the polypeptide for which it codes will likewise be altered. Such an alteration may affect the operation of the resulting enzyme, preventing it from properly catalyzing its reaction and thus preventing a trait from being expressed by the cell. The majority of gene mutations are harmful because they result in the cell being impaired from performing some specific task. In most cells, the corresponding allele continues to function, and the cell continues its activities unaffected. In rare cases, a mutation may result in a lethal gene that kills the cell either by producing a substance toxic to the cell or by failing to produce a protein of vital importance to the cell.

Gene mutations are passed on to every cell that arises from the mutated cell. If the mutation occurs in **somatic** (body) tissues, its effect is limited to the tissues immediately surrounding the mutated cell. If the mutation occurs in a **primary sex cell**, it can be passed on to the offspring that result from fertilization of or by the gametes produced from that primary sex cell. In the latter case, a mutation can enter the gene pool of a population and be passed on to succeeding generations. This is now known to be a cause of variation in a species.

Amino Acids

The work of the cell is carried out by the many different types of molecules it assembles, mostly proteins. Protein molecules are long, usually folded chains made from 20 different kinds of amino acids in a specific sequence. This sequence influences the shape of the protein. The shape of the protein, in turn, determines its function. Although there are 20 different types of amino acids, they all share a common general structure. All amino acids contain an amino group ($-NH_2$) and a carboxyl group ($-COOH$) attached to a central carbon atom. A third attached group (radical) varies and gives each type of amino acid its unique properties. Two amino acid molecules may be joined together chemically by dehydration synthesis to form a dipeptide. The term "peptide" refers to the name of the actual chemical bond between carbon and nitrogen atoms that joins the two amino acid units together. As more and more amino acids link together by dehydration synthesis, an amino acid chain, known as a **polypeptide**, is formed. Polypeptide chains form the basis of protein molecules. Because of the almost endless variations in which amino acids may be arranged in a polypeptide, proteins are formed in thousands of different forms. The extreme variability of proteins is thought to be responsible for the individual variations in living things. (See page 103 for chemical structure of amino acids.)

The functions of proteins may be classified as either structural (forming a part of the cell material) or functional (having a role in the chemistry of the cell). A few common examples of proteins are insulin (a hormone), hemoglobin (an oxygen-carrying pigment in red blood cells), and enzymes (organic catalysts).

Offspring resemble their parents because they inherit similar genes that code for the production of proteins that form similar structures and perform similar functions. We have learned that DNA's principal role in the cell is to code for the production of specific polypeptides. Each enzyme or structural protein synthesized by a cell has a specific function to perform. These functions comprise the characteristics, or traits, of the cell. It follows logically, then, that the individual traits that a cell displays are a function of the particular combinations of DNA found in its cells. The particular combination of these traits gives an organism its individuality that enables us to recognize them as individual organisms. The following diagram will help to illustrate this concept.

DNA ⟶ gene ⟶ polypeptide ⟶ enzyme ⟶ reaction ⟶ reaction product ⟶ trait

Within this individuality, certain traits resemble those of the parent organisms, illustrating the genetic continuity that links one generation to the next. Species traits are similar enough for us to recognize an individual organism as part of a species group but particular enough for us to recognize family similarities and individual characteristics.

Differentiation

The many body cells in an individual can be very different from one another. This occurs even though they are all descended from a single cell and thus have essentially identical genetic instructions. This is because different parts of these instructions are used in different types of cells and are influenced by the cell's environment and past history. All cells in an individual multicellular organism arise by mitotic cell division from a single parent cell, the **zygote**. All cells in the body therefore receive the same genetic information. Despite this, the body contains many specific tissue types, many highly specialized to perform a narrow range of tasks. The process at work to produce different cells that perform different functions is **differentiation**. Differentiation is how cells become specialized into tissues to perform the various tasks needed to keep the body functioning. This will be discussed in more depth in the section "Mechanisms of Reproduction" (pages 159–171). It is mentioned here to illustrate the hypothesis that a gene will operate in the cell only when its polypeptide is required by the activities of the cell and that at other times the gene is switched off. In fact, considerable evidence has accumulated that points to just such a concept. Only the genes specific to the activities of a specific cell type are activated in that cell.

This concept maximizes the efficiency of the cell. The cell's protein synthesis mechanisms are not overloaded producing enzymes and structural proteins that have no direct function in the specialized cells in which they operate. For example, human muscle tissues must produce the proteins actin and myosin that make up the muscle cell's contractile apparatus. Therefore, the genes for the production of these proteins are switched on in muscle cells. At the same time, the genes that code for the production of hemoglobin are switched off in muscle cells since hemoglobin, which is required in red blood cells, has no direct function in muscle cells. See "Stem Cell Research," page 170.

Genetic Engineering

Essential Questions:

- *What effect will genetic engineering have on the future of genetic inheritance?*
- *What cautions should be considered in the use of genetic engineering?*

Performance Indicator 4.2.2 *The student should be able to explain how the technology of genetic engineering allows humans to alter the genetic makeup of organisms.*

For thousands of years, new varieties of cultivated plants and domestic animals have resulted from **selective breeding** for particular traits. Geneticists have applied many of the theoretical concepts of the science of genetics to the practical areas of plant and animal breeding. Many of the most productive plant and animal breeds raised for human uses have been developed through careful artificial selection of breeding individuals, crossbreeding (hybridization) between related varieties, and inbreeding of perfected strains to maintain established traits. Such traits as reproductive potential, resistance to disease, adaptability to climate, and productivity, among others, are considered desirable traits for human uses. These and other traits are purposely bred into grains (corn, wheat), flowers (roses, orchids), fruits (apples, oranges), cattle, horses, dogs, and many other species. They are maintained by careful attention to breeding patterns.

In the past, crude methods were used to mate members of a species displaying desirable traits in the hopes that some of the offspring of such a mating would display enhanced characteristics of the desired type. When desirable offspring were obtained, the hybrid animals or plants were crossbred in an attempt to produce a pure-breeding strain of the desired variety.

Genetic Manipulation

In recent years, new varieties of farm plants and animals have been engineered by manipulating their genetic instructions to produce new characteristics. Modern methods of **gene manipulation** have been developed that make isolating and inserting specific genes for particular traits within the genome of a species possible. Once inserted, perpetuating the desired trait through traditional methods of hybrid crosses is possible. **Genetic engineering** refers to the series of techniques used to transfer genes from one organism to another. It involves removing a small piece of DNA from a cell and adding it to the gene structure of another cell. The new DNA that results is known as recombinant DNA. The spliced gene in the recombinant DNA will continue to produce its polypeptide product in the new cell, thus transferring to that cell a genetic ability it lacked before.

These techniques have been used to produce new, improved varieties of certain farm plants (wheat, corn, coffee) and animals (cattle, fish). Some of these species have been bred to display enhanced characteristics such as resistance to disease. Others have been bred to produce human proteins needed for research or for medical applications.

An already successful medical technique that holds great promise for the future is the production of human hormones and other proteins in the bodies of laboratory animals or bacterial colonies. Human genes have been removed from healthy human cells and added to the gene sequences of bacteria that are then cultured in large quantities in the laboratory. The substances produced in this manner are considered to be safer than those produced by traditional methods, including those drawn from human subjects. Some examples of this process are as follows.

- Bacteria colonies have been genetically engineered to produce human insulin, interferon, hepititis B vaccine, and human growth hormone. These proteins were gathered and purified for use in treating various human disorders.
- Potato plants have been altered to produce serum albumin.
- Tobacco plants with spliced human genes produce antibodies and melanin.

Different enzymes can be used to cut, copy, and move segments of DNA. Characteristics produced by the segments of DNA may be expressed when these segments are inserted into new organisms, such as bacteria. In the mid-20th century, research with certain bacteria led to the discovery of enzymes capable of snipping DNA molecules at particular nitrogenous base sequences. Such enzymes are known as **restriction enzymes**. This knowledge was used to develop laboratory methods by which the desirable genes of one species may be snipped out of the genome of that species and inserted into the genome of a different species. Once part of the genome of the host species, the gene can function to produce the hormone or other protein coded by the inserted gene. In this way, it is possible for a human trait to be displayed by a different species, including bacteria. The DNA that is altered by this method is known as **recombinant DNA**.

Inserting, deleting, or substituting DNA segments can alter genes. An altered gene may be passed on to every cell that develops from it. Any altered sequence of DNA in a living cell can be passed along to succeeding generations of the cell by means of mitotic cell division. Like mutation, genetic engineering can result in altered sequences of nitrogenous bases on a molecule of DNA. When nitrogenous base sequences are added, deleted, or substituted in a DNA molecule, these changes are inheritable. If the genetically engineered cell is a primary sex cell, the gametes resulting from that primary sex cell may inherit the altered gene sequence, leading to the possibility that an artificial gene may be passed along from generation to generation in a host species that does not normally display the characteristic associated with the species.

Genetic engineering involves removing a desirable gene from a cell, adding it to the gene structure of a bacterial cell, and replacing the recombinant DNA into the bacterial cell. The bacteria then have the capability to produce the protein produced by the original animal or plant cell.

Genetic Engineering

Gene Therapy

Knowledge of genetics is making possible new fields of health care. For example, finding genes that may have mutations that can cause disease will aid in the development of preventive measures to fight disease. Substances, such as hormones and enzymes, from genetically engineered organisms may reduce the cost and side effects of replacing missing body chemicals. The term **gene therapy** has become a commonly used term in biomedical research. This process involves the splicing of functional genes into cells that contain defective, nonfunctional genes for a particular trait. The clinical techniques used to accomplish gene therapy are still being worked out. However, they involve research into methods of introducing the functional genes into the person or organism affected by the defective gene. Cystic fibrosis is one human disorder successfully treated in clinical trials using this technique.

Cloning

A scientific curiosity just a few years ago, cloning has become an increasingly common method for the production of plants and animals used as food. Cloning in plants involves the growth and differentiation of undifferentiated tissues, resulting in the formation of a set of offspring with characteristics identical to those of the donor plant. In animals, cloning is accomplished by replacing the haploid nucleus of an egg cell with a diploid nucleus of a somatic (body) cell of a donor animal. Stimulated to divide, the egg produces an embryo that matures into an adult animal with characteristics identical to those of the donor animal—a clone.

This method of reproduction is controversial because it places the creation of new life in the hands of laboratory scientists who make decisions about what organisms will be selected for cloning. A potential dilemma surrounds the cloning of humans, a currently banned practice that could have widespread social and ethical implications. Students of today will need to face this and many other ethical decisions about how our expanding knowledge of genetics should be used to impact the human condition. See "Stem Cell Research," page 170.

QUESTION SET 2.5—GENETIC CONTINUITY
(ANSWERS EXPLAINED, P. 297)

1. Living things contain units of structure and function that arise from pre-existing units. This statement best describes the
   (1) cell theory
   (2) lock-and-key model of enzyme activity
   (3) concept of natural selection
   (4) heterotroph hypothesis

2. Which statement most accurately compares mitotic cell division in plant and animal cells?
   (1) It is exactly the same in plant and animal cells.
   (2) The walls of plant cells pinch in, but the membranes of animals do not.
   (3) Most plant cells use centrioles, but most animal cells do not.
   (4) In both plants and animals, the daughter cells are genetically identical to the original cell.

3. What are the normal chromosome numbers of a sperm, egg, and zygote, respectively?
   (1) haploid, haploid, and haploid
   (2) haploid, diploid, and diploid
   (3) diploid, diploid, and diploid
   (4) haploid, haploid, and diploid

4. Which statement describes the work of Gregor Mendel?
   (1) He developed some basic principles of heredity without having knowledge of chromosomes.
   (2) He explained the principle of dominance on the basis of the gene-chromosome theory.
   (3) He developed the microscope for the study of genes in the garden pea.
   (4) He used his knowledge of gene mutations to help explain the appearance of new traits in organisms.

5. Genes carried on only an X chromosome are said to be
   (1) hybrid   (3) autosomal
   (2) codominant   (4) sex-linked

6. In which situation could a mutation be passed on to the offspring of an organism?
   (1) Ultraviolet radiation causes skin cells to undergo uncontrolled mitotic division.
   (2) The DNA of a human lung cell undergoes random breakage.
   (3) A primary sex cell in a human forms a gamete that contains 24 chromosomes.
   (4) A cell in the uterine wall of a human female undergoes a chromosomal alteration.

7. The diagram below illustrates what happens to the fur coloration of a Himalayan hare after exposure to a low temperature.

Ice pack

Before   After

This change in fur coloration is most likely due to
(1) the effect of heredity on gene expression
(2) the arrangement of genes on homologous chromosomes
(3) environmental influences on gene action
(4) mutations resulting from a change in the environment

8–11. Base your answers to questions 8 through 11 on the diagram below and on your knowledge of biology. The diagram represents molecules involved in protein synthesis.

8. In plant cells, molecule 1 is found in the
   (1) centriole
   (2) nucleus
   (3) cell wall
   (4) lysosome

9. The building blocks of molecule 3 are known as
   (1) amino acids
   (2) DNA molecules
   (3) fatty acids
   (4) RNA molecules

10. Where do the chemical reactions coded for by molecule 2 take place?
    (1) in the vacuole
    (2) on the plasma membrane
    (3) in the lysosome
    (4) at ribosomes

11. Molecule 3 is formed as a result of
    (1) deamination
    (2) dehydration synthesis
    (3) enzymatic hydrolysis
    (4) oxidation

12–15. Base your answers to questions 12 through 15 on the reading passage below and on your knowledge of biology.

Female or Male, Which Will It Be?

After fertilization, *all* human embryos begin forming the basic female reproductive structures. These structures are present by the time the embryo has toes, fingers, eyes, and a heart at 35–40 days into gestation. If the egg was fertilized by a sperm containing a Y chromosome, a series of changes occurs that will produce a male.

Recent research has isolated a genetic switching mechanism that is part of the process that determines sex in humans. The Y chromosome contains a trigger factor known as the *SRY* gene, which activates the male pattern of development after 35–40 days of gestation. The *SRY* gene causes testes to develop. These, in turn, produce testosterone, which causes the development of male characteristics such as the penis, masculine muscles, and eventually, facial hair.

At this stage of development, the embryo has both male and female potential. However, the *SRY* gene sends a chemical message to another gene known as *MIS*. The *MIS* gene causes the developing female organs in the embryo to disappear. The combined action of the *SRY* and *MIS* genes results in the change of the embryo from female to male.

12. During the first five weeks after a human egg is fertilized, the embryo develops
   (1) only toes, fingers, eyes, and a heart
   (2) male reproductive structures and other organs
   (3) female reproductive structures and other organs
   (4) male or female reproductive structures depending on whether the egg was fertilized by an X-bearing or a Y-bearing sperm

13. The male pattern of development is activated by the
   (1) *SRY* gene
   (2) entire Y chromosome
   (3) entire X chromosome
   (4) *MIS* gene

14. The *MIS* gene is activated by
   (1) the X chromosome
   (2) a chemical message
   (3) the presence of male reproductive structures
   (4) the presence of female reproductive structures

15. An embryo is changed from a female to a male by the action of
    (1) two X chromosomes
    (2) all the genes on a Y chromosome
    (3) an *MIS* gene, only
    (4) *SRY* and *MIS* genes

---

16. The code of a gene is delivered to the enzyme-producing region of a cell by a
    (1) hormone
    (2) nerve impulse
    (3) messenger RNA molecule
    (4) DNA molecule

17. Which process could be used by breeders to develop tomatoes with a longer shelf life and to develop cows with increased milk production?
    (1) natural selection
    (2) sporulation
    (3) genetic engineering
    (4) chromatography

18. What would most likely result if mitosis was not accompanied by cytoplasmic division?
    (1) two cells, each with one nucleus
    (2) two cells, each without a nucleus
    (3) one cell with two identical nuclei
    (4) one cell without a nucleus

19. Breeders have developed a variety of chicken that has no feathers. Which methods were most likely used to produce this variety?
    (1) artificial selection and inbreeding
    (2) grafting and hybridization
    (3) regeneration and incubation
    (4) vegetative propagation and binary fission

20. In fruit flies with the curly wing mutation, the wings will be straight if the flies are kept at 16°C but curly if they are kept at 25°C. The most probable explanation for this is that
    (1) fruit flies with curly wings cannot survive at high temperatures
    (2) the environment influences wing phenotype in these fruit flies
    (3) high temperatures increase the rate of mutations
    (4) wing length in these fruit flies is directly proportional to temperature

21–23. Base your answers to questions 21 through 23 on the diagram below and on your knowledge of biology.

21. Structure *A* contains a
    (1) genetic code
    (2) single nucleotide only
    (3) messenger RNA molecule
    (4) small polysaccharide

22. Structure *B* represents
    (1) a ribosome
    (2) transfer RNA
    (3) recombinant DNA
    (4) a male gamete

23. The technique illustrated in the diagram is known as
    (1) cloning
    (2) genetic engineering
    (3) protein synthesis
    (4) *in vitro* fertilization

VI. ORGANIC EVOLUTION

> **KEY IDEA 3—ORGANIC EVOLUTION** Individual organisms and species change over time.

Essential Question:

- *What has caused the wide variety of living things that have inhabited Earth over the past 3.5 billion years?*

Little scientific doubt exists that the physical forms of nearly all living things have changed over time. Genetic and environmental factors have combined in the past to make these changes happen. These same forces are still at work, selecting traits that better adapt species to Earth's ever-changing environment. Organic evolution specifically refers to the mechanisms thought to govern the structural, behavioral, and functional changes in living species over geologic time. These changes may include variation within a species or the production of new species.

Theory of Natural Selection

The unifying theory underlying the modern concept of evolution is the theory of natural selection. First advanced by Charles Darwin in the 19th century, this theory is based on scientific observation coupled with an understanding of how living things respond to environmental pressures. The basic features of the **theory of natural selection** include observations of overproduction of offspring in naturally occurring species, the existence of variation among offspring, and the struggle for survival. The theory also includes logical assumptions concerning the selection of individuals whose variations provide adaptive value and the subsequent survival and reproduction of the best-adapted individuals. To Darwin's original construct has been added a wealth of new observations and a more complete understanding of the genetic basis of variation. See pages 151–152.

Many scientific observations, some only peripherally related to the science of evolution, have lent support to the concepts embodied in the theory of natural selection. The sciences of geology (study of rocks) and paleontology (study of fossils) have demonstrated the extreme age of the Earth and the existence of fossilized organisms different from, but still similar to, organisms that exist today. The sciences of anatomy (study of structure) and embryology (study of reproduction and development) have enabled scientists to classify organisms based on similarities of structure and developmental patterns. The sciences of cytology (study of cells) and biochemistry (study of chemical reactions in living things) have pointed to the basic similarity of all living things. The relative relatedness of living things shows close structural similarities.

Process of Evolution

Essential Question:

- *How do new species arise from previously existing species?*

Performance Indicator 4.3.1 *The student should be able to explain the mechanisms and patterns of evolution.*

Evolution is a natural force thought to be responsible for the dizzying array of different life-forms that have inhabited our planet. A survey of living things reveals that approximately 1.9 million different species of every description exist on Earth. Classification provides scientists with the means for sorting and grouping these organisms for easier study. The basis of biological classification is physical structure, although other criteria, such as embryonic, genetic, and biochemical similarities, are also used. Organisms that are similar in their physical traits are usually similar in other ways. Because of this fact, the characteristics of a large grouping of similar organisms can be learned by studying a few representatives of the group. When a new organism is discovered, it can be readily grouped with other, similar organisms when only a few of its characteristics are known. It is assumed that organisms sharing many traits in common are likely to share a common ancestry as well.

Scheme of Classification

The most widely accepted scheme of classification places every known organism into one of five large groupings known as kingdoms. The organisms within a particular kingdom share many broad characteristics in common, although there can be considerable diversity (difference) of form among them. The five kingdoms are as follows.

- Monera—unicellular forms having a primitive cell structure
- Protista—unicellular organisms with plantlike or animal-like characteristics
- Fungi—unicellular, colonial, or multicellular saprophytic organisms
- Plant—multicellular, photosynthetic organisms
- Animal—multicellular, heterotrophic organisms

Within each kingdom, organisms having greater similarity to each other than to the organisms in other groups are classified together into phyla. Therefore, each phylum contains groups of organisms showing characteristics distinctly different from those of other phyla. Phyla are subdivided into still narrower groupings of organisms including classes, orders, and families, and finally into groups showing high degrees of similarity known as genera.

The members of each genus are so similar that they might easily be mistaken for each other by most people. Each genus is further broken down into species. The members of a species are so similar biologically that they can share genetic information and reproduce more individuals like themselves. Fertile offspring result from reproductive activities between members of the same species.

Kingdom	Phylum	Characteristics	Examples
Monera	Bacteria Blue-green algae	Have primitive cell structure lacking a nuclear membrane.	E. coli
Protista	Protozoa Algae	Are predominantly unicellular organisms with plant-like and/or animal-like characteristics.	Paramecium Ameba Spirogyra
Fungi		Cells are usually organized into branched, multi-nucleated filaments that absorb digested food from their environment.	Yeast Bread mold Mushroom
Plant		Are multicellular, photosynthetic organisms.	
	Bryophytes	Lack vascular tissues; have no true roots, stems, or leaves.	Moss
	Tracheophytes	Possess vascular tissue; have true roots, stems, and leaves.	Geranium Fern Bean Trees (maples, oaks, pines, etc.) Corn
Animal		Are multicellular, heterotrophic organisms.	
	Coelenterates	Have two cell layers, hollow body cavity.	Hydra Jellyfish
	Annelids	Have segmented body walls.	Earthworm Sand worm
	Arthropods	Have jointed appendages, exoskeleton.	Grasshopper Lobster Spider
	Chordates	Possess dorsal nerve cord, internal skeleton.	Shark Frog Human being

This classification scheme is used by scientists to organize living things into easily understood groupings. To keep track of the more than 1.9 million different species that have been discovered and classified, a system of naming (nomenclature) has been developed. This system uses two names in much the same way that most people have at least two names to help others tell them apart. This naming system, known as **binomial nomenclature** (two-name naming), was first devised by Carl Linnaeus in the eighteenth century. The two names are the genus name (always capitalized) and the species name (always written in lowercase). The language used in the system of binomial nomenclature is Latin. Some examples of this system are as follows.

Genus	Species	Common Name
Homo	sapiens	human being
Canis	familiaris	dog
Canis	lupis	wolf
Felis	domestica	cat
Felis	leo	lion

Evolutionary Assumptions

The basic theory of biological evolution states that Earth's present-day species developed from earlier, distinctly different species. The currently accepted theories of organic evolution assume that modern life-forms have evolved from previously existing life-forms. One source of this assumption is the logical conclusions drawn from a knowledge that living things arise from other living things by reproduction. Just as present-day organisms have been produced by the generations that immediately proceeded them, so modern-day species must have developed from the ancient species of eons past. In addition to this logical conclusion, a great deal of scientific evidence supports this assumption provided by studies of geology and paleontology.

No direct evidence indicates the exact date of Earth's formation. By studying the indirect evidence drawn from the radioactive dating of Earth's rocks, however, geologists have estimated the age of Earth to be between 4.5 and 5.0 billion years. In arriving at this estimation, scientists have assumed that Earth is at least as old as the oldest rocks so far discovered. Fossils of early life-forms have been found preserved in certain of these rocks and are dated by scientists using various techniques. It is logical to conclude that the fossils are as old as the rock layers in which they are embedded.

Genetic Mechanisms of the Cell

New inheritable characteristics can result from new combinations of existing genes or from mutations of genes in reproductive cells. Mutation and the sorting and recombining of genes during meiosis and fertilization result in a great variety of possible gene combinations. We have already learned that the source of individual traits is found within the genetic mechanisms of the cell. These mechanisms include the cellular processes of mutation, meiosis, fertilization, and protein synthesis.

Mutation is important in providing the new genes that may lead to the production of new genetic traits. Although the majority of such mutations are harmful or neutral, a small percentage may provide significant adaptive advantages to a species. It is important to recognize that in sexually reproducing species, only mutations in gametes can be passed on to succeeding generations. Protein synthesis is the basic chemical process in the cell that results in production of new proteins that function in the cell and result in new variations.

Within the reproductive process, meiosis and fertilization provide the mechanism by which new combinations of both old and new traits may be tried out as new varieties within a species. These can be counted in the millions of possible individual combinations. Successful combinations are perpetuated in the species through the processes of sexual reproduction. Unsuccessful combinations are removed from the population's gene pool by the selective forces of the environment.

Gene Mutations

Mutations occur as random events. Gene mutations can also be caused by such agents as radiation and chemicals. When they occur in sex cells, the mutations can be passed on to offspring. If they occur in other cells, the mutations can be passed on to other body cells only. Gene mutations involve changes in the chemical nature of the gene. The active chemical in the gene is DNA. When this material undergoes chemical alteration, its control over cell activities and cell characteristics changes, causing alterations in the phenotype of the organism. Although these changes are likely to be small and difficult to detect, they may occasionally be great enough to be easily noticed or even to cause death. Albinism (lack of skin pigment in humans) is an example of a human trait caused by a single gene mutation whose effects are quite obvious and dramatic. The phenotype (albino) that results from the homozygous recessive allelic combination is characterized by pale white skin, yellow hair, and pink eyes.

Although most gene mutations cause changes that are neutral or harmful, occasionally their effect is beneficial. A beneficial mutation, such as one that produces a useful enzyme, causes a phenotypic change that in some way gives an organism an advantage in its environment over other organisms of the same species. An extremely beneficial gene mutation can cause major shifts in the species' genetic characteristics, a phenomenon considered by many scientists to be a major driving force in evolution.

Gene mutation can occur spontaneously by random chemical alteration of DNA during replication. In addition, a number of naturally occurring and human-produced phenomena are known to cause or accelerate gene mutation. These phenomena include:

- Radiation—Sources of radiation include cosmic rays, radon, X rays, ultraviolet radiation, radioactive radiation, and electromagnetic radiation; and
- Chemicals—Sources include benzene, formaldehyde, asbestos, dioxin, and tobacco residues.

Geologic Record

Natural selection and its evolutionary consequences provide a scientific explanation for the fossil record of ancient life-forms as well as for the molecular and structural similarities observed among the diverse species of living organisms. Found frequently within certain types of rocks are **fossils**, the preserved direct or indirect evidence of organisms that lived in the past. Fossils are most commonly discovered embedded in sedimentary rock, such as sandstone or limestone. However, the remains of organisms may also be found preserved in ice or permanently frozen soil or in naturally occurring tars, salts, or other chemical deposits. Knowing the age of the rock layers in which fossils are embedded enables scientists to determine with reasonable accuracy the age of those fossils. Fossils have been discovered that have been dated by scientific methods to be more than 3 billion years old.

In undisturbed layers (strata) of sedimentary rock (such as sandstone, limestone, slate, and shale), the lowest layers were laid down first, the middle layers next, and the topmost layers last. It follows logically, then, that fossils found embedded in the lower strata are older than those in the upper strata. In fact, deeper layers of such rock are known to contain fossils of older, simpler life-forms. Strata formed near the surface contain younger and generally more complex forms. Note that these sedimentary layers may be disturbed by seismic activity that folds, inverts, subtends, or factures them. They may also be disturbed by erosion cycles that can wash fossils out of the sediments that originally encased them.

Geologic Record

Scientists have been able to identify a certain degree of continuity among the fossils in consecutive layers of fossil-bearing rock. It is possible to find fossils in upper strata that resemble those in lower strata, even though they are clearly different species. This fact lends support to the theory that genetic links exist between modern life-forms and ancient forms. It also suggests that genetic links exist among diverse modern life-forms by virtue of their common links to ancestral species. For example, all vertebrate species, from fish to mammals, share a chordate **common ancestor** that lived millions of years in the past. Modern species having similar structures share these ancestral forms in common. Even diverse species displaying few obvious similar-

147

ities, such as earthworms and mollusks, are thought to share distant common ancestors. The concept of common ancestry, in which two divergent forms can trace their lineage to a single preexisting life form, is central to understanding the science of evolution.

Comparative Sciences

Added to evidence supplied through the fossil record is that stemming from the sciences of comparative cytology, anatomy, biochemistry, and genetics. The cell as a structural and functional unit is a feature that all known living things share in common. The organelles located within these cells function in much the same way in the cells of every organism. Despite the basic similarity of all cells, certain differences among cells of different species are known to exist. Organisms with a very similar cell structure are usually considered to be more closely related than organisms whose cells show many differences.

Homologous Forelimbs and a Comparison of Vertebrate Brains

The determination of similarities in anatomic (structural) features is perhaps the most common method of demonstrating biological relationships among organisms. This method provides the basis for biological classification, in which an organism is placed into a kingdom, phylum, class, order, family, genus, and species, based on its degree of structural similarity with other members of those groups. Similar organisms have limbs, internal organs, or other structures that are constructed similarly. Such structures, known as **homologous structures**, are believed to have originated from common ancestral forms of the same structures.

We learned that each polypeptide in the cell is coded by a unique strand of DNA. We also learned that the ability to produce such polypeptides may be passed from generation to generation through the processes of reproduction and genetic inheritance. Related organisms, therefore, having inherited their characteristics from common ancestors, may be expected to share many genes and their corresponding enzymes in common. Biochemical analysis of enzymes and other proteins shows that a great deal of similarity exists in the biochemical makeup of organisms known to be related genetically. For example, the complex protein hemoglobin is found in the blood of many vertebrate species, whereas it is less common among invertebrates. Generally, the more closely related two organisms are, the more similar is their biochemical makeup. Likewise, organisms that are not as closely related share fewer biochemical similarities.

Origin of Life

Billions of years ago, life on Earth is thought to have begun as simple, single-celled organisms. About a billion years ago, increasingly complex multicellular organisms began to evolve. A question that has challenged scientists and nonscientists alike is the question of the origin of life. The cell theory assumes that all cells arise from previously existing cells. However, what gave rise to the first cell? Scientists have proposed the **heterotroph hypothesis** to help explain the origin of the first primitive life-forms on the ancient Earth. This scientific hypothesis assumes that the first primitive life-forms were not able to manufacture their own food (were heterotrophic). The heterotroph hypothesis is consistent with much of the currently accepted scientific theory on the origins of the universe and with current understandings of the sciences of biology and biochemistry. However, like many hypotheses developed to explain phenomena that cannot be directly observed and measured, the heterotroph hypothesis is based on extensions of basic assumptions about Earth's origins.

The first cell is theorized to have been an aggregate of simple organic molecules that were produced in the primitive oceans from the chance combinations of inorganic molecules under extremely energy-rich conditions. Experiments conducted in the laboratory by Stanley Miller, Sidney Fox, and other scientists have confirmed that this chance aggregation is possible, even in relatively short periods of time, given the right set of conditions.

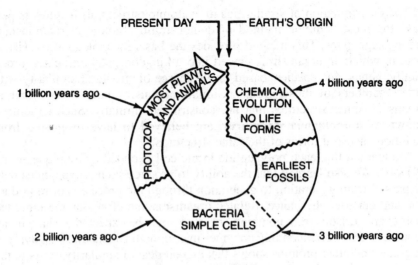

The geologic clock shows all of earth's history from its formation 5 billion years ago to the present. Human beings appeared at "11:59," "1 minute" before the present!

Geologic Clock

It is thought that the earliest living cells obtained their energy by means of a cellular process similar to fermentation, which has the natural by-product carbon dioxide. This carbon dioxide, it is believed, built up in concentration in Earth's atmosphere as a result of widespread fermentative activity. Certain early cells, having spontaneously evolved the ability to use the newly introduced gaseous carbon dioxide to manufacture their own organic foods, became Earth's first food producers. Extensive autotrophic nutritional activity, similar to photosynthesis, added free molecular oxygen to Earth's atmosphere. This new environmental condition likely proved toxic to many of Earth's newly evolved species. A few species, having spontaneously evolved an ability to use this molecular oxygen in the respiratory process, became Earth's first aerobic organisms.

The first cell-like aggregations are assumed to have appeared about 3.5 billion years ago. Unicellular protists exist in the fossil record from about 1.8 billion years ago. The first simple multicelled organisms appear in the record approximately 1.0 billion years before the present.

Theories of Evolution

Species evolve over time. Evolution is the consequence of the interactions of (1) the potential for a species to increase its numbers, (2) the genetic variability of offspring due to mutation and recombination of genes, (3) a finite supply of the resources required for life, and (4) the ensuing selection by the environment of those offspring better able to survive and leave offspring. In the 19th century in England, a naturalist named Charles Darwin devised a theory of evolution based on variation and **natural selection**. This theory forms the basis of the modern theory of evolution. The theory of evolution by natural selection has been modified over the past 100 years to include the following points.

- *Overproduction*—Scientists observe that naturally occurring species have a tendency to produce far more offspring than can possibly survive to become reproducing adults. In fact, if all these offspring did survive, Earth's habitats and available resources would be quickly used up.
- *Competition*—Despite the tendency to overproduce, the number of individuals in natural populations tends to remain relatively constant over many generations. This suggests that within each species, there is a struggle for survival that eliminates many individuals before they reach reproductive maturity. This struggle may be termed intraspecies (within a species) competition.
- *Variation*—For centuries, scientists have been aware of the extreme variability that exists between and within species. Scientists in the time of Darwin were unaware of the mechanisms that produced this variation. This knowledge would have to wait until scientists such as Mendel, DeVries, Morgan, and others had conducted their groundbreaking work in the science of genetics. We know now that this variability results from genetic and reproductive forces. As the chemical composition of DNA changes in the cell via mutation, the resulting new traits that develop enter the gene pool and are subjected to selection pressures that either increase or decrease the frequency of the variation in the species population.

Adaptive Advantage

Each individual in a species population is engaged in a struggle for existence in its own environment. The individuals that survive (are selected) are assumed to be those best adapted to survive under the particular set of environmental conditions in question. The individuals that perish are considered to be those less well adapted for survival. When the survivors later reproduce, they tend to pass on the genes associated with their adaptive advantages. In this way, nature provides selection pressures that limit or eliminate traits that do not promote individual survival. This is the primary role of natural selection in the process of evolution.

The frequencies of these favorable genes, then, increase in the gene pool relative to the frequencies of these genes controlling less favorable traits. This shift of gene frequencies in the gene pool of a species population is thought to constitute the mechanism of evolution.

The variation of organisms within a species increases the likelihood that at least some members of the species will survive under changed environmental conditions. It is important to recognize that the adaptations in a population that are favorable under one set of environmental conditions may prove to be highly unfavorable under a different set of environmental conditions. At the same time, traits present in the gene pool of a population that have low or neutral survival value may markedly increase in value if the environment changes or if the population moves to a new environment.

Some examples of this concept are as follows.

- In the past 100 years, the environments of many insect pests (for example, houseflies, mosquitoes, roaches, weevils) changed when new chemical pesticides such as DDT were introduced into their environments. Most of the insects that came into contact with the insecticide died since they were not genetically resistant to it. A small number of the organisms were genetically resistant to the chemicals, however, and survived to reproduce offspring that were also genetically resistant. Today, such resistant strains present a problem to chemists attempting to develop other new insecticides to deal with insect infestations. In this case, the insecticide has acted as an agent of natural selection. (Note that chemical pesticides are now considered to be an environmental threat to many species other than the target insect pests, including beneficial insects and humans.)

- A similar situation has occurred in the evolution of antibiotic-resistant strains of bacteria in some hospital environments. As in the example above, newly introduced antibiotics such as penicillin represented a change in the environment of disease-causing bacteria. Nonresistant bacteria died when they came into contact with the antibiotic, but a few resistant bacteria survived. The resistant bacteria gave rise to entire strains of the bacterial species that are resistant to the antibiotic. The selecting agent in this case is the antibiotic.

Behaviors have also evolved through natural selection. The broad patterns of behavior exhibited by organisms are those that have resulted in greater reproductive success. Although structural and biochemical adaptations are those most frequently considered in discussions of genetically controlled characteristics in species, behavioral patterns are increasingly being recognized as having a genetic basis. Examples of genetically programmed behaviors include bird migration patterns, fish spawning behaviors, bird imprinting behaviors, and a host of others. The science of ethology studies the simple and complex behavior patterns of animals. Many of these behaviors promote reproductive success, whether to ensure that species members meet in favorable nesting environments or to ensure that only members of reproductively compatible species attempt mating.

Patterns of Evolution

Evolution does not result in long-term progress in some set direction. Evolutionary changes appear to be like the growth of a bush. Some branches survive from the beginning with little or no change, many die out altogether, and others branch repeatedly, sometimes giving rise to more complex organisms. The processes of mutation, allelic recombination, and natural selection are constantly at work in the production and selection of new adaptations in all species. When enough unique adaptations have been accumulated in a species population so that it becomes distinct from other populations of the same species, it may be classified as a new variety of the species. Under the right set of selection pressures, a species variety may accumulate so many variations that it becomes reproductively isolated from other varieties of the species. At this point, it has become a distinct species. The process by which new species arise from parent species is known as **speciation**. Speciation may be accelerated by factors that isolate one variety from another. These include geographic isolation and reproductive isolation.

Most scientists generally agree that the mechanisms controlling the evolutionary process are similar to those outlined above. However, considerable debate still exists concerning the time frame in which this mechanism operates. Two theories about the time frame question have been put forth.

- **Gradualism**—This theory assumes that evolutionary change is slow, gradual, and continuous. A gradualistic view of evolution is supported by fossil records of a species that display slight changes in each sedimentary layer, leading to a significant divergence between specimens found in the bottom and top layers.
- **Punctuated equilibrium**—This theory assumes that species experience long geologic periods of stability (of a million years or more) in which little or no significant change takes place. This stability is punctuated by brief periods (of a few thousand years) in which dramatic changes occur within species. During these brief periods of change, many species are thought to evolve very quickly from parent species, while other species become extinct on a massive scale. Such a view of evolution is supported by fossil evidence in which little change is noted between most sedimentary layers but sudden bursts of change (that may coincide with dramatic and cataclysmic geological events such as meteor strikes and volcanic eruptions) are evident in the fossils of a few sedimentary layers. The diagram below provides models of these two theories that may be useful in helping you understand the differences and similarities between them. The models demonstrate how an ancestral species (X) might have evolved into modern species (A, B, and C) within the context of these theories.

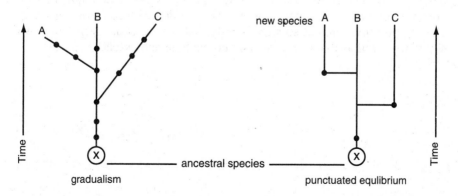

These evolutionary events are at best incompletely understood. How they actually occurred and under what circumstances they took place can only be a matter of speculation supported by scientific evidence. What is known, however, is that modern species show a wide divergence (difference) in form and function. Modern species may be autotrophic or heterotrophic. They may be aerobic or anaerobic. They may reproduce sexually or asexually. They may differ from other species in countless ways. These varieties are thought to have come about via the evolutionary processes previously described, filling the Earth's available environments with countless species able to survive the various physical conditions they encountered.

Extinction

Extinction of a species occurs when the environment changes and the adaptive characteristics of a species are insufficient to allow its survival. Fossils indicate that many organisms that lived long ago are extinct. Extinction of species is common; most of the species that have lived on Earth no longer exist. A natural component of evolution is the **extinction** of species whose adaptations no longer suit them to the environmental conditions that they encounter.

The natural environment is always changing in subtle ways. As long as this change is slow and gradual, most species can adapt to it or migrate to nearby habitats in which the favored conditions can be reestablished. However, when environmental change is rapid or when species cannot escape gradual change by migrating to new environments, the extinction of non-adaptable species is inevitable.

Much speculation has been made about the rapid extinction of most dinosaur species within a relatively short period of time. One theory holds that an asteroid strike of sufficient magnitude to be classified as an extinction level event may have created massive environmental changes that lead to the dinosaurs' extinction. Smaller, more adaptable species are thought to have evolved rapidly during this period, including primitive bird species from ancestral dinosaur species and mammals from primitive reptilian species.

In modern times, environmental changes introduced by human activities have greatly accelerated the rate of species' extinction. Although far from normal extinction, this situation should be regarded with concern that we are altering our environment so significantly that we are endangering our own survival as well as that of many other compatible and essential species.

1. The diagram below represents undisturbed rock strata in a given region. A representative fossil of an organism is illustrated in each layer.

Which statement best describes a relationship among these representative organisms?
(1) Organism A was probably more structurally advanced than organism B and organism C.
(2) Organism C probably gave rise to organism A and organism B.
(3) All of these organisms probably evolved at the same time.
(4) Organism A was probably more primitive than organism B and organism C.

2. In the early stages of development, the embryos of birds and reptiles resemble each other in many ways. This resemblance suggests that they
(1) belong to the same species
(2) are adapted for life in the same habitat
(3) share a common ancestry
(4) are both animal-like protists

3. In addition to the basic ideas of Darwin, the modern theory of evolution includes a concept that
(1) variations result from mutations and gene recombination
(2) overproduction of organisms leads to extinction
(3) variations exist only in large populations
(4) competition occurs only between members of the same species

4. The diagrams below represent some structural changes that occurred over time, resulting in the development of the modern horse.

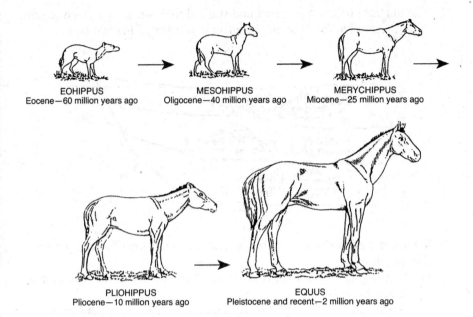

EOHIPPUS
Eocene—60 million years ago

MESOHIPPUS
Oligocene—40 million years ago

MERYCHIPPUS
Miocene—25 million years ago

PLIOHIPPUS
Pliocene—10 million years ago

EQUUS
Pleistocene and recent—2 million years ago

This sequence of structural changes best illustrates the concept of
(1) organic evolution
(2) ecological succession
(3) intermediate inheritance
(4) geographic isolation

5. Some species undergo long periods of stability interrupted by geologically brief periods of significant change. During these brief periods, new species may evolve. This pattern of evolution is part of the concept of
(1) use and disuse
(2) reproductive isolation
(3) homologous structures
(4) punctuated equilibrium

6. The diagram below illustrates the change that occurred in the frequency of phenotypes in an insect population over 10 generations.

First generation

Tenth generation

10 years later

Key:

= AA, Aa

= aa

A probable explanation for this change would be that over time there was

(1) a decrease in the adaptive value of gene *a*
(2) an increase in the adaptive value of gene *a*
(3) an increase in the population of this insect
(4) a decrease in the mutation rate of gene *A*

7. According to Darwin's theory of evolution, differences between species may result from

(1) the disuse of body structures
(2) the transmission of acquired characteristics
(3) natural selection
(4) mutagenic agents

8. The concept that new varieties of organisms are still evolving is best supported by the

(1) increasing need for new antibiotics
(2) increasing number of individuals in the human population
(3) decreasing number of new fossils discovered in undisturbed rock layers
(4) decreasing activity of photosynthetic organisms due to warming of the atmosphere

9. Two nucleotide sequences found in two different species are almost exactly the same. This suggests that these species
   (1) are evolving into the same species
   (2) contain identical DNA
   (3) may have similar evolutionary histories
   (4) have the same number of mutations

10. The theory that evolutionary change is slow and continuous is known as
    (1) punctuated equilibrium
    (2) geographic isolation
    (3) speciation
    (4) gradualism

11–12. Base your answers to questions 11 and 12 on the information below and on your knowledge of biology.

   Before the Industrial Revolution, a light-colored variety of peppered moth was well camouflaged among light-colored lichens that grew on the bark of trees around London. A dark-colored variety of the peppered moth probably existed but was rarely observed because it was so easily seen by birds and eaten. When industry was introduced in London, soot killed the pollution-sensitive lichens, exposing dark tree bark. As a result, the dark-colored variety of the moth became the better camouflaged of the two moth varieties.

11. In this situation, what is the relationship between the birds and the moths?
    (1) producer-consumer
    (2) predator-prey
    (3) parasite-host
    (4) autotroph-heterotroph

12. Identify one way in which humans influenced the change in the populations of the peppered moth.

___

13. When Charles Darwin was developing his theory of evolution, he considered variations in a population important. However, he could not explain how the variations occurred. Name *two* natural processes that can result in variation in a population and explain how these processes actually cause variation. [2]

VII. CONTINUITY OF LIFE BY MEANS OF REPRODUCTION

KEY IDEA 4—REPRODUCTIVE CONTINUITY The continuity of life is sustained through reproduction and development.

Essential Question:

- *What mechanisms are at work to create and perpetuate life on Earth?*

Species produce more of their own kind through the process of reproduction. In a sense, species are able to transcend the life spans of individual members of the species by passing along the genetic information contained in their reproductive cells. In asexual reproduction, the parent organism passes an exact copy of its genetic makeup to the offspring. In sexual reproduction, the father and mother each contribute 50 percent of the genetic information needed to produce the new individual. For this reason, the offspring of sexual reproduction resemble, but are not exactly like, the parents. A **zygote** (fertilized egg) contains all the genetic information needed to produce a complete organism. Development of the embryo is a highly regulated process that involves mitosis and differentiation. It is sensitive to the environmental conditions in which it is occurring. Reproductive technologies such as cloning, genetic engineering, and stem cell research have medical, agricultural, and ecological applications.

Mechanisms of Reproduction

Essential Question:

- *How do living things produce more living things of the same kind?*

Performance Indicator 4.4.1 *The student should be able to explain how organisms, including humans, reproduce their own kind.*

Reproduction and development are necessary for the continuation of any species. Cells have finite life expectancies. After formation by mitosis or fertilization, each new cell undergoes a period of growth followed by mitotic cell division. Growing evidence points to the fact that such cells have a genetically programmed limit to the number of times they can reproduce before they age and die. When cells specialize, their reproductive potential becomes significantly limited and aging becomes a more influential factor in their life expectancy. For all cells and all organisms, death is a certainty.

Therefore, the only way that a species can be continued is through the reproductive process. The living material and genetic information passed from one generation to the next literally perpetuate the species.

Asexual Reproduction

Some organisms reproduce asexually with all the genetic information coming from one parent. Other organisms reproduce sexually with half the genetic information typically contributed by each parent. Cloning is the production of identical genetic copies. All types of asexual reproduction have in common the production of new organisms from a single parent organism. However, the process is carried on differently by different living things. Some types of asexual reproduction are as follows.

Binary fission is accomplished when a single cell undergoes mitosis followed by equal cytoplasmic division. This forms two daughter cells each having roughly the same size and shape and containing identical genetic information. Binary fission is carried on by many species of unicellular organisms, including the paramecium and the ameba. Bacteria are also known to reproduce in this manner.

Binary Fission in Ameba (above) and Paramecium (below)

Budding is accomplished when mitosis is followed by unequal cytoplasmic division. This results in daughter cells of unequal size that contain identical genetic information. The larger of the two cells may divide rapidly several more times, producing a chain or colony of daughter cells. Budding of this sort occurs commonly in yeast. A second form of budding occurs in certain simple multicelled animals, including the hydra. Undifferentiated cells begin to divide rapidly, forming a new, smaller organism from the tissues of the parent. This bud, genetically identical to the parent, then separates from the parent and begins to undergo rapid growth as an independent organism.

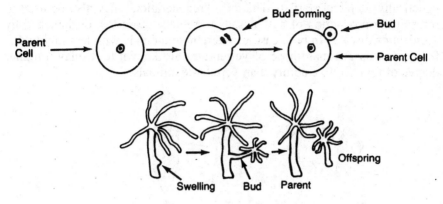

Budding in Yeast (above) and Hydra (below)

Sporulation is the formation of specialized reproductive cells, known as spores, within the parent organism. Each spore contains a nucleus surrounded by cytoplasm. When the spores are released from the parent plant and land in an environment containing conditions favorable to their growth, they begin to undergo mitotic division. The spores of most species require moisture and warmth to germinate. The mitotic divisions result in the formation of a new multicellular organism genetically identical to the original parent organism. This type of sporulation is carried on by fungi, mosses, and ferns.

Regeneration involves the production of one or more new organisms from the severed parts of a single parent organism. The pieces of cut-up planaria worm may frequently grow back lost tissues to produce as many as four or five new, identical planaria worms. Starfish cut into pieces that contain part of the central disk area may undergo regeneration that produces several new, genetically identical starfish. The term "regeneration" may also be used to refer to the production of new tissues to replace those lost or damaged by accident or disease. In both cases, invertebrate animals with less highly differentiated tissues than those of vertebrates are known to display a higher degree of regenerative ability than vertebrate animals.

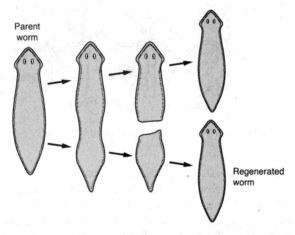

Vegetative propagation is a general term referring to any of several forms of asexual reproduction carried on by multicellular plants. An aspect common to all types of vegetative propagation is the production of new plant organisms from the leaves, stems, or roots (vegetative parts) of the parent plant rather than from the flower (reproductive part). As in other forms of asexual reproduction, the new organisms produced by vegetative propagation are genetically identical to the parent. Examples of vegetative propagation include cuttings, bulbs, tubers, runners, and grafting.

162

Cloning is a term that refers to the production of a group of genetically identical offspring from the cells of a single parent organism that would normally reproduce by sexual means. Cloning has been attempted with varying success with a number of different organisms, both plant and animal. Although the process remains experimental in animals, it has proven to be quite successful in plants. It is used extensively in the production of certain commercial crops. Its main advantage is that organisms with desirable combinations of traits, which would otherwise be changed in sexual reproduction, can be reproduced rapidly, with no alteration of their phenotypic combinations. Each of the genetically identical offspring is known as a clone.

The diagram below represents the process used in 1996 to clone the first mammal, a sheep named Dolly.

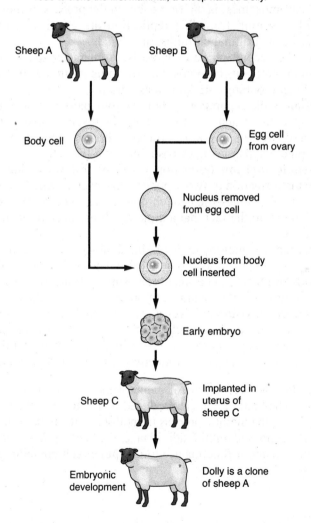

Sheep A

Sheep B

Body cell

Egg cell from ovary

Nucleus removed from egg cell

Nucleus from body cell inserted

Early embryo

Sheep C

Implanted in uterus of sheep C

Embryonic development

Dolly is a clone of sheep A

Sexual Reproduction

The processes of meiosis and fertilization are key to sexual reproduction in a wide variety of organisms. The process of meiosis results in the production of eggs and sperm that each contain half of the genetic information. During fertilization, gametes unite to form a zygote, which contains the complete genetic information for the offspring. Sexual reproduction involves the production and fusion of sex cells (**gametes**) of two parent organisms. This process ensures variations in a species. Key to this process is the production of **haploid** (*n*) gametes from the diploid (*2n*) cells of the parent, using the process of **meiotic cell division**. Meiotic cell division, also known as meiosis, occurs in distinct stages.

- Replication of the cell's single-stranded chromosomes, forming double-stranded chromosomes, is the first event in the process of meiosis. The chromatids resulting from this replication are chemical and genetic duplicates of the original chromosome strands.
- **Synapsis**, the second phase of meiosis, is characterized by the close pairing of homologous chromosomes, forming groupings of four **chromatids**. Such groupings are known as tetrads.
- **Disjunction** is the separation of homologous pairs into two groups as the cell enters the actual division phase. The result of disjunction is the formation of two sets of double-stranded chromosomes. These migrate along a spindle apparatus to opposite poles of the cell.
- **First meiotic division** (reduction division) follows disjunction and results in the formation of two separate cells, each of which contains the haploid set of double-stranded chromosomes. The term "**reduction division**" refers to the fact that a diploid cell has been divided into two haploid cells.
- Soon after the first meiotic division, the double-stranded chromosomes line up in the center of the two new cells. The centromeres joining the chromatids replicate. The resulting single-stranded chromosomes migrate along a spindle apparatus to opposite poles of the cell.
- **Second meiotic division** involves a second cytoplasmic division and the formation of four **daughter nuclei**, each of which contains a haploid set of single-stranded chromosomes. The cells that result from this process will mature into specialized reproductive cells known as **gametes**.

Spermatogenesis is a specific type of gametogenesis carried on in the male gonad, or testis. The kind of gamete produced by the **testis** is the sperm cell. In spermatogenesis, the meiotic process normally results in the production of four haploid nuclei housed within individual cells. Each of these cells has the potential to mature into a functional, motile **sperm cell** containing its own haploid nucleus.

Oogenesis is a specific type of gametogenesis carried on in the female gonad, or ovary. The kind of gamete produced by the **ovary** is the **egg cell**. In oogenesis, the meiotic process normally results in the production of four haploid nuclei housed within individual cells. Only one of these cells has the potential to mature into a functional egg cell containing a haploid nucleus. The other three cells, known as **polar bodies**, degenerate and are eventually reabsorbed by the body.

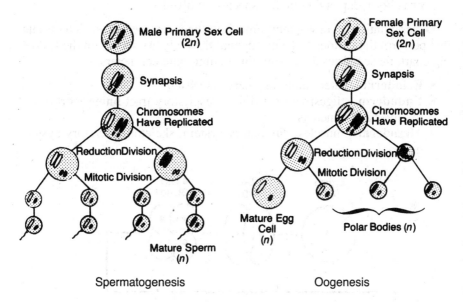

Spermatogenesis   Oogenesis

During **fertilization**, fusion of a haploid sperm cell and a haploid egg cell results in the formation of a diploid **zygote**, or fertilized egg. The restoration of the diploid condition allows the zygote to undergo mitotic divisions that eventually lead to the production of a new organism. Because the genes of the father and mother are combined in the offspring, the offspring resembles both parents but is not identical to either. In reference to meiotic cell division, this union of male and female genes is known as **recombination**.

Early Development

The zygote may divide by mitosis and differentiate to form the specialized cells, tissues, and organs of multicellular organisms. After fertilization, the zygote begins to undergo a series of rapid mitotic divisions. This process, known as **cleavage**, increases the number of cells in the growing cell mass. Little cell growth accompanies cleavage, resulting in a reduction of cell size with each division. A zygote undergoing cleavage is an **embryo**. As cleavage continues, distinct stages of development may be noted that differ with different species:

- **Blastula**—a hollow ball of cells during which the basic body plan becomes established.
- **Gastrula**—an indented blastula in which cells begin to move into the positions they will have in the adult offspring. Basic cell differentiation is thought to occur as this phase is reached.
- **Embryo**—a stage of development in which the new living organism takes on the final shape and differentiation it will have when it hatches or is born, depending on the species involved.

Differentiation is how embryonic cells become specialized into tissues that perform the various tasks throughout the body. As shown on the following chart, these tissues arise from three embryonic cell layers:

- **Ectoderm**—outer epidermis, nervous system
- **Endoderm**—digestive tract lining, respiratory tract lining, portions of the liver and pancreas
- **Mesoderm**—muscles, circulatory system, skeleton, excretory system, gonads

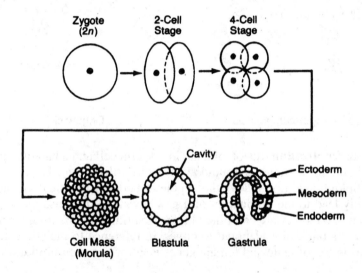

Cleavage and Differentiation

Human Reproduction

Human reproduction and development are influenced by factors such as gene expression, hormones, and the environment. The reproductive cycle in both males and females is regulated by hormones such as testosterone, estrogen, and progesterone. Development of a human embryo is governed by many factors. Principal among these factors is the precise genetic information provided to the embryo by the parents. The physical and functional traits inherited by the embryo will comprise a combination of the traits of the father and the mother.

Evidence supports the probability that basic behavioral traits are inherited in a similar fashion.

Hormones produced by the embryo as it develops, as well as those of the mother, will also help to shape the embryo. The environment of the embryo is also of extreme importance in determining the characteristics of the child. Growing evidence suggests that the health, diet, and use of drugs, tobacco, and alcohol by the mother during pregnancy can have a powerful negative effect on the developing embryo and subsequently on the child after birth.

The reproductive process in adult humans, like most physiological activities, is governed by hormones. **Testosterone** in males is important in the production of male secondary sex characteristics, including the production of healthy sperm cells. In females, the hormones **estrogen** and **progesterone**, among others, are essential in the production of female secondary sex characteristics, including the production and maturation of healthy egg cells.

Human Female   The structures and functions of the human female reproductive system, as in almost all other mammals, are designed to produce gametes in ovaries, allow for internal fertilization, support the internal development of the embryo and **fetus** (late-stage embryo) in the uterus, provide essential materials through the placenta, and provide nutrition through milk for the newborn. In human females, ova (egg cells), the female gametes, are produced by meiosis in the **ovary**. Two ovaries are located in the lower abdomen. After it has been formed, the egg cell is stored in the follicle of the ovary until it matures and is released. Once released, the egg cell travels along a short tube, the oviduct (fallopian tube), to the uterus.

The **menstrual cycle**, which begins at puberty, is a hormone-controlled process in the human female that is responsible for the monthly release of mature eggs. The average duration of the menstrual cycle is 28 days. The four stages of the menstrual cycle are as follows.

- **Follicle stage**—As the cycle begins, a single egg matures within the ovarian follicle under the influence of follicle stimulating hormone (FSH) produced by the pituitary gland. The ovary begins to produce estrogen, which stimulates the lining of the uterus to thicken and vascularize. This portion of the cycle takes approximately 14 days to complete.
- **Ovulation stage**—This portion of the cycle involves the release of the mature egg from the follicle into the oviduct. On average, this event occurs about day 14 of the cycle.
- **Corpus luteum stage**—After ovulation, the cells that made up the ovarian follicle begin to change under the influence of lutenizing hormone (LH), which is produced by the pituitary gland. The resulting structure, known as the corpus luteum, secretes the hormone progesterone, which helps to ready the uterine lining for the possible implantation of a fertilized egg. This portion of the cycle continues for about 8 to 10 days.

- **Menstruation stage**—If no fertilized egg is received in the uterus within a few days after ovulation, the uterine lining begins to break down. The disintegrated tissue and blood are expelled from the body via the vaginal canal during menstruation. This aspect of the cycle normally occurs over a few days.

The lower end of the female reproductive tract consists of the **vagina**, which functions to receive sperm cells during intercourse and later as the birth canal. **Internal fertilization** normally occurs in the **oviduct**, where sperm cells and the egg cell meet. The **uterus** acts as the internal development chamber after implantation of a fertilized egg in its lining. The lower end of the uterus is bounded by the muscular **cervix**.

Human Reproductive System: Female

Prenatal development includes all the processes of embryonic development that occur before birth. Prenatal development begins as soon as the egg is fertilized. The resulting zygote begins to undergo cleavage. Within 10 days after fertilization, the ball of cells resulting from cleavage, now known as an embryo, implants itself in the lining of the uterus. Once embedded in the uterine lining, the embryo undergoes the process of gastrulation. At the same time, the cells of the embryo begin to differentiate into the specialized tissues and organs of the adult organism.

The tissues of the embryo and the mother grow together to form the **placenta**, the connection that allows nutrients and oxygen to pass from mother to embryo during the 9-month gestation period. Wastes such as carbon dioxide and urea pass by diffusion from the embryo's blood to that of the mother for excretion through the mother's excretory processes. The fetus (late-stage embryo) is connected to the placenta by the **umbilical cord**. The fetus is enclosed in an **amnion** sac filled with amniotic fluid.

Birth occurs after a period of development, known as the **gestation** period, of approximately 9 months. Strong contractions of the muscular uterus force the baby headfirst through the cervix and vagina to the outside of the body, where the baby begins to breathe, eat, and excrete wastes on his/her own. Milk produced by the mammary glands provides food specialized for newborns.

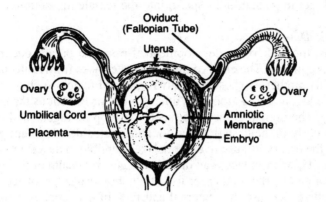

Placental Development

Human Male   The structures and functions of the human male reproductive system, as in other mammals, are designed to produce gametes in **testes** and make possible the delivery of these gametes for fertilization. Two testes are located in an internal pouch, the **scrotum**, which extends from the wall of the lower abdomen. Temperatures in the scrotum are generally 1°C to 2°C cooler than normal body temperature. This reduced temperature is optimum for sperm production. When first formed, the sperm are inactive and are stored in the testis until activated and released. Once released, the sperm travel along a series of tubes, the **vas deferens**, to the exterior of the body. Along this route, a number of glands add various fluids to the sperm cells to activate them and increase their volume. This mixture of sperm cells and fluid is known as **semen**.

Human Reproductive System: Male

169

The **urethra** is a tube that carries the activated sperm along the last portion of its journey in the body. The urethra extends from the urinary bladder, at its connection with the vas deferens, to the exterior of the body through the penis. In human beings, the **penis** is the structure that permits internal fertilization through direct implantation of sperm into the female reproductive tract.

Embryonic Development

In humans, the embryonic development of essential organs occurs in early stages of pregnancy. The embryo may encounter risks from faults in its genes and from its mother's exposure to environmental factors such as inadequate diet, use of alcohol/drugs/tobacco, other toxins, or infections throughout her pregnancy. The human embryo begins to develop its organs and organ systems very early in the gestation period. During this development period, the tissues of the embryo/fetus are extremely susceptible to the effects of outside influences. These influences can include the general health of the mother, the diet maintained by the mother, or use of alcohol, drugs, or tobacco products by the mother. Because the maternal and fetal blood supplies exchange soluble materials through the tissues of the placenta, any substances that enter the mother's blood can easily diffuse across the placental membranes to affect the fetal tissues. Substances easily tolerated by adult tissues can cause severe damage to the developing fetal tissues. A direct connection has been drawn between these environmental influences and learning disabilities, attention deficits, health problems, and physical deformities experienced by infants and young children.

Stem Cell Research

In recent years, much scientific attention has been paid to the area of **stem cell research**. Stem cells are undifferentiated cells that reside in the tissues of the body and give rise to new, specialized cells. Scientists are studying stem cells to determine how their characteristics can be used to treat a variety of conditions in the body, ranging from cancer and other diseases, to damaged tissues and organs, to the effects of aging.

Scientists are working with both embryonic and adult stem cell lines to find out how to induce these tissues to replace heart muscle tissues damaged by heart disease, spinal cord tissues damaged by physical trauma, and brain tissues lost due to stroke. Other disorders being researched include Parkinson's and Alzheimer's diseases, diabetes, and arthritis. It may also prove useful in replacing skin tissues damaged by burns and in stabilizing or reversing disorders caused by birth defects.

The promise of human health benefits resulting from stem cell research is enormous. However, the nature of this research makes it controversial and the focus of political debate. This controversy includes the sources of human stem cells and the ethical dilemma surrounding the applications of the research findings. Should stem cell lines be developed from embryonic tissues, from adult tissues only, or not at all? Who will be able to receive the therapies resulting from this research? Who will make such decisions? These are ethical questions that will be answered by today's generation of young people.

QUESTION SET 2.7—CONTINUITY OF LIFE
BY MEANS OF REPRODUCTION
(ANSWERS EXPLAINED, P. 310)

1. Which diagram represents binary fission?

(1)   (3)

(2)   (4)

2. The process of meiotic cell division in a human male usually forms, from each primary sex cell,
   (1) one diploid cell, only
   (2) four diploid cells
   (3) one haploid cell, only
   (4) four haploid cells

3. Which sequence represents the correct order of events in the development of sexually reproducing animals?
   (1) fertilization → cleavage → differentiation → growth
   (2) cleavage → fertilization → growth → differentiation
   (3) growth → cleavage → fertilization → differentiation
   (4) fertilization → differentiation → cleavage → growth

4–7. Base your answers to questions 4 though 7 on the diagrams below and on your knowledge of biology.

4. Which structures secrete hormones that regulate the development of secondary sex characteristics?
   (1) A and J
   (2) D and H
   (3) F and I
   (4) E and G

5. After sperm cells are deposited inside the female, the pathway they follow to reach the egg is from
   (1) H to I to K
   (2) J to K to H
   (3) K to I to H
   (4) G to H to I

6. Gametogenesis occurs within structures
   (1) A and J
   (2) E and G
   (3) B and I
   (4) D and H

7. Which structures are directly affected by hormones involved in the menstrual cycle?
   (1) C and E
   (2) A and D
   (3) G and I
   (4) I and J

8–10. Base your answers to questions 8 through 10 on the information in the chart below and on your knowledge of biology.

STAGES OF THE MENSTRUAL CYCLE

Stage	Event
A	Periodic shedding of the thickened uterine lining
B	Release of the egg
C	Production of progesterone by tissue in a follicle
D	Maturation of the egg and secretion of estrogen

8. Which structure does the egg released in stage *B* normally enter first?
   (1) cervix
   (2) vagina
   (3) uterus
   (4) oviduct

9. Which stage is represented by letter *A*?
   (1) ovulation
   (2) menstruation
   (3) follicle
   (4) corpus luteum

10. Which sequence best represents the order of stages in the menstrual cycle?
    (1) $D \rightarrow B \rightarrow C \rightarrow A$
    (2) $A \rightarrow B \rightarrow D \rightarrow C$
    (3) $C \rightarrow A \rightarrow B \rightarrow D$
    (4) $A \rightarrow B \rightarrow C \rightarrow D$

---

11. In human males, sperm cells are suspended in a fluid medium. The main advantage gained from this adaptation is that the fluid
    (1) removes polar bodies from the surface of the sperm
    (2) activates the egg nucleus so that it begins to divide
    (3) acts as a transport medium for sperm
    (4) provides currents that propel the egg down the oviduct

12. Substances can diffuse from the mother's blood into the fetal blood through the structure known as the
    (1) amnion
    (2) fallopian tube
    (3) yolk sac
    (4) placenta

173

13–15. Base your answers to questions 13 through 15 on the diagram below and on your knowledge of biology.

13. Which diagram shows the first appearance of the distinct layer of cells that will form the muscular, skeletal, and circulatory systems?
    (1) 11   (3) 6
    (2) 8   (4) 4

14. If stages 1 through 4 represent developmental stages of a human, where in the human female would these stages normally occur?
    (1) ovary   (3) oviduct
    (2) vagina   (4) uterus

15. Which events must occur immediately before the sequence represented in the diagrams can take place?
    (1) gametogenesis and fertilization
    (2) menstruation and menopause
    (3) prenatal development and gestation
    (4) placental formation and metamorphosis

16–18. Base your answers to questions 16 through 18 on the information and data tables below and on your knowledge of biology. Use one or more complete sentences to answer each question.

Drinking alcohol during pregnancy can cause the class of birth defects known as fetal alcohol syndrome (FAS). Scientists do not yet understand the process by which alcohol damages the fetus. Evidence, however, shows that the more a pregnant woman drinks, the greater the chances that the child will be affected and the birth defects will be serious. Some evidence indicates that even low levels of alcohol consumption can cause intellectual and behavioral problems.

INFANT CHARACTERISTICS

Characteristics (Average)	Alcohol Use During Pregnancy	
	Drinker	Nondrinker
Weeks of development before birth	36.9	38.7
Birth weight (g)	2,555	3,094
Birth length (cm)	46.8	50.1
Head circumference (cm)	32.1	34.5

PHYSICAL ABNORMALITIES DETECTED IN INFANTS AT BIRTH

Physical Abnormalities	Alcohol Use During Pregnancy	
	Drinker (Percentage of 40 Infants)	Nondrinker (Percentage of 80 Infants)
Low birth weight	73	12
Small brain	33	0
Flattened nasal bridge	8	0
Abnormal facial features	15	0
Spinal defects	8	0
Heart defects	8	0

16. Do the data in the tables justify scientists' conclusions that alcohol causes physical abnormalities at birth by interfering with the normal development of the fetus? Defend your position with supporting data.

17. What additional data would be needed to better support the scientists' conclusions?

18. Explain why alcohol consumption by the mother is especially harmful during the early stages of pregnancy.

---

19. Write one or more paragraphs that compare the two methods of reproduction, asexual and sexual. Your answers must include: [4]

 • *one* similarity between the two methods [1]
 • *one* difference between the two methods [1]
 • *one* example of an organism that reproduces by asexual reproduction [1]
 • *one* example of an organism that reproduces by sexual reproduction [1]

VIII. DYNAMIC EQUILIBRIUM IN LIVING THINGS

> ## KEY IDEA 5—DYNAMIC EQUILIBRIUM AND
> ## HOMEOSTASIS Organisms maintain a dynamic equilibrium
> that sustains life.

Essential Question:

- *How do living things create a balanced steady state that maintains the living condition?*

Energy and materials must be constantly provided to living systems to keep them operating. Within the cell, biochemical processes occur within a very narrow range of conditions. These conditions are maintained by the very biochemical processes that require their stability. This maintenance is accomplished by the responsiveness of living systems to the environmental changes that occur around the cell. These responses can range from simple biochemical reactions to complex behaviors and result in a **dynamic equilibrium** known as **homeostasis**. When the feedback mechanisms that create and maintain homeostasis fail, the living condition is threatened, and disease or death can result.

Biochemical Processes and Homeostasis

Essential Question:

- *What biochemical reactions are common in living cells and promote the living condition?*

Performance Indicator 4.5.1 *The student should be able to explain the basic biochemical processes in living organisms and their importance in maintaining dynamic equilibrium.*

Photosynthesis
The energy for life comes primarily from the Sun. Photosynthesis provides a vital connection between the Sun and the energy needs of living systems. The Sun, actually a medium-sized star, is the ultimate source of all energy on Earth. It produces radiant energy in many forms, some of which is detectable by humans as light. Visible sunlight appears white but actually contains the wavelengths needed to produce all colors. In **photosynthesis**, the energy of sunlight is trapped and converted into the chemical bond energy of organic

compounds such as sugar. Once trapped in the chemical bonds of these compounds, the energy of sunlight is made available to all living things in a more stable form. Green plants provide a vital link between the Sun's energy and the life processes of all living things.

Plant cells and some one-celled organisms carry on photosynthesis. An organelle found in the cells of green plants and algae is the **chloroplast**. The chloroplast absorbs sunlight and converts the energy to chemical bond energy. **Chlorophyll**, a green pigment found in the chloroplast, absorbs light energy. In the presence of the proper enzymes, the atoms of carbon dioxide (CO_2) and water (H_2O) are rearranged to form the more complex organic molecule glucose ($C_6H_{12}O_6$). A by-product of this process is molecular oxygen (O_2).

Chloroplast

A more detailed study of the photosynthetic process reveals that it actually consists of two separate processes. Each is characterized by its own set of chemical reactions.

Photochemical (light) reactions Stacked layers of chlorophyll-containing membranes within the chloroplast, known as **grana**, contain the enzymes that catalyze this process. The light energy absorbed by the chlorophyll and other pigments on these grana is used to split water molecules into their component elements, hydrogen and oxygen. This process is known as **photolysis** (splitting with light) and involves the production of an energy-carrying molecule known as adenosine triphosphate (ATP). Atoms of oxygen recombine to form molecular (atmospheric) oxygen, which is released as a gas. The hydrogen released is transferred to the next phase of the process by a hydrogen carrier compound.

Carbon-fixation (dark) reactions A second set of photosynthetic reactions combines the released hydrogen atoms with the atoms making up carbon dioxide. An intermediate product of this reaction is a carbohydrate-like, three-carbon compound known as **phosphoglyceraldehyde (PGAL)**. This compound may be used by the cell to synthesize several other compounds such as glucose. The name carbon-fixation reactions is derived from the fact that carbon atoms are fixed in place in a stable form as a result of this process. This set of reactions occurs in the **stroma** of the chloroplast. The stroma lies between the grana.

In all organisms, organic compounds can be used to assemble other molecules such as proteins, DNA, starch, and fats. The chemical energy stored in bonds can be used as a source of energy for life processes. The compound glucose, which results from photosynthesis, may be used in many ways by the cell. Some of these include the following.

- Direct use as a source of energy in cellular respiration, which is carried on by all living things, plants as well as animals.
- Conversion, for purposes of storage, into more complex forms of carbohydrates, such as starch, by the process of dehydration synthesis. Such storage carbohydrates must be reconverted to simpler molecules by intracellular digestion before they can be used in metabolic processes or transported to other sites in the organism.
- Conversion into other types of metabolic compounds, such as proteins, lipids, and nucleic acids.

Photosynthesis is perhaps the single most significant biochemical process carried on by living things. Significant results of the photosynthetic reactions include the following.

- Carbohydrates manufactured as a result of photosynthesis are used by virtually all living things as a source of cellular energy. This energy is released in a controlled fashion in the process of cellular respiration.
- Oxygen gas released as a by-product of the photosynthetic reactions is the principal source of atmospheric oxygen required by most living things for cellular respiration. Reductions in photosynthetic rate can have disastrous effects on such life-forms by reducing the concentration of oxygen in the air or water environment.

Respiration

In all organisms, the energy stored in organic molecules may be released during cellular respiration. This energy is temporarily stored in ATP molecules. The chemical bonds between atoms of food molecules provide the energy used by all living organisms to sustain life. Respiration is the biochemical mechanism by which this energy is released for use in the cell. Once released, it becomes available to the cell for use in all other life functions. Respiration takes place in cell organelles known as the **mitochondria**.

Cellular respiration is a biochemical process that includes the reactions used by cells to release energy from organic molecules such as glucose. The energy released as a result of these reactions is temporarily stored in the bonds of molecules of **adenosine triphosphate (ATP)**. This energy may be released for use in cell processes when ATP is converted to adenosine diphosphate (ADP) and phosphate (P). Like other reactions that occur in the cell, the reactions of respiration are controlled by enzymes. The following equation illustrates the reactions involving the formation of ADP and the release of cell energy:

$$H_2O + ATP \xrightarrow{\text{ATPase}} ADP + P + energy$$

The two forms of cellular respiration carried on by living things are known as **aerobic** (with oxygen) and **anaerobic** (without oxygen).

1. **Anaerobic respiration** is a form of cellular respiration carried on by cells in the absence of molecular oxygen. This process operates by means of chemical reactions catalyzed by enzymes located in the cytoplasm of the cell. Organisms typically employing this form of cellular respiration include certain bacteria and fungi (yeasts). Human beings have learned to take advantage of this process by using these organisms in the manufacture of foods such as cheese, buttermilk, and yogurt. These organisms also play an important role in the baking, wine-making, and brewing industries. Two principal types of anaerobic respiration occur.

   • **Alcoholic fermentation** is so called because one of the major by-products produced in this process is ethyl alcohol:

   $$glucose \xrightarrow{\text{enzymes}} 2 \text{ ethyl alcohol} + 2\ CO_2 + 2\ ATP$$

   • **Lactic acid fermentation**—which produces lactic acid as one of its by-products:

   $$glucose \xrightarrow{\text{enzymes}} 2 \text{ lactic acid} + 2\ ATP$$

   Both types of anaerobic respiration are considered to be relatively inefficient. The end products of both (ethyl alcohol and lactic acid) contain considerable amounts of energy that are not effectively released for use by the cell.

2. **Aerobic respiration** is a form of cellular respiration carried on by certain cells in the presence of molecular oxygen. This process operates by means of chemical reactions catalyzed by enzymes located primarily in the mitochondria of the cell. As in anaerobic respiration, the principal product of aerobic respiration is the energy released from organic molecules for use by the cell in other processes. Unlike anaerobic respiration, however, aerobic respiration uses molecular oxygen to release substantially more energy per glucose molecule metabolized. For this reason, aerobic respiration is a relatively more efficient form of respiration than anaerobic respiration. In general, aerobic respiration may be illustrated as follows.

$$\text{glucose + 6 oxygen} \xrightarrow{\text{enzymes}} \text{6 water + 6 carbon dioxide + 36 ATP}$$

Like anaerobic respiration, aerobic respiration is characterized by a series of enzyme-catalyzed reactions that progressively convert glucose to end products and energy. These reactions are conceptually divided into two phases, known as the anaerobic phase and the aerobic phase.

- **Anaerobic phase**—Glucose molecules are converted to two molecules of **pyruvic acid**. The process is initiated through release of the energy stored in two molecules of ATP. Four molecules of ATP are formed in this conversion, resulting in a net gain of two molecules of ATP. A summary equation of this phase is as follows:

$$\text{glucose + 2 ATP} \xrightarrow{\text{enzymes}} \text{2 pyruvic acid + 4 ATP}$$

- **Aerobic phase**—The cell uses molecular oxygen to break down pyruvic acid further to form energy and the waste end products water and carbon dioxide. In this part of the process, an additional 34 molecules of ATP are formed. A summary equation of this phase is as follows:

$$\text{2 pyruvic acid + 6 oxygen} \xrightarrow{\text{enzymes}} \text{6 water +}$$
$$\text{6 carbon dioxide + 34 ATP}$$

The net gain of ATP molecules in aerobic respiration is 36, making it 18 times more efficient than anaerobic respiration. The significance of aerobic respiration is that usable energy is stored in the chemical bonds of the ATP molecules.

The energy from ATP is used by the organism to obtain, transform, and transport materials and to eliminate wastes. Cellular operations, primarily biochemical reactions, are energy-consuming activities. The energy in ATP is released to initiate these reactions if they are exothermic (energy-releasing) or to provide sustained energy if they are endothermic (energy-consuming). Active transport, in which materials are moved into or out of the cell against the concentration gradient, is a good example of an energy-consuming process. By using this process, the cell can obtain and concentrate needed materials, move them around the cell interior, and expel potentially toxic wastes to the exterior environment.

Enzyme Action

Biochemical processes, both breakdown and synthesis, are made possible by a large set of biological catalysts called enzymes. Enzymes can affect the rates of chemical change. The rate at which enzymes work can be influenced by internal environmental factors such as pH and temperature. The controlled nature of cellular chemical reactions is made possible through the actions of cellular enzymes. Enzymes regulate the rate at which the cell's chemical reactions occur. In this role, enzymes function as organic catalysts. Each chemical reaction in the cell requires its own, specific type of enzyme in order to operate.

The rate at which enzymes catalyze their reactions changes as conditions inside the cell change. Such conditions as the temperature of the cell, the relative concentrations of enzyme and substrate within the cell, and the pH of the cell can alter the rapidity with which enzymes work.

Temperature The cell and its enzymes are very sensitive to temperature conditions. Extreme cold can slow enzyme action nearly to a halt. As the temperature of the cell rises, its enzymes begin to operate more and more rapidly. An optimum (best or most efficient) temperature allows the most rapid reaction rate. In humans this optimum temperature is 98.6°F (38°C). Extreme heat—at or above 104°F (40°C)—can halt enzyme action by deforming the molecular shape of the enzyme. This distortion is known as **denaturation** and is a permanent condition that makes the enzyme incapable of further catalytic action. The effect of temperature on a typical human enzyme is shown below.

Enzyme Activity Versus Temperature

Concentration of enzyme and substrate In order to operate, enzyme molecules require the presence of substrate. In a system in which the concentration of enzyme is constant, increasing the substrate concentration from zero will result in a steady rise in reaction rate. This rise continues until the concentration roughly equals that of the enzyme. At this point, the enzyme and substrate are in equilibrium and the rate of increase slows and levels off. Increases in substrate concentration beyond this equilibrium point have little or no effect on the rate of enzyme action.

Cell acidity (pH) The cell environment may be characterized by its level of acidity or alkalinity. To measure this characteristic accurately, scientists have devised methods of measuring the relative concentration of hydrogen ion (H^+), which is given off by acids in solution with water. The pH scale has been devised to indicate the relative hydrogen-ion concentration. This scale runs from 0 to 14, 0 being extremely acidic and 14 being extremely alkaline. A pH of 7 indicates a neutral condition (neither acidic nor alkaline). A pH less than 7 indicates an acidic condition; a pH greater than 7, an alkaline condition.

Most human enzymes seem to work best at or near a pH of 7 (neutral). Despite this fact, there are many enzymes for which the optimum pH is highly acidic (stomach enzymes) or highly alkaline (intestinal enzymes). In general, however, as the cell's acidity increases above 7 or decreases below 7, most enzymes show a decrease in activity.

Enzyme Rate Versus pH

Enzyme Structure Enzymes and other molecules, such as hormones, receptor molecules, and antibodies, have specific shapes that influence both how they function and how they interact with other molecules. **Enzymes** are protein molecules. They are both extremely complex and extremely variable. Certain enzymes are also known to contain nonprotein components known as coenzymes; vitamins frequently function as coenzymes. Enzymes are usually named for the particular chemical (substrate) whose reactions they catalyze; enzyme names end in the suffix -*ase*.

Enzyme molecules are thought to contain specific areas responsible for linking to the substrate molecules. These reacting areas are known as **active sites**. They can be thought of as pockets or slots on the enzyme molecule into which the substrate molecules fit during the reaction catalyzed by the enzyme.

A = Enzyme Molecule (Unchanged).
B = Substrate Molecule.
C & D = Products of Hydrolysis of Molecule B.

Catalytic Action of an Enzyme

The catalytic action of enzymes may be depicted in many ways. However, the following model, known as the **lock-and-key model**, is widely accepted as being a reasonable summary of this action.

- To begin the process, the enzyme molecule must link with the substrate in a temporary, close physical association known as the **enzyme-substrate complex**.
- This association is thought to occur at the enzyme's **active site**. The chemical fit between this active site and the substrate must be exact in order for the desired reaction to occur, in much the same way that a key must fit a lock in order for the lock to be opened (hence the name "lock-and-key" model).
- During the existence of the enzyme-substrate complex, the reaction involving the substrate takes place. The substrate is chemically altered in this reaction, and products are formed. The enzyme molecule, however, remains unchanged by the reaction.
- When the reaction is completed, the enzyme and product(s) separate. The enzyme molecule, being unchanged, is free to form additional complexes and to catalyze additional reactions of this substrate.
- A single enzyme molecule may catalyze millions of reactions in this manner during its lifetime. Eventually, however, the structure of the enzyme molecule begins to deteriorate. The molecule must then be replaced.

Disease as a Failure of Homeostasis

Essential Questions:

- *How does the body defend itself against disease?*
- *What happens when a living thing gets sick?*

Performance Indicator 4.5.2 *The student should be able to explain disease as a failure of homeostasis.*

We have learned that **homeostasis** refers to the normal condition of balance (or equilibrium, steady state, stability) that exists within cells and organisms. This equilibrium depends on the coordination of thousands of chemical reactions occurring at the same time within the cells and that may be easily upset by any change of the cell's physical or chemical environment. Examples of homeostasis include the maintenance of body temperature and blood sugar levels in human beings.

This balance of life is fragile at best. Living things are constantly subjected to changes in their environments that threaten to interfere with homeostatic balance. Living systems have been developed that provide feedback mechanisms to help the regulatory system monitor the interior and exterior environment and send commands to the body to adjust to changes it detects. If these feedback mechanisms fail to provide the appropriate responses, then cell functions could become less efficient than normal, which in turn could lead to illness or death.

Organisms That Interfere with Cell Activity

Viruses, bacteria, fungi, and other parasites may infect plants and animals and interfere with normal life functions. In addition to internal disruptions in the cell that can trigger an imbalance, other organisms and materials can also disrupt the cell's operation. Many human diseases are caused by viruses, bacteria, fungi, and parasites of various types. These organisms interfere with cell activity in a number of ways.

Viruses enter the body by inhalation of airborne viruses, consumption of contaminated water or foods, or through direct contact with infected tissues. They act by commandeering the cell's protein synthesis mechanism, redirecting it to produce many copies of the virus that go on to infect other cells. The host cells are destroyed in the process, impacting the operation of the affected tissues and organs. Examples of viral diseases in humans include the common cold, measles, hepatitis, polio, smallpox, rabies, viral pneumonia, yellow fever, herpes, and AIDS.

Bacteria invade healthy tissues through breaks in the skin, through consumption of contaminated food or water, or by inhalation of airborne bacteria. They begin to reproduce rapidly within the host. Bacterial infections are dangerous because they produce toxins that can kill healthy host tissues. Some bacteria actually enter host cells and consume them from within. Examples of bacterial diseases in humans include Lyme disease, giardiasis, whooping cough, tetanus, typhoid fever, diphtheria, bacterial pneumonia, bubonic plague, tuberculosis, gonorrhea, chlamydia, and syphilis.

Fungi feed on healthy tissues of the body by sending rootlike tendrils into the tissues and digesting the tissues extracellularly. Common examples of fungal infections in humans include athlete's foot and ringworm.

Parasites of various kinds can invade the body through the skin via insect bites or through consumption of contaminated water or foods. Some parasites bore directly through the exposed skin to reach internal tissues. Examples of parasitic diseases in humans include hookworm, tapeworm, trichinosis, malaria, typhus, and African sleeping sickness.

Plants and other animals can be similarly infected by disease-causing agents that are specific to the tissues of those host organisms. Scientists are constantly at work developing new treatments for the diseases caused by these agents, both to protect human health and to safeguard the health of agricultural products such as grains and livestock.

The Immune System

The immune system protects against antigens associated with pathogenic organisms or foreign substances and against some cancer cells. A special function of the blood is to defend the body against disease, to provide **immunity**. In addition to the other functions associated with the transport system, the blood provides the immune response to help it react to foreign invaders. Such invaders may include the viruses, bacteria, fungi, and parasites discussed above. They may also include the chemical antigens and toxins produced by those organisms. Special cells in the body can also be important in defending the body against its own cells, when those cells become abnormal through cancer.

Some white blood cells engulf invaders. Others produce antibodies that attack the invaders or mark them for killing. Some specialized white blood cells will remain, able to fight off subsequent invaders of the same kind. White blood cells exist in several different forms with several different functions. Two such types are described here, both important in the control of disease in humans.

- **Phagocytes** engulf and destroy bacteria that enter the bloodstream. These phagocytic white blood cells gather in large numbers at sites of bacterial infection in the body.
- **Lymphocytes** produce **antibodies** specifically designed to fight off particular types of foreign proteins, known as antigens, that enter the bloodstream by various routes.

Once these cells react to a foreign invader, the immune system develops a chemical memory of the invader's proteins (antigens). When this happens, an **active immunity** develops, which is a long-lasting protective condition created when the body recovers from an infectious disease. The lymphocytes carry their chemical memory of that antigen and churn out specific antibodies at the first sign of its return to the bloodstream.

Vaccination uses weakened microbes (or parts of them) to stimulate the immune system to react. This reaction prepares the body to fight subsequent invasions by the same microbes. Many years ago, it was discovered that inoculating people with solutions containing dead or weakened disease microbes (or fragments of these microbes) would impart the same level of immunity to the inoculated person as does recovery from the disease but without many of the risks associated with contracting the full-blown disease. This inoculation of dead or weakened microbes is known as vaccination. Active immunity acquired through vaccination usually results in the same long-lasting immunity as does the body's own immune reaction.

Some viral diseases, such as AIDS, damage the immune system, leaving the body unable to deal with multiple infectious agents and cancerous cells. AIDS is an acronym for acquired immune deficiency syndrome and is caused by the HIV virus. As its name implies, AIDS leads to a reduced efficiency of the immune system, leaving its victims unable to ward off disease-causing agents that enter the body. Certain types of cancer can also gain a foothold under these conditions. As the body's ability to fight off infectious diseases diminishes, an AIDS victim can contract multiple diseases, leaving him/her in a progressively weakened condition. If untreated, this situation can gradually lead to death. There is currently no cure for AIDS, although treatments have been developed that slow the disease's progress.

Some **allergic reactions** are caused by the body's immune responses to usually harmless environmental substances. Under certain circumstances the immune system may attack some of the body's own cells or transplanted organs. Allergies are the result of the body's reaction to the chemical composition of such materials as pollen, dust, animal dander, insect saliva, foods, drugs, molds, and many other substances. Many of these materials are harmless allergens. Sensitive people produce antibodies or antitoxins to these allergens as though they were the antigens of disease-causing organisms. An additional bodily response to the antigens is the production of histamines, which may cause irritation and swelling of the mucous membranes, a symptom typical of allergic reactions that may be more life-threatening than the allergens themselves.

Organ and tissue transplants, including blood transfusions, can be safely accomplished only if the antigen types of the donor and recipient are the same or very similar. If a close match does not exist between the recipient's and the donor's antigens, the recipient's body will react to the transplanted tissues as though they were foreign, disease-causing organisms and produce antibodies. The result of such a reaction is the rejection of the transplanted tissue or organ.

Other Disorders

Disease may also be caused by inheritance, toxic substances, poor nutrition, organ malfunction, and some personal behaviors. Some effects show up right away; others may not show up for many years. There are other causes of diseases and disorders. Some examples are as follows.

- **Inheritance:**

  - ➤ sickle cell anemia—abnormal red blood cells are crescent-shaped and hemoglobin is defective
  - ➤ Tay-Sachs disease—fatty deposits build up around nerves, leading to mental retardation
  - ➤ hemophilia—blood does not clot due to its inability to produce clotting factors
  - ➤ arthritis—cartilage breaks down and skeletal joints deteriorate and/or fuse together
  - ➤ phenylketonuria (PKU)—mental retardation caused by inability to metabolize phenylalanine

- **Toxic substances:**

  - ➤ ulcers—erosion of the stomach/intestine from irritation by alcohol or other substances
  - ➤ asthma—narrowing of the bronchial tubes due to swelling caused by allergic reaction
  - ➤ emphysema—a general deterioration of the lung structure that results from exposure to air pollutants or tobacco

- **Poor nutrition:**

  - ➤ goiter—enlargement of the thyroid caused when that gland fails to produce sufficient quantities of thyroxin due to a lack of iodine in the diet
  - ➤ gallstones—deposits of hardened cholesterol that lodge in the gallbladder and cause pain
  - ➤ anemia—a reduced ability of the blood to carry oxygen that results from an iron-deficient diet

- **Organ malfunction:**

  - ➤ constipation—the large intestine absorbs too much water from feces, leading to difficult elimination
  - ➤ diarrhea—large intestine fails to absorb sufficient water from feces, leading to watery elimination
  - ➤ coronary thrombosis—blockage of the coronary artery, which feeds the heart muscle
  - ➤ angina—chest pain caused by gradual deterioration of coronary circulation

➤ gout—painful inflammation of the skeletal joints caused by the buildup of uric acid

➤ stroke—condition caused when a portion of the brain fails due to inadequate or blocked blood supply

➤ diabetes—inability to maintain proper sugar balance in the blood due to failure of the pancreas to produce sufficient quantities of insulin

- **Personal behavior:**

➤ high blood pressure—caused by fatty buildup in arteries, is aggravated by stress, smoking, and diet, among other activities

➤ tendinitis—inflammation of the tendon-bone junction caused by repeated physical stress on the affected part

Cancer

Gene mutations in a cell can result in uncontrolled cell division, known as **cancer**. Exposure of cells to certain chemicals and radiation increases mutations and thus increases the chance of cancer. Normal mitotic cell division results in the production of new cells for growth and for the repair of damaged or worn out body tissues. It is controlled within the cell itself and occurs countless times in every living cell without flaw. However, in some cells under certain conditions, the mitotic process appears to break down and begins to occur so rapidly that insufficient time is available for normal replication and chromosome separation. This rapid, abnormal cell division is known as cancer. In a very short time, cancer may produce a very large number of such abnormal cells, which begin to crowd out the normal tissues. This results in damage to these tissues and often in the death of the host organism.

Research to Control and Eliminate Disease

Biological research generates knowledge used to design ways of diagnosing, preventing, treating, controlling, or curing diseases of plants and animals. A large and increasingly important branch of biology deals with the study and control of disease in humans and other organisms. Some scientists study the bacteria and viruses that cause disease in an attempt to determine how these organisms work within the body to disrupt systems. Other scientists are hard at work decoding the genomes of these organisms to learn how their genetic makeup affects other living things. The human genome is also being mapped and decoded, enabling scientists to trace the root causes of genetically based diseases and to design gene therapy techniques to treat them. Branches of science deal with finding effective treatments for disease in the form of drugs and antibiotics. These and other research activities are providing medical and agricultural science with a sophisticated and effective arsenal to help us to combat diseases in living things.

Body Systems and Homeostasis

Essential Question:

- *How do the activities that sustain the life of a cell contribute to the survival of a multicellular organism?*

Performance Indicator 4.5.3 *The student should be able to relate processes at the system level to the cellular level in order to explain dynamic equilibrium in multicelled organisms.*

Dynamic equilibrium results from detection of and response to stimuli. Organisms detect and respond to change in a variety of ways both at the cellular level and at the organismal level. A **stimulus** is any change that occurs in the environment of an organism that elicits a response in that organism. Such stimuli may be external or internal to the body of the organism. Examples of stimuli include light, sound, and chemical stimuli. A **response** is a reaction that an organism makes to a specific stimulus. Responses may include physical movements or glandular secretions. By reacting to its environment, an organism is able to adjust to changing conditions with the result that its systems remain in a **dynamic equilibrium** (energy-charged and adapting steady state). This equilibrium is an important prerequisite to homeostasis.

A unique feature of living matter is its sensitivity to its environment and its ability to adjust to changes both inside and outside of itself. As is the case with all life functions, this sensitivity has its basis at the cellular level. Independent one-celled organisms display sensitivity appropriate to their lifestyle and environment. This phenomenon is known as cytoplasmic sensitivity. It allows the cell to receive environmental stimuli and perform simple actions, such as avoidance behaviors, in response to them. Some one-celled organisms have specialized sensory areas that can detect environmental stimuli, such as light or vibration. Unicellular organisms use simple structures such as cilia or flagella to effect their responses to these stimuli.

In multicelled organisms, a variety of specialized adaptations assist the organism to detect stimuli and to effect responses to those stimuli. In simple organisms such as the hydra, these adaptations are limited to specialized cells that sense stimuli and communicate to other cells (nerve net) and other cells that produce crude movement (contractile fibers) or release stinging barbs (nematocysts). In more complex organisms such as the human, specialized sensing organs take in a wide range of environmental stimuli and send complex information, via nerves, to the central-processing centers of the brain and spinal cord. After analysis, this information provides the basis for complex responses including movements and secretions. The effectors of these responses (muscles and glands) receive their commands from the brain. Constant feedback allows the system to adjust its reactions based on the effects of the initial response.

In addition to sophisticated nervous responses, the bodies of complex multicelled organisms use a system of endocrine glands to control the metabolic activities of specific tissues, organs, and systems. By controlling the body's metabolism, the balance needed to sustain life is maintained.

Feedback mechanisms have evolved that maintain homeostasis. Examples include the changes in heart rate or respiratory rate in response to increased activity in muscle cells, the maintenance of blood sugar levels by insulin from the pancreas, and the changes in openings in the leaves of plants by guard cells to regulate water loss and gas exchange.

The term **feedback mechanism** refers to any process through which the body is able to adjust its actions in response to internal stimuli. The examples given represent just a few ways that the body monitors its own internal environment and adjusts to the changes it detects. Examples include:

- As the body becomes physically active, the muscle cells carry on respiration at an elevated rate. This produces additional energy but also elevated levels of carbon dioxide. As this carbon dioxide enters the bloodstream, its concentration increases and blood chemistry changes as a result. This change in the body's internal environment is detected by nerve centers in the brain and spinal cord. These nerve centers initiate commands to the breathing apparatus and heart that cause these organs to increase their rates of activity. The body's elevated heart rate causes the blood to circulate more rapidly around the body, enabling it to carry more oxygen to the muscle cells and carry away more carbon dioxide from these tissues. Breathing rate increases to accelerate the rate of gas exchange in the lungs. This response reduces carbon dioxide levels and increases oxygen levels in the blood. When carbon dioxide levels drop to a sufficiently low level, this change is also detected by the nerve centers. They then issue commands to the breathing apparatus and heart muscle to slow their rates of activity, returning the body to its resting levels. This is a clear example of a feedback loop.

- In a like manner, the body monitors the concentration of sugar (glucose) in the blood. When blood sugar levels become elevated, as they would after a person consumes and digests a meal containing carbohydrates, the pancreas is stimulated to release the hormone insulin. Insulin causes the liver to increase the rate at which blood sugar is converted to glycogen (animal starch). A second hormone, glucagon, is released from the pancreas when blood sugar concentrations drop too low, as they would long after a meal has been consumed. Glucagon stimulates the conversion of glycogen to glucose, which then enters the bloodstream to increase the blood sugar concentration. This is a clear example of a feedback loop.

- Stomates are tiny openings on the leaves of green plants that allow gas exchange between the environment and the photosynthetic cells of the leaf interior. These openings are regulated by pairs of **guard cells** that flank each stomate. Guard cells contain chloroplasts. When photosynthetic activity is high, the concentration of sugar builds up in the guard cells. This change causes the guard cells to absorb additional water in an effort to maintain a constant sugar concentration in the cytoplasm. The absorption of this water causes the guard cells to become turgid, opening the stomate and promoting rapid gas exchange. When photosynthetic activity is low, the guard cells lose sugar and water, causing them to become limp. In this condition, the guard cells are no longer able to keep the stomates open, so gas exchange is minimized.

Both single-celled organisms and multicelled organisms sense and respond appropriately to their environments. The responses of a multicelled organism are a function of the responses of the many individual cells that make it up.

QUESTION SET 2.8—DYNAMIC EQUILIBRIUM IN LIVING THINGS
(ANSWERS EXPLAINED, P. 316)

1. What is a major distinction between living and nonliving matter?
   (1) Living matter is unable to diffuse materials.
   (2) Living matter is able to control chemical activities with organic catalysts.
   (3) Living matter is able to create energy.
   (4) Living matter is unable to use energy for metabolic activities.

2–3. Base your answers to questions 2 and 3 on the diagram below and on your knowledge of biology. The diagram represents some processes occurring in the leaf of a plant.

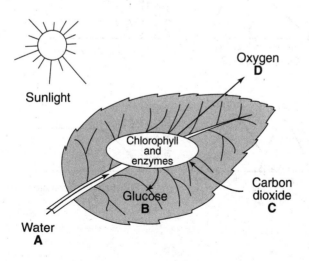

2. Which equation illustrates a process of nutrition carried out within the leaf?

(1) $B + D \rightarrow A + C$ (3) $B + C \rightarrow A + D$

(2) $A + C \rightarrow A + B + D$ (4) $A + B + D \rightarrow B + C$

3. Which letters indicate substances needed by the leaf to carry out the process of aerobic cellular respiration?

(1) A and C   (3) B and C

(2) C and D   (4) B and D

---

4. All producers and consumers use the chemical process of respiration to synthesize

(1) $C_6H_{12}O_6$ (3) alcohol

(2) ATP   (4) oxygen

5. Brine shrimp live in shallow coastal waters or near the surface of the ocean where light penetrates. Under laboratory conditions, brine shrimp are attracted to areas with the greatest light intensity and avoid areas of low light intensity. The movement of the brine shrimp to bright light is an example of

(1) negative feedback   (3) a stimulus

(2) a response   (4) active transport

6–7. Base your answers to questions 6 and 7 on the graphs below and on your knowledge of biology. The graphs represent human enzyme activity.

A

C

B

D

6. Human enzymes would most likely begin to denature at a
   - (1) temperature of 40°C
   - (2) temperature of 23°C
   - (3) pH of 3
   - (4) pH of 2

7. Certain enzymes work best within an acidic or a basic environment. This concept is illustrated in graph
   - (1) A
   - (2) B
   - (3) C
   - (4) D

---

8. The equation below summarizes some of the reactions involved in a specific biochemical process.

   hydrogen + carbon dioxide → carbohydrate + water

   The source of the hydrogen in the equation is most likely
   - (1) $C_6H_{12}O_6$
   - (2) H_2O
   - (3) PGAL
   - (4) ATP

9. In which structure of the cell shown below do photolysis and carbon-fixation reactions occur?

   - (1) A
   - (2) B
   - (3) C
   - (4) D

10–11. For each statement in questions 10 and 11, select the immune response, chosen from the list below, that is most closely associated with that statement. Then record the name of the immune response in the space provided.

<u>*Immune Response*</u>

Active immunity
Passive immunity
Allergies
Tissue rejection

10. A vaccine containing a weakened disease-causing organism is injected into the body.

　　_____

11. Chemicals known as histamines are released as a result of antibody production.

　　_____

12. An insufficient amount of hemoglobin is most closely associated with the disorder known as
    (1) angina
    (2) anemia
    (3) coronary thrombosis
    (4) high blood pressure

13. In a certain community, a number of humans have an abnormally enlarged structure under the skin of the lower front side of their necks. The cause of this condition is most likely
    (1) an excess of calcium in the diet, which has caused a muscle deformity
    (2) deposits of fat under the skin caused by a vegetable diet
    (3) inherited neck deformities caused by elevated environmental temperatures
    (4) a lack of iodine in the diet, which has caused the development of a goiter

14. What does photosynthesis produce?
    (1) starch, which is metabolized into less complex molecules by dehydration synthesis
    (2) protein, which is metabolized into less complex molecules by dehydration synthesis
    (3) glycerol, which is metabolized into more complex molecules by dehydration synthesis
    (4) glucose, which is metabolized into more complex molecules by dehydration synthesis

15. Which statement best describes the enzymes represented in the graphs below?

(1) This enzyme works best at a temperature of 35°C and a pH of 8.
(2) This enzyme works best at a temperature of 50°C and a pH of 12.
(3) Temperature and pH have no effect on the action of this enzyme.
(4) This enzyme works best at a temperature above 50°C and a pH above 12.

16–19. Base your answers to questions 16 through 19 on the reading passage below and on your knowledge of biology.

Take Two and Call Me in the Morning

Hippocrates observed that pain could be relieved by chewing the bark of a willow tree. We now know that this bark contains salicylic acid, which is similar to acetylsalicylic acid, the active ingredient in aspirin. Over 2,300 years after this observation by Hippocrates, scientists have learned how aspirin works.

When people get the flu or strain their backs, the body responds by making prostaglandins (PG), a group of hormone-like substances. The presence of certain prostaglandins may result in fever, headaches, and inflammation. Scientists have determined that aspirin interferes with prostaglandin H2 synthase (PGHS-2), an enzyme that the body uses to make pain-causing prostaglandins. In 1994, the structure of this enzyme was found to be a crystal with a tube running up the middle of it. Raw materials move through this tunnel to reach the core of the enzyme, where they are transformed into prostaglandin molecules. Research has shown that aspirin blocks this tunnel. Part of the aspirin molecule attaches to a particular place inside the tunnel, preventing the raw materials from passing through the tunnel. This blockage interferes with the production of prostaglandins, thus helping to prevent or reduce fever, headaches, and inflammation.

The body makes two forms of the enzyme. PGHS-1 is found throughout the body and has a variety of uses, including protecting the stomach. PGHS-2 usually comes into play when tissue is damaged or when infections occur. Its action results in pain and fever. Aspirin plugs up the tunnel of PGHS-1 completely and often causes stomach irritation in some people. Aspirin plugs up the tunnel partially in PGHS-2, thus helping to relieve pain and fever.

Perhaps further research could result in a drug targeting PGHS-2 but not PGHS-1, relieving the aches, pains, and fever but not irritating the stomach as aspirin does now.

16. How does aspirin relieve the symptoms of the flu?
    (1) It forms a barrier around the outer surface of PGHS-2 molecules, separating them from the prostaglandins.
    (2) It dissolves the crystal of the enzyme, preventing it from producing prostaglandins.
    (3) It is an acid that dissolves the prostaglandins that cause the symptoms.
    (4) It reduces the amount of raw material reaching the active site of the enzyme that produces prostaglandins.

17. Why does aspirin irritate the stomach of some people who take it?
    (1) It interferes with the activity of an enzyme that helps to protect the stomach.
    (2) It is the only acid in the stomach and eats away at the stomach lining.
    (3) It stimulates prostaglandin production in the stomach.
    (4) It is obtained from willow bark, which cannot be digested in the stomach.

18. By using one or more complete sentences, describe the molecular structure of prostaglandin H2 synthase.

19. By using one or more complete sentences, explain why chewing the bark of a willow tree could help relieve the symptoms of headache and fever.

___

20. Which human disorder is characterized by a group of abnormal body cells that suddenly begin to undergo cell division at a very rapid rate?
    (1) albinism   (3) hemophilia
    (2) cancer   (4) color blindness

21. An allergic reaction characterized by the constriction of the bronchial tubes is known as
    (1) coronary thrombosis
    (2) arthritis
    (3) asthma
    (4) emphysema

22. In some regions of the world, children suffer from a protein deficiency known as kwashiorkor. This deficiency occurs when a child's diet is changed from high-protein breast milk to watery cereal. Even though the child is receiving calories, the child becomes sick and less active, and growth ceases. These symptoms are probably due to
    (1) too many nucleic acids in the diet
    (2) an overconsumption of complete protein foods
    (3) not enough carbohydrates in the diet
    (4) a lack of essential amino acids in the diet

23. A photograph of a slide of human blood taken from a healthy individual is shown below.

Which statement best describes the change that would be observed if the slide contained blood from an individual with anemia?
    (1) Cell type A would be fewer in number and larger in size.
    (2) Cell type B would be fewer in number and lighter in appearance.
    (3) Cell type B would be larger in size and greater in number.
    (4) Cell type A would be larger in size and darker in appearance.

24–26. Base your answers to questions 24 through 26 on the information below and on your knowledge of biology. Use one or more complete sentences to answer each question. [3]

When a drug manufacturer develops a new drug to treat some form of disease, the drug should be tested to ensure that it does what it is supposed to do. Usually, the drug is tested on animals. If these tests are successful, it is then tested on humans.

A drug called Lowervil was developed by a drug company to lower blood pressure. Lowervil has been tested successfully on animals, and the drug company is now ready to test it on humans. The drug company claims that one dose of Lowervil per day will decrease blood pressure in individuals experiencing high blood pressure.

A researcher has been hired to determine whether or not Lowervil lowers blood pressure. Answer the following questions related to the experimental testing of the new drug Lowervil.

24. How should the experimental group and control group be treated differently? [1]

25. Why would using a large number of people be important in this experiment? [1]

26. How could the researcher determine if the drug is effective in reducing blood pressure? [1]

27–30. Base your answers to questions 27 though 30 on the information below and on your knowledge of biology.

Organ Transplants of the Future

While most people take good health for granted, thousands of others desperately need to replace a failing organ with one that is healthy. Most healthy organs come from people who agreed to donate them upon their death, although it is possible to remove some tissue and organs (such as kidneys and bone marrow) from living donors. Unfortunately, organs for transplant are in short supply. As of 1992, over 22,000 Americans were waiting for a transplant.

Although increasingly common, transplants are risky procedures. During the operation, veins and arteries must be blocked to prevent blood loss. This deprives parts of the body of oxygen and nutrients and may result in permanent damage. In addition, the body may recognize the transplanted organ as foreign and mount an immune response in which specialized white blood cells (T cells) attack the transplanted organ.

Drugs called immunosuppressants are given to transplant patients to prevent their immune system from rejecting the transplanted organ. However, these drugs weaken the ability of the body to fight disease and leave the patient less able to fight infection.

Scientists are exploring new technology for producing transplanted tissues and organs. Unspecialized cells called stem cells are removed from the patient and then grown in a laboratory. Treating stem cells with the appropriate chemicals causes them to differentiate into various specialized tissues. In the future, scientists hope to develop chemical treatments that will cause stem cells to grow into complete organs needed for transplants. Transplants produced by this process would not be foreign material and, therefore, would not be rejected by the immune system of the patient.

27. Explain why a transplant might be dangerous to the health of a patient.

28. State one reason that transplant patients might take an immunosuppressant drug.

29. State one specific disadvantage of taking an immunosuppressant drug.

30. Explain why doctors would consider using tissues or organs grown from stem cells.

IX. INTERDEPENDENCE OF LIVING THINGS

> ## KEY IDEA 6—INTERDEPENDENCE OF LIFE
> Plants and animals depend on each other and their physical environment.

Essential Questions:

- *How do species of different kinds depend on each other and contribute to each other's survival?*
- *How do conditions in the nonliving environment contribute to the survival of plant/animal communities?*

Living things in the environment interact with each other, developing an interdependence crucial to the survival of these populations and to the overall health of the environment. The energy and materials needed to drive this living system derive from the nonliving, abiotic factors present in the environment. Living systems serve to cycle materials and provide a mechanism to absorb and transfer energy among organisms.

Populations of organisms interact with each other in many ways in the natural environment, including competition, symbiosis, and nutritional relationships. Each species population in a natural community has a role to play in the environment, further enhancing the interdependence of species.

The Ecosystem

Diversity and Ecosystem Stability

Essential Question:

- *How do the varied species that make up an ecological community affect environmental stability?*

Performance Indicator 1.1 *The student should be able to explain how diversity of populations within ecosystems relates to the stability of ecosystems.*

When studying ecological interactions, the basic unit of study is the **ecosystem**. The ecosystem is the lowest level of ecological organization in which all environmental factors, both living and nonliving, are represented and interact freely. Earth contains many different ecosystems that provide habitats for millions of different species of plants, animals, and simpler life-forms. The diversity of life directly results from the vast variety of different ecosystems that exist on Earth.

Organisms can be categorized by the functions they serve in an ecosystem. Each species in an ecosystem has a role for which it is best suited. The environmental role of an organism is known as its **niche**. In general, no two species inhabit the same niche in an ecosystem. When two species attempt to inhabit the same niche, **interspecies competition** between them will normally result in the elimination of one of the two from the ecosystem. This fact allows different species to coexist successfully and helps maintain the stability of the ecosystem. Producers, consumers, and decomposers exist within food webs in these ecosystems, carrying out either autotrophic nutrition or heterotrophic nutrition. Each species population in an ecosystem has a different role to play in the transfer of energy and the cycling of elements in the environment.

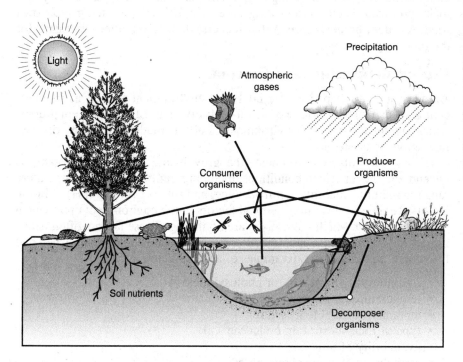

Balanced Ecosystem

Producers are the green plants and algae (the autotrophs) in the community responsible for trapping the Sun's radiant energy and using it to manufacture organic compounds that are used for their own consumption, as well as that of animals. **Consumers** include **herbivores**, **carnivores**, **omnivores**, and **decomposers** (the heterotrophs). Herbivores are **primary**

consumers because they are the first consumers to tap the energy trapped by the producers. Carnivores are **secondary consumers** because they do not tap plant energy directly but obtain it through their consumption of primary-consumer organisms. Omnivores may be either primary consumers or secondary consumers, depending on whether they consume plant matter or animal matter, respectively. **Decomposers** include saprophytic fungi and bacteria responsible for breaking down the complex structure of the bodies of living things into simpler forms that can be used by other living things. In a sense, decomposers are responsible for operating a recycling system that reuses the chemical substances of life over and over. This action is essential to the continued functioning of the ecosystem.

Organisms capable of producing their own food are known as **autotrophs** (self-feeders). Green plants and algae, the principal autotrophic organisms, manufacture organic molecules that serve as food for all the other organisms in the environment. This food is the main source of energy and structural components for all living things. Heterotrophic organisms, particularly animals, protozoa, bacteria, and fungi, are incapable of producing their own food. As such, **heterotrophs** (other feeders) depend on other organisms for the food they consume.

Ecosystem Structure and Function

An ecosystem is shaped by the nonliving environment as well as its interacting species. The world contains a high diversity of physical conditions, which creates a variety of environments. Earth is made up of many diverse, interactive ecosystems.

In any particular environment, the growth and survival of organisms depend on the physical conditions including light intensity, temperature range, mineral availability, soil/rock type, and relative acidity (pH). **Abiotic** factors are the physical and chemical factors in the environment upon which life depends, but which themselves are nonliving. These factors often determine what type of plant and animal community can become established and thrive in a particular area. Examples of abiotic factors include:

- light intensity available for photosynthesis
- temperature range
- amount of available moisture
- type of rock substratum under the soil
- availability of minerals
- availability of atmospheric gases
- relative acidity (pH) of the system

The abiotic conditions determine to a large extent what plant species can exist in a particular environment. In turn, the animal populations that exist in an environment are determined largely by the plant community. For example, a dry, sunlit field with a clay-loam soil in a temperate zone might support a wide diversity of wildflower species, such as goldenrod, aster, and wild daisy, that would thrive under these conditions. On the other hand, species such as mosses and ferns would have a difficult time existing in such an environment because of the need of these species for moist, shady conditions such as those found in a forest environment.

The living species that are successful in any given ecosystem are those best adapted to survive in that set of environmental conditions. The living, or **biotic**, factors abide by natural biological laws that govern all life-forms. For example, all known life-forms are composed of cells that serve as the organisms' functional unit. All living things depend on their ability to carry out basic life functions. Living things depend on the presence and activities of other living things to provide them with nutrients and other essential components. Relationships between organisms may be negative, neutral, or positive. Some organisms may interact with one another in several ways. They may be in a producer/consumer, predator/prey, or parasite/host relationship; or one organism may cause disease in, scavenge, or decompose another. Biotic factors include all the living components of the environment that affect the ecological community, either directly or indirectly, and help to limit the species that inhabit an area. Examples of biotic factors include:

- the population levels of an individual species
- the particular set of food requirements of a species
- the interactions that a species has with other species
- the wastes produced by the members of a species

Nutritional relationships between species involve the transfer of nutrient materials between one organism and another within the environment. Organisms may be classified as follows in terms of the type of nutritional relationships they have with other organisms.

Symbiotic relationships between species involve the ways that different types of organisms can live together in close physical association. Symbiosis is the term used to describe such relationships. Types of symbiosis include parasitism, commensalism, and mutualism.

Biomes
The Earth's natural environments are organized into broad climatic zones known as **biomes**. Each biome is categorized by a particular range of abiotic conditions, such as temperature, rainfall, solar radiation, and altitude. Each biome is also known by the type of dominant, or **climax**, vegetation that establishes itself in the region as well as by the typical, or index, animal species it supports. The biome that makes up the majority of New York State is the temperate deciduous forest biome. It has moderate temperature and rainfall, abundant solar radiation, climax vegetation made up of a mixture of broadleaf hardwood trees such as oaks, maples, and beeches, and index animals such as the deer, squirrel, and hawk. Other terrestrial biomes found on Earth include the tundra, taiga, grassland, desert, and tropical rain forest biomes. The chart that follows summarizes the characteristics of these terrestrial biomes.

CHARACTERISTICS OF TERRESTRIAL BIOMES

Biome	Abiotic Features	Climax Vegetation	Index Animals
Tundra	Permanently frozen subsoil	Mosses, lichens	Caribou, owl
Taiga	Long, severe winters, thawing subsoil	Pine, spruce, cedar	Moose, bear, eagle
Temperate deciduous forest	Moderate precipitation, cold winters, warm summers	Oak, hickory, maple, beech trees	Squirrel, fox, deer, hawk
Grassland	Considerable variation in rainfall and temperature, strong prevailing winds	Grasses, tumbleweed	Prairie dog, antelope, bison
Desert	Sparse rainfall, extreme variation in temperature	Cactus, mesquite	Roadrunner, snake, lizard
Tropical rain forest	Abundant rainfall, constant warmth	Broad-leaved plants, palm trees	Monkey, parrot, jaguar

Interaction of Populations and the Environment

In all environments, organisms compete for vital resources. The linked and changing interactions of populations and the environment compose the total ecosystem. A vital characteristic of all natural ecosystems is the interaction of species populations found within them. A **population** is defined as all the members of a species inhabiting a particular place at a particular time. In an ecosystem, some species take on the role of producers, others the role of consumers. Among the consumers, some species are predators, others prey. Still others may be scavengers or decomposers that break down complex organic matter into simpler components usable by other living things. Some species produce oxygen, while others consume it. The wastes left by one type of living thing may be vitally important to the life functions of another. The complexity of these interactions is extreme and the dynamic balance to which they contribute is delicate. Specific types of interactions based on nutritional relationships are material cycles, food chains, and food webs.

Competition

When different species living in the same environment (habitat) use the same limited resources, **competition** occurs. These resources may include requirements such as food, space, light, water, oxygen, and minerals. The more similar the requirements of the competing species, the more intense the competition is likely to be. Of course, when such resources are abundant, different species may share them without creating significant competition. Only when the resources become scarce does the competition become intense. Such competition might be termed **interspecies competition** (competition *between* species) since it involves members of different species. This type of competition should not be confused with the concept of **intraspecies competition** (competition *within* a species).

As an example, consider the variable growth patterns of goldenrod plants growing in different but neighboring environments. Goldenrod is a species of wildflower that commonly inhabits sunny fields. It is a hardy plant that competes well against other field plants, often growing to heights of five or six feet and crowding out other light-loving species. Goldenrod seeds carried into a nearby forest may germinate and produce young plants, but these plants grow to be only a few inches tall. Their stems are thin and weak, their leaves are small and sparse, and their flowers, if they form at all, are under-developed. How can one species of plant, inhabiting the same general area, have such different growth patterns? The key is in the different amounts of light available in the forest and the field. As long as light is abundant, as it is in the field, the goldenrod is a good competitor. When light becomes scarce, as it is in the forest, goldenrod does not compete well, and will eventually be eliminated from that environment altogether.

The interdependence of organisms in an established environment often results in approximate stability over hundreds and thousands of years. For example, as one population increases, it is held in check by one or more environmental factors or another species. Over a period of many years, a steady cycle of interspecies competition will result in a relatively stable environment in which populations coexist under sets of environmental conditions that will support their long-term survival. In this sense, a successful plant species is one whose adults can grow and thrive and whose seeds can germinate and survive to adulthood. A plant species that dominates as an adult population in an environment, but whose seedlings cannot grow to adulthood in that environment, will not continue to be a dominant species in the area for more than a few generations.

Cyclic Change

Ecosystems, like many other complex systems, tend to show cyclic changes around a state of approximate equilibrium. Ecosystems are dynamic and complex. The many biotic and abiotic factors that characterize a given environment interact in complex ways that create a relatively stable system. This stable state is said to be at **equilibrium**. These factors balance each other. When one factor changes, all the remaining factors are affected. Periodic changes in these factors can set off repeating cycles of change that become part of the dynamic state of the environment. Such periodic changes can include food availability, seasonal temperature change, cyclic rainfall patterns, and population reductions caused by disease, among many others.

Perhaps the easiest cyclic change to understand is that caused by seasonal change. As spring arrives in most of New York State, natural ecosystems emerge from a dormant state to an actively growing one. Seeds lying in the moist earth germinate, producing new annual and perennial plants. The buds of mature trees and shrubs swell with new growth and open to produce new leaves and flowers. Ice, snow, and ground frost give way to warming temperatures. The abundance of liquid water, warm temperatures, and new plant food provide an inviting environment that supports herbivores. The presence of herbivores attracts populations of carnivores. The environment continues to warm and stabilize throughout the summer months, bringing still more change. As these periodic fluctuations occur, they affect the ability of other species populations in the area to thrive or survive, including migratory birds and other species. Swings in population numbers represent another aspect of the changes in the environment. Such changes do not always promote the survival of individuals; death is a normal part of the cycle of change.

Another example of cyclic change is the predator-prey relationship. Each environment has a carrying capacity for its animal populations. The carrying capacity of herbivores, such as rabbits, is the abundance and type of plant matter (such as clover) available in the environment. If left unchecked, the rabbit population would increase as a function of the amount of such food until all clover plants were consumed. Obviously, the destruction of the food supply would cause the destruction of the rabbit population as well. In a balanced environment, predators such as foxes exist that feed on the populations of herbivores, reducing their number. This promotes the growth of clover as the numbers of rabbits decrease. As more rabbits are consumed, the well-fed fox population increases. As the rabbit population declines due to hunting by the foxes, the fox population soon also declines. As the fox population declines, hunting pressure on the rabbits is diminished, so there is a resurgence in the rabbit population. Clover populations are subsequently thinned by the resurgent rabbit population. This cycle repeats itself as the populations of clover, rabbits, and foxes respond to increases or decreases in the other populations. Such cyclic and interrelated changes ultimately lead to relatively stable populations of all three species over time.

This graph illustrates how populations of predators and their prey vary in a balanced environment. Note that increases in prey population usually precede increases in predator populations, indicating that high numbers of prey directly stimulate the growth of predator populations.

Ironically, the result of these small periodic changes is normally the perpetuation of a relative stability in the environment. Each of the small changes contributes to a dynamic tension to which other environmental factors respond. The cumulative result of all these changes is ecosystem stability.

Direct and Indirect Interactions

Every population is linked, directly or indirectly, with many others in an ecosystem. Disruptions in the numbers and types of species and environmental changes can upset ecosystem stability. The presence or absence of other species populations in an environment can affect the stability of an individual species. Species populations interact in many ways, sometimes directly and sometimes indirectly. Since most organisms depend on other organisms for food, the absence of a species that is used as a food source by other species will affect the ability of those other species to survive. For example, the Karner Blue butterfly, a species unique to New York pine barrens, depends on the growth of a plant, known as lupine, for food and breeding. Lupine, in turn, can tolerate only a narrow range of soil conditions. If these conditions change, the population of lupine may be adversely affected. If the availability of lupine decreases, the survival of the Karner Blue is also impacted adversely.

Some direct relationships are **nutritional**. A nutritional relationship is where one organism uses the bodies of other organisms for food. In nature, all organisms fit into food chains that involve patterns of consumption. For example, grass is a common plant that is used by many herbivorous species for food. An example of a grass-eating herbivore is a grasshopper. Grasshoppers may serve as a food source for a carnivore, such as a frog. The frog, in turn, may become a food source for a raccoon. (See page 210). An outside force that affects any of the populations in a food chain will affect all other populations in the food chain. A blight that kills the grass will cause the grasshoppers to die of starvation. The lack of grasshoppers will drive frogs to new environments to find food. Without the frogs, the raccoons will have to seek out alternative food sources, a step that puts them into direct competition with other top predators for limited food resources. Food chains that intersect are known as food webs. (See page 210.)

Other direct relationships are **symbiotic**. Symbiosis refers to interacting species populations where some aspect of their lives is shared. Common symbiotic relationships include parasitism, mutualism, and commensalism.

In addition to these direct relationships, many interrelationships in an ecosystem are indirect. Sometimes, many steps intervene in the connection between two organisms, but a connection nearly always exists. When no direct link is obvious, it is easy to dismiss such interactions as being of little consequence to human survival. However, the potential impact of species destruction on human survival should not be discounted. There is every reason to assume that when the environment becomes inhospitable to one species, it becomes inhospitable to all similarly situated species, including humans.

Limiting Factors

Limiting factors are those conditions in an ecosystem that impose caps on the sizes of the populations that inhabit it. Limiting factors generally fall into two categories: materials and energy.

Essential Question:

- *What forces of nature manage the balance of species populations?*

Performance Indicator 4.6.1 *The student should be able to explain factors that limit the growth of individuals and populations.*

Materials as Limiting Factors

The chemical components that make up living matter are limited. In order for living things to survive over millions of years of Earth's history, these chemicals must be recycled between the living and the nonliving world. The atoms and molecules on the Earth cycle among the living and nonliving components of the biosphere. For example, carbon dioxide and water molecules used in photosynthesis to form energy-rich organic compounds are returned to the environment when the energy in these compounds is eventually released by cells. Continual input of energy from sunlight keeps the process going.

Material cycles function in nature to make chemical substances available to living things for their continued growth and reproduction. Material cycles consist of sequential chemical reactions that allow for such periodic recycling. Some examples of material cycles are as follows.

Carbon-hydrogen-oxygen cycle—In previous sections, we studied the processes of respiration and photosynthesis and discovered that they are quite similar in terms of the chemical substances involved. In photosynthesis, carbon dioxide and water combine with the aid of solar energy, producing glucose and oxygen gas. In respiration, glucose and oxygen combine to produce carbon dioxide and water, releasing cellular energy. One process produces waste products that serve as raw materials for the other. In this way, environments that contain balanced communities of plants and animals should be self-sustaining in terms of the supply of the elements carbon, hydrogen and oxygen.

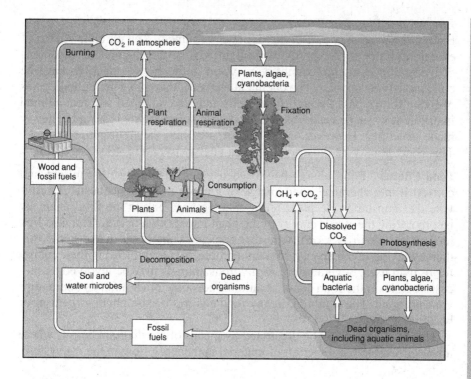

Nitrogen cycle—Nitrogen is an elemental component of the class of compounds known as proteins. The nitrogen cycle makes nitrogen available to organisms for use in protein synthesis. As in the carbon cycle, living organisms are an important part of the nitrogen cycle. Decomposers and other soil bacteria are essential in converting nitrogenous wastes of animals and plants into a form of nitrogen usable by plants as fertilizer. Animals of various types round out the nitrogen cycle as consumer organisms in the food web.

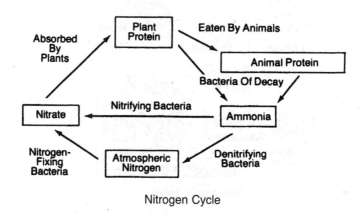

Nitrogen Cycle

Energy as a Limiting Factor

Energy is a necessary part of the living environment. The ultimate source of energy on Earth is the energy of sunlight. Unlike the chemical components of living things, energy cannot be recycled in the environment, so must be continuously resupplied. The chemical elements that make up the molecules of living things pass through **food webs** and are combined and recombined in different ways. At each link in a food web, some energy is stored in newly made structures. However, much is dissipated into the environment as heat. This concept may be illustrated with an **energy pyramid**. (See page 211.)

Food Chains Food chains begin when green plants absorb sunlight and convert it into chemical bond energy of glucose in photosynthesis. A herbivore that consumes the plant represents the next link in the food chain by incorporating the stored energy of the plant into its own tissues. The next link in the food chain is a carnivore that consumes the body of the herbivore, thereby taking in the energy-containing molecules and releasing their energy for its own uses. Several more carnivorous animals may be involved as successive links in the food chain as animals consume other animals and in turn are consumed. The final link in any food chain is a saprophytic organism or bacteria of decay.

This food chain may be illustrated by means of the diagram below, where arrows illustrate the direction of material and energy flow in the ecosystem:

grass ⟶ grasshopper ⟶ frog ⟶ raccoon ⟶ bacteria of decay

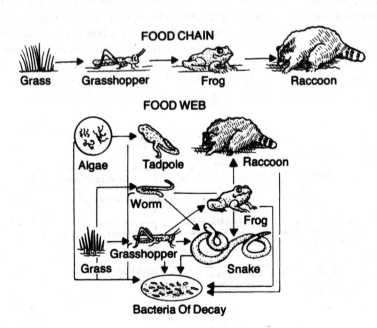

FOOD CHAIN

Grass   Grasshopper   Frog   Raccoon

FOOD WEB

Algae   Tadpole   Raccoon

Worm

Frog

Grasshopper

Grass   Snake

Bacteria Of Decay

Food Chain and Food Web

Food Webs   To illustrate more realistically the complex nature of nutritional relationships in a natural community, the food web is used. The food web concept recognizes that many plant species are present in any ecological community, all producing energy-rich organic compounds for consumption by many different species of herbivore. At the same time, multiple combinations of carnivorous and omnivorous species interact to consume the herbivorous species. Saprophytes of many different varieties are responsible for consuming the decaying bodies of plants and animals alike. In fact, different species may compete strongly for the same type of food available in the ecosystem. When the names of all the species present in a community are written onto a sheet of paper and lines are drawn between the species for which a nutritional relationship exists, a pattern resembling a web emerges.

   The solar energy available to the producer organisms in a food chain is considerable. At each step of the food chain, however, some of this energy is lost. Some is used in the life processes, some is radiated as heat, some is lost in excretory waste. In any food chain, the producer level contains the greatest amount of energy. Primary consumers contain only about 10 percent of the energy found in the producers. In turn, secondary consumers contain only about 10 percent of the energy housed in the bodies of the primary consumers and only about 1 percent of the energy in the producers. Since each feeding level contains less energy than the level below it, this phenomenon can be illustrated as a pyramid known as the **pyramid of energy**, also known as a **food pyramid**. Eventually, all energy received by the ecosystem will be lost and will radiate as heat into the atmosphere. A constant resupply of energy from the Sun is necessary to sustain the ecosystem.

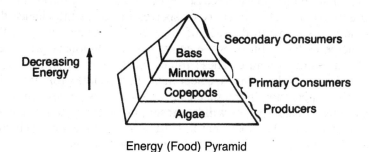

Energy (Food) Pyramid

Carrying Capacity
The number of organisms any habitat can support is limited by the available energy, water, oxygen, and minerals and by the ability of ecosystem to recycle the residue of dead organisms through the activities of bacteria and fungi. The **carrying capacity** of an environment is the number of organisms of different species that can be supported by that environment. Various factors, including available sunlight, moisture, minerals, oxygen, organic food, and

other abiotic factors, are instrumental in determining the number of each species that can be supported by that environment. For example, in a grassland environment, the number of buffalo that can be supported is determined by the available food, water, and other components required by the buffalo. If these requirements are limited by natural disaster or human intervention, then the number of buffalo that can be supported may be severely limited.

Also critical to the carrying capacity of the environment is the rate of turnover of the material components of the ecosystem. This turnover is accomplished by organisms of decay—fungi and bacteria. These organisms use their specific characteristics to convert the complex polymers characteristic of most plant and animal life to simpler components that can be absorbed by plants and other organisms capable of synthesizing organic molecules. These activities are directly related to the material cycles discussed previously.

Diversity of Species and Habitat

Preservation of Species Diversity and Habitat

Essential Question:

- *Why is it important to protect endangered and threatened species and habitats?*

Performance Indicator 4.6.2 *The student should be able to explain the importance of preserving diversity of species and habitats.*

As a result of evolutionary processes, there is a diversity of organisms and roles in ecosystems. This diversity of species increases the chance that at least some will survive in the face of large environmental changes. **Biodiversity** increases the stability of the ecosystem. Evolutionary processes are constantly at work in nature. These forces create new varieties, test them against environmental pressures, select those best adapted for survival, and ensure their perpetuation in the species gene pool. This process ensures that many varieties of each species will be present in the ecosystem at any given time, increasing the likelihood that at least some members of each species will survive in the event of a drastic change in the environment.

Environmental Niche

Each species of organism has a role to play in the environment. This role is the organism's **niche** in the environment. Although the niches of some organisms may, on the surface, seem insignificant to human survival, scientists have come to understand that all populations in a natural community depend on all other populations to provide a balanced set of conditions conducive to the survival of all members of the community. Removal of any one species degrades the quality of the ecosystem because its elimination removes the role that species plays and makes it that much more difficult for the environment to recover from drastic change. The elimination of any species reduces biodiversity and makes survival of other species in that environment problematic. See page 229 for a more complete discussion.

Putting Ecological Principles to Work

Scientists have learned much about the interactions among diverse species in the ecosystem and the effects of removing species from a balanced system. Today's system of identifying and protecting endangered and threatened species is a manifestation of our growing understanding of the vital roles that all species, from snails to whales, play in our global environment. We are beginning to understand that an environment hostile to any natural species is potentially hostile to our own species, as well.

Biodiversity also ensures the availability of a rich variety of genetic material that may lead to future agricultural or medical discoveries with significant value to humankind. As diversity is lost, potential sources of these materials may be lost with it. Medical science is developing techniques to analyze and exploit the genetic makeup of animals and plants to treat diseases in humans and other species. An essential aspect of this research is the discovery of new species. Each new discovery is studied in order to determine whether it might contain genetic information leading to the production of antibiotics and other medicinal biochemicals. These newly discovered species can then be cultured in the laboratory in order to maximize the production of the desired chemical. The specific genes responsible for the manufacture of these biochemicals may also be isolated and spliced into the genome of common bacteria in order to reduce the time and expense of this production.

Ecosystem Formation

Ecosystem Change

Essential Question:

- *How do Earth's environments affect, and how are they affected by, the living community?*

Performance Indicator 4.6.3 *The student should be able to explain how the living and nonliving environments change over time and respond to disturbances.*

Ecosystem Stability

An individual species' reproductive rate may be limited by the microclimate created by the dominant community. It may also be limited by the microclimate created by such factors as moisture, temperature, light, latitude, and altitude. For example, the seedlings of hardwood trees such as oak and hickory grow well in environments characterized by low altitudes, sandy soils, and high light intensities. As these species mature, they create a microenvironment that is so deeply shaded that their own seedlings can no longer thrive. Instead, the seedlings of beech and maple, which tolerate low-light conditions, can successfully grow until they replace the populations of oak and hickory that once dominated. However, these same beech and maple seedlings do not compete well at higher altitudes and can easily be crowded out by tree species, such as spruce and hemlock, that are more tolerant of the conditions found at high altitudes. At very high altitudes, tree species in general do not thrive and are replaced by alpine wildflowers and other extreme altitude plant populations.

A stable ecosystem can be altered, either rapidly or slowly, through the activities of organisms (including humans) or through climatic changes or natural disasters. The altered ecosystem can usually recover through gradual changes back to a point of long-term stability. A climax community will normally remain intact as long as conditions in the environment do not change appreciably. If, however, the environment changes drastically because of some catastrophic event, then the climax community may die out or be swept away, making way for a new succession of communities. If the alteration is temporary, the same climax community may result. If the change is permanent, then a new group of climax organisms may be favored.

Ecological Succession
The term **ecological succession** refers to a process in which an established ecological community is gradually replaced by another. It, in turn, is replaced by other communities until a stable, self-perpetuating community is formed.

The beginning stages of an ecological succession frequently occur in barren, almost lifeless environments that may have been swept clean by glaciation, erosion, fire, volcanic eruption, or some other destructive event. The first living things to invade such an area and establish themselves in it are known as **pioneer organisms**. A typical pioneer organism populating bare rock is lichen, a fungus-alga symbiotic association that can tolerate this harsh, dry, soilless environment. A pioneer organism such as lichen, although small, can significantly alter the environment it has invaded. The lichen produces a mild acid that acts on the rock surface and erodes it into grains of sand. At the same time, the material comprising the lichen adds organic matter to the sand grains, producing a crude form of soil. This soil gradually fills the rock crevices, providing favorable areas for seed germination and thereby paving the way for later stages of ecological succession, including mosses and grasses.

The germinating spores of mosses and seeds of grasses and other herbs give rise to the second succession stage. As this stage dominates over several generations, its members contribute organic matter to the soil. The root system protects the thin soil layer from erosion. Weathering and root pressure continue to fragment the rock substratum, adding still more substance to the thickening soil layer. Eventually, the soil layer is able to support the growth of shrubs and other woody plant species. When soil depth and quality are sufficient, varieties of trees and the animal species associated with them begin to invade and then dominate the area.

In this way, each successive community of plants and animals modifies the environment to make it less favorable for its own offspring but more favorable for the establishment of the next succession stage. In such a pattern, plant species tend to dominate since they provide the basis of the food chain for the area. The animal species that can live in the area are normally those that depend on the plant community. A succession stage is normally named for the dominant plant types in the environment since these species exert the most influence on the environmental conditions of the area.

Eventually in each succession, a community becomes established that is self-perpetuating and relatively stable. Such a community is known as a **climax community**. A climax community remains the dominant community for an indefinite period, maintaining a relatively stable set of environmental conditions for both plant and animal species. The predominant climax community in New York State is the oak-hickory forest.

Forest Succession

Examples of this phenomenon can be readily observed in nature in a variety of settings. Keen observation and careful identification and inventorying of plant species in an area will provide evidence that succession is an ongoing biological activity in many ecosystems of the world. A few examples include the following.

- An abandoned farm field undergoes rapid succession, first playing host to wild grasses and wildflowers, then to invading shrubs and light-loving tree species such as aspen and white pine. After a relatively short period, the former field resembles a dense, young forest and the seedlings of climax stage trees are sprouting and competing with the early tree species. Often, such a succession pattern requires only a few decades to revert to the low-elevation oak-hickory climax forest characteristic of much of New York State.

- A burned mountain forest tract quickly sprouts new pitch pine and other fire-resistant species. The seed cones of pitch pine are specially adapted to protect the seeds of this species during a fire but to release them rapidly after the ground has cooled. The new pine seedlings act to stabilize the soil, reducing erosion. Seed of other tree species, such as birch and aspen, from nearby unburned forest are transported to the burn area and quickly germinate to help stabilize the soil still further. Eventually, the seeds of slow-growing hemlock, beech, and maple trees take root and reestablish this high-elevation climax community typical of the Catskill and Adirondack Mountains of New York State.

- A stream valley flooded by a landslide that dams the stream quickly takes on the characteristics of a shallow lake. Over a period of many years, sedimentation and decay gradually fill the lake until marsh plants such as cattail and sphagnum moss gain a foothold in the wet shoreline. Over a period of years, the filling continues and the shoreline moves farther and farther out into the lake. Eventually, the marsh area is invaded by red maples and other wet-soil plants. These species stabilize the marsh mat and enable climax species to repopulate the valley, converting the lake to a forest stream once again.

- A volcanic eruption, such as the one observed on Mount St. Helens in Washington State (1980), completely covers the mountainside forest environment with lava flows and volcanic ash. No soil remains, but the seeds of pioneer plants blown in from surrounding areas take root in the unstable ash deposits and in crevasses in the lava rock. Over a period of as little as a decade, signs of recovery from the complete devastation were evident. Many decades later, the forest ecosystem will undergo succession in the area until a stable, self-perpetuating climax community is established. However, the mineral content of the volcanic ash may lead to a climax community different from the one swept away by the eruption.

QUESTION SET 2.9—INTERDEPENDENCE OF LIVING THINGS
(ANSWERS EXPLAINED, P. 324)

1. Which statement best describes some organisms in the food web shown below?

(1) Minnows and fish are primary consumers.
(2) Algae and floating plants are decomposers.
(3) Aquatic crustaceans are omnivorous.
(4) Raccoons, fish, and ducks are secondary consumers.

2. The graph below shows the changes in two populations of herbivores in a grassy field.

A possible reason for these changes is that
(1) all of the plant populations in this habitat decreased
(2) population B competed more successfully for food than population A did
(3) population A produced more offspring than population B did
(4) population A consumed the members of population B

3–5. For each description in questions 3 through 5, select the biome, chosen from the list below, that best fits that description. A biome may be used more than once or not at all.

Biomes

(1) Tundra
(2) Taiga
(3) Temperate deciduous forest
(4) Grassland
(5) Desert
(6) Tropical forest

3. Lichens and mosses are present; subsoil permanently frozen.

4. Constant, warm temperature; abundant rainfall.

5. Wide variation in daily temperature; little rainfall.

6. The diagram below shows a milkweed plant and some of the insects that live on it or visit it.

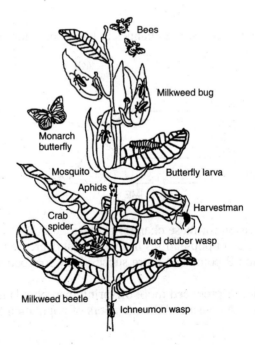

Which term best describes the group of organisms in the diagram?
(1) biosphere   (3) habitat
(2) community   (4) biome

7. Hawks and owls living in the same area compete for the same type of mouse for food. Which situation would lead to the greatest problem in the food supply?
   (1) an increase in the owl population
   (2) an increase in the mouse population
   (3) a decrease in the hawk population
   (4) a decrease in the owl population

8. Which group represents a population?
   (1) all the vertebrates living in New York State
   (2) all the *Homo sapiens* living in New York State
   (3) all the plant and animal species found in New York State
   (4) all the flowering plants found in New York State

9. A student measured some abiotic factors present in an aquarium in a biology laboratory. Which data did the student most likely record?
   (1) the weight and color of each type of scavenger
   (2) the number of each type of green plant and each type of snail
   (3) the size and number of each species of fish
   (4) the temperature and oxygen content of the water

10. Energy stored in organic molecules is passed from producers to consumers. This statement best describes an event in
    (1) the process of photosynthesis
    (2) natural selection
    (3) a food chain
    (4) ecological succession

11. Which material cycle relies *least* on the processes of photosynthesis, transpiration, evaporation, respiration, and condensation?
    (1) oxygen cycle   (3) water cycle
    (2) nitrogen cycle   (4) carbon cycle

12–14. For each symbiotic relationship in questions 12 through 14, select the type of symbiosis, chosen from the list below, that best identifies that relationship. A number may be used more than once or not at all.

Types of Symbiosis

(1) Commensalism
(2) Mutualism
(3) Parasitism

12. A tapeworm lives in the digestive tract of a human.

13. Nitrogen-fixing bacteria live in the nodules on the roots of legumes.

14. A flea sucks blood from the skin of a dog.

---

15. By starting on bare rock, what is the usual ecological succession of organisms?
    (1) grasses → shrubs → lichens → trees
    (2) lichens → shrubs → grasses → trees
    (3) grasses → lichens → shrubs → trees
    (4) lichens → grasses → shrubs → trees

16. An ecosystem is represented in the diagram below.

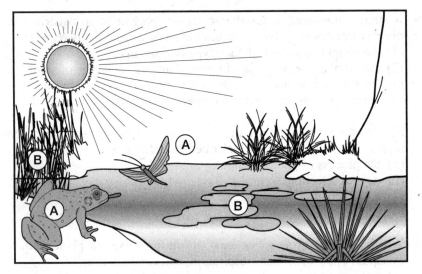

This ecosystem will be self-sustaining if
    (1) the organisms labeled A outnumber the organisms labeled B
    (2) the organisms labeled A are equal in number to the organisms labeled B
    (3) the type of organisms represented by B are eliminated
    (4) materials cycle between the organisms labeled A and the organisms labeled B

17. A certain plant requires moisture, oxygen, carbon dioxide, light, and minerals in order to survive. This statement shows that a living organism depends on
    (1) biotic factors
    (2) abiotic factors
    (3) symbiotic relationships
    (4) carnivore-herbivore relationships

18. Events that take place in a biome are shown in the diagram below.

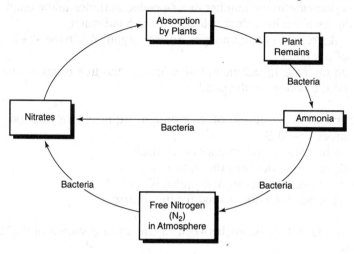

Which information is represented in the diagram?
(1) Respiration and photosynthesis are interrelated.
(2) Transpiration and condensation are related to the water cycle.
(3) Decomposers release a material that is acted on by other organisms.
(4) Predators and their prey are involved in many interactions.

19. The diagram below represents an energy/biomass pyramid.

Which statement concerning the energy in this pyramid is correct?
(1) The producer organisms contain the least amount of stored energy.
(2) Stored energy decreases from consumer 2 to consumer 3.
(3) Consumer 3 contains the greatest amount of stored energy.
(4) Stored energy increases from the producer to consumer 1.

20. Which statement concerning the climax stage of an ecological succession is correct?
(1) It changes rapidly.
(2) It persists until the environment changes.
(3) It is the first community to inhabit an area.
(4) It consists entirely of plants.

21. In a pond, which change would most likely lead to terrestrial succession?
    (1) a decrease in the number of suspended particles in the pond water
    (2) an increase in current velocity of the pond water
    (3) a decrease in the number of diverse organisms in the shallow water of the pond
    (4) an increase in sediment, fallen leaves, and tree limbs accumulating on the bottom of the pond

22. The most likely result of a group of squirrels relying on limited resources would be
    (1) an increase in the number of squirrels
    (2) competition between the squirrels
    (3) increased habitats for the squirrels
    (4) a greater diversity of food for the squirrels

23. Nutritional relationships between organisms are shown in the diagram below.

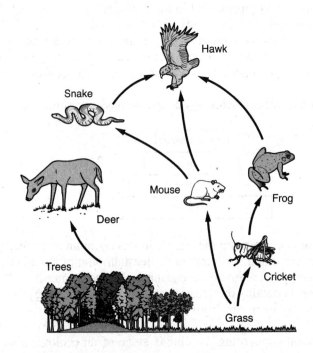

Which organisms are primary consumers?
    (1) mouse, snake, and hawk
    (2) snake, hawk, and frog
    (3) cricket, frog, and deer
    (4) mouse, deer, and cricket

24. Information relating to an ecosystem is contained in the diagram shown below.

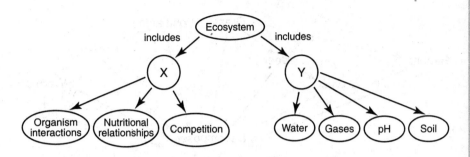

Which information belongs in areas X and Y?
(1) X—biotic factors; Y—abiotic factors
(2) X—ecological relationships; Y—biotic relationships
(3) X—abiotic factors; Y—interacting populations
(4) X—energy flow; Y—biotic factors

25. The American dogwood, a flowering tree of New York State's woodlands, has been attacked by a fungal disease specific to this tree species. Many dogwoods have died because fungicides have not proven effective in fighting the spread of this disease. Which term best describes the relationship between the dogwood trees and the fungus?
(1) commensalism
(2) mutualism
(3) parasitism
(4) saprophytism

26. The food web below shows some of the relationships that exist between organisms in a field and pond ecosystem.

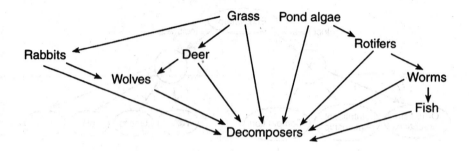

a   Write one or more paragraphs describing some of the relationships in this food web. In your answer, be sure to:
- identify a carnivore from the food web
- describe the complete path of energy from the Sun to that carnivore
- explain why decomposers are necessary in this food web

b   A significant decrease in the wolf population occurs. After a period of one year, what change in the grass population would most likely be observed?

c   A farmer sprayed pesticides onto a field next to the pond. By using one or more complete sentences, explain why several years later the fish population would contain higher pesticide levels than any other pond organisms would contain.

X. HUMAN IMPACT ON ECOSYSTEMS

> ### KEY IDEA 7—HUMAN IMPACT ON THE ENVIRON-
> **MENT** Human decisions and activities have had a profound impact on the physical and living environment.

Interdependence of Human and Natural Systems

Essential Questions:

- *What is the proper role of the human species in the ecological community?*
- *What responsibility does the human species have to protect the natural ecosystem?*
- *How do humans positively and negatively affect the ecosystem?*

Performance Indicator 4.7.1 *The student should be able to describe the range of interrelationships of humans with the living and nonliving environment.*

The human species is just one of Earth's 1.9 million living life forms. It has arisen by the same mechanisms as other species. It has the same physical requirements as other, similar species. It is dependent for its survival on its successful interactions with Earth's plant and animal species. Ultimately, it is subject to the same limitations to growth as any other species.

However, humans are also unlike any other species because of their niche as thinking, planning, and technological beings. Due to their ability to use technology, the human population is growing virtually unchecked by the natural factors that limit other species populations. Human technologies have had significant impacts on the natural world by producing materials that pollute the air, water, and soil. Its activities are increasingly displacing or destroying natural habitats and their ecological communities, reducing biodiversity and endangering the survival of many, if not all, of Earth's living things.

As a species, it is essential that we understand the necessity of preserving the natural environment and its living species as a means of ensuring our own survival. While some progress has been made in correcting certain environmental problems, much remains to be done. Education and environmental awareness on a global level is essential. Governments, industries, and the general public must come to grips with the long range impact of human activity that destroys the very fabric of biological life on Earth.

Technological Oversight and Human Population Growth

Essential Question:

- *How has human technology and the growth of the human population affected the quality of life for other species?*

Performance Indicator 4.7.2 *The student should be able to explain the impact of technological development and growth in the human population of the living and nonliving environment.*

Human Population Growth

Human population growth, unlike that of all other species, has risen at a rapid rate over the past two centuries. Human inventiveness has allowed our species to avoid or extend the limiting factors that keep natural species populations in check. For example, the development of medical technology has reduced the incidence of disease in many parts of the world, modern agriculture has reduced the likelihood of starvation in developed nations, construction technologies have allowed humans to avoid exposure to the elements and to extend their range into regions that would be otherwise uninhabitable by our species. This rapid increase in human population and the technology that supports that growth has put extreme pressures on the natural world, threatening the survival of natural species and habitats.

In many less developed areas of the world, the human population has grown faster than the food supply, causing widespread famine and threatening to eliminate large portions of the populations. Poverty and hunger in these areas has led to the resurgence of diseases that have been all but eradicated in the world's developed nations. These same conditions have forced people to abandon their homes in search of food and to be exposed to the elements, further reducing their ability to survive. Such occurrences are becoming more frequent and severe. It remains to be seen when similar collapses will occur in developed nations. Apparently, the human species is rapidly approaching a point where it will be unable to sustain continued growth.

Human Population Growth

Technological Oversight

Humans have used their ability to alter the environment to support their continued survival on Earth. Technologies have been developed to assist humans to this end. Inadvertent or purposeful misuse of technology in regard to the natural world is known as **technological oversight** and is illustrated in the following paragraphs.

In creating their technologies, humans have used natural materials at an alarming rate with little concern for conservation or reuse. Many of these resources are **nonrenewable**, meaning that their supply on Earth is finite and, once used up, cannot be replaced. Nonrenewable resources that have been mined or pumped from the ground to serve human needs include metals (e.g., gold, silver, copper, iron), salts (e.g., limestone, granite, seasalt), petroleum (e.g., oil, coal, natural gas), and fresh water. **Renewable** resources are those resources that can, with proper care and attention to the future, be replaced. Timber (e.g., wood, paper, cardboard) and other plant and animal resources are considered to be renewable, though positive action by humans is needed to expedite that renewal.

Environmental Pollution and Destruction

Technological and industrial processes have led to the production of chemicals and by-products that are harmful (toxic) to living things. Such toxic chemicals have contributed to global **environmental pollution**. Examples of pollutants include acidic ions, pesticides, hydrocarbons, growth hormones, pharmaceuticals, and radioactive particles. When released into the environment, these pollutants contaminate water, air, and/or soil, and make survival of natural species difficult or impossible. To the extent that humans are exposed to these pollutants, we are also subject to their toxic effects. In New York State in the early 20th century, the agricultural use of the pesticide DDT negatively impacted many native bird species by interfering with their reproductive cycle. Today, industrial by-products such as dioxin pose similar dangers to many species, including humans. The combining of acidic ions with atmospheric water has resulted in acid precipitation (acid rain), which has acidified lakes and disrupted natural ecosystems downwind of industrial sites.

Acid Precipitation

Habitat Destruction

In order to support the rapid growth in their numbers, humans have used their technologies to remove trees and other plant life from wide areas of natural habitat for the purpose of creating agricultural lands, industrial zones, roadways, or residential and commercial developments. Humans have drained, filled, or redirected wetlands, ponds, streams, and other natural water habitats for these same purposes. Collectively known as **habitat destruction**, these activities have had the effect of displacing plant and animal species important to the maintenance of a balanced ecosystem. Virtually all present-day developed lands in New York State have come about through human technologies that have resulted in habitat destruction.

Human agriculture has produced a variety of negative effects due to inattention to its impact on the natural environment. Failure to use **cover crops** (crops that protect soil from erosion) between planting cycles has exposed bare soil to erosion, resulting in loss of topsoil and sedimentation of water sources. **Overcropping** is the failure to allow soil to recover nutrients and organic matter (humus) content between plantings. **Overgrazing** is the practice of allowing large numbers of domestic animals to graze an area too small to support them.

Non-native Animals and Plants

Humans' desire to possess non-native exotic animals and plants or their products has been enabled by modern transportation, communication, and economic technologies. This has resulted in the **exploitation** of exotic species of primates such as the spider monkey, big cats such as the ocelot, birds such as the Columbian parrot, and others as pets. In addition, rhinoceroses have been killed for their horns, elephants for their tusks, tigers for their reproductive organs, and uncounted mammal species (e.g., bison, fox, beaver) for their pelts. Each species loss has impacted its ecosystem in a negative way. Each exploited species is, to one degree or another, in danger of extinction because of this exploitation.

Transportation technology has accelerated the movement of non-native species of plants and animals from foreign countries to our own. Such organisms are known as **invasive species**, and enter our environment either through intentional introduction or inadvertently on the bottoms of ships. Because they often have no natural enemies in their adopted environment, these newly introduced organisms have disrupted local ecosystems and in some cases have totally eliminated native species. Examples of invasive species in New York include the Japanese beetle, gypsy moth, zebra mussel, Asian clam, European starling, Eurasian milfoil, purple loosestrife, chestnut blight, and the virus that causes Dutch elm disease.

Diminishing Biodiversity
In an activity known as **direct harvesting**, humans have removed plants from the ecosystem for their economic value without regard for the effect of this removal on the natural ecosystem. This removal has negatively impacted the ecosystem by decreasing **biodiversity** (variety of life) in these habitats. In the 19th century in New York State, forest trees were directly harvested from the Adirondack and Catskill mountains to provide lumber for building purposes. New York's mountains and mountain valleys were denuded in the process, allowing fragile soils to be washed away. This activity resulted in the destruction of entire mountain habitats. Internationally, direct harvesting activities for exotic hardwoods in tropical rainforests has resulted in similar habitat destruction and loss of biodiversity.

When this concept is discussed in relation to animal species, the term **overhunting** is often used to describe it. During the 17th and 18th centuries in New York State, beavers were trapped from the Hudson, Mohawk, Genesee, and Susquehanna river watersheds in large numbers for the purpose of obtaining their pelts for export. This activity nearly drove the beaver to extinction in these areas; the beaver population only began to recover in the mid-late 20th century. In the 19th century, bounties imposed on the hunting of gray wolves and mountain lions effectively removed these species from New York. Internationally, species such as the passenger pigeon, the great auk, and the dodo were hunted to extinction in this same period. Today, the overhunting of several species of whale in the open ocean has threatened to drive them into extinction. Whenever a species is eliminated from its native habitat, the species' environmental niche is vacated and ecosystem balance is disrupted.

For hundreds if not thousands of years, the community of plants and animals on the Kaibab plateau in northern Arizona remained in balance. Each species filled its appropriate niche and interacted as predator, prey, producer, or decomposer. Misguided conservation policies introduced in the early 1900s placed a bounty on mountain lions and other natural predators of the Kaibab deer. As hunters reduced the numbers of predatory species on the plateau, the deer population, with fewer natural enemies, exploded. The thousands of additional deer that resulted quickly outstripped the food supply of this fragile environment. As edible plants were consumed and root systems destroyed by the foraging deer, the deer population starved to death in great numbers.

Warning Signs of Environmental Contamination

Many warning signs exist that indicate that a dangerous disequilibrium is being created in our ecosystems. Some researchers suspect that toxic chemical by-products, heavy metals, pesticides, hormones, pharmaceuticals, and other contaminants are responsible for an increased frequency in deformation of, illness in, and death of natural populations worldwide. For example, dramatic declines and deformations in frog populations have been observed in aquatic ecosystems around the world. The culprits suspected by some scientists are hormone-disrupting chemicals in agricultural runoff. Other examples include:

- reproductive cycle disruption in birds as a result of high concentrations of DDT and other pesticides;
- deaths and population declines of birds and fish linked to contamination from oil spills;
- deaths and disorientation of marine mammals (porpoises, whales, and manatees), possibly due to coastal pollutant runoff;
- high concentrations of polychlorinated biphenyls (PCBs) in fish taken from contaminated waters.

In our own species, adverse health effects linked to chemical exposure are increasingly prevalent among children. Infancy and early childhood are periods of vulnerability to environmental pollution. Because they expend more energy and require more food, water, and oxygen per unit of mass than adults, young children are likely to take in and store more toxic chemicals per pound than do adults. In recent years, the incidence of birth defects and diseases linked to environmental pollution has risen among children living in environmentally-contaminated areas of the world. Childhood cancer has become the second leading cause of childhood deaths. Other health problems being experienced today by children living in contaminated communities include chronic lung disease, childhood asthma, attention deficit disorder, and too-early or delayed sexual maturation. Chemical pollutants such as PCBs, dioxin, lead, pesticides, and solvents are increasingly suspected as contributors to these issues among children.

Energy Consumption

Industrialization has brought an increased demand for and use of energy resources. This usage has had both positive and negative effects on humans and ecosystems. The world's industrialized nations use tremendous quantities of energy to fuel their economies. Countries such as the United States and England have long been the major consumers of energy. China, formerly a light user of energy resources, has recently become a major consumer of petroleum as a source of energy to operate its growing economy.

Over time, the source of this energy has changed, with traditional water, wood, coal, and oil gradually being replaced with natural gas, wind, wave/tidal, nuclear, solar, geothermal, and fuel cell technologies. Each power source has its advantages and disadvantages for application in any particular industry. However, the cost per unit and availability of these power sources have always been prime determiners of the choices made. The development of these energy sources often uses valuable agricultural lands for the construction of extraction, storage, and transport of the energy they produce.

Energy Consumption History and Outlook, 1950–2025

Fossil Fuels

Fossil fuels, commonly used as an energy source in industrial processes for the past 200 years, include naturally occurring materials such as oil, coal, and natural gas. These fuels are drawn from underground deposits formed millions of years ago by biological and geological processes. Fossil fuel deposits are not being replenished by these same processes today. As a result, the supply of these fuels is finite and will someday be depleted. Fossil fuels are burned to power internal combustion engines or to produce steam for running electrical turbines. Their hydrocarbon chemistry results in the emission of carbon dioxide (CO_2) and carbon monoxide (CO) gases when they are burned. Impurities in these fuels may also result in the emission of particulate matter, sulfur oxides, and nitrogen oxides. The drilling and mining operations used to obtain these fuels is also worthy of consideration, as the technology surrounding these operations can have negative consequences for the natural environment, as well as for humans.

Hydraulic fracturing, commonly known as **hydrofracking**, is a process that has been used, with variable success, to release natural gas from certain geologic shale formations (such as the Marcellus Formation underlying much of New York's southern tier). Because some of these shale formations are found in south/central New York State, hydrofracking has become an important point of discussion for many New Yorkers. The most prominent issue under discussion involves the trade-off between energy independence, economic development, and jobs vs. protection of the quality of aquatic ecosystems for native fish species and groundwater for human consumption. Are the short-term advantages of economic gain worth the long-term dangers of environmental disruption? This is one of many questions facing New Yorkers regarding the pros and cons of economic development vs. environmental quality.

The chemical by-products of fossil fuel combustion are released into the atmosphere and contribute to the worldwide air pollution problem as well as a phenomenon known as acid precipitation, or acid rain. **Acid precipitation** results from the combining of sulfur and nitrogen oxide ions with atmospheric water to produce sulfuric and nitric acids. When rain, snow, and sleet containing these acids falls on aquatic ecosystems they can significantly alter the acid/base chemistry of these ecosystems. This alteration often leads to stress on, or elimination of, acid/base sensitive species of fish and amphibians in those ecosystems. The burning of fossil fuels has also been linked to the phenomenon known as global warming (see page 235).

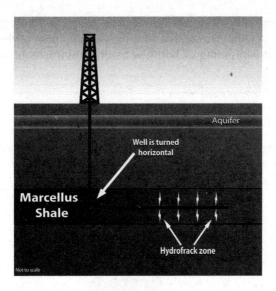

Hydrofracking

Nuclear fuel, in the form of plutonium, is a highly refined and concentrated form of the radioactive mineral uranium. In a containment vessel specially designed to promote it, plutonium emits high-energy particles that collide with atoms, forcing them to emit neutrons that collide with other atoms in a chain reaction. This reaction produces a tremendous quantity of heat, which is used to boil water to make steam for running electrical turbines. The wastes of this process are heat released into aquatic ecosystems and radioactive emissions released into the atmosphere from depleted fuel rods. Disposal of these wastes can be disruptive to natural environments and dangerous to human health. Waste heat can disrupt water environments by raising the temperature of those ecosystems to unnatural levels, killing fish and other aquatic life in the process. Radiation can contaminate soils, water, and air and can kill or cause mutations and cancers in living organisms including humans. For example, radioactive waste is currently being stored in outdoor containment areas in New York at Lake Ontario's nuclear power plants; its radioactivity is likely affecting communities downwind (southeast) of these facilities.

Alternative energy sources, such as wind, wave/tidal, solar, fuel cell, and geothermal, are in their developmental stages. Developed properly, these alternative energies have the potential to provide energy far in excess of the current annual energy needs of the United States, a fact that could lead to energy-independence, new jobs, and an improved economy. Because these energy sources are renewable, they have the advantage that they will not be depleted as will fossil fuels, allowing them to carry our energy needs far into the future. These technologies are also relatively pollution-free, making them a safer alternative to fossil and nuclear energy technologies. It benefits us as a species to support research into the development of these alternative energies, for the well-being of both the natural environment and ourselves.

Wind Energy

Tidal Energy

Global Warming

The negative effects of 200 years of worldwide industrial and technological development have recently manifested themselves as the phenomenon known as **global warming**. Although the signs of this phenomenon have been recognized by competent scientists for many years, it has taken industrialists, politicians, and the general public much longer to recognize the threat to the living environment posed by its effects. In the mid-20th century, atmospheric scientists and environmental biologists pointed out that increasing concentrations of carbon dioxide and methane, known collectively as **greenhouse gases**, had the effect of trapping solar radiation in Earth's atmosphere in the form of heat. These scientists warned that, if left unchecked, this process would lead to global warming sufficient to melt polar ice caps and glaciers, expand deserts, raise ocean levels, and significantly disrupt global weather patterns. They further warned that an increase of only a few degrees in average temperature would disrupt ecological communities in virtually all parts of the world, accelerating the pace of extinction of plant and animal species and threatening human survival. Recent estimates of the rate of global warming predict that Earth's average temperatures will rise by as much as 11°F (–12°C) by the year 2100. That much warming will create environmental effects that will devastate the entire human population.

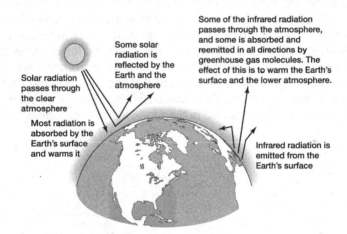

Solar radiation passes through the clear atmosphere

Some solar radiation is reflected by the Earth and the atmosphere

Some of the infrared radiation passes through the atmosphere, and some is absorbed and reemitted in all directions by greenhouse gas molecules. The effect of this is to warm the Earth's surface and the lower atmosphere.

Most radiation is absorbed by the Earth's surface and warms it

Infrared radiation is emitted from the Earth's surface

These scientists urged governments worldwide to impose controls on industries and technologies known to emit greenhouse gases into the atmosphere, but little was done at the time to curb these emissions. Recently, the serious effects of global warming have begun to capture worldwide attention. In 2007, the United Nations scientific panel on climate change noted that the evidence of global warming was "unequivocal" and that human technological oversight was the "likely" culprit. Still, the international community has been slow to respond with meaningful environmental controls of greenhouse gas emissions. Unfortunately for all of us, the actions taken may be "too little, too late." Adding to the problem is the repeated failure of vocal and influential politicians and corporate leaders to acknowledge that the phenomenon exists at all or, more specifically, to acknowledge that it is most likely a human-made problem requiring a human-made solution.

Reports of accelerated glacial melting, shrinking polar caps, epic droughts, intense and destructive weather patterns, catastrophic crop failures, disappearing fisheries, and rising ocean level are more and more frequent. If immediate steps are not taken to reduce greenhouse gas emissions, it is estimated that **ocean levels** will rise as much as 2 feet by the year 2050. This amount of sea level rise will drown coastal habitats and subject cities such as New York, Boston, and Philadelphia to massive flooding. It will also push salt water far up tidal estuaries (such as the Hudson River from New York to Albany), destroying freshwater habitats and flooding riverside agricultural lands and communities along the way. It will have a profound negative effect on the world's economic situation and will only get worse over the next several decades.

Weather patterns that in the past have been mild and predictable have become increasingly violent and unpredictable. Just weeks apart in 2005, hurricanes Katrina and Rita devastated the Gulf coast, taking many lives and destroying billions of dollars of property in Texas, Louisiana, Alabama, and Mississippi. In 2011 and 2012, respectively, New Yorkers felt the effects of hurricanes Irene and Sandy, also with the significant loss of life and property. Hurricane Irene destroyed communities and natural habitats along the Hudson and Mohawk river valleys to as far north as the Adirondack Mountains, as well as throughout New England. Hurricane Sandy subjected lower Manhattan and coastal areas of New York and New Jersey to high winds, rain, and massive storm surge flooding. There is little doubt that the intensity and destructive power of these storms were greatly enhanced by the effects of global warming. More such storms can be expected in the future as Earth continues to warm.

Another serious example of global warming's effect on weather patterns involves significant shifts in **temperature and rainfall patterns** across the world's most productive agricultural zones. Between the years 2000 and 2012, increasing temperatures in the "breadbasket" of the United States have had a significant impact on the production of wheat and other grains. An increase of just 1°F (0.56°C) in average temperature during this period has resulted in a 5% decrease in wheat production in North Dakota alone. Throughout the breadbasket, these changes have significantly shifted the areas where wheat can be grown at all, gradually pushing them northward and westward. Grains such as wheat, corn, and oats are essential crops needed to feed the country's population and/or its domestic meat-producing animals. Drastically decreased rainfall in this same area has resulted in record drought conditions that further threaten the production of food crops in the future. Similar effects have been noted in agricultural zones of Canada and Australia, as well as throughout Europe and Asia. Continued increases in temperature and decreases in rainfall threaten to seriously reduce or completely eliminate agricultural activities in these areas. The implications for human survival may be disastrous if the situation is not reversed.

In a perverse feedback effect of global warming, the melting of **polar ice and arctic permafrost** caused by global warming has been shown to be responsible for the release into the atmosphere of huge quantities of methane, a greenhouse gas that has been trapped in frozen soils and sea bottom muck for tens of thousands of years. The rapid addition of this methane to Earth's atmosphere is having the effect of compounding the global warming problem, which will have the effect of releasing still more methane into the atmosphere as increased temperatures promote still more ice/permafrost thawing.

Retreat of Northern Polar Ice Cap

Polar Cap Extent 2002

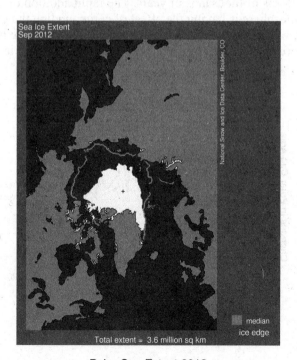

Polar Cap Extent 2012

While alternating **cycles of global warming and cooling** are known to have occurred many times over the four billion years of Earth's history, the current shift is happening with unprecedented speed. Whereas past temperature shifts have taken place over hundreds of thousands or millions of years, this one is taking place over mere decades or centuries. This means that we will see changes in our lifetimes that in past cycles would only have occurred over hundreds of human generations. Though difficult to observe from the perspective of a single human lifetime, these changes have been studied and predicted by competent scientists throughout the world and must be addressed by the present generation if we hope to survive as a species.

Need for Action

It is imperative that we, as a people, act together to make the regulatory and behavioral changes necessary to turn back the tide of global warming. Each of us has a personal obligation to reduce the amount of carbon dioxide we add to the atmosphere. We can do this by reducing personal energy consumption, buying locally produced foods and products that do not need to be shipped long distances, and minimizing the use of synthetic materials that require energy to produce. This is sometimes referred to as "**reducing our carbon footprint**." But we also must demand that our political leaders enact effective legislation to enforce reduction of greenhouse gas emissions from vehicles, factories, power plants, and public buildings. We must insist that our federal government agree to enforce international treaties aimed at reducing energy use and environmental pollution, as well as those designed to protect fragile ecosystems and endangered species. To do less will only serve to hasten our extinction.

Individual Choices and Social Actions

Essential Questions:

- *How can humans take positive action to reverse the degradation of the natural environment caused by human technology?*
- *What trade-offs must be considered when making economic and political decisions that affect the environment?*

Performance Indicator 4.7.3 *The student should be able to explain how individual choices and social actions can contribute to improving the environment.*

Societies must develop procedures that ensure that we can prove the safety and effectiveness of new technologies before implementing them. This concept is sometimes referred to as the "**precautionary principle**," and is meant to help protect human health from reckless technological implementation. Individual voters need to make personal decisions that will help them assess the risks, costs, benefits, and trade-offs of such new technologies. Until it can be proven that a particular technology poses no significant risk to the natural environment and to human health, it should not be allowed to operate, no matter what the economic benefit to individuals, corporations, or governments. Developing the political will to do so will help ensure the survival of our environment, and of our species as well.

We must constantly keep in mind that we share this small planet with diverse species that together with us make up the living community. We must acknowledge that the **maintenance of biodiversity** promotes the general health and welfare of both human society and of the ecosystem. We cannot afford to lose a single species to extinction caused by environmental abuse, lest we threaten our own existence. We must remain vigilant that we do not revert to the dangerous and destructive practices of the past but continue to improve the ways in which we care for and understand our natural environment.

This requires that we become informed about environmental issues so we can apply the precautionary principle by asking critical questions about the potential impact of new technologies on these issues. We must also support research aimed at studying the effects of these technologies on human health as well as the health of other species. This information should then be weighed against the benefits that this technology would bring to society as a whole. This thought process is sometimes referred to as **examining the trade-offs**. If the benefits outweigh the risks, then the technology should be implemented with appropriate precautions to guard against unanticipated negative effects. If, on the other hand, the risks outweigh the benefits, then the technology should be set aside until adequate safeguards are developed.

Recently, some forward-thinking groups, governments, communities, and companies have made efforts to establish procedures that allow development without damaging the environment. Examples of such **positive environmental actions** include:

- Local governments requiring that developers replace wetlands destroyed by commercial and residential development activities;
- State and county governments requiring that logging industries replant native species in clear-cut tracts;
- State and federal governments requiring that coal and oil-burning power plants install smokestack scrubbers to remove greenhouse emissions and acid-producing ions from stack gases;
- Commercial companies and schools making the decision to build "green," energy-efficient buildings to reduce their carbon footprint;
- Auto manufacturers designing and marketing electric, hybrid, and other fuel-efficient vehicles that help conserve fossil fuels and reduce greenhouse gases released into the atmosphere;
- Schools, businesses, individuals, and governments installing geothermal, solar panel, and/or wind turbine energy as alternatives to traditional fossil fuel energy sources;
- Communities and schools developing and participating in recycling efforts to reduce energy use and materials depletion;

- States and local governments reducing the amount of road salt applied to highways during winter months in environmentally sensitive areas;
- State and local laws requiring that public landfills be properly constructed so as to minimize the likelihood of groundwater contamination and methane release into the atmosphere;
- Private institutions and foundations creating preservation lands for the conservation of habitat for endangered species;
- Federal and state governments setting aside wilderness areas for the preservation of natural ecosystems;
- Federal and state governments requiring industrial companies to clean PCBs and other toxic pollutants from ecosystems contaminated by the activities of these companies;
- Citizens' cooperatives being founded to promote the use of locally grown agricultural foods and other products to reduce the carbon footprint of the local economy;
- Environmental activists demanding that state and local governments investigate incidents of environmental contamination and prosecute violators of environmental and public health laws;
- Voters demanding that state governments require that energy companies prove the safety and effectiveness of hydrofracking and other energy-extraction technologies prior to implementation.

Environmental Responsibility

The decisions of one generation provide both the limit and range of possibilities open to the next generation. The technological advances of human beings have often had a negative impact on the environment, but we need not accept these past decisions going forward. Increasing awareness of the role of humans and all the other organisms in the ecosystem has begun to reverse the negative trend of the past. Only through continued efforts to protect wild species, conserve resources, preserve natural habitats, control human population growth, and value all life forms as essential contributors to the maintenance of a healthy environment will we, as a global community, survive and provide suitable living conditions for future generations.

As it becomes increasingly clear that the solution to pollution is to avoid creating hazardous wastes in the first place, today's generation is trying to open up more possibilities for those that follow. We are examining manufacturing processes more closely, demanding that industry use as few toxic inputs as possible and produce few or no toxic wastes. Grassroots activists are making it so difficult and expensive for industry to dump its hazardous wastes that cleaning up the production processes becomes a more cost-effective option for industries to consider. Consumers are beginning to show a preference for products that create less waste by demanding local manufacture, small carbon footprint, minimal packaging, and reusable vs. disposable products. In today's global economy, these ideals are sometimes hard to come by, but are an essential component in our economic and ecological recovery.

How will you respond to this environmental imperative? Will you shop at "big box" stores whose products have been manufactured and shipped from halfway around the world, giving them a large carbon footprint, or will you shop at local stores for products that have been produced locally? Will you insulate your home and set its programmable thermostat at a minimal level to conserve energy, or will you set it to a higher level without considering insulation merely to provide physical comfort? Will you allow local governments to issue building permits in your town without regard to their environmental impact, or will you become involved by demanding environmental review of such decisions? Will you be a citizen, only, or will you be an active voter, participant, and leader in the governance of your community? These are only a few of the questions that you should ask yourself as an emerging adult member of your town, county, state, nation, and world community. It will be your responsibility to make the best choices for the furtherance of our environment, our economy, and our species.

QUESTION SET 2.10—HUMAN IMPACT ON ECOSYSTEMS (ANSWERS EXPLAINED, P. 332)

1. The use of ladybugs and preying mantises to consume insect pests in gardens is an example of
   (1) biological control of insect pests
   (2) exploitation of insect pests
   (3) abiotic control of insect pests
   (4) use of biocides to control insect pests

2. The creation of wildlife refuges and the enforcement of game laws are conservation measures that promote increased
   (1) use of biocides
   (2) preservation of species
   (3) use of biological controls
   (4) exploitation of species

3. One practice that has successfully increased the number of bald eagles in the United States is the
   (1) protection of natural habitats
   (2) importation of food to their nesting sites
   (3) preservation of other eagle species that occupy the same niche
   (4) increased use of pesticides

4. Which human activity would be more likely to have a negative impact on the environment than the other three?
   (1) using reforestation and cover cropping to control soil erosion
   (2) using insecticides to kill insects that compete with humans for food
   (3) developing research aimed toward the preservation of endangered species
   (4) investigating the use of biological controls for pests

5. Which human activity would most likely result in the addition of an organism to the endangered species list?
   (1) cover cropping
   (2) use of pollution controls
   (3) use of erosion controls
   (4) habitat destruction

6–10. Base your answers to questions 6 through 10 on the passage below and on your knowledge of biology.

The Mystery of Deformed Frogs

Deformities, such as legs protruding from stomachs, no legs at all, eyes on backs, and suction cup fingers growing from sides, are turning up with alarming frequency in North American frogs. Clusters of deformed frogs have been found in California, Oregon, Colorado, Idaho, Mississippi, Montana, Ohio, Vermont, and Quebec.

Scientists in Montreal have been studying frogs in more than 100 ponds in the St. Lawrence River Valley for the past four years. Normally, less than 1 percent of frogs are deformed. In ponds where pesticides are used on surrounding land, as many as 69 percent of the frogs were deformed. A molecular biologist from the University of California believes that the deformities may be linked to a new generation of chemicals that mimic growth hormones. The same kind of deformities found in the ponds have been replicated in laboratory experiments.

Some scientists have associated the deformities with a by-product of retinoid, which is found in acne medication and skin rejuvenation creams. Retinoids inside a growing animal can cause deformities. For this reason, pregnant women are warned not use skin medicines that contain retinoids. Recent laboratory experiments have determined that a pesticide can mimic a retinoid.

A developmental biologist from Hartwick College in Oneonta, New York, questioned whether a chemical could be the culprit because no deformed fish or other animals were found in the ponds where the deformed frogs were captured. He believes parasites are the cause. When examining a three-legged frog from Vermont, the biologist found tiny parasitic flatworms packed into the joint where a leg was missing. In a laboratory experiment, he demonstrated that the invasion of parasites in a tadpole caused the tadpole to sprout an extra leg as it developed. Scientists in Oregon have made similar observations.

6. Why are pregnant women advised not to use skin medicines containing retinoids?
   (1) Retinoid by-products may cause fetal deformities.
   (2) Retinoid by-products cause parasites to invade developing frogs
   (3) Retinoid by-products mimic the effects of pesticides on fetal tissue.
   (4) Retinoid by-products reduce abnormalities in maternal tissue.

7. Some scientists argue that pesticides may not be the cause of the frog deformities because
   (1) pesticide use has decreased over the last four years
   (2) new pesticides are used in skin-care products
   (3) other animals in the ponds containing deformed frogs did not have abnormalities
   (4) laboratory experiments have determined that a pesticide can mimic retinoids

8. A possible reason for the absence of deformed fish in the ponds that contained deformed frogs is that
   (1) fish can swim away from chemicals introduced into the pond
   (2) parasites that affect frogs usually do not affect fish
   (3) fish cannot develop deformities
   (4) frogs and fish are not found in the same habitat

9. Which inference can be made from the information in the passage?
   (1) Only a few isolated incidents of frog deformities have been observed.
   (2) If frog parasites are controlled, all frog deformities will stop.
   (3) Deformities in frogs are of little significance.
   (4) Factors that affect frogs may also affect other organisms.

10. By using one or more complete sentences, describe how pesticides could cause deformities in frogs.

   _____

11. Habitat destruction is an environmental problem that affects our own generation and will affect future generations if not solved. Write an essay in which you identify a habitat that is being destroyed and explain how the destruction of this habitat relates to humans and the overall ecosystem. Your essay must include at least:

- *two* human activities that contribute to the destruction of this habitat
- *three* ways the destruction of this habitat has affected plants, humans, and other animals
- *two* ways to limit further destruction of this habitat

12–13. Base your answers to questions 12 and 13 on the information below and on your knowledge of biology.

In July 1997, about 25,000 *Galerucella pusilla* beetles were released at Montezuma National Wildlife Refuge in western New York State. These beetles eat purple loosestrife, a beautiful but rapidly spreading weed that chokes wetlands. Purple loosestrife is native to Europe, but here it crowds out native wetland plants, such as cattails, and does not support wildlife the way that native plants do. Purple loosestrife grows too thick to allow birds to nest. Most native insects do not eat it, leaving little for insect-eating birds to feed on. Bernd Blossey, a professor at Cornell University, spent 6 years in Europe trying to find out what limited the loosestrife population there.

12. By using one or more complete sentences, explain why the introduction of the beetle is an advantage over the use of herbicides to control the purple loosestrife population.

_____

13. By using one or more complete sentences, describe one possible environmental problem that may result from the introduction of this beetle.

_____

14–16. Base your answers to questions 14 through 16 on the information below.

In a rural area is a wetland with a large population of mosqui- toes. Nearby residents are concerned because the mosquitoes are always annoying and occasionally carry diseases. The community decides to have an insecticide sprayed from an airplane onto the area during the prime mosquito season. Whenever they stop spraying, the mosquito population quickly rebounds to a higher level than existed before the spraying program began. After 10 years, the spraying became much less effective at reducing the mosquito population. Higher doses of insecticide were required to accomplish the same population decreases.

14. State *one* possible *disadvantage* of spraying the insecticide from the air- plane.

_____

15. State *one* alternative method of mosquito control that may have a more lasting impact on the mosquito population.

_____

16. Give *one* positive effect or *one* negative effect, other than killing mos- quitoes, of the alternative method of mosquito control you stated in question 15.

_____

APPENDIX

Answer Key

UNIT ONE

Question Set 1.1
1. 2
2. 2
3. 1
4. 1
5. 2
6. 3
7. 4
8. 2
9. 4
10. see Answers Explained
11. 3
12. 1
13. 2
14. 1
15. 1

Question Set 1.2
1–4. see Answers Explained
5. 4
6. 1
7. 4
8–10. see Answers Explained
11. 1
12. 4
13. 2
14. see Answers Explained
15. 3

Question Set 1.3
1. see Answers Explained
2. 1
3. 3
4. 1
5. 4
6. 2
7. see Answers Explained
8. see Answers Explained
9. see Answers Explained
10. 4
11. 4
12. see Answers Explained

Question Set 1.4

1. **2**	17. **1**	30. see Answers
2. **B**	18. **3**	Explained
3. **C**	19. **2**	31. see Answers
4. **D**	20. **8.9 cm**	Explained
5. **4**	21. **1**	32. see Answers
6. **3**	22. **4**	Explained
7. see Answers	23. see Answers	33. see Answers
Explained	Explained	Explained
8. **1**	24. see Answers	34. **4**
9. **4**	Explained	35. **1**
10. **2**	25. **2**	36. **2**
11. **2**	26. **3**	37. **3**
12. **4**	27. **4**	38. **3**
13. **3**	28. see Answers	39. **3**
14. **1**	Explained	40. **4**
15. **4**	29. see Answers	41. **4**
16. **1**	Explained	42. **2**

Reading, Writing, and Current Events in Science

1. **3**	8. see Answers	12. see Answers
2. **1**	Explained	Explained
3. **1**	9. see Answers	13. see Answers
4. **2**	Explained	Explained
5. **3**	10. see Answers	14. see Answers
6. see Answers	Explained	Explained
Explained	11. see Answers	
7. see Answers	Explained	
Explained		

UNIT TWO

Question Set 2.1

1. **1**	5. **1**	9. **2**
2. **4**	6. **4**	10. **1**
3. **4**	7. **3**	11. **4**
4. **4**	8. **1**	

Question Set 2.2

1. **4**	5. **1**	9. **4**
2. **1**	6. **4**	10. **2**
3. **2**	7. **3**	
4. **2**	8. **2**	

Question Set 2.3

1. **1**	5. **1**	10. **4**
2. **3**	6. **3**	11. **2**
3. **3**	7. **2**	12. **1**
4. see Answers	8. **3**	
Explained	9. **3**	

Question Set 2.4

1. **1**	6. **3**	11. **2**
2. **2**	7. **3**	12. **1**
3. **3**	8. **3**	13. **4**
4. **4**	9. **2**	
5. **3**	10. **3**	

Question Set 2.5

1. **1**	11. **2**	21. **1**
2. **4**	12. **3**	22. **3**
3. **4**	13. **1**	23. **2**
4. **1**	14. **2**	
5. **4**	15. **4**	
6. **3**	16. **3**	
7. **3**	17. **3**	
8. **2**	18. **3**	
9. **1**	19. **1**	
10. **4**	20. **2**	

Question Set 2.6

1. **4**	6. **2**	11. **2**
2. **3**	7. **3**	12. see Answers
3. **1**	8. **1**	Explained
4. **1**	9. **3**	13. see Answers
5. **4**	10. **4**	Explained

Question Set 2.7

1. **1**	7. **3**	13. **1**
2. **4**	8. **4**	14. **3**
3. **1**	9. **2**	15. **1**
4. **4**	10. **1**	16–19. **see Answers**
5. **3**	11. **3**	**Explained**
6. **2**	12. **4**	

Question Set 2.8

1. **2**	12. **2**	21. **3**
2. **2**	13. **4**	22. **4**
3. **4**	14. **4**	23. **2**
4. **2**	15. **1**	24–30. **see Answers**
5. **2**	16. **4**	**Explained**
6. **1**	17. **1**	
7. **2**	18. **see Answers**	
8. **2**	**Explained**	
9. **4**	19. **see Answers**	
10–11. **see Answers**	**Explained**	
Explained	20. **2**	

Question Set 2.9

1. **4**	10. **3**	19. **2**
2. **2**	11. **2**	20. **2**
3. **1**	12. **3**	21. **4**
4. **6**	13. **2**	22. **2**
5. **5**	14. **3**	23. **4**
6. **2**	15. **4**	24. **1**
7. **1**	16. **4**	25. **3**
8. **2**	17. **2**	26. **see Answers**
9. **4**	18. **3**	**Explained**

Question Set 2.10

1. **1**	5. **4**	9. **4**
2. **2**	6. **1**	10–16. **see Answers**
3. **1**	7. **3**	**Explained**
4. **2**	8. **2**	

Answers Explained

UNIT ONE

Question Set 1.1 (p. 16)

1. **2** The biologist should observe *a large number of frogs in their natural habitat*. The large sample will help to ensure the probability that the sample will be representative of the species's behavior. Observing the frogs in their natural habitat will ensure that no unusual conditions exist that may influence how the frogs react to each other.

Wrong Choices Explained
(1) Observing *a small number of frogs in their natural habitat* will not ensure that a representative sample has been taken. If a few of the frogs in a small sample behave in an unusual way, they will represent a larger percentage of a small sample than they would of a large sample.
(3), (4) Observing *several groups of frogs maintained in different temperatures in the laboratory* or *several groups of frogs maintained on different diets in the laboratory* will introduce variables that interfere with the natural behaviors of the frogs. Changes in temperature or diet may affect the frogs' behavior greatly, making the observations dependent on these experimental conditions instead of the natural factors that affect the frogs' behavior.

2. **2** Counting the number of lines on the cylinder between the meniscus at A and the meniscus at B yields the number 4. Since a total of 8 milliliters was removed, the graduations on the cylinder must be *2 milliliters* each.

Wrong Choices Explained
(1) Graduations of *1 milliliter* would yield eight lines with the removal of 8 milliliters from the cylinder.
(3) Graduations of *8 milliliters* would yield one line with the removal of 8 milliliters from the cylinder.
(4) Graduations of *4 milliliters* would yield two lines with the removal of 8 milliliters from the cylinder.

3. **1** The statement *environmental conditions affect germination* represents a hypothesis. This statement contains a belief about the outcome of an investigation in which a dependent variable (germination) is changed by an experimental variable (environmental conditions). Although very general, the statement still predicts an outcome and therefore meets the definition of a hypothesis.

Wrong Choices Explained

(2) The statement *boil 100 milliliters of water, let it cool, and then add ten seeds to the water* represents an experimental procedure, not a hypothesis.

(3) The statement *is water depth in a lake related to available light in the water?* represents an experimental question, not a hypothesis.

(4) The statement *a lamp, two beakers, and elodea plants are selected for the investigation* represents a materials list for an experiment, not a hypothesis.

4. **1** The *millimeter* is the unit of measure most appropriate to perform this measurement. About 25.4 millimeters make up one inch.

Wrong Choices Explained

(2) The micrometer is a unit of measure used to describe the sizes of very small objects, such as cell organelles. It is not an appropriate unit of measure for an object of this size.

(3), (4) The foot and the meter are too large to be used to measure an object of this size, since the measurements would be fractional.

5. **2** A millimeter is made up of 1,000 micrometers. The number of whole millimeters shown in the diagram is 3 (3,000 micrometers). The amount of distance left to the right of the third millimeter is a little larger than half of a millimeter (about 70 micrometers). The total distance across the diameter, therefore, is approximately *3,700 micrometers*. (HINT: remember to count as 1,000 micrometers the distance from the left edge of one millimeter mark to the left edge of the next millimeter mark.)

Wrong Choices Explained

(1), (3), (4) Each of these distracters has been chosen to catch the unwary student. Count carefully and mark on the test booklet if that will assist you.

6. **3** *It differs in the one variable being tested* is the way that the control setup differs from the experimental setups in the same experiment. In all other respects, it is identical to the experimental setups. For example, in an experiment testing how varying concentrations of enzymes affect the rate of hydrolysis of a substrate, an appropriate control would be a setup containing water, only, but identical to the experimental tubes in all other respects. This control assures that the water in which the enzyme is dissolved has no effect on the hydrolytic action.

Wrong Choices Explained

(1), (2), (4) The statements *it tests a different hypothesis, it has more variables*, and *it utilizes a different method of data collection* are not true

252

of a control group in a properly designed experiment. The experimental design should seek to eliminate the differences between the experimental setups and the control setup. The more variables and differences between the control and experimental setups, the less reliable the results of, and the conclusions drawn from, the experiment.

7. **4** The experimental variable is the condition changed (varied) by the experimenter. In this experiment, the condition varied is the *distance of the plant from light*.

Wrong Choices Explained
(1) No measure is being made of the *concentration of gas in the water*. If this condition is changing, it is incidental to the experiment.
(2) The *aquatic plant type*, in this case elodea, is a *constant* in the experiment as described.
(3) No measure is made of the *amount of water in the test tube*. If this condition is changing, it is incidental to the experiment.

8. **2** *Lugol's solution* will test for the presence of starch in foods. It turns blue black in the presence of starch.

Wrong Choices Explained
(1) *pH paper* is used to test how acidic or alkaline a solution is.
(3) *Methylene blue* is a commonly used stain in preparing wet-mount microscope slides.
(4) *Bromthymol blue* is used to detect the presence of carbon dioxide in a solution.

9. **4** Before a scientist begins to study a natural phenomenon, he/she must first pose a question to be answered or a *problem to be solved*. If this step is skipped, much time is wasted in the laboratory studying factors that may have no direct bearing on the object of study.

Wrong Choices Explained
(1), (2), (3) An appropriate sequence for conducting a scientific investigation is as follows:

— state the problem
— *formulate a hypothesis* (educated guess)
— define the experimental method
— *perform the experiment*
— gather experimental data
— *analyze the experimental data*
— draw inferences

10. *Goggles should be worn.* OR *Gloves should be worn.* OR *The student should take care not to touch the sharp scalpel to his skin.*
(NOTE: The preceding answers represent acceptable responses. Other complete-sentence responses are acceptable as long as they give correct information.)

11. **3** The *control* group in an experiment duplicates all aspects of the experiment except the independent variable (the variable being tested). In this case, the drug is the independent variable, so the group given pills without the drug, but with glucose, is considered the control group.

Wrong Choices Explained
   (1) The *experimental group* is the group that receives pills containing the drug rather than pills containing glucose.
   (2) The term *limiting factor* relates to environmental conditions that limit the survival of individuals or populations.
   (4) An *indicator* is a chemical that helps to determine the chemical characteristics of biological samples. Examples of indicators include bromthymol blue and litmus.

12. **1** Of those given, the hypothesis *light intensity affects the growth of algae* best explains the observations reported. The intensity of natural light affects the depth to which it can penetrate the lake water. Algae can photosynthesize only in the presence of adequate light intensity. Therefore, we are led to the conclusion that the presence (or growth) of algae depends on the degree to which light is available (its intensity) in the environment.

Wrong Choices Explained
   (2) The hypothesis *wind currents affect the growth of algae* has no basis for support in the reported observations. Not only is wind not mentioned as a variable, but logical analysis of the situation would indicate that wind is unlikely to directly affect algae 1 to 6 meters below the surface of the lake.
   (3) The hypothesis *nitrogen concentration affects the growth of algae* has no basis for support in the reported observations. Nitrogen concentration is not mentioned as a variable in this experiment.
   (4) The hypothesis *precipitation affects the growth of algae* has no basis for support in the reported observations. Precipitation is not mentioned as a variable in this experiment.

13. **2** *Diagram 2* illustrates how the hydra will appear to move under the compound light microscope. The optics of this instrument invert the image of what is seen in the field of view, so that everything appears upside down. As the hydra moves right, its image will appear to move left. If the hydra moves away from the observer, its image will appear to move toward the observer, and so on.

Wrong Choices Explained
(1), (3), (4) These diagrams all represent incorrect illustrations of how the image will move in the field of view under the conditions stipulated.

14. **1** The approximate length of a nucleus of one of these cells is *100 μm*. Two nuclei are visible as dark ovals within each of the two cells illustrated. Care should be taken to read the question carefully so that the correct object is measured.

Wrong Choices Explained
(2) *500 μm* is the approximate length of one cell.
(3) *1,000 μm* is the approximate length of both cells end to end.
(4) *1,500 μm* is the approximate diameter of the field of view (actually closer to 1,600 μm).

15. **1** The difference in length between leaves *A* and *B* is closest to *20 mm*. Leaf *A* is approximately 55 mm long, while leaf *B* is approximately 33 mm long. The difference, 22 mm, is closest to 20 mm.

Wrong Choices Explained
(2) *20 cm* is ten times the correct measurement. Students should take care that the correct units are used in determining the lengths of these specimens.
(3) *0.65 m* is a measurement equal to 650 cm or 6,500 mm (more than 2 feet) in length. These specimens are much smaller than this measurement.
(4) *1.6 μm* is a measurement equal to 0.0016 mm or 0.00016 cm. These specimens are much larger than this measurement.

Question Set 1.2 (p. 26)

1–3.

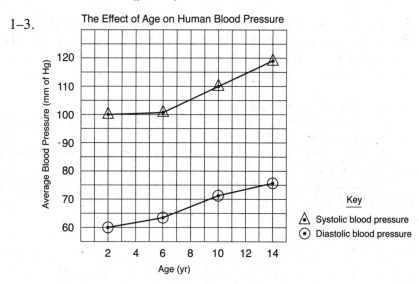

The Effect of Age on Human Blood Pressure

Y-axis: Average Blood Pressure (mm of Hg)
X-axis: Age (yr)

Key
△ Systolic blood pressure
⊙ Diastolic blood pressure

4. *Systolic pressure is higher than diastolic pressure.* OR *Both systolic pressure and diastolic pressure increase between the ages of 2 and 14.* (NOTE: Other correct, complete-sentence responses are acceptable.)

5. **4** Brook trout growth is represented by the lighter solid line in the graph. By tracing along the horizontal axis to 16°C and then tracing up to the point at which the brook trout line is encountered, we see that the growth rate is between 80% and 100% but closer to 100%. Estimate about *95%*.

Wrong Choices Explained
    (1) Brook trout growth rate is at 30% when the water temperature is about 18°C. Largemouth bass growth rate is about 30% at 16°C.
    (2) None of the species indicated have growth rates of 55% at 16°C. Brook trout growth rate is at 55% at about 17.5°C.
    (3) Brook trout growth rate falls to 75% at about 17°C. The growth rate of northern pike is at 75% at about 16°C.

6. **1** The diagram represents an experiment in cloning. Since the nucleus (which contains the genetic material) used in this experiment comes from a tadpole, the egg will produce new cells that have *genetic characteristics identical to those of the tadpole.*

Wrong Choices Explained
    (2) It is unclear from this diagram how frog *A* came into being.
    (3) This conclusion is refuted by the results of this experiment; a new frog was created without the use of fertilization.
    (4) This conclusion is refuted by the results of this experiment. The tadpole nucleus was taken from an intestinal cell, which is a body cell.

7. **4** The chart on the left shows that a man 6'0" tall with a small frame has an ideal weight range of 149–160 pounds. Of the choices given, the closest comparison can be made to the ideal weight range of a *6'0" woman with a medium frame* at 148–162 pounds.

Wrong Choices Explained
    (1) The weight range shown for a *5'10" man with a medium frame* is 151–163 pounds.
    (2) The weight range shown for a *5'9" woman with a large frame* is 149–170 pounds.
    (3) The weight range shown for a *6'0" man with a medium frame* is 157–170 pounds.

8.

Color of Light	Wavelength of Light (nm)	Percent Absorption by Spinach Extract
Violet	412	49.8
Blue	457	49.8
Green	533	17.8
Yellow	585	25.8
Orange	616	32.1
Red	674	41.0

9–10.

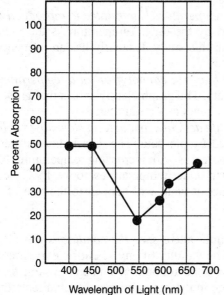

11. **1** The statement *photosynthetic pigments in spinach plants absorb blue light and violet light more efficiently than red light* is a valid conclusion that can be drawn from the data. The high point of the chart/graph data is very clearly shown to be over the blue and violet wavelengths of light.

Wrong Choices Explained
(2) The statement *the data would be the same for all pigments in spinach plants* is not supported by the results of this experiment. The chart/graph data show considerable variation in the experimental results as the wavelength of light varies.
(3) The statement *green and yellow light are not absorbed by spinach plants* is not supported by the results of this experiment. Although the chart/graph data show a lower absorption rate in these wavelengths, there is still some absorption in this range.

(4) The statement *all plants are efficient at absorbing violet and red light* is not supported by the results of this experiment. The experimental data is limited to the absorption of light by pigments found in one type of plant. This data cannot be extended to all plants unless all other types of plants are tested under the same experimental conditions and the results are found to be similar.

12. **4** Of those given, *refrigeration will most likely slow the growth of these bacteria* is the most reasonable inference that can be made from the graph data. The graph clearly shows a slower rate of growth (reproduction) at 5°C than at either 10°C or 15°C.

Wrong Choices Explained
(1) The inference that *temperature is unrelated to reproductive rate of bacteria* is not supported by the data. Temperature is the independent (experimental) variable in this study. It clearly has an influence on the bacterial reproductive rate.
(2) The inference that *bacteria cannot grow at a temperature of 5°C* is not supported by the data. The graph clearly shows that growth at this temperature, while slow, occurs at a steady pace.
(3) The inference that *life activities in bacteria slow down at high temperatures* is not supported by the data. The data indicates that, if anything, bacterial activity increases with increasing temperature. No data is shown for bacterial growth under temperatures above 15°C, so we cannot draw any inference about what will happen to the rate of bacterial growth at these extremes.

13. **2** These researchers should *make sure the conditions are identical to those in the first study*. The validity of any scientific experiment can be verified only if the same results are obtained under the same experimental conditions. Any change in these conditions will invalidate the results of the verification study.

Wrong Choices Explained
(1) If the researchers *give growth solution to both groups*, there will be no control group against which to compare the experimental group. The results of the verification study will be invalid since experimental conditions will have changed.
(3) If the researchers *give an increased amount of light to both groups of plants*, the original experimental method will be altered. The results of the verification study will be invalid since experimental conditions will have changed.
(4) If the researchers *double the amount of growth solution given to the first group*, the original experimental method will not be followed. The results of the verification study will be invalid since experimental conditions will have changed.

14. *130. The number of turns in the waggle dance decreases as the distance of the food supply from the hive increases.* OR *The closer to the hive the food source is located, the more turns there are in the waggle dance.* (NOTE: Any correct, complete-sentence answer is acceptable.)

15. **3**   Other biologists in other laboratories should be able to perform *the same experiment and obtain the same results* if the experimental results are valid. Any experimental results obtained by one scientist must be validated by independent research by other scientists following the same procedures.

    Wrong Choices Explained
       (1), (4) If different scientists perform *an experiment with a different variable and obtain the same results* or *an experiment under different conditions and obtain the same results*, they will have neither validated nor invalidated the results of the original experiment. All variables and conditions must be kept the same if the experimental results are to be properly tested.
       (2) If different scientists perform *the same experiment and obtain different results*, the original experimental results will have been invalidated.

Question Set 1.3 (p. 37)

1. Two credits are awarded for a correct two-part response that correctly identifies the finch species most affected and the reason for this effect. Acceptable responses may include:

    - *The small ground finch [1] because the sharp-billed ground finch looks for the same kinds of food. [1]*
    - *The native finch species most probably strongly affected will be the small ground finch. [1] The reason for this is that both the small ground finch and the sharp-billed ground finch depend on the same food source (small seeds). [1]*
    - *My hypothesis is that the small ground finch will be most affected by the introduction of the sharp-billed ground finch. [1] The chart shows that both species are ground-feeding seedeaters and have similar bill shapes, indicating that they will compete for the same types of available seeds. [1]*

2. **1**   The *large ground finch* will have the greatest adaptive advantage under the changed environmental conditions described. According to the chart, the large ground finch is well-adapted for crushing the shells of large, thick-walled seeds. Since these seeds will be in abundance in the changed environment, it is expected that the large ground finch will thrive compared to the other species native to the island.

Wrong Choices Explained

(2) The *small ground finch*, with its small beak, is specifically adapted for picking up and eating small seeds. It is poorly adapted for eating large, thick-walled seeds. It is likely that this species will not thrive under the conditions of environmental change described.

(3), (4) According to the chart, the *large tree finch* and the *small tree finch* are adapted to feed on insects in an environment different from that of the other native finch species. It is unlikely that this environmental change will either significantly advantage or disadvantage insect-eating finch species.

3. **3** *Interspecies competition* occurs when one species tries to gain an advantage over another species in the same environment. Interspecies (between-species) competition may occur when any shared resource, including food, territorial area, nesting sites, or other habitat, is in short supply.

Wrong Choices Explained

(1) *Variation within a species* is not represented by one finch species gaining an advantage over other species in the same habitat. This term refers to the concept that each naturally occurring species contains differences (variations) among its individual members. Such variations are thought to be the basis of the ongoing evolution in all species by means of natural selection.

(2) *Environmental change* is not represented by one finch species gaining an advantage over other species in the same habitat. This term refers to any alteration (such as rainfall or temperature) in the set of environmental conditions that exist in a particular habitat.

(4) *Mutagenic agents* are not represented by one finch species gaining an advantage over other species in the same habitat. This term refers to forces in nature (such as radiation and chemicals) that are able to alter the genetic makeup of living things.

4. **1** *CUU-GUU-ACU-GAA-CAU-CAU-CCC-GUU* is the mRNA codon sequence that will result from this DNA sequence. In the formation of mRNA, ribonucleotides position themselves on the DNA template such that complementary nitrogenous bases link together. DNA bases C, G, A, and T link with RNA bases G, C, U, and A, respectively.

Wrong Choices Explained

(2), (3), (4) *GAA-CAA-UGA-CUU-GUA-GUA-GGG-CAA, CTT-GTT-ACT-GAA-CAT-CAT-CCC-GTT*, and *UGG-AGG-CUG-ACC-UCG-UCG-UUU-AGG* are not mRNA codon sequences that will result from this DNA sequence. Each of these sequences violates the rules of complementarity referenced above or includes the nitrogenous base T (thymine) instead of U (uracil) in its structure.

5. **4** *Leu-Val-Thr-Glu-His-His-Pro-Val* is the amino acid sequence that will result from this DNA sequence. In the synthesis of a protein strand on the ribosome, each amino acid is positioned according to the sequence of codons on the mRNA strand (which in turn derived its sequence from the DNA template in the nucleus). For example, the amino acid Leu (leucine) attaches at the mRNA codon *CUU* or *CUG* (which in turn is encoded by the DNA template codon *GAA* or *GAC*). Following this example through the entire DNA codon sequence leads to the sequencing of amino acid units as shown above.

Wrong Choices Explained
(1), (2), (3) *Leu-Glu-His-Pro-Val-Val-Pro-Val, Leu-Val-Thr-Leu-Glu-Glu-His-Pro*, and *Leu-Pro-His-Val-Thr-Thr-Glu-Pro* are not amino acid sequences that will result from this DNA sequence. Each of these sequences violates the rules of amino acid encoding referenced above and data shown in the Partial Genetic Code Chart.

6. **2** *Species X* produces a protein segment most similar to that produced by *S. hunta*. Because each of the amino acids in this segment can be encoded by more than one DNA code, the slight variations in this gene between the two species still result in the same amino acid sequence (*Leu-Val-Thr-Glu-His-His-Pro-Val*).

Wrong Choices Explained
(1), (2), (3) *Leu-Glu-His-Pro-Val-Val-Pro-Val, Leu-Val-Thr-Leu-Glu-Glu-His-Pro*, and *Leu-Pro-His-Val-Thr-Thr-Glu-Pro* are not amino acid sequences that will result from this DNA sequence. Each of these sequences violates the rules of amino acid encoding referenced above and data shown in the Partial Genetic Code Chart.

7. One credit is awarded for a correctly completed histogram, as shown below:

Average Pulse Rate Range (beats per minute)

8. One credit is awarded for a complete sentence that correctly describes one pattern that is evident in the data that would help someone to understand how pulse rate is distributed among these 20 students. Answers may include:

- *Nearly two-thirds (65%) of this class has resting heart rates between 61 and 80 beats per minute.*
- *The average (mean) pulse rate for this class is 73.35 beats per minute.*
- *The distribution of heart rates for this class follows a bell curve that skews slightly to the left.*
- *Generally, the female members of this class have lower pulse rates than the males.*

9. One credit is awarded for a complete sentence that correctly states an experimental question about pulse rate that could be answered by further study of these 20 students. Answers may include:

- *What is the average pulse rate for males versus females in this class?*
- *How do the distribution curves differ for resting heart rate in females versus males?*
- *What is the relationship between resting pulse rate and body mass index (BMI)?*
- *What is the relationship between heart rate and amount of daily exercise among these students?*
- *What are the three-trial average pulse rates of these students after 5 minutes of moderate exercise?*

10. **4** The difference between the fluids in the initial and final setups is that *glucose has diffused from the fluid inside the dialysis tubing, across the membrane, and into the beaker fluid*. In the initial stage chemical test with Benedict's solution, the dialysis tubing was shown to contain glucose, whereas the beaker water was not. When the same test was performed on the setup in the final stage, glucose was detected in both the dialysis tubing and the beaker water. Since the dialysis tubing is a semipermeable membrane, it can be concluded that the small molecular size of glucose allowed it to diffuse through the membrane from a region of higher concentration (inside the tubing) to a region of lower concentration (outside the tubing).

Wrong Choices Explained
(1) *Lugol's solution has been converted to Benedict's solution* does not describe the difference between the initial and final setups. Lugol's solution is a mixture of potassium salt, iodine, and water. Benedict's solution is a mixture of various sodium and copper salts dissolved in

water. It is chemically impossible for Lugols's solution to be converted to Benedict's solution.

(2) *Starch has been converted to glucose by the indicator solutions* does not describe the difference between the initial and final setups. The conversion of starch to glucose requires the catalytic action of the enzymes amylase and maltase. Lugol's and Benedict's solutions contain no such enzymes.

(3) *Starch has diffused from the fluid inside the dialysis tubing, across the membrane, and into the beaker fluid* does not describe the difference between the initial and final setups. Starch is a polymer whose molecules are too large to diffuse through the dialysis tubing. Had this occurred, the final setup beaker water would have tested positive for the presence of starch (turned blue-black) in the presence of Lugol's solution.

11. **4** *More water moving out of the cell than into the cell* is the most likely cause of the change in appearance illustrated in diagram *B*. Under normal conditions, the cell cytoplasm and the watery environment have nearly identical salt concentrations, allowing water molecules to diffuse into and out of the cytoplasm at equal rates, as shown in *A*. When salt is added to the watery medium, the relative concentration of water decreases compared to the water concentration inside the cell. Under these conditions, water molecules diffuse out of the cell more rapidly than they diffuse into the cell, causing the cytoplasm to shrink, as shown in *B*.

Wrong Choices Explained

(1), (2) *More salt moving out of the cell than into the cell* and *more salt moving into the cell than out of the cell* are not probable causes of the condition shown in diagram *B*. Salt does not normally diffuse into or out of the cell. Rather, water is the substance that moves by osmosis in response to the concentration gradient.

(3) *More water moving into the cell than out of the cell* is not a probable cause of the condition shown in diagram *B*. If this were the case, the cytoplasm inside the cell would increase, rather than decrease, in volume.

12. Four credits are awarded for a paragraph, containing complete sentences, that correctly describes the biological explanation for the farmer's observations of the corn plants in field *B*. Sample paragraph:

When the farmer irrigated field B with the 10% salt solution on day 1, the ground around the roots became saturated with it. This created a situation in which the concentration of water inside the corn root cells was higher (>99%) than the concentration of water in the ground outside the corn root cells (90%). [1] As a result, water moved by osmosis (diffusion of water from a region of high relative concentration toward a region of low relative concentration) [1] from the inside to the outside of the corn plants.

On day 2, the rainfall washed the salt away from the soil around the roots of the corn plants sufficiently to allow the equilibrium (equal concentrations of water both inside and outside the cell) [1] to become reestablished, enabling the corn plant root cells to absorb water and the plants to regain their healthy appearance. [1]

Question Set 1.4 (on Laboratory Skills) (p. 49)

1. **2** The student's suggestion is considered to be *a hypothesis*. A hypothesis is an "educated guess" as to the scientific reason behind an observed phenomenon. The student is making this suggestion based on a hunch rather than on direct observation. To test the hypothesis, the student would need to design an experiment, collect data, analyze the data, and draw inferences and conclusions concerning the results.

Wrong Choices Explained
(1) The student's suggestion is not considered to be *a control*. A control is a laboratory setup created by the researcher in a scientific experiment that duplicates the experimental setup in every way except for the variable being tested.

(3) The student's suggestion is not considered to be *an observation*. An observation is an action taken by the researcher during a scientific experiment that normally results in the generation of data.

(4) The student's suggestion is not considered to be *a variable*. A variable is a condition that changes in a scientific experiment either because it is manipulated by the researcher or because it changes in response to the manipulated variable.

2. **B** A student testing for the presence of a carbohydrate in food should select set *B*. This laboratory setup contains two indicators (Benedict's solution and iodine) that are specialized to indicate the presence of specific carbohydrates (simple sugars and complex starches, respectively) in foods. The assorted glassware is needed to hold the foods being tested, and the heat source is used to heat the Benedict's solution in the presence of simple sugars. Goggles protect the eyes from accidental exposure to heated liquids.

Wrong Choices Explained
(A) The student should not select set *A*, which would probably be best used to determine the effect of different colors (wavelengths) of light on the rate of photosynthesis in a water plant.

(C) The student should not select set *C*, which would probably be best used to dissect a biological specimen in order to facilitate a study of gross anatomical features.

(D) The student should not select set D, which would probably be best used to study the microscopic features of very small biological specimens such as the organelles of a cell.

3. **C** A student determining the location of the aortic arches of the earthworm should select set C. This laboratory setup contains dissection instruments and a wax pan in which dissection of an earthworm can be carried out. The stereomicroscope is used to magnify the small anatomical features of the earthworm (in this case, aortic arches) for closer study. Goggles protect the eyes from accidental exposure to preservative chemicals and sharp instruments.

Wrong Choices Explained
(A) The student should not select set A, which would probably be best used to determine the effect of different colors (wavelengths) of light on the rate of photosynthesis in a water plant.
(B) The student should not select set B, which would probably be best used to determine the presence of various carbohydrates in foods.
(D) The student should not select set D, which would probably be best used to study the microscopic features of very small biological specimens such as the organelles of a cell.

4. **D** A student observing the chloroplasts in elodea should select set D. This laboratory setup contains a compound light microscope used to magnify the microscopic features of cells for closer study. The slide and water are used to provide a transparent viewing surface, and the forceps are used to manipulate the very small elodea leaf.

Wrong Choices Explained
(A) The student should not select set A, which would probably be best used to determine the effect of different colors (wavelengths) of light on the rate of photosynthesis in a water plant.
(B) The student should not select set B, which would probably be best used to determine the presence of various carbohydrates in foods.
(C) The student should not select set C, which would probably be best used to dissect a biological specimen in order to facilitate a study of gross anatomical features.

5. **4** The corresponding control group would most likely consist of truck drivers who drove *daily for 8 hours in very light traffic*. A control is a setup created by the researcher in a scientific experiment that duplicates the experimental setup in every way except for the variable being tested. Because the variable being tested is the stress of driving for 8-hour periods in heavy traffic, the stress factors that would contrast with these would be either reduced exposure to time driving in heavy traffic (not offered as a choice) or reduced exposure to traffic stresses for the same time duration (choice 4).

Wrong Choices Explained

(1) Truck drivers who drove *daily for 12 hours in very heavy traffic* is not the likely control group for this experiment. Because the experimental group is already considered by the researcher to be stressed with 8 hours in very heavy traffic, the control group would not be set up to receive even more time exposure to this stressful situation.

(2), (3) Truck drivers who drove *every third day for 8 hours in very heavy traffic* or who drove *every other day for 12 hours in very light traffic* are not the likely control groups for this experiment. The introduction of a third variable (intermittent exposure) in these groups would tend to confuse the research and raise questions about which variable of the three is really responsible for the stresses being observed in the experimental group.

6. **3** *Temperature* is a variable in this experiment. A variable is a condition that changes in a scientific experiment either because it is manipulated by the researcher or because it changes in response to the manipulated (experimental) variable. Of the conditions listed, only temperature changes in the experiment. The researcher places each of the five test tubes in a different temperature-controlled environment (0°C, 10°C, 20°C, 30°C, and 40°C), and so temperature is the experimental variable. The other variable in the experiment is the amount of egg white digested in each tube.

Wrong Choices Explained

(1), (2), (4) *Gastric fluid*, *length of glass tubing*, and *time* are not variables (conditions that change) in this experiment. These conditions remain constant throughout the experiment for all five laboratory setups.

7. (1) *The coarse adjustment knob is used to provide general focus on the specimen being viewed.*

(2) *The fine adjustment knob is used to adjust focus on an object more precisely and in finer detail than can be obtained under coarse adjustment.*

(3) *The objective contains optical lenses that permit magnification of the image of the specimen.*

(4) *The diaphragm is used to adjust the amount of light passing through the specimen to the objective.*

8. **1** The student should have focused under *low power using the coarse and fine adjustments, and then under high power using only the fine adjustment*. Following this procedure will allow the student to locate the portion of the specimen to be studied and then to focus on and study a small part of the specimen in detail. A compound light microscope should never be focused with the coarse adjustment when the high-power objective is in place. Even minor adjustments of the coarse adjustment knob under these conditions could crush the specimen and break the slide.

Wrong Choices Explained

(2) The student should not have focused under *high power first, then under low power using only the fine adjustment*. The high power objective should not be used to locate the general area of the specimen to study, but rather the low-power objective should be used first for this purpose.

(3) The student should not have focused under *low power using the coarse and fine adjustments, and then under high power using the coarse and fine adjustments*. A compound light microscope should never be focused with the coarse adjustment when the high-power objective is in place. Even minor adjustments of the coarse adjustment knob under these conditions could crush the specimen and break the slide.

(4) The student should not have focused under *low power using the fine adjustment, and then under high power using only the fine adjustment*. The fine adjustment is designed to move the objective very small distances and is not effective when used on low power unless the coarse adjustment has been used previously to gain general focus at this level of magnification.

9. **4** The diameter of the low-power field of vision is *2,000* micrometers (μm). By definition, a micrometer is one/one millionth (1/1,000,000) of a meter (m). Because there are 1,000 millimeters (mm) in a meter, each millimeter must contain 1,000 micrometers (1,000 mm × 1,000 μm/mm = 1,000,000 μm = 1 m). Because there are 2 millimeters that span the low-power field of view illustrated in this diagram, the total low-power-field diameter must be 2,000 micrometers (2 mm × 1,000 μm/mm = 2,000 μm).

Wrong Choices Explained

(1), (2), (3) The diameter of the low-power field in the illustration is not *1* micrometer, *2* micrometers, or *1,000* micrometers. See explanation above.

10. **2** The diameter of the high-power field of vision in this illustration would be closest to *0.5 mm* (millimeters). The ratio of magnification between the low-power and high-power objectives is 1:4. With the formula below, the high-power-field diameter can be calculated to be one fourth of the 2-mm low-power-field diameter, or 0.5 mm:

$$\frac{\text{Low-power objective magnification } (10\times)}{\text{High-power objective magnification } (40\times)} = \frac{\text{High-power-field diameter } (D \text{ mm})}{\text{Low-power-field diameter } (2 \text{ mm})}$$

$$40D = 2 \text{ mm} \times 10 = 20 \text{ mm}$$

$$D = 20 \text{ mm}/40 \text{ mm} = 1/2 \text{ mm} = 0.5 \text{ mm}$$

Wrong Choices Explained

(1), (3), (4) The diameter of the low-power field is not *0.05 mm*, *5 mm*, or *500 mm*. Each of these answers is a multiple of the correct answer. See explanation on page 267.

11. **2** *Iodine solution* is the substance that, when added to a wet mount containing starch grains, would react with the starch grains and make them more visible. Also known as Lugol's solution, this tan indicator turns blue-black when it contacts a substance containing starch.

Wrong Choices Explained

(1) *Litmus solution* would not react with the starch grains by changing color. Litmus changes color in reaction to acidic and basic solutions, not to solutions containing starch. Litmus turns blue in the presence of bases and red/pink in the presence of acids.

(3) *Distilled water* would not react with the starch grains by changing color. Distilled water is a neutral substance that normally does not react with any substance to produce a color change.

(4) *Bromthymol blue* would not react with the starch grains by changing color. Bromthymol blue is an indicator that turns brick red when heated in the presence of simple sugars such as glucose.

12. **4** The formation of bubbles on slide *A* could have been prevented by *bringing one edge of the coverslip into contact with the water and lowering the opposite edge slowly*. This technique allows air to escape from under the coverslip as it is lowered onto the slide, preventing the formation of air bubbles.

Wrong Choices Explained

(1) *Using a thicker piece of potato and less water* would not prevent the formation of air bubbles on slide *A*. The thicker specimen would tend to suspend the coverslip off the water surface and could actually increase the likelihood of bubble formation.

(2) *Using a longer piece of potato and a coverslip with holes in it* would not prevent the formation of air bubbles on slide *A*. The length of the specimen would not affect the formation of bubbles. However, a coverslip with holes in it would allow air to enter the area between the coverslip and slide, leading to the formation of air bubbles.

(3) *Holding the coverslip parallel to the slide and dropping it directly onto the potato* would not prevent the formation of air bubbles on slide *A*. This technique would not allow trapped air to escape from under the coverslip, increasing the likelihood that air bubbles will form and remain trapped.

13. **3** *It allows the stain to penetrate the potato tissue without the removal of the coverslip* is the statement that best describes the purpose of the procedure. This technique takes advantage of capillary forces to pull the liquid stain under the coverslip as water is being removed by the absorbent paper on the opposite side of the coverslip. When the stain comes into contact with the potato tissue, some of it is absorbed by the cells of the tissue, staining it for study.

Wrong Choices Explained

(1) *It prevents the stain from getting on the ocular of the microscope* is not the statement that best describes the purpose of the procedure. The ocular lens should never be exposed to chemical stains, as a stain can coat the lens, obscuring visual observation of the specimen by the researcher. This procedure should be conducted away from the proximity of the ocular lens.

(2) *It prevents the water on the slide from penetrating the potato tissue* is not the statement that best describes the purpose of the procedure. By drawing the stain under the coverslip, this technique helps to ensure that this penetration will occur.

(4) *It helps to increase the osmotic pressure of the solution* is not the statement that best describes the purpose of the procedure. Although the active ingredient of many stains is a form of salt, and although this salt can have the effect of altering the osmotic gradient in a system, the main purpose of the procedure is to introduce the stain into proximity of the tissues to be stained, not to increase the osmotic pressure of the solution.

14. **1** *Onion epidermal cells* most likely make up the tissue represented in the drawing. The blocklike arrangement of the cells and the wall-like divisions between them are typical of a plant tissue. Iodine stain would tend to darkly stain the starch in vacuoles and the cellulose in the cell walls of these cells.

Wrong Choices Explained

(2) *Ciliated protists* do not most likely make up the tissue represented in the drawing. Most ciliated protests are not colonial but free-living, whereas this tissue clearly shows a structure of contiguous blocks. Cilia are not apparent on any of the cells in the drawing.

(3) *Cardiac muscle cells* do not most likely make up the tissue represented in the drawing. Such cells are shaped like interlocking spindles and show a prominent striped pattern within each cell. Muscle cells lack cell walls, whereas the cells in the diagram clearly contain cell walls.

(4) *Blue-green algae* do not most likely make up the tissue represented in the drawing. Blue-green algae are among the most primitive of life forms, lacking organized nuclei. Structure *A* in the diagram is most likely an organized nucleus.

15. **4** The organelle labeled *B* in the diagram is most likely a *cell wall*. The thick, well-defined, rigid cell boundary illustrated in the diagram is typical of the cell walls of typical plant cells.

Wrong Choices Explained

(1) The organelle labeled *B* is not likely a *mitochondrion*. Mitochondria are too small to be clearly visible under typical laboratory high-power (400×–430×) magnification.

(2) The organelle labeled *B* is not likely a *centriole*. Centrioles are too small to be clearly visible under typical laboratory high-power (400×–430×) magnification. Centrioles are rod-shaped and are found only in animal cells.

(3) The organelle labeled *B* is not likely a *lysosome*. Lysosomes are specialized vacuoles that contain digestive enzymes used to break down complex molecules into small, soluble molecules. Lysosomes are irregularly spherical in shape.

16. **1** *The tissue is composed of more than one layer of cells* is the best conclusion to be made from these observations. In diagram *A*, the microscope is focused on the cell walls and nuclei of one cell layer, whereas the second layer remains out of focus in the foreground. As the fine adjustment knob is turned slowly, the second layer is brought into sharp focus, while the first layer is put out of focus in the background. This is illustrated in diagram *B*.

Wrong Choices Explained

(2) *The tissue is composed of multinucleated cells* is not the best conclusion to be made from these observations. A characteristic of a multinucleated cell is that it contains two or more nuclei. None of the cells in the diagrams have more than one nucleus.

(3) *The cells are undergoing mitotic cell division* is not the best conclusion to be made from these observations. In tissues undergoing mitotic cell division, additional cells are created. The diagrams do not show additional cells being created.

(4) *The cells are undergoing photosynthesis* is not the best conclusion to be made from these observations. The cells in the diagrams may or may not be carrying on photosynthetic activity, but this activity would not be responsible for the different views illustrated in diagrams *A* and *B*.

17. **1** The presence of carbon dioxide would be indicated by *a color change in the solution*. Bromthymol blue is blue in a slightly basic solution but turns yellow in a slightly acidic solution. When the student exhales through a straw into the bromthymol blue solution, carbon dioxide molecules in the exhaled air combine with water to form weak carbonic acid. As the acidity of the solution increases, the indicator reacts by changing from blue to yellow.

Wrong Choices Explained

(2) *A change in atmospheric pressure* would not indicate the presence of carbon dioxide in the exhaled air. This phenomenon occurs for meteorological reasons independent of the laboratory.

(3) *The formation of a precipitate in the solution* would not indicate the presence of carbon dioxide in the exhaled air. This is not a characteristic of the indicator bromthymol blue.

(4) *The release of bubbles from the solution* would not indicate the presence of carbon dioxide in the exhaled air. Bubbles of any gas that is less dense than water will rise to the surface when released in the water. This is not a characteristic of the indicator bromthymol blue.

18. **3** *Benedict's solution* is the indicator used to detect the presence of glucose and other simple sugars. When heated in the presence of glucose, blue Benedict's solution changes to brick red.

Wrong Choices Explained

(1) *Iodine solution* would not indicate the presence of glucose in the container. Iodine (Lugol's) solution is used to detect the presence of starch in foods.

(2) *Bromthymol blue* would not indicate the presence of glucose in the container. Bromthymol blue is used to detect the presence of carbon dioxide in solution.

(4) *pH paper* would not indicate the presence of glucose in the container. pH paper is used to detect the presence of acids or bases.

19. **2** The total volume of water indicated in the diagram is *11 mL* (milliliters). When reading the volume of liquid in a graduated cylinder, the observer should note the mark that corresponds to the bottom of the meniscus. In the diagram, the meniscus falls on the 11-mL mark.

Wrong Choices Explained

(1), (3), (4) The total volume indicated is not *10 mL, 12 mL,* or *13 mL.* Each of these readings falls either below or above the 11-mL mark indicated at the bottom of the meniscus in the diagram.

20. *8.9 cm* (or *89 mm* or *0.089 m*) In the illustration, the earthworm is laid out along the metric ruler so that the proboscis is positioned at the 2.0-cm mark. The posterior of the specimen is shown to extend to the 10.9-cm mark on the ruler. Subtracting 2.0 cm from 10.9 cm provides the specimen's total length of 8.9 cm.

21. **1** *Micrometer, millimeter, centimeter, meter* is the group of measurement units correctly arranged in order of increasing size. A meter contains 1,000,000 micrometers, 1,000 millimeters, or 100 centimeters.

Wrong Choices Explained

(2), (3), (4) The groups *millimeter, micrometer, centimeter, meter; meter, micrometer, centimeter, millimeter;* and *micrometer, centimeter, millimeter, meter* do not correctly arrange these units in order of increasing size. See explanation on page 271.

22. **4** The *nerve cord* is ventral to the esophagus in earthworms and grasshoppers. The ventral surface of a bilaterally symmetrical animal is the surface normally oriented downward when the animal is in a horizontal position (except in humans and other upright-walking animals where it is located on the "front" of the body). This question requires a knowledge of the anatomy of these organisms gained through dissection or other means.

Wrong Choices Explained

(1) The *gizzard* of the earthworm and grasshopper is located posterior, not ventral, to the esophagus.

(2) The *brain* of the earthworm and grasshopper is located anterior and dorsal, not ventral, to the esophagus.

(3) The *intestine* of the earthworm and grasshopper is located posterior, not ventral, to the esophagus.

23. *Anther* The stamen is the male part of the flower and is made up of the filament and the anther. The anther contains tissues that produce pollen. This question requires a knowledge of the anatomy of flowers gained through dissection or other means.

24. A scientifically accurate, complete-sentence response is required. Examples of acceptable responses include:

- *This structure functions in the formation of pollen.*
- *The anther contains tissues that produce pollen.*

25. **2** The student should *discontinue heating and report the defect to the instructor.* A crack in any piece of laboratory glassware is a sign of imminent failure. If the beaker had shattered while the student was near it, she could have been spattered and burned by the hot water in the beaker and burned or cut by hot broken glass. The instructor should remove and discard the beaker from the student's lab station and replace it with a new one for her use.

Wrong Choices Explained

(1), (3), (4) The student should not *discontinue heating and attempt to seal the crack, discontinue heating and immediately take the beaker to the instructor,* or *continue heating as long as fluid does not seep from the crack.* These actions are unsafe for both the student and others in the lab-

272

oratory, as each could result in injury from hot broken glass and hot water.

· 26. **3** *Rinse the eyes with water; then notify the teacher and ask for advice* is the proper laboratory procedure to follow if some laboratory chemical splashed into a student's eyes. Because of the sensitivity of the eyes and skin surface to chemical burns, the student should not wait for direction from the teacher but should immediately flood the exposed areas with clean water. Each school science laboratory should have an eyewash station for this purpose. The student should then immediately report the incident to the teacher. The teacher in turn should immediately refer the student to the school nurse for examination and treatment as needed.

Wrong Choices Explained

(1) The student should not *send someone to find the school nurse* before thoroughly rinsing the exposed area and informing the teacher of the incident. This delay could cause severe damage to the eyes that could result in blindness.

(2) The student should not *rinse the eyes with water and do not tell the teacher because he or she might become upset*. Chemical exposures should always be treated by qualified medical personnel. If the exposed tissues are not thoroughly rinsed and examined, the chemical could cause severe damage to the eyes that could result in blindness.

(4) The student should not *assume that the chemical is not harmful and no action is required*. Until it is determined otherwise, all chemicals used in the laboratory should be considered dangerous. Chemical exposures should always be treated by qualified medical personnel. If the exposed tissues are not thoroughly rinsed and examined, the chemical could cause severe damage to the eyes that could result in blindness.

27. **4** When heating a liquid in a test tube, the student should always keep the mouth of the tube *aimed away from everybody*. Doing so will minimize the risk that someone will be injured by the hot liquid in the test tube. If heated too quickly, the liquid in the tube could splash or boil from the open end of the tube and burn anyone close enough to be spattered by it.

Wrong Choices Explained

(1) The student should not keep the mouth of the tube *corked with a rubber stopper*. This action could result in the formation of hot, high-pressure gases inside the test tube that could eject the stopper at high speed or shatter the tube. Either of these events could cause injury to the student or someone standing nearby.

(2) The student should not keep the mouth of the tube *pointed toward the student*. This action may result in injury to the student. If heated too

quickly, the liquid in the tube could splash or boil from the open end of the tube and burn the student.

(3) The mouth of the tube should not be *allowed to cool*. Good laboratory practice suggests that the student should heat the tube evenly along its entire length so as to avoid uneven expansion of the glass walls of the tube that could result in cracking. This practice also helps to ensure that the liquid in the tube will heat evenly, minimizing the probability of uncontrolled boiling and splashing.

28. One credit is allowed for appropriately labeling column III of the chart. Acceptable responses include:

    - *Height of Bean Plants (cm)*
    - *Bean Plant Height in Centimeters*

29. One credit is allowed for appropriately organizing data on the chart. Students should complete the chart so that the values entered in column II are arranged in increasing order from top to bottom and that the corresponding bean plant measurement is entered in column III.

DATA TABLE

I	II	III
Group	Concentration of Substance X (%)	Height of bean plants (cm)
B	0	28.7
C	2	29.4
E	4	31.5
A	6	32.3
D	8	37.1
F	10	30.7

30. One credit is allowed for marking an appropriate scale on the axis labeled "Number of Members of Each Population." Students should be sure to start numbering from zero at the lower left corner of the grid and enter values along the vertical axis in equal increments up to 150, a value that just exceeds the highest recorded moose population.

31. One credit is allowed for marking an appropriate scale on the axis labeled "Year." Students should be sure to start numbering from 1970 at the lower left corner of the grid and enter values along the horizontal axis in equal increments up to 1980.

32. One credit is allowed for correctly plotting the data for the wolf population, surrounding each point with a small triangle, and connecting the points.

33. One credit is allowed for correctly plotting the data for the moose population, surrounding each point with a small circle, and connecting the points.

34. **4** According to the chart, *lake trout* can tolerate the highest level of acidity in their water environment. The chart shows that the survival rate for lake trout is strong at pH values ranging from 7.0 (neutral) to 5.5 (moderately acidic). At a pH value of 5.5, the survival rate for this species begins to decrease; at pH values below 5.0, the survival rate drops off drastically and reaches zero at a pH value of approximately 4.4.

Wrong Choices Explained
   (1) *Mussels* do not tolerate the highest levels of acidity in their water environment. The chart shows that survival rates for this species begin to decrease at a pH value of 6.5 and reach zero at a pH value of approximately 5.9.
   (2) *Smallmouth bass* do not tolerate the highest levels of acidity in their water environment. The chart shows that survival rates for this species begin to decrease at a pH value of 5.8 and reach zero at a pH value of approximately 5.3.
   (3) *Brown trout* do not tolerate the highest levels of acidity in their water environment. The chart shows that survival rates for this species begin to decrease at a pH value of 5.5 and reach zero at a pH value of approximately 4.9.

35. **1** According to the chart, *mussels* would have been the first species eliminated from these lakes during the period from 1880 to 1980. The chart shows that survival rates for this species begin to decrease at a pH value of 6.5 and reach zero at a pH value of approximately 5.9. Assuming that lake acidification occurred at a constant rate throughout the period, by about 1930 a pH value of 5.9 would have been reached and mussels would have been eliminated from these lakes.

Wrong Choices Explained

(2) *Smallmouth bass* would not have been the first species to be eliminated from these lakes. The chart shows that survival rates for this species reach zero at a pH value of approximately 5.3. Assuming that lake acidification occurred at a constant rate throughout the period, by about 1960 a pH value of 5.3 would have been reached and smallmouth bass would have been eliminated.

(3) *Brown trout* would not have been the first species to be eliminated from these lakes. The chart shows that survival rates for this species reach zero at a pH value of approximately 4.9. Assuming that lake acidification occurred at a constant rate throughout the period, by about 1980 a pH value of 4.9 would have been reached and brown trout would have been eliminated.

(4) *Lake trout* would not have been the first species to be eliminated from these lakes. The chart shows that survival rates for this species reach zero at a pH value of approximately 4.4. As of 1980, the pH value in these lakes was approximately 4.9, and lake trout had presumably not yet been eliminated.

36. **2** The total change in the pH value of rainwater between 1880 and 1980 was *1.7*. In 1880, the pH value was approximately 5.7; in 1980, the pH value was approximately 4.0. The numerical difference between these values is found by carrying out the mathematical operation $5.7 - 4.0 = 1.7$.

Wrong Choices Explained

(1), (3), (4) The total change in the pH value of rainwater between 1880 and 1980 was not *1.3, 5.3,* or *9.7*. These changes are incorrect according to the method described above.

37. **3** The total change in mass of the apple piece in the 10% sugar solution was *a decrease of 0.30 gram*. We assume that the natural sugar concentration of the apple slices is approximately 3%, a point at which the gain/loss in mass is at 0.0 gram on the graph. At a sugar concentration of 10%, the mass change shown on the graph is –0.3 gram.

Wrong Choices Explained

(1) A mass change is not shown as *a decrease of 0.45 gram*. This degree of mass change is indicated on the graph as corresponding to a sugar concentration of approximately 15%.

(2) A mass change is not shown as *an increase of 0.30 gram*. This degree of mass change is indicated on the graph as corresponding to a sugar concentration of approximately 1%.

(4) A mass change is not shown as *an increase of 0.10 gram*. This degree of mass change is not indicated on the graph as corresponding to

any specific sugar concentration but can be inferred to occur at a sugar concentration of approximately 2%.

38. **3**   The apple slices should neither lose nor gain weight at a sugar concentration of approximately *3%*, a point at which the gain/loss in mass is at 0.0 gram on the graph. This value can be determined by drawing a curved line connecting the points on the graph and approximating the sugar concentration at the point where this line crosses the 0.0 mass change line. ·

Wrong Choices Explained
(1), (2), (4) According to the data shown in the graph, at sugar concentrations of *6%, 10%,* and *20%* the apple slices will lose total mass. All these values exceed the equilibrium concentration of 3%.

39. **3**   Based on the graph, it can be correctly concluded that the pituitary extract *affects the growth of rats* over a 200-day period. The graph shows data on their growth in grams of two groups of rats: one group (the control group) untreated and the other group (the experimental group) treated with anterior pituitary extract. The two groups of rats appear to grow at the same rate until day 70, when the treated (experimental) group begins to show an increased growth rate compared to the untreated (control) group.

Wrong Choices Explained
(1) It cannot be correctly concluded that the pituitary extract *is essential for life*. It is obvious from the data that the untreated rats have survived and grown over the 200-day period of this experiment, a fact indicating that the lack of extract does not lead to the death of these individuals.

(2) It cannot be correctly concluded that the pituitary extract *determines when a rat will be born*. There is no data on the graph concerning a measurement of the birthrates of rats.

(4) It cannot be correctly concluded that the pituitary extract *affects the growth of all animals*. Only rats, not "all animals," are being tested in this experiment. Further experiments involving other animal species may be conducted to provide this information, but the data from this experiment provides information only about the effect of anterior pituitary extract on rats.

40. **4**   Based on the graph, it can be correctly concluded about plant hormones that *they stimulate maximum root and stem growth at different concentrations*. The graph shows data on the effect of relative plant hormone concentration on the relative degree of growth of roots and stems. The two types of plant tissues respond differently to the same concentrations of hormones. In general, root tissues are stimulated by low con-

centrations of hormones and inhibited by high concentrations of hormones, whereas stem tissues are stimulated by moderately high hormone concentrations and inhibited by very low and very high concentrations of hormones.

Wrong Choices Explained

(1) It cannot be correctly concluded that *they stimulate maximum root growth and stem growth at the same concentration*. It is obvious from the data that the relative degree of stimulation or inhibition is different in roots than in stems at any particular concentration (other than at the point where the graph lines cross).

(2) It cannot be correctly concluded that *they stimulate maximum stem growth at low concentrations*. The data in the graph show that moderately high hormone concentrations (up to a point) are needed to stimulate the highest degree of stem growth.

(3) It cannot be correctly concluded that *they most strongly inhibit root growth at low concentrations*. The data in the graph show that relatively low hormone concentrations are needed to stimulate the highest degree of root growth.

41. **4** Based on the data, it can be correctly concluded that *as temperature increases from 15°C to 30°C, the number of compound eye sections in male Drosophila with bar-eyes decreases*. The chart presents data on the effect of incubation temperature on the number of compound eye sections in male bar-eye *Drosophila*. The data show that, as the incubation temperature increases from 15°C to 30°C, the number of eye sections decreases from 270 to 74 in males of this specific species and variety.

Wrong Choices Explained

(1) *The optimum temperature for culturing Drosophila is 15°C* is not the best conclusion to be drawn from analysis of the data. The chart contains no general information concerning the effect of temperature on the culturing of this or other varieties of *Drosophila*.

(2) *Drosophila cultured at 45°C will show a proportionate increase in the number of compound eye sections* is not the best conclusion to be drawn from analysis of the data. The chart contains no information concerning the effect of an incubation temperature of 45°C on the number of eye sections in this or other varieties of *Drosophila*.

(3) *Temperature determines eye shape in Drosophila* is not the best conclusion to be drawn from analysis of the data. The chart contains no information concerning the effect of incubation temperature on the expression of eye shape in varieties of *Drosophila* other than bar-eye.

42. **2** Based on the diagram, it can be correctly concluded that *nuclei are more dense than mitochondria*. At the very high spin rates of an ultra-centrifuge, suspended organelles from broken cells are forced to separate

into layers based on their relative densities. The more dense the organelle, the lower in the tube it concentrates. The diagram shows nuclei concentrated in a lower layer than mitochondria, indicating that nuclei are more dense than mitochondria.

Wrong Choices Explained

(1), (3) *Ribosomes are more dense than mitochondria* and *mitochondria and ribosomes are equal in density* are not the best conclusions to be drawn from analysis of the data. The diagram shows that ribosomes concentrate in a higher layer than mitochondria, indicating that ribosomes are less dense than mitochondria.

(4) *The cell consists of only solid components* is not the best conclusion to be drawn from analysis of the data. The diagram shows that the highest (least dense) layer in the ultracentrifuge tube contains cell fluid, indicating that cells contain a liquid component.

Question Set on Reading, Writing, and Current Events in Science (p. 63)

1. **3** *There is not sufficient information given in the paragraph.* The first section of the paragraph mentions that scientists are now able to splice human genes into bacteria, but there is no indication of the technical difficulties involved in this process.

2. **1** *This statement is true according to the paragraph.* The first sentence of the second section of the paragraph states that "bacteria reproduce rapidly under certain conditions."

3. **1** *This statement is true according to the paragraph.* The second sentence of the third section of the paragraph states that "continued use of [animal] insulin can trigger allergic reactions in some humans."

4. **2** *This statement is false according to the paragraph.* The first sentence of the fourth section of the paragraph states that "the bacteria used for these experiments are . . . common to the digestive systems of many humans."

5. **3** *There is not sufficient information given in the paragraph.* The fourth section of the paragraph mentions that scientists have used *E. coli* for these experiments, but no information is given concerning the use or nonuse of other bacterial species.

6. One credit is allowed for a scientifically correct, complete-sentence response that describes a difference between gradualism and punctuated equilibrium. Acceptable responses include:

- *Gradualism proposes that evolutionary change is continuous and slow, whereas punctuated equilibrium proposes that species undergo long periods without change and then undergo considerable change quickly.*
- *The theory of gradualism involves slow, gradual change, whereas the theory of punctuated equilibrium involves rapid, infrequent change.*
- *Gradualism is supported by the presence of transitional fossils, and punctuated equilibrium is supported by the lack of transitional fossils.*

7. One credit is allowed for a scientifically correct, complete-sentence response that states what may result from the accumulation of small variations. Acceptable responses include:

- *Gradualism theorizes that new species evolve as a result of the accumulation of many small genetic changes.*
- *The theory of gradualism states that the accumulation of many small genetic changes will eventually result in the formation of new species.*

8. One credit is allowed for a scientifically correct, complete-sentence response that states what fossil evidence supports the theory of gradualism. Acceptable responses include:

- *Gradualism is supported by fossil evidence that shows transitional forms.*
- *The theory of gradualism is supported by fossil evidence that includes transitional forms within an evolutionary line.*

9. Three credits are allowed for listing three human structures and describing in a scientifically correct response the structure's adaptive value to the human body. Acceptable responses include:

- *SWEAT GLAND—This structure regulates body temperature by releasing moisture from the blood to the skin surface where it can evaporate.*
- *PANCREAS—This organ produces digestive enzymes used in the digestive function.*
- *LIVER—This organ filters the blood of dead blood cells and toxic materials.*
- *EPIGLOTTIS—This structure prevents swallowed food from entering the trachea.*
- *CAPILLARY—This structure enables the blood to reach all body tissues.*
- *VILLUS—This structure permits the absorption of digested food from the intestine into the bloodstream.*
- *KIDNEY—This organ filters certain toxic materials from the blood and excretes them from the body.*
- *PLATELET—This structure promotes blood clotting by releasing enzymes into the blood.*

10. **4**   Acid rain has a pH closest to that of *vinegar*. The second paragraph of the selection states that acid rain has a pH value between 3.0 and 5.0. The scale presents data on the pH of several common substances, one of which is vinegar with a pH value of approximately 3.0.

Wrong Choices Explained
    (1) *Ammonia* does not have a pH closest to that of acid rain. According to the scale, ammonia has a pH value of approximately 13.0.
    (2) *Tap water* does not have a pH closest to that of acid rain. According to the scale, tap water has a pH value of approximately 6.5.
    (3) *Baking soda* does not have a pH closest to that of acid rain. According to the scale, baking soda has a pH value of approximately 8.2.

11. **2**   The most likely source of acid rain in New York State is the *midwestern United States*. The first paragraph of the selection states that this acid rain's principal sources are smokestack gases released by coal-burning facilities located in the midwestern states.

Wrong Choices Explained
    (1), (3), (4) The areas of the *far western United States, far eastern Canada*, and *far western Europe* are not the most likely sources of acid rain in New York State. Although smokestack gases are produced in these regions to a greater or lesser extent than in the midwestern United States, the prevailing westerly winds would be unlikely to carry them from these regions to the atmosphere over New York State.

12. **2**   The food chain *algae → aquatic insect → trout → otter* would most immediately be affected by acid rain. These organisms all live in or near water and depend on clean, unpolluted water for survival.

Wrong Choices Explained
    (1), (3), (4) The food chains *grass → rabbit → fox → decay bacteria, shrub → mouse → snake → hawk*, and *tree → caterpillar → bird → lynx* are all terrestrial food chains. Although all would eventually be affected by acid rain, they would not be the most immediately affected.

13. **3**   Acid rain can most appropriately be characterized as a type of *technological oversight*. The technologies of steam and electrical power generation have had a positive effect on our economy and the growth of our nation but have had an unintended negative impact on the natural environment. The gases produced by coal and oil-fueled power plants and factories have become a problem in that the acid precipitation that results has killed fish and aquatic insects that form an important part of the aquatic ecosystem. The destruction of these ecosystems in turn has profound negative implications for human health and survival.

Wrong Choices Explained

(1) Acid rain is not most appropriately characterized as a type of *biological control*. A biological control is an environmentally sound method for controlling unwanted insects and other pests by using their natural biological enemies to exert this control.

(2) Acid rain is not most appropriately characterized as *conservation of resources*. Conservation of resources involves using the minimum amount of our scarce natural resources, as well as recycling materials to reclaim these resources for reuse.

(4) Acid rain is not most appropriately characterized as a type of *land-use management*. Land-use management involves using the minimum amount of our dwindling land resources, preserving open space and arable land for enjoyment by future generations.

14. **3** A strain of fish that could survive under conditions of increased acidity could best be obtained by *selective breeding*. Fish of all species are organisms that reproduce by means of sexual reproduction, the foundation of selective breeding techniques. These techniques allow breeders to mate pairs of organisms that display the traits desired for breeding (in this case, fish with a tolerance for elevated levels of acidity). The offspring of these breeding pairs have an increased likelihood of displaying the desired trait and forming a self-reproducing variety of the species over time.

Wrong Choices Explained

(1), (2), (4) *Binary fission*, *vegetative propagation*, and *budding* are not reproductive processes that would be likely to produce a strain of fish that could survive under conditions of increased acidity. These reproductive processes are all examples of asexual reproduction that are not used by any species of fish as its primary method of reproduction. In addition, as a reproductive process, asexual reproduction does not normally result in a wide variety of traits. Rather, it tends to maintain genetic traits intact and unchanging over many generations.

Question Set 2.1 (p. 77)

1. **1** The human digestive system is characterized by *A, B,* and *C*. Peristalsis (*A*) is a wave of muscular contractions that moves food through the food tube. In humans, this occurs in the esophagus, stomach, and intestines. In humans, smooth (involuntary) muscle tissues (*B*) are responsible for peristaltic movement. Accessory organs (*C*) present in humans include the liver, gallbladder, and pancreas, which are not part of the food tube but that produce or store substances important to the digestive process.

Wrong Choices Explained

(2), (3), (4) The combinations of characteristics *A only*, *B only*, and *B and C only* do not occur in humans or in other commonly studied animals.

2. **4** *It has two atria and two ventricles, and it pumps blood directly into arteries* is a correct statement about the human heart. The human heart, like that of other mammal species, is a four-chambered vessel whose design helps to separate oxygenated and deoxygenated blood. These chambers include two thin-walled atria, which receive blood from veins, and two muscular ventricles, which pump blood out to the body through arteries.

Wrong Choices Explained

(1) *It has two atria and one ventricle, and it pumps blood directly into veins* is not a correct statement about the human heart. A three-chambered heart is characteristic of amphibians such as frogs but not humans. Blood is not pumped into veins from the heart.

(2) *It has one atrium and one ventricle, and it is composed of cardiac muscle* is not a correct statement about the human heart. Cardiac muscle is the muscle type found in human blood vessels, but the heart in humans is not two-chambered.

(3) *It has one atrium and two ventricles, and it is composed of visceral muscle* is not a correct statement about the human heart. This is a nonsense distracter since no known organism has this combination of heart chambers. Visceral muscle is not found in the human heart.

3. **4** *Filtering of bacteria from the lymph* is a process that would be affected by a malfunction of the lymph nodes. This process is one of the essential functions of the lymphatic system.

Wrong Choices Explained

(1), (2), (3) These are all nonsense distracters, since none of these processes is carried out in the lymph. *Release of carbon dioxide* and *oxygen* are associated with the respiratory system. *Filtering of glucose* may be carried out by the excretory system under certain circumstances.

4. **4** Of the choices given, *decreased consumption of complex carbohydrates* (for example, starches) is least likely to have the effect of promoting heart attack. By decreasing the consumption of starchy foods, there may even be a positive effect on the body's ability to ward off heart disease.

Wrong Choices Explained

(1), (2), (3) *An increase in arterial blood pressure*, *oxygen deprivation of cardiac muscle*, and *narrowing of the arteries transporting blood to the heart muscle* are all potential contributors to heart disease and may help to set up the conditions for heart attack in certain people. All place additional strain onto the heart muscle, making failure under stress more likely.

5. **1** Structure *A* (mouth) is the site in which the initial hydrolysis of carbohydrates occurs. Saliva secreted into the mouth contains digestive enzymes that hydrolyze starch into disaccharides.

Wrong Choices Explained
(2) Structure *E* (small intestine) is the organ where the digestion of all food types is completed and the end products of digestion are absorbed into the blood.
(3) Structure *C* (stomach) is the site in which the chemical digestion of protein begins.
(4) Structure *D* (pancreas) is an accessory organ that produces a variety of digestive juices important for the complete hydrolysis of carbohydrates, proteins, and lipids.

6. **4** Structure *E* (small intestine) is the organ where the digestion of all food types is completed and the end products of digestion are absorbed into the blood.

Wrong Choices Explained
(1) Structure *F* (large intestine) is where solid wastes accumulate and water is reabsorbed.
(2) Structure *H* (liver) is an accessory organ that manufactures bile and stores glycogen.
(3) Structure *C* (stomach) is the site in which the chemical digestion of protein begins.

7. **3** Structure *C* (stomach) is the site in which the chemical digestion of proteins begins.

Wrong Choices Explained
(1) Structure *G* (gallbladder) is the site where bile is stored after it is manufactured by the liver.
(2) Structure *B* (esophagus) is the structure that moves food from the mouth to the stomach.
(4) Structure *E* (small intestine) is the organ where the digestion of all food types is completed and the end products of digestion are absorbed into the blood.

8. **1** *An artery*, with its muscular walls, is the blood vessel most likely to produce a detectable pulse. The pulse is a wave of muscular contractions that help blood to move through the circulatory system.

Wrong Choices Explained

(1) *A vein*, by lacking muscular walls, is not likely to produce a detectable pulse. Veins serve as a return system for blood to the heart.

(2) *A capillary* is only one-cell thick and contains blood under very low pressure. Pulses would not be detected in the capillary.

(3) *A lacteal* is a microscopic extension of the lymphatic system found within a villus in the small intestine. There is no blood pressure in a lacteal.

9. **2** Structures *B, F, and I* are those most likely to contain deoxygenated blood in the human heart. Deoxygenated blood enters the heart from the body in structure *I* (left atrium) and then passes to structure *F* (left ventricle), where it is pumped through structure *B* (pulmonary artery) to the lungs.

Wrong Choices Explained

(1) Structures *A, B, and C* is not the correct choice. Structure *A* is the aorta, which carries oxygenated blood to the body. Structures *C* are the pulmonary veins, which carry oxygenated blood from the lungs to the heart.

(3) Structures *C, D, and E* is not the correct choice. Structures *C* are the pulmonary veins, which carry oxygenated blood from the lungs to the heart. Structure *D* is the right atrium, which receives oxygenated blood from the pulmonary veins. Structure *E* is the right ventricle, which pumps oxygenated blood into the aorta.

(4) Structures *D, H, and I* is not the correct choice. Structure *D* is the right atrium, which receives oxygenated blood from the pulmonary veins. Structure *H* is the vena cava, which carries deoxygenated blood from the body to the heart.

10. **1** Structure *G* is the left atrioventricular (A-V) valve that opens when blood passes from structure *I* to structure *F*, but then closes to prevent the blood from flowing back into structure *I* (left atrium).

Wrong Choices Explained

(2) Structure *B*, the aorta, is the major artery leading from the heart to all parts of the body. It does not control the backflow of blood into an atrium.

(3) Structures *C* are the pulmonary veins. These blood vessels carry oxygenated blood from the lungs to the heart and do not control the backflow of blood into an atrium.

(4) Structure *H* is the vena cava, which carries deoxygenated blood from the body to the heart. It is not involved in the control of backflow of blood.

11. **4** *Lymph vessels* have specialized regions, known as nodes, where bacteria are attacked by phagocytic white blood cells and removed from the bloodstream.

Wrong Choices Explained
(1) *Arteries* are muscular blood vessels specialized for moving blood from the heart to the body extremities and internal organs.
(2) *Capillaries* are microscopic blood vessels that connect arteries to veins. Most molecular exchange occurs in the capillaries.
(3) *Veins* are valved blood vessels that carry blood back to the heart from all around the body.

Question Set 2.2 (p. 83)

1. **4** This question requires a knowledge of the structural design of the human respiratory system. Air passing out of the body during exhalation would start in the alveoli of the lungs and then proceed in order through the bronchioles, bronchi, trachea, *pharynx*, and oral/nasal cavity to the environment.

Wrong Choices Explained
(1), (2), (3) These distracters all reference parts of the human respiratory system. However, during exhalation, air passes through the *alveoli*, *bronchioles*, and *bronchi* before reaching the trachea.

2. **1** In humans, blood is cooled as it passes through capillaries surrounding *sweat glands* in the skin. Heat (along with excess salt and water) is absorbed from the blood and transferred to the glands as sweat. As sweat droplets evaporate from the skin surface, heat dissipates with it and the body cools.

Wrong Choices Explained
(2) The *nephron* is the functional unit of the human kidney. Although it is associated with removal of materials (urea, water, and salts) from the blood, it is not directly involved in temperature regulation.
(3) The *liver* is involved with filtering the blood; it is responsible for removing worn-out red blood cells and recycling their components in the body. The liver also removes a variety of toxic substances, including alcohol, from the blood. The liver is not directly involved in temperature regulation.
(4) The *urinary bladder* is responsible for collecting and storing urine passing from the kidneys. It is not directly involved in temperature regulation.

3. **2** This question requires a knowledge of the structural and functional design of the human excretory system. In humans, the urethra (*X*) is responsible for the *transport of urine out of the body* to the environment.

Wrong Choices Explained

(1) The *filtration of cellular wastes from the blood* is performed by the paired kidneys, shown at the top of the diagram.

(3) The *storage of urine* is a function performed by the urinary bladder, shown immediately above the urethra in the diagram.

(4) The *secretion of hormones* is not a function performed by the human excretory system. However, the adrenal glands, which are a part of the human regulatory system, are located near the kidneys (not shown in the diagram).

4. **2** The region labeled *X* is most likely an *alveolus* within the human lung. The diagram shows oxygen (O_2) diffusing into the blood from the alveolus and carbon dioxide (CO_2) diffusing out of the blood and passing into the alveolus.

Wrong Choices Explained

(1) *A glomerulus* is a portion of the nephron within the human kidney. Glomeruli are not associated with gas exchange in humans.

(3) *A villus* is a microscopic projection of the inner lining of the human small intestine. Villi are not associated with gas exchange in humans.

(4) *The liver* is an organ in humans specialized for the removal and storage of toxic materials from the blood. The liver is not associated with gas exchange in humans.

5. **1** *Anaerobic respiration in muscle cells, forming lactic acid* is a condition arising from the continual use of muscles during prolonged physical exercise such as that encountered in a marathon. In the absence of sufficient oxygen, the muscle cells carry on lactic acid fermentation to produce energy. The resulting lactic acid buildup produces a sensation of pain in the muscle, commonly called muscle fatigue.

Wrong Choices Explained

(2), (3), (4) *Aerobic respiration in muscle cells, generating glycogen*; *anaerobic respiration in liver cells, producing glucose*; and *aerobic respiration in liver cells, synthesizing alcohol* are nonsense distracters. No type of respiration in humans produces glycogen, glucose, or alcohol.

6. **4** The *nasal cavity* in humans is lined with a ciliated mucous membrane that warms, moistens, and filters air. The cilia move dust-laden mucus to

the outside, while tissues warm and moisten the air before it enters the lung interior.

Wrong Choices Explained

(1) The *pharynx* is a cavity at the back of the throat, where the mouth cavity and the nasal cavity join. No ciliated tissues are in the pharynx.

(2) The *alveolus* is the respiratory sac in the lung where gas exchange occurs between the environment and the blood. The air reaching the alveolus must already be warmed and moistened to prevent drying of the lung tissues.

(3) The *epiglottis* is a small flap of tissue that closes the trachea while food is being swallowed. The epiglottis helps to prevent choking.

7. **3** The ureter transports urine from the *kidney to the urinary bladder*. From the urinary bladder, the urine passes through the urethra to the environment.

Wrong Choices Explained

(1) From the *blood to the kidney* is incorrect. This is accomplished in the nephron within the kidney.

(2) From the *liver to the kidney* is incorrect. This is accomplished only by the interaction of the bloodstream, as the waste products of protein deamination are removed for filtering in the kidney.

(4) From the *urinary bladder to outside the body* is incorrect. This is accomplished by way of the urethra.

8. **2** *Constriction of the bronchial tubes and wheezing* are symptoms commonly associated with asthma. In its acute state, asthma can be life threatening, as the victim's breathing tubes narrow and close.

Wrong Choices Explained

(1) *Enlargement and degeneration of the alveoli* may be a symptom of the human disorder known as emphysema.

(3) *Inflammation and swelling of the epiglottis* is not associated with any known human disorder.

(4) *Constriction of the nasal cavity and watery eyes* may be a symptom of an allergic reaction in humans.

9. **4** Structure *D*, the diaphragm, is a sheet of muscle tissue that contracts, causing a pressure change in the chest cavity during breathing. When the diaphragm presses downward, pressure in the chest cavity is reduced and air is forced into the lungs by atmospheric pressure. When the diaphragm presses upward, pressure in the chest cavity is increased and air is expelled to the atmosphere.

Wrong Choices Explained

(1) Structure *A* is the trachea, which is a rigid tube that conducts air from the mouth and nasal cavities downward toward the lung.

(2) Structure *B* is the bronchus, one of two extensions of the trachea leading deep into the lung tissues.

(3) Structure *C* is the lung, which contains thousands of small air sacs (alveoli) that facilitate the absorption of oxygen from the atmosphere.

10. **2** The statement *it regulates the chemical composition of the blood* is true of the main function of the nephron. Blood passes through capillary networks where materials diffuse out of the blood fluid and into the nephron interior. Metabolic wastes, such as urea, salts, and water, are removed from the blood. However, beneficial materials, such as dissolved food molecules, are returned to the blood. With this mechanism, the chemical composition of the blood is maintained.

Wrong Choices Explained

(1), (3) The statements *it breaks down red blood cells to form nitrogenous wastes* and *it forms urea from the waste products of protein metabolism* are true of the liver but not of the nephron.

(4) The statement *it absorbs digested food from the contents of the small intestine* is true of the villus but not of the nephron.

Question Set 2.3 (p. 92)

1. **1** This process is an example of *maintenance of homeostasis*. As the runner exercises, an increase in metabolism produces heat. This heat is absorbed by the blood and is carried to sweat glands in the skin, where it is transferred to droplets of perspiration. As these droplets evaporate from the skin surface, the blood is cooled, helping to maintain a stable body temperature.

Wrong Choices Explained

(2) *Antigen-antibody reactions* involve the body's defense mechanism against disease. By itself, it is an example of maintenance of homeostasis. However, the description given does not match this particular process.

(3) *Acquired characteristics* is a concept contained in the evolutionary theories of Jean Lamarck. These theories have been largely discredited by modern science.

(4) *Environmental factors affecting phenotype* is a concept discussed in modern genetic theory. It states that the expression of certain genetic traits can be altered by particular environmental conditions.

2. **3** *Neurotransmitters*, such as acetylcholine, are secreted by the terminal branches of neurons. These chemical neurotransmitters diffuse across the synapse and cause subsequent neurons to fire nerve impulses. In this manner, nerve impulses are transferred from neuron to neuron until the impulse reaches the central nervous system.

Wrong Choices Explained
(1) *Antibodies* are produced by the blood in response to the detection of foreign proteins (antigens) in the blood.
(2) *Antigens* are foreign proteins that may enter the blood and stimulate the production of antibodies.
(4) *Lipids* are biochemicals that take on a wide variety of particular shapes and characteristics. Lipids are composed of a glycerol molecule bonded chemically to three fatty acid molecules.

3. **3** *C* represents the most logical choice of those given. The statement should read as follows: "The *endocrine* glands produce *hormones*, which are transported by the *circulatory* system." Endocrine glands lack ducts and so depend on the bloodstream to carry hormones throughout the body.

Wrong Choices Explained
(1), (2), (4) These nonsense distracters indicate combinations of glands, products, and transport systems that do not exist in nature.

4. **Definitions** (acceptable sample responses):
- Coordination—regulates the body's activities through the use of nerve impulses and hormones
- Excretion—gets rid of cellular wastes from the body
- Digestion—breaks complex foods down into soluble subunits that can be absorbed by the cell
- Circulation—moves materials through the body from one place to another
- Synthesis—manufacture of complex molecules from simpler subunits

Explanations of interaction (acceptable sample responses):
- Circulation/excretion—The circulatory system moves cell wastes to places where they can be removed from the body.
- Coordination/digestion—The nervous system coordinates the peristaltic movement of food through the food tube so that digestion and absorption of the food can occur.

5. **1** In humans, muscle is anchored to bones by connective tissues known as tendons. *Cartilage* does not perform this function in humans. (HINT: Always pay close attention to exam questions that use the italicized word *not*.)

Wrong Choices Explained
(2), (3), (4) These distracters all reference functions characteristic of human cartilage.

6. **3** This question requires a knowledge of the structural and functional design of the human nervous system. An *interneuron* is shown at X lying within the spinal cord and linking a sensory neuron (*below*) with a motor neuron (*above*). The motor neuron in turn enervates a muscle effector. A reflex arc typically involves this structural pattern, designed to provide rapid responses (reflexes) to potentially dangerous situations in nature.

Wrong Choices Explained
(1) The *effector* in this diagram is the muscle, which enables a movement response.
(2) The *motor neuron* is shown linking the interneuron (X) with the muscle.
(4) The *receptor* (not shown) could be any of the body's sense organs or internal receptors.

7. **2** *Production of blood cells* is a function of the human skeletal system. Both red and white blood cells are produced in bone marrow.

Wrong Choices Explained
(1) *Transmit impulses* is a function of the regulatory system in humans.
(3) *Produce lactic acid* occurs in human muscle under anaerobic conditions.
(4) *Store nitrogenous wastes* is a function of the excretory system in humans.

8. **3** The *pituitary gland* is responsible for the elongation of structures A and B in the photograph. Structures A and B are long bones of the arm and leg. Pituitary growth hormone produced by the pituitary gland is responsible for the elongation of these bones.

Wrong Choices Explained
(1) The *islets of Langerhans* in the pancreas are responsible for producing the hormone insulin; insulin regulates the concentration of sugar in the blood.
(2) The *liver* acts as a reservoir for glycogen produced when the body stores excess sugar. The liver produces no hormones and does not directly influence the growth of bone tissue.
(4) *Striated muscles* produce no hormones and do not directly affect the elongation of these bones.

9. **3** *Both the central and peripheral nervous systems* are brought into play in the example given. When the child hears the clap of thunder, both the central and peripheral nervous systems are involved in the initial startle reflex (reflex arc). The central nervous system is involved in the child's decision to find his/her mother for comfort (cerebrum). The peripheral nervous system is involved in the child's feelings of fear (endocrine response) and the movement of muscles needed to allow the child to run to his/her mother (locomotion). Coordination of those muscles is a central nervous system function (cerebellum).

Wrong Choices Explained

(1) *The central nervous system only* is not correct. The peripheral nervous system is needed to relay central commands to the muscles and glands.

(2) *The peripheral nervous system only* is not correct. Central control of a complex response is necessary for survival.

(4) *Neither the central nor the peripheral nervous system* is not correct. Both systems are needed to effect a coordinated response.

10. **4** *Ligaments*, strong elastic connective tissues, are responsible for attaching bone to bone at joints in the skeleton, including the shoulder and elbow joints.

Wrong Choices Explained

(1) *Cartilage* is a type of connective tissue found in many flexible parts of the human body. Cartilage acts as a cushion, not a connector, between bones.

(2) *Tendons* are another form of elastic connective tissue in the body. Tendons are specialized for connecting muscle to bone, not bone to bone.

(3) *Extensors* are muscles that extend joints. They are not associated with linkages between bones.

11. **2** The *adrenal* gland secretes the hormone adrenaline in times of emergency. Adrenaline speeds up metabolic rate, enabling the body to respond to the emergency.

Wrong Choices Explained

(1) The *thyroid* gland secretes the hormone thyroxin, which regulates the body's general metabolic rate.

(3) The *islets of Langerhans*, embedded within the pancreas, secrete the hormone insulin, which regulates the concentration of sugar in the blood.

(4) The *parathyroid* gland secretes the hormone parathormone, which regulates the metabolism of calcium in the body.

12. **1** The *thyroid* gland secretes the hormone thyroxin, which regulates the body's general metabolic rate. Dietary iodine is necessary for the proper synthesis of thyroxin.

Wrong Choices Explained
(2), (3), (4) The *adrenal* gland, *islets of Langerhans*, and *parathyroid* gland do not require dietary iodine to produce their respective hormones.

Question Set 2.4 (p. 115)

1. **1** The *compound light microscope* typically has a low-power magnification of 100× and relatively limited resolution. At this level of magnification and resolution, viewing cells and some of the larger cell organelles (such as *cell walls, chloroplasts*, and *nuclei*) is posssible, especially if stains are added to enhance the color or contrast of these features.

Wrong Choices Explained
(2), (3), (4) The items mentioned in these distracters are generally too small or too difficult to stain to be viewed under a compound light microscope on low power. Under conditions of higher magnification and specialized specimen preparation (such as those used in electron microscopy), viewing *ribosomes, endoplasmic reticula, lysosomes,* and *mitochondria* would be possible. *Genes* and *nucleotides*, being molecular in size, would be difficult or impossible to view using known methodologies.

2. **2** In a living cell, *ribosomes* provide the site for the manufacture of new polypeptide molecules (such as enzymes and structural proteins) from free amino acid molecules in the cytoplasm. While the precise role of the ribosomes in protein synthesis is not completely understood, they are known to be necessary for its successful completion. If ribosomes were to become disabled in the cell, the synthesis of essential enzymes would theoretically cease.

Wrong Choices Explained
(1) *Uncontrolled mitotic cell division* is a characteristic of many cancers. It is associated with changes that are thought to occur within the cell's nucleus. Ribosomes are not directly involved in this process.

(3) *Antibodies* are substances produced by specialized cells in response to the antigens of a foreign invading cell or virus. Antibodies are thought to combine to specific antigens in the body, rendering them harmless. Ribosomes are essential to the production of antibodies, which are proteins. If the ribosomes were not functioning, antibody production would cease, not increase.

(4) The *transport of glucose* and other small molecules in a cell depends on the cell membrane and cytoplasm of the cell. Ribosomes are not directly involved in this process.

3. **3** The diagram most likely represents *a protozoan ingesting food during heterotrophic nutrition*. The protozoan illustrated is most likely an ameba carrying out the process of phagocytosis to ingest a food particle found in its water environment. The food particle is first detected (illustration 1), then engulfed by cytoplasmic extensions known as pseudopods (illustration 2), and finally enclosed in a food vacuole for digestion (illustration 3).

Wrong Choices Explained
    (1) The diagram does not represent *a virus destroying a cell by extracellular digestion*. An illustration of this process would show a crystalline virus particle attaching to the exterior of a cell, injecting its genetic material into the cell interior, producing a number of new viruses from the cell's genetic material, and bursting the cell in order to release the new virus particles.
    (2) The diagram does not represent *a member of the bryophyte phylum performing intercellular digestion*. Members of the bryophyte phylum include mosses, which are autotrophic and whose cells are surrounded by cell walls. The cell illustrated is clearly not a plant cell carrying on photosynthesis.
    (4) The diagram does not represent *a lysosome egesting a food particle into the cytoplasm*. Lysosomes are specialized vacuoles in the cell that contain digestive enzymes. During the nutritional process, lysosomes combine with food vacuoles to facilitate chemical digestion. They are not directly involved with the process of egestion.

4. **4** The arrows in this diagram represent the movement of the indicated substances from the interior of a unicellular protist (a paramecium) to the environment. The substances indicated are carbon dioxide (CO_2) and ammonia (NH_3), which are both *metabolic wastes* produced in the cell. These substances are being *excreted* from the cell in this illustration.

Wrong Choices Explained
    (1) Certain types of *anaerobic respiration* may result in the production of carbon dioxide, but this process is not carried on in paramecia. Ammonia is never a product of anaerobic respiration.
    (2) Certain types of *autotrophic nutrition* may use (but not release) carbon dioxide. Ammonia is never a product of autotrophic nutrition. Paramecia do not carry on autotrophic nutrition.
    (3) *Deamination of amino acids* may occur in a paramecia and may result in the production of ammonia as a metabolic waste. However, carbon dioxide is never a product of deamination.

5. **3** In this diagram, structure A illustrates a flagellum of an alga cell. Like the flagella of protists, these structures are primarily used to allow the organism to move about in its watery environment. This movement is classified as a type of *locomotion*.

Wrong Choices Explained

(1), (2) The structure most closely associated with the life functions of *excretion* and *transport* in this organism is the cell membrane, which is not labeled.

(4) The structure most closely associated with the life function of *reproduction* in this organism is the nucleus, which is not labeled.

6. **3** $C_{12}H_{22}O_{11}$ illustrates the organic compound maltose. The presence of the key elements carbon (C), hydrogen (H), and oxygen (O) clearly indicate that this compound is organic.

Wrong Choices Explained

(1) $Mg(OH)_2$ is the chemical formula for magnesium hydroxide, an inorganic base. No carbon is present in the formula of this compound; therefore it cannot be organic.

(2) $NaCl$ is the chemical formula for sodium chloride, common table salt, which is an inorganic salt. No carbon is present in the formula of this compound, therefore it cannot be organic.

(4) NH_3 is the chemical formula for ammonia, an inorganic material resulting from protein metabolism in certain bacteria. No carbon is present in the formula of this compound, therefore it cannot be organic.

7. **3** *Diffusion* is the process that results in the movement of molecules from a region of higher concentration (area A) to a region of lower concentration (area B). This net movement will occur until the concentrations of molecules have reached equilibrium between area A and area B.

Wrong Choices Explained

(1), (2) *Phagocytosis* and *pinocytosis* are processes by which certain protists engulf their food and enclose it within a vacuole for digestion.

(4) *Cyclosis* refers to the streaming of cytoplasm in the cell, a simple form of intracellular transport.

8. **3** This diagram illustrates a molecule of lipid. Lipids have *a high energy content*. The many carbon-hydrogen bonds present in this molecule indicate that a large amount of energy will be derived from its complete breakdown.

Wrong Choices Explained

(1) The statement *it has the ability to control heredity* is false. A molecule that controls heredity is the nucleic acid DNA.

(2) The statement *it has the ability to control reactions* is false. Molecules capable of controlling cell reactions are known as catalysts and enzymes.

(4) The statement *it is involved in photosynthesis* is false. Photosynthesis is a chemical process involving principally carbon dioxide, water, ATP, PGAL, glucose, and oxygen, as well as a number of enzymes.

9. **2** The molecule shown, which may also be represented by the empirical formula $C_3H_5(OH)_3$, is a molecule of glycerol. Glycerol is one of two products that result from the chemical digestion of a triglyceride lipid, the other being fatty acid.

Wrong Choices Explained

(1) The molecule shown, which may also be represented by the empirical formula CO_2, is a molecule of carbon dioxide. Carbon dioxide is not an end product of any known chemical digestion but is an end product of aerobic respiration.

(3) The molecule shown, which may also be represented by the empirical formula H_2O, is a molecule of water. Water is a reactant, not an end product, of chemical digestion.

(4) The molecule shown, which may also be represented by the empirical formula CH_4, is a molecule of methane. Methane is not an end product of any known chemical digestion but is an end product of decomposition.

10. **3** *A carboxyl group* is illustrated in box *Y*. This group is common to organic acids, including amino acids, and is often written as –COOH.

Wrong Choices Explained

(1) *An amino group*, also common to amino acids, is written $-NH_2$.

(2) *A variable group*, more commonly known as a radical group, refers to the chemical composition of individual amino acid molecules. Each type of amino acid contains a unique chemical composition in the radical group position.

(4) *A peptide group* is a nonsense distracter. No known chemical group carries this name.

11. **2** There are two peptide bonds illustrated in this diagram. These bonds are the C–N bonds that link one amino acid to the next prior to hydrolysis.

Wrong Choices Explained

(1), (3), (4) These bond counts are not correct for the molecule illustrated.

12. **1** Diagram A illustrates a group of plant cells whose cell membranes are pressed tightly against the cell wall by cytoplasm containing a relatively high concentration of water. Diagram B illustrates the same cells after the addition of *salt water*. Note that the cytoplasm has shrunk in volume as water has moved out of the cell by osmosis (diffusion of water) from a region of high concentration of water (the cell interior) toward a region of lower concentration of water (the cell exterior).

Wrong Choices Explained

(2) The addition of *distilled water* to this system would have the opposite effect from that illustrated; the cytoplasm would swell with the rapid inward movement of water, perhaps to the point of bursting.

(3), (4) Most common sources of *pond water* or *tap water* contain dissolved mineral concentrations similar to that found in the natural habitats of most terrestrial or aquatic species. The addition of water from these sources would probably have little or no effect on the cells' appearance.

13. **4** *Structure X transports mateirals for metabolic activities* is the statement that is valid inference concerning structure X represented in the diagram. Structure X represents a vein of a maple tree leaf. The vein in a leaf contains vascular tissues that are specialized to transport water, sugar, and other materials resulting from, or needed for, the plant's metabolic activities.

Wrong Choices Explained

(1) *Structure X contains guard cells that regulate glucose intake* is *not* the statement that is valid inference concerning structure X represented in the diagram. Guard cells are specialized cells that flank the stomates of leaves and regulate the passage of atmospheric gases into and out of the leaf. Stomates and guard cells in plants such as the maple tree are located in the underside of flat areas, not in the veins, of their leaves.

(2) *Structure X carries out heterotrophic nutrition* is *not* the statement that is valid inference concerning structure X represented in the diagram. Heterotrophic nutrition is a feeding mechanism in which an organism consumes and metabolizes preformed organic matter for food. Green plants such as the maple tree are autotrophic, not heterotrophic, organisms.

(3) *Structure X produces gametes for asexual reproduction* is *not* the statement that is valid inference concerning structure X represented in the diagram. Gametes (pollen and ova) in flowering plants such as the maple tree are produced in their flowers, not in their leaves.

Question Set 2.5 (p. 136)

1. **1** The *Cell Theory* is one of the central principles of modern biology. The "units of structure and function" are the cells that make up the bod-

ies of all known living organisms, from amebas to humans. The "functions" referenced are the chemical reactions that constitute the cell's metabolic activities. These, in turn, enable the life functions of living organisms. A unique feature of living cells is their ability to produce new living cells like themselves (cells "arising from preexisting units").

Wrong Choices Explained

(2) The *lock-and-key model of enzyme activity* is a theoretical concept used to explain the specificity of enzymes involved in the cell's metabolism. This concept is much more limited in scope than cell theory. Enzymes are important in the functioning of a cell but are not structural units and cannot arise from preexisting enzymes.

(3) *Natural selection* is a term used to describe the mechanism by which favorable variations within a species are thought to be selected by natural forces to increase in frequency (and by which unfavorable traits are gradually reduced in frequency) in a species population. Natural selection deals with forces outside the organism and therefore cannot be described as making up the structure of living things.

(4) The *heterotroph hypothesis* is a theory proposed by scientists to explain the origin of life on Earth from primitive organic molecules floating in the oceans of the early Earth. Because the heterotroph hypothesis deals with events that theoretically took place before true cells existed, it is not used to describe the structure and function of modern life-forms.

2. **4** The statement *"in both plants and animals, the daughter cells are genetically identical to the original cell"* is true. Mitotic cell division ensures this continuity by replicating the genetic material prior to its separation into the two new daughter cells. This process occurs in all organisms employing mitotic cell division as a mechanism of growth and repair of body tissues.

Wrong Choices Explained

(1) The statement *"it is exactly the same in plant and animal cells"* is not correct since minor differences in the mechanics of cell division are evident when comparing plant and animal cells.

(2) The statement *"the walls of plant cells pinch in, but the membranes of animal cells do not"* is not correct. In plant cells, cell division is associated with the formation of the cell plate, which divides the cytoplasm and separates the daughter nuclei. In animal cells, the cell membrane pinches in to perform the same function.

(3) The statement *"most plant cells use centrioles, but most animal cells do not"* is not correct. Centrioles are commonly found in animal cells undergoing mitotic cell division. This process occurs in plant cells without the presence of centrioles.

3. **4** Sperm cells and egg cells are *monoploid* (*n*) gametes formed during the process of meiotic cell division. A zygote results from the fusion of two monoploid nuclei in fertilization, so it must be *diploid* (*2n*) in chromosome number.

Wrong Choices Explained
   (1), (2), (3) Each of these distracters contains an incorrect combination of choices (see above).

4. **1** Gregor Mendel performed his experiments in the 1800s, *prior to modern scientific research findings concerning chromosome structure or function*. He described observed characteristics of and performed controlled breeding experiments with garden peas that led to the development of theories concerning the inheritance of characteristics. He postulated the existence of physical structures, which he referred to as factors (today's genes) that are passed from generation to generation in the reproductive process.

Wrong Choices Explained
   (2) Mendel developed the principle of *dominance*, but he did so without reliance on or knowledge of the *gene-chromosome theory*.
   (3) The modern compound light *microscope* was developed over time through the contributions of many scientists; Mendel is not known for his contributions in this area. He would not have studied *genes* with a microscope since he was unaware of the genes' existence except as theoretical factors.
   (4) Mendel had no direct knowledge of the existence of *gene mutations* and so would not have been able to use such mutations as an explanation for variation.

5. **4** The X and Y chromosomes carry genes that determine maleness and femaleness in humans. Genes located on either of these two chromosomes are said to be *sex-linked*.

Wrong Choices Explained
   (1) The term *hybrid* refers to the heterozygous condition of an allele pair.
   (2) The term *codominant* refers to a type of intermediate inheritance.
   (3) The term *autosomal* refers to chromosomes other than the X and Y chromosomes.

6. **3** Mutations can be passed on to offspring only if they occur in cells (*primary sex cells*) that give rise to gametes. Mutations in other (somatic) cells may cause cancerous conditions in their recipients, but these conditions cannot be passed on to succeeding generations.

Wrong Choices Explained

(1), (2), (4) *Skin cells*, *lung cells*, and *uterine* cells are all examples of somatic cells. Mutations in somatic cells cannot be passed on to off-spring.

7. **3** Fur color in the Himalayan hare depends on the temperature of the skin during fur growth. The diagram illustrates an experiment that shows that artificial cooling stimulates the skin to produce dark fur. Since temperature is an environmental condition, this experiment shows that certain *genetic traits can be influenced by environmental conditions*.

Wrong Choices Explained

(1) Heredity always influences *gene expression*. The experiment is not focused on this area.

(2) Genes are arranged on *homologous chromosomes* in essentially the same order. The experiment is not focused on this area.

(4) It is unlikely that this experiment would have resulted in *gene mutations* so extensive as to be immediately observable as a change in hair color.

8. **2** Molecule 1 represents a DNA molecule. The presence of units labeled T (for thymine) confirms this fact. In the living cell, DNA is found in the *nucleus*, associated with the chromatin.

Wrong Choices Explained

(1) The *centriole* is an organelle found in animal cells that serves as a point of attachment for the spindle fibers during mitotic cell division.

(3) The *cell wall* is an organelle that bounds and gives mechanical support to animal cells.

(4) The *lysosome* is an organelle that houses hydrolytic enzymes used in the digestion of complex food molecules in the cell.

9. **1** Molecule 3 represents a polypeptide (protein) molecule composed of four units of *amino acid*. Each of the amino acids is coded by a unique combination of three nitrogenous bases in DNA, as translated by mRNA molecules (molecule 2).

Wrong Choices Explained

(2) *DNA* is a very large polymer composed of repeating units of nucleic acid. DNA does not serve as a building block for other molecules.

(3) *Fatty acids* serve as components of lipid molecules, not of protein molecules.

(4) *RNA*, like DNA, is a very large polymer composed of repeating units of nucleic acid. RNA does not serve as a building block for other molecules.

10. **4** Molecule 2 represents a molecule of messenger RNA (mRNA). This molecule is manufactured from free nucleic acids, using DNA as a template. mRNA then migrates to the *ribosome* to perform the task of protein synthesis.

Wrong Choices Explained

(1) Cell *vacuoles* are specialized for storing various materials, including ingested foods, water, and toxic wastes.

(2) The *plasma membrane* provides the outer covering of the cell. It is specialized to regulate the entry and exit of materials into and out of the cell.

(3) The *lysosome* is an organelle that houses hydrolytic enzymes used in the digestion of complex food molecules in the cell.

11. **2** *Dehydration synthesis* (joining by removing water) is a process by which complex molecules are formed by the chemical combination of two or more simpler subunits. By using this reaction, two amino acid molecules might be joined to synthesize one molecule of a dipeptide. Water is a by-product of this reaction.

Wrong Choices Explained

(1) *Deamination* is a process by which amino acids are broken down into their component parts for conversion into urea.

(3) *Enzymatic hydrolysis* is sometimes referred to as splitting with water. A complex molecule is split into two simpler molecules by adding the elements making up a water molecule to the bond that used to join the molecules. It is the chemical opposite of dehydration synthesis.

(4) *Oxidation* is a general biochemical term that relates to chemical reactions in which oxygen is chemically added to an element or compound.

12. **3** In the first paragraph, reference is made to the fact that *all embryos develop female sex organs* during the first 35–40 days (5–6 weeks) after fertilization.

Wrong Choices Explained

(1) Although these structures are formed during this period, they are not the only structures to be formed.

(2), (4) According to the passage (second paragraph), sex differentiation occurs after 35–40 days, not within the first 5 weeks.

13. **1** The passage identifies the *SRY* gene as the determiner of maleness.

Wrong Choices Explained

(2) The *SRY* gene is a single locus on the *Y* chromosome; the remainder of the chromosome is evidently not critical to this process.

(3) The *X* chromosome does not carry the *SRY* gene, and therefore cannot determine maleness.

(4) The *MIS* gene is activated by the *SRY* gene therefore the *MIS* gene cannot activate the process.

14. **2** The *MIS* gene is activated by a *chemical* produced by the *SRY* gene, which stimulates it to cause the female organs to disappear.

Wrong Choices Explained

(1) The *X* chromosome is not involved in the production of maleness in humans.

(3), (4) The presence of reproductive structures, male or female, is not cited in this passage as having anything to do with the stimulation of the *MIS* gene.

15. **4** Both the *SRY and the MIS genes* are involved in this process, the *SRY* gene acting as a trigger factor and the *MIS* gene performing the task of tissue resorption.

Wrong Choices Explained

(1) The *X* chromosome is not involved in the production of maleness in humans.

(2) The *SRY* gene is a single locus on the *Y* chromosome; the remainder of the chromosome is evidently not critical to this process.

(3) The *MIS* gene is activated by the *SRY* gene, therefore the *MIS* gene cannot activate the process.

16. **3** *Messenger RNA* (mRNA) is manufactured from free nucleic acids, using DNA as a template to produce a specific coded message concerning the design of an enzyme molecule. mRNA then migrates to the enzyme-producing region (ribosome) to perform protein synthesis.

Wrong Choices Explained

(1) *Hormones* are chemical substances produced by endocrine glands that are responsible for the chemical regulation of the body. They are not directly involved in protein synthesis.

(2) *Nerve impulses* are carried along neurons as electrochemical changes on the cell membrane. They are not directly involved in protein synthesis.

(4) *DNA molecules* provide the code for protein synthesis. They do not directly participate in protein synthesis.

17. **3** *Genetic engineering* refers to a set of laboratory techniques that involve the artificial translocation of genes from one cell to the chromosomes of other host cells. The host cells may then be used to produce

specific enzymes or other materials unique to the translocated gene. The examples given are two of the economic benefits that might be derived from such activities.

Wrong Choices Explained

(1) *Natural selection* is a process occurring in nature by which favorable varieties of a species may be selected by natural forces for survival and then pass their favorable traits on to their offspring. It is unlikely that natural selection would lead to the kinds of traits referenced in the question.

(2) *Sporulation* is a reproductive process carried out by certain molds and fungi. As an asexual reproductive process, it is unlikely to result in unique new varieties such as those referenced.

(4) *Chromatography* is a laboratory technique used to separate and study complex organic compounds found in cells. It does not relate directly to the production of new genetic varieties.

18. **3** *One cell with two identical nuclei* would result from this situation. Mitosis is a process by which a cell nucleus first replicates its genetic material and then undergoes a nuclear division that results in two identical daughter nuclei. Cytoplasmic division normally accompanies mitosis, separating the daughter nuclei into two separate daughter cells.

Wrong Choices Explained

(1) *Two cells, each with one nucleus* would be the result of normal mitotic cell division.

(2) *Two cells, each without a nucleus* could not result since mitosis normally results in the formation of identical daughter nuclei. Also, two cells could not result in the absence of cytoplasmic division.

(4) *One cell without a nucleus* could not result since mitosis normally results in the formation of identical daughter nuclei.

19. **1** *Artificial selection and inbreeding* are techniques that may be used to develop animals with certain desirable characteristics. The breeder selects individuals with evidence of desirable traits to be used as breeding stock. Once a favorable line has been established, it is maintained in future generations by inbreeding related individuals within the breeding population.

Wrong Choices Explained

(2) *Grafting and hybridization* are techniques normally associated with plant reproduction. Grafting refers to a process by which a plant slip (scion) is forced to grow into a stem (stock) of a different plant of the same species.

(3) *Regeneration and incubation* are processes associated with repair of damaged or lost tissues (regeneration) or the care by parents of eggs in a nest (incubation).

(4) *Vegetative propagation and binary fission* are methods of asexual reproduction carried out by plants and ameba, respectively.

20. **2** *The environment influences wing phenotype in these fruit flies.* For this particular mutation, the expression of the curly wing phenotype depends not only on the homozygous recessive genotype but also on the presence of high environmental temperatures (25°C) during development. Since the same genotype expresses a different phenotype (straight wing) at lower temperatures (16°C), we say that the trait is influenced by environmental conditions.

Wrong Choices Explained

(1) The statement *fruit flies with curly wings cannot survive at high temperatures* is not supported by the information given. If anything, the question leads us to believe that high temperatures support the curly wing condition. ·

(3) The statement *high temperatures increase the rate of mutations* is not supported by the information given. There is no evidence derived from the question that this mutant condition was caused by the environmental condition of temperature but instead, only influenced by it.

(4) The statement *wing length in these fruit flies is directly proportional to temperature* is not supported by the information given. The question implies an all-or-none condition with respect to the expression of this trait. No gradations in wing length are referenced, so wing length cannot be described as being proportional.

21. **1** A *genetic code* is contained within structure *A*, which represents a chromosome inside the nucleus of an animal cell. This genetic code is specific for the production of one or more proteins in the cell.

Wrong Choices Explained

(2) A *single nucleotide only* is not found in structure *A*. A chromosome is made up of hundreds, it not thousands, of nucleotides for each strand of DNA contained within it.

(3) A *messenger RNA molecule* is not found in structure *A*. Messenger RNA molecules are formed next to DNA molecules but then migrate from the nucleus to the ribosome.

(4) A *small polysaccharide* is not found in structure *A*. Polysaccharides, such as starch, are not normally associated with chromosome structure.

22. **3** *Recombinant DNA* is represented by structure *B*. Genes from an animal chromosome (structure *A*) have been spliced into the genome of a bacterial cell, forming a new genetic combination. The bacterial cell will now be capable of producing the protein or proteins coded for by the animal DNA.

Wrong Choices Explained

(1) *A ribosome* is not represented by structure *B*. Ribosomes are small organelles containing RNA and are located on the endoplasmic reticulum of most cells.

(2) *Transfer RNA* is not represented by structure *B*. The diagram indicates that structure A is composed of DNA, not RNA.

(4) *A male gamete* is not represented by structure *B*. Male gametes are sperm cells. Although sperm cells contain DNA, they are not found in the nuclei of animal cells.

23. **2** The technique illustrated is *genetic engineering*. Genetic engineering involves using various laboratory procedures to move genes from one cell to another.

Wrong Choices Explained

(1) *Cloning* is a technique in which the undifferentiated cells of an organism are used to produce a new organism with the same set of characteristics.

(3) *Protein synthesis* is a natural cellular process involving DNA and RNA molecules where amino acids are joined in particular sequences to produce specific proteins for use by the cell.

(4) *In vitro fertilization* is a laboratory technique in which mature eggs are fertilized outside the mother's body and the resulting zygote is reintroduced into the mother's uterus for development.

Question Set 2.6 (p. 155)

1. **4** *Organism A was probably more primitive than organism B and organism C* is a correct statement about the relationship among these organisms. By assuming that the fossils illustrated in the diagram are of sequentially evolved species, it can be inferred from evolutionary theory that the most primitive fossil is in the lowest (oldest) rock layer (A), and that the most advanced fossil is in the upper (youngest) rock layer (C).

Wrong Choices Explained

(1) *Organism A was probably more structurally advanced than organism B and organism C* is incorrect because it is unlikely that an older form of organism is more structurally advanced than a more recent form of that organism.

(2) *Organism C probably gave rise to organism A and organism B* is not a correct statement about the relationship among these organisms. Since the rock layers are "undisturbed," the top layer must have been the last one laid down. For this reason, it would be impossible for an organism whose fossils are found in layer C to have lived before, and given rise to, organisms whose fossils are found in deeper strata (A or B).

(3) *All of these organisms probably evolved at the same time* is not a correct statement about the relationship among these organisms. Given the long time frame required for the deposition of sedimentary rock, it is unlikely that all three of these organisms could have lived at the same time.

2. **3** Comparative embryology is often used to assess the degree of similarity between two species. The more similar two organisms are in terms of their embryological development, the more closely related they are assumed to be. This assessment can, and often does, point to the likelihood that they *share a common ancestry*.

Wrong Choices Explained

(1) Birds and reptiles are assigned to different genera. Therefore, they cannot *belong to the same species*.

(2) Species do not have to be closely related to be *adapted for life in the same habitat*. An example of this phenomenon is cacti and snakes that live successfully in the desert.

(4) The embryos of birds and reptiles show marked similarities. Neither is an *animal-like protist*. Rather, they are members of closely related vertebrate genera.

3. **1** Modern evolutionary theory borrows a great deal from the work of Charles Darwin. In addition, it uses the discoveries of geneticists of the 20th century in the areas of genetic inheritance and gene mutation to shed light on the mechanisms of the evolutionary process. As a result of this work, we now know that *variations are the result of mutations and gene recombination*.

Wrong Choices Explained

(2) *Overproduction of organisms leads to extinction* is not a concept included in modern theories of evolution. Overpopulation can be a serious problem for species. However, by itself, it rarely results in the extinction of a species since natural checks and balances tend to keep species population growth under control.

(3) *Variations exist only in large populations* is not a correct statement. Variations are not limited to large populations but exist in all populations, small and large.

(4) *Competition occurs only between members of the same species* is not a correct statement. Competition for environmental resources is not limited to members of the same species but occurs both between different species and within a single species.

4. **1** The concept of *organic evolution* is illustrated by these diagrams. Organic evolution refers to the mechanism thought to govern the changes in living things over geologic time. The diagram illustrates the physical changes that have occurred within the horse family over 60 million years, based on the fossil record.

Wrong Choices Explained

(2) *Ecological succession* refers to the changes that occur in ecological communities over long periods of time. It is not used to describe evolutionary change within a species and is not referenced in this diagram.

(3) *Intermediate inheritance* is used to describe a pattern of genetic inheritance in which the contrasting traits controlled by an allele pair blend to produce an intermediate phenotype. No reference is made to this pattern of inheritance in the diagram.

(4) *Geographical isolation* is used to describe situations in which species populations are separated by physical barriers (such as oceans, mountains, or canyons). This separation tends to promote variation leading to speciation. No geographical isolation can be inferred from this diagram.

5. **4** When long periods of evolutionary stability (equilibrium) are interrupted by relatively short periods of rapid evolutionary change (punctuation), the pattern of evolution is referred to as *punctuated equilibrium*.

Wrong Choices Explained

(1) The term *use and disuse* refers to the evolutionary theory of Jean Lamarck, which has been largely discredited as a viable theory of evolutionary change.

(2) *Reproductive isolation* refers to behavioral mechanisms that can isolate two varieties of a species from interbreeding even though they may inhabit the same geographical area.

(3) The term *homologous structures* refers to the study of comparative anatomy that postulates that similarities in structure, regardless of function, signal common ancestry between two or more different species.

6. **2** The phenomenon illustrated in the diagram is *an increase in the adaptive value of gene a.* We are led to believe that selective pressures have been brought to bear on the test species over 10 generations such that the lighter phenotype is negatively selected. This selection pressure gradually eliminates organisms displaying the light phenotype, eliminating with them a large proportion of the dominant allele A. As the frequency of allele A diminishes, the frequency of allele a increases proportionately, resulting in more organisms displaying the darker phenotype controlled by genotype aa.

Wrong Choices Explained

(1) *A decrease in the adaptive value of gene a* would result in an increase in the proportion of light phenotypes in this population rather than a decrease.

(3) A quick comparison of the branch at generations 1 and 10 reveals the same relative density of this population. No evidence indicates *an increase in the population of this insect* in this diagram.

(4) No information is given in the diagram that would lead us to believe that *a decrease in the mutation rate of gene A* has occurred. This situation might be caused by an increased mutation rate of *A* but not a decreased rate.

7. **3** *Natural selection* is the centerpiece of Darwin's theory of evolution. Darwin theorized that natural conditions (selection pressures) favor some varieties of a species over other varieties of the same species. As the favorable variety survives these pressures at a higher rate, its members are more readily available to produce offspring expressing the same favorable variety. In this way, Darwin thought, new and favorable varieties of a species are selected by nature to become common in the species. When carried to an extreme, these varieties can become different species surviving under different sets of environmental conditions.

Wrong Choices Explained

(1), (2) Evolution by means of the use and *disuse of body structures* and *the transmission of acquired characteristics* were central ideas expressed in the evolutionary theory of Jean Lamarck. Lamarck's theories have been largely disproved by modern science.

(4) *Mutagenic agents* and gene mutation as a means of creating new varieties in species were concepts unknown at the time that Darwin developed his evolutionary theory. These concepts have been developed in the context of modern science.

8. **1** The *increasing need for new antibiotics* to kill new, antibiotic-resistant strains of bacteria is often used as a clear example of evolution at work in the present day. Bacteria, like other living species, express many genetic variations, including resistance to the effects of antibiotics. In an environment where the presence of antibiotics is common (such as a hospital), the antibiotics act as a selection pressure that eliminates most (but not all) members of the bacterial species. A single bacterium that contains a mutation for resistance to a particular antibiotic can reproduce millions of bacterial cells like itself in a very short time. New types of antibiotics must constantly be produced to combat these resistant strains.

Wrong Choices Explained

(2) *An increasing number of individuals in the human population* is an environmental problem not directly supporting the concept that evolution is occurring in the present.

(3) *A decreasing number of new fossils discovered in undisturbed rock layers* is not directly related to evolution occurring in living species. The discovery and study of fossils provides evidence that evolution occurred in the past.

(4) *Decreasing activity of photosynthetic organisms due to warming of the atmosphere* references an environmental problem not directly supporting the concept that evolution is occurring in the present.

9. **3** Similar nucleotide sequences in the genetic material of separate species provide evidence that these species *may have similar evolutionary histories*. Genetic material, and particularly gene sequences, are passed from generation to generation by reproduction. The more similar these sequences are, and the more sequences there are that are similar, the more closely related the species are considered to be.

Wrong Choices Explained

(1) It is extremely unlikely that these two species *are evolving into the same species*. The process of evolution tends to lead to greater, not less, diversity among living things.

(2) No information is presented in the question that would point to the possibility that these two species *contain identical DNA*. If this were the case, then they would be the same, not different, species.

(4) No information is presented in the question that would indicate that these two species *have the same number of mutations*. Mutations are thought to occur randomly in nature, so the likelihood that two separate species have the same number of mutations is small.

10. **4** *Gradualism* is a theory of evolutionary change that assumes that mutations occur at a predicable pace and the environmental pressures stay relatively constant over time. These two factors, then, result in a relatively constant rate of change of the gene pool of most species, leading to slow, gradual evolution.

Wrong Choices Explained

(1) The term *punctuated equilibrium* refers to a theory of evolutionary change in which long periods of relatively slow change are interrupted by relatively short periods of rapid evolution.

(2) *Geographic isolation* is a term relating to a condition (physical separation) that can promote the development of new species varieties as well as new species.

(3) *Speciation* is a term that relates to the change of separate varieties of a single species into two separate and distinct species.

11. **2** The relationship between the bird and moth is *predator-prey*. The bird is the predator of the moth and other insects that constitute its prey.

Wrong Choices Explained
   (1) A *producer-consumer* relationship would have to involve a plant (the producer) and a herbivore (the consumer). No such relationship is indicated in the passage.
   (3) A *parasite-host* relationship is a symbiosis between an organism (the parasite) that lives in or on another (the host) and harms it in the process. The bird and moth live separate existences in this case.
   (4) An *autotroph-heterotroph* relationship would have to involve a plant (the autotroph) and a herbivore (the heterotroph). No such relationship is indicated in the passage.

12. *They introduced industry to the area.* OR *They built factories that produced pollution.*
   (NOTE: Other correct, complete-sentence responses are acceptable.)

13. *Mutations change DNA, resulting in new traits.* OR *Crossing-over during meiosis may produce new gene combinations.* OR *Fertilization involves the union of sex cells from each of two parents, resulting in offspring different from either parent.*
   (NOTE: Other correct, complete-sentence responses are acceptable.)

Question Set 2.7 (p. 171)

1. **1** Of those shown, diagram 1 most clearly illustrates the process of binary fission. The diagram shows an ameba-like organism undergoing the cytoplasmic division stage of mitotic cell division. Clearly visible in each of the two forming daughter cells is a newly formed nucleus.

Wrong Choices Explained
   (2) Diagram 2 illustrates a yeast cell undergoing budding. Budding is a form of asexual reproduction characterized by mitosis followed by unequal cytoplasmic division.
   (3) Diagram 3 illustrates a strawberry plant reproducing by runners. While mitotic cell division is involved in this form of reproduction, it is not binary fission.
   (4) Diagram 4 illustrates a bread mold in the process of spreading spores. Spore formation is a form of asexual reproduction, but does not involve binary fission.

2. **4** *Four monoploid cells* normally form in a male when a primary sperm cell undergoes meiotic cell division. Meiotic cell division involves two distinct phases, the first of which reduces the number of chromosomes from diploid $(2n)$ to monoploid (n). The second division separates homologous chromosomes into gamete cells known as sperm cells.

Wrong Choices Explained
(1), (2), (3) Each of these is a nonsense distracter. Normal cell division does not result in the combinations indicated.

3. **1** The sequence that represents the correct order of events in the development of sexually reproducing animals is *fertilization* → *cleavage* → *differentiation* → *growth*. Fertilization refers to the fusion of gametes to form a zygote. Cleavage is the rapid mitotic cell division of the zygote to form a cell mass. Differentiation is a process by which the new cells begin to specialize into embryonic tissues. Growth is the increase in cell number and organism size that leads to the formation of an adult organism.

Wrong Choices Explained
(2), (3), (4) These sequences are each out of order.

4. **4** Structures *E and G* secrete hormones that regulate the development of secondary sex characteristics. Structure *E* is the male testis; structure *G* is the female ovary. These structures produce hormones that regulate the development of secondary sex characteristics such as production of body hair, breast enlargement, and voice and muscle tone changes in the adult human.

Wrong Choices Explained
(1) Structures *A and J* represent the male urinary bladder and the female urinary bladder, respectively. These structures serve to store urine prior to its controlled elimination.
(2) Structures *D and H* represent the male vas deferens and the female oviduct, respectively. The primary role of these structures is to transport gametes to the site of fertilization.
(3) Structures *F and I* represent the male scrotum and the female uterus, respectively. The scrotum surrounds the testes and provides conditions conducive to the formation and storage of sperm cells. The uterus functions to receive a fertilized egg and provide a stable environment for its development.

5. **3** The pathway followed by sperm cells implanted inside the female is *K to I to H*. *K* represents the vagina, which serves as the site for the implantation of sperm during intercourse. *I* represents the uterus,

311

through which the sperm must travel to reach the egg. *H* represents the oviduct, within which the sperm meet the mature egg and where fertilization normally occurs.

Wrong Choices Explained
(1), (2), (4) These sequences are each out of order.

6. **2** Gametogenesis occurs within structures *E and G*. Structure *E* is the male testis; structure *G* is the female ovary. These structures produce sperm cells and egg cells, respectively, which are the male and female gametes involved in the process of sexual reproduction.

Wrong Choices Explained
(1) Structures *A and J* represent the male urinary bladder and the female urinary bladder, respectively. These structures serve to store urine prior to its controlled elimination.

(3) Structures *B and I* represent the male urethra and female uterus, respectively. In males, the urethra serves to guide the release of sperm cells into the female reproductive tract. The urethra also conducts urine from the urinary bladder to the environment. The uterus functions to receive a fertilized egg and provide a stable environment for its development.

(4) Structures *D and H* represent the male vas deferens and the female oviduct, respectively. The primary role of these structures is to transport gametes to the site of fertilization.

7. **3** Structures *G and I* are directly affected by hormones involved in the menstrual cycle. Structure *G* is the female ovary, which produces eggs throughout the period of sexual maturity in females. Structure *I* is the uterus, which functions to receive a fertilized egg and provide a stable environment for its development.

Wrong Choices Explained
(1) Structures *C and E* represent the male penis and the male testis, respectively. These structures are not affected by hormones produced in the female menstrual cycle.

(2) Structures *A and D* represent the male urinary bladder and the male vas deferens, respectively. These structures are not affected by hormones produced in the female menstrual cycle.

(4) Structures *I and J* represent the female uterus and the female urinary bladder, respectively. While the uterus is affected by the production of hormones, the urinary bladder is not similarly affected.

8. **4** When a human egg is released from the ovary, it passes directly into the *oviduct* for transport to the uterus.

Wrong Choices Explained

(1) The *cervix* is a muscular structure at the base of the uterus that keeps the developing embryo and its associated membranes within the uterus during gestation.

(2) The *vagina* serves as the site for the implantation of sperm during intercourse as well as the birth canal following gestation.

(3) The *uterus* is the organ in which the embryo implants and the placenta forms during gestation.

9. **2** The uterus periodically readies itself for implantation of a fertilized egg. The duration of the menstrual cycle is approximately 28 days. If no fertilized egg is received by the 28th day of the cycle, the uterine lining is shed in a process known as *menstruation*.

Wrong Choices Explained

(1) *Ovulation* is the process by which a mature egg cell is released from the ovarian follicle once every 28 days in the menstrual cycle.

(3) The *follicle* stage of the menstrual cycle is the 14-day period, immediately prior to ovulation, in which the egg undergoes maturation inside the ovarian follicle.

(4) The *corpus luteum* stage of the menstrual cycle is the 8- to 10-day period, immediately following ovulation, in which the cells of the follicle transform into the corpus luteum for the production of progesterone.

10. **1** The sequence $D \rightarrow B \rightarrow C \rightarrow A$ provides for follicle stage (D), followed by ovulation (B), followed by corpus luteum stage (C), followed by menstruation (A).

Wrong Choices Explained

(2), (3), (4) These are nonsense distracters whose sequences do not coincide with any known reproductive cycle.

11. **3** The fluid *acts as a transport medium for sperm*. This fluid is known as semen. Its primary function is to provide a protective watery medium for the sperm cells as they enter the female reproductive tract.

Wrong Choices Explained

(1) *Removes polar bodies from the surface of the sperm* is a nonsense distracter. Polar bodies are not associated with sperm production.

(2) *Activates the egg nucleus so that it begins to divide* is not an advantage of this fluid medium. The egg is stimulated to divide by the act of fertilization. Semen is not directly involved in this process.

(4) *Provides currents that propel the egg down the oviduct* is not an advantage of this fluid medium. Cilia that line the oviduct are responsible for establishing fluid currents that both carry the egg downward

toward the uterus and carry sperm upward toward the ovary. Semen is not directly involved in this process.

12. **4** The *placenta* is the structure through which substances can diffuse from the mother's blood into the fetal blood. This structure forms in the uterus and establishes a physical link between the maternal and fetal tissues.

Wrong Choices Explained
(1) The *amnion* is a membrane that immediately surrounds the developing embryo and provides a watery environment during development.

(2) The *fallopian tube*, also known as the oviduct, is the site of fertilization and early development (cleavage).

(3) The *yolk sac* provides nutrition for the human embryo during the very early stages of development, before the placenta is formed. It deteriorates as soon as the placental connection is established.

13. **1** Stage *11* shows the first appearance of the mesoderm layer, which can be seen as a group of cells lying between the outer ectoderm layer and inner endoderm layer. The mesoderm will eventually give rise to many of the body's internal organs, including muscle, bone, and circulatory structures.

Wrong Choices Explained
(2) Stage *8* represents the blastula stage of embryonic development, characterized by a hollowing of the cell mass created by the cleavage divisions. No cell specialization is present at this stage.

(3) Stage *6* represents the cell mass resulting from five consecutive cleavage divisions. No cell specialization is present at this stage.

(4) Stage *4* represents the cell mass resulting from three consecutive cleavage divisions. No cell specialization is present at this stage.

14. **3** The *oviduct* is normally the site of the first several cleavage divisions. Cleavage follows fertilization and precedes implantation of the embryo in the uterus.

Wrong Choices Explained
(1) The *ovary* is the site of egg formation and the process of ovulation. The released egg is carried along the oviduct toward the uterus.

(2) The *vagina* is the site of sperm implantation during intercourse. Sperm cells must swim through the female tract until they contact the egg within the oviduct.

(4) The *uterus* is the site of embryo implantation and fetal development.

15. **1** *Gametogenesis and fertilization* most immediately precede the sequence of developmental stages represented in the diagram.

Gametogenesis is the process by which sperm cells and egg cells are produced in the testes and ovaries, respectively. Fertilization is the fusion of egg cell and sperm cell nuclei within the female tract.

Wrong Choices Explained

(2) *Menstruation and menopause* do not most directly precede this sequence of developmental stages. Menstruation is the shedding of the uterine lining that occurs when an egg is not fertilized. Menopause is the cessation of the menstrual cycle that occurs in most females at about age 50.

(3) *Prenatal development and gestation* do not most directly precede this sequence of developmental stages. Prenatal development refers to the entire developmental process, encompassing this sequence of stages up to birth. Gestation refers to the 9-month period during which prenatal development occurs in humans.

(4) *Placental formation and metamorphosis* do not most directly precede this sequence of developmental stages. Placental formation is an event that follows, rather than precedes, this sequence of stages. The placenta forms a connection between mother and fetus during gestation. Metamorphosis is a term relating to development in insects, not humans.

16. *The data in the tables do support the scientists' claim. The measurements of newborns of drinkers are uniformly lower than those of nondrinkers, including shorter gestation periods (4.7% shorter), lower birth weight (17.4% lower), less length (6.6% shorter), and smaller head size (7.0% smaller).*

    (NOTE: Any correct, complete-sentence response is acceptable.)

17. *The data are based on a relatively small sample size of 40 drinkers, so additional data is needed.*

    (NOTE: Any correct, complete-sentence response is acceptable.)

18. *Embryological tissues are very sensitive to chemical changes in the uterus. The younger and more highly undifferentiated fetal tissues are, the more they can be affected by chemicals in their environment.*

    (NOTE: Any correct, complete-sentence response is acceptable.)

19. Sample essay: *"Sexual and asexual reproduction are alike in that they both result in the production of new organisms from a parent or parents. They are unlike in that asexual reproduction results in offspring that have no genetic variation from the parent, while sexual reproduction allows a mixing of new traits each time an offspring is produced. This is due to the fact that in sexual reproduction, two monoploid sex cells fuse, each carrying half of each parent's genetic information. The resulting zygote contains a unique combination of traits.*

"*Examples of organisms that reproduce asexually include the ameba, hydra, yeast, and planaria worm. Organisms that reproduce sexually include earthworms, humans, and flowering plants.*"

(NOTE: Any essay that correctly addresses the points required in the question is acceptable.)

Question Set 2.8 (p. 191)

1. **2** *Living matter is able to control chemical activities with organic catalysts.* The controlled chemical activity in living matter is one of the distinguishing characteristics of life. Enzymes (organic catalysts) are principally responsible for controlling this chemical activity, enabling reactions to occur under conditions found in the cell.

Wrong Choices Explained
(1) *Living matter is unable to diffuse materials* is not a true statement. Diffusion is a process by which molecules of a substance move from a region of higher concentration to a region of lower concentration of that substance. It occurs naturally in many living and nonliving systems and does not depend on living things for its operation.

(3) *Living matter is able to create energy* is not a true statement. Scientists theorize that the amount of energy in the universe is fixed and is neither created nor destroyed but only transferred from one form to another. Energy, although essential to the survival of living things, cannot be created by them.

(4) *Living matter is unable to use energy for metabolic activities* is not a true statement. The life function of respiration is involved with the transfer of chemical bond energy in foods (such as glucose) to molecules of ATP. The energy in ATP is subsequently used by the cell as a controlled source of energy to run cellular reactions (metabolism). Respiration is common to all known life-forms.

2. **2** The nutritional process illustrated by the equation $A + C \rightarrow A + B + D$ is photosynthesis. The carbon-fixation reactions combine the atoms of hydrogen, oxygen, and carbon found in water (A) and carbon dioxide (C) into molecules of glucose (B); Water (A) and oxygen (D) are byproducts of this autotrophic nutritional process. Water (A) is both a reactant in and a product of this process.

Wrong Choices Explained
(1) The equation $B + D \rightarrow A + C$ illustrates the process of aerobic respiration, which involves the same substances as does photosynthesis but is its chemical opposite.

(3), (4) These equations, $B + C \rightarrow A + D$ and $A + B + D \rightarrow B + C$, are nonsense equations that cannot occur either in nature or in the laboratory.

3. **4** *B and D* represent chemicals needed for cellular respiration. Glucose (B) is combined with oxygen (D) to release energy for use in the cell. This is a process carried on by both animals and plants.

Wrong Choices Explained

(1) The substances water (A) and carbon dioxide (C) are reactants in the process of photosynthesis, the chemical opposite of aerobic respiration.

(2), (3) The combinations of substances in these distracters are nonsense combinations that would rarely, if ever, occur in nature.

4. **2** Respiration involves chemical reactions in which food molecules (such as glucose) are broken down and molecular energy is released to form molecules of adenosine triphosphate (*ATP*). This process is carried on by all known living organisms as a means of making energy available for cellular processes.

Wrong Choices Explained

(1) $C_6H_{12}O_6$ is the chemical formula for glucose, which is synthesized in the process of photosynthesis, not respiration.

(3) *Alcohol* is a by-product of fermentation, a type of anaerobic respiration. While it is a product of the respiratory process, it is not produced by all organisms but rather by a few specialized yeasts.

(4) Molecular *oxygen* is a by-product of photosynthesis, not respiration.

5. **2** When an organism reacts to a change in environmental conditions, its reaction is known as a *response*. The brine shrimp discussed in this paragraph respond to a change in light conditions by moving toward the light, a nervous response. Response to environmental stimuli is a characteristic of regulation.

Wrong Choices Explained

(1) *Negative feedback* is a concept normally associated with hormonal regulation in animals. Hormonal negative feedback is not normally associated with simple nervous responses such as the one discussed in the paragraph.

(3) A *stimulus* is defined as an environmental change that results in a response by an organism. In this paragraph, the stimulus is the change in light conditions.

(4) *Active transport* is a process by which materials may be actively moved into or out of a cell through the expenditure of energy. The life function associated with active transport is that of transport.

6. **1** An examination of graph A shows that, from a temperature of 0°C to a temperature of 38°C, enzyme action tends to increase at a fairly steady pace. Once the temperature of *40°C* is reached, however, the rate of enzyme action begins to decrease rapidly, indicating that the enzyme molecules in the system are being destroyed through the irreversible process of denaturation.

Wrong Choices Explained ·
(2) Graph A shows that the rate of enzyme action is approximately 50% of optimum at *23°C*.
(3), (4) Graph B shows that, at least for the human stomach enzyme pepsin, the rate of enzyme action is near optimum at *pH 2* and *pH 3*.

7. **2** The measure of acidity or basicity used in the biology laboratory is pH (concentration of hydrogen ion). Graph B is clearly labeled as a representation of the effect of pH on enzyme action.

Wrong Choices Explained
(1) Graph A represents the effect of temperature on the rate of enzyme action.
(3) Graph C represents the effect of enzyme concentration on the rate of enzyme action.
(4) Graph D represents the effect of substrate concentration on the rate of enzyme action.

8. **2** The equation represents those steps in photosynthesis known as the carbon-fixation reactions. Prior to these reactions, water (H_2O) is split into hydrogen and oxygen in the reaction known as photolysis, making water the most likely source of hydrogen for this equation.

Wrong Choices Explained
(1), (3) *PGAL* (phosphoglyceraldehyde) is the intermediate and glucose ($C_6H_{12}O_6$) the final molecular product resulting from this process. These molecules are the recipients, not the sources, of the hydrogen represented in the equation.
(4) *ATP* (adenosine triphosphate) is a specialized molecule that provides energy for many cellular reactions, including this one. However, it does not serve as a source of hydrogen for these reactions.

9. **4** The cell organelle indicated as structure *D* is a chloroplast. The chloroplast is the site of the photosynthetic reactions, including photolysis and carbon-fixation reactions.

Wrong Choices Explained
(1) Structure *A* represents a plant cell nucleus, which is involved in cell regulation.

(2) Structure *B* represents the cytoplasm, which provides a watery medium for cell activities.

(3) Structure *C* represents a food vacuole, which is involved in storage of manufactured starch.

10. In *active immunity*, the body is stimulated to mount its own defenses against foreign antigens in the blood. These foreign antigens can be acquired naturally, by contracting a disease, or introduced artificially by inoculation with a vaccine. Active immunity is usually long lasting.

Wrong Choices Explained
   Passive immunity is acquired when antibodies produced in another organism are introduced into the blood via injection. Passive immunity is usually short-lived.

   Allergies arise in response to specific foreign antigens. The body responds to these antigens by releasing chemicals known as histamines, which in turn affect the body by causing headaches, runny nose, watery eyes, and a variety of other symptoms.

   Tissue rejection is used to describe the reaction of a host organism to a transplanted tissue or organ from a donor organism. If the antigens of the host and those of the donor do not closely match, the host may reject the transplanted tissues in a process similar to that associated with natural active immunity.

11. *Allergies* arise in response to specific foreign antigens. The body responds to these antigens by releasing chemicals known as histamines, which in turn affect the body by causing headaches, runny nose, watery eyes, and a variety of other symptoms.

Wrong Choices Explained
   See question 10.

12. **2** The human disorder known as *anemia* is associated with insufficient amounts of hemoglobin, the oxygen-carrying compound in red blood cells. Anemia sufferers usually tire quickly due to insufficient amounts of oxygen reaching body tissues.

Wrong Choices Explained
   (1) *Angina* pectoris refers to chronic chest pain caused by insufficient oxygen reaching the heart muscle. Angina is usually associated with narrowing of the coronary arteries due to cholesterol buildup on the artery walls.

   (3) *Coronary thrombosis*, commonly known as heart attack, is an acute disorder of the circulatory system resulting from complete blockage of the coronary artery.

(4) *High blood pressure* is a chronic disorder of the circulatory system, which may result from the gradual narrowing of arterial vessels throughout the body.

13. **4** *A lack of iodine in the diet, which has caused the development of a goiter* is the correct response. Goiter is often associated with diets insufficient in iodine, an element essential to the operation of the thyroid gland located in the throat. Without iodine, the thyroid enlarges, producing the swelling described in the question.

Wrong Choices Explained
    (1), (2), (3) *An excess of calcium in the diet, which has caused a muscle deformity, deposits of fat under the skin caused by a vegetable diet,* and *inherited neck deformities caused by elevated environmental temperatures* are all nonsense distracters that relate to no known human disorders.

14. **4** *Glucose* is the principal product of photosynthesis. Glucose may be *metabolized into more complex carbohydrates by dehydration synthesis,* a chemical process occurring in plant cells that produces starch, cellulose, and other complex carbohydrates.

Wrong Choices Explained
    (1), (2) *Starch* and *protein* can result from dehydration synthesis of smaller molecular units. Dehydration synthesis results in more complex, not less complex, molecular by-products.
    (3) *Glycerol,* a component of certain fat molecules, cannot be synthesized into complex carbohydrates.

15. **1** The graphs indicate that *this enzyme works best at a temperature of 35°C and a pH of 8.* The temperature graph clearly shows the high point of the graph directly at the midpoint between 30°C and 40°C, while the pH graph shows the optimum pH at 8.

Wrong Choices Explained
    (2), (4) The statements *this enzyme works best at a temperature of 50°C and a pH of 12* and *this enzyme works best at a temperature above 50°C and a pH above 12* are not supported by the information in the graphs. The graphs show that the rate of enzyme activity drops off drastically above 40°C and above pH 9.
    (3) The statement *temperature and pH have no effect on the action of this enzyme* is incorrect. The graphs clearly show that the rate of enzyme activity varies greatly when these two variables are manipulated.

16. **4** Of those given, the statement *it reduces the amount of raw material reaching the active site of the enzyme that produces prostaglandins* most

correctly describes how aspirin acts to reduce flu symptoms. In the second paragraph, the article describes how aspirin blocks a critical area of the enzyme that manufactures prostaglandins, preventing the raw materials used in its manufacture from reaching the active site of the enzyme.

Wrong Choices Explained
(1), (2), (3) These statements are not supported by the information given in the article. Close reading of the article is necessary to eliminate these distracters.

17. **1** Of those given, the statement *it interferes with the activity of an enzyme that helps to protect the stomach* most accurately describes how aspirin can irritate the stomach of some people. In the third paragraph, aspirin's effect on the stomach-protecting enzyme PGHS-1 is discussed. By completely blocking this enzyme, stomach-protecting substances cannot be manufactured, and stomach irritation can result.

Wrong Choices Explained
(2), (3), (4) These statements are not supported by the information given in the article. Close reading of the article is necessary to eliminate these distracters.

18. *The molecule is a crystal with a tube running up the middle of it.*
(Other correct, complete-sentence responses are acceptable.)

Wrong Choices Explained
Students should take care to give their responses in biologically accurate, complete-sentence format, including a subject, predicate, and appropriate punctuation. Responses that do not meet these basic criteria cannot be considered for credit.

19. *The willow tree bark contains salicylic acid, which is similar to acetylsalicylic acid, the active ingredient in aspirin.*
(Other correct, complete-sentence responses are acceptable.)

Wrong Choices Explained
Students should take care to give their responses in biologically accurate, complete-sentence format, including a subject, predicate, and appropriate punctuation. Responses that do not meet these basic criteria cannot be considered for credit.

20. **2** *Cancer* is characterized by abnormal cells that suddenly begin to undergo cell division at a very rapid rate. A cancerous cell can produce a mass of abnormal cells in a relatively short time, crowding out normal tissues.

Wrong Choices Explained

(1) *Albinism* is a genetic condition, controlled by a single mutant somatic gene, in which the skin lacks the ability to produce pigments.

(3) *Hemophilia* is a genetic condition, controlled by a single mutant sex-linked gene, in which the blood lacks the ability to produce clotting factors.

(4) *Color blindness* is a genetic condition, controlled by a single mutant sex-linked gene, in which the eyes lacks the ability to produce proteins necessary for recognizing certain colors.

21. **3** *Asthma* is a disorder in humans characterized by constricted airways due to the presence of an irritant. Asthma can be life-threatening if not properly treated.

Wrong Choices Explained

(1) *Coronary thrombosis* is a human disorder in which the heart muscle becomes damaged due to blockage of the coronary artery.

(2) *Arthritis* is a human disorder in which skeletal joints become inflamed, swollen, and painful.

(4) *Emphysema* is a human disorder in which lung tissue deteriorates, leaving the lung inefficient at absorbing oxygen.

22. **4** *A lack of essential amino acids in the diet* is the most likely cause of the condition described in the passage. Because breast milk contains proteins produced by the mother's own body, it will contain the correct balance of amino acids needed by the infant for proper growth. When the diet is changed to cereal, mainly carbohydrates, these proteins are no longer available to the child. As a result, the child is not able to obtain essential amino acids for its own cells to synthesize proteins.

Wrong Choices Explained

(1) *Too many nucleic acids in the diet* is not the likely cause of these symptoms. Nucleic acids are manufactured in the cell nucleus from free nucleotides. They are not found in large quantities in cereal grains.

(2) *An overconsumption of complete protein foods* is not the likely cause of these symptoms. The disorder is described as a protein deficiency in the passage. Overconsumption of complete protein foods would lead to an abundance, not a deficiency, of protein in the body.

(3) *Not enough carbohydrates in the diet* is not the likely cause of these symptoms. Cereal grains contain large concentrations of carbohydrates, which provide an energy source but relatively small quantities of protein.

23. **2** *Cell type B would be fewer in number and lighter in appearance* is the statement that best describes the changes observed in the blood of a

person with anemia, assuming all other variables are kept constant. The cells labeled *B* are red blood cells, which contain the red oxygen-carrying pigment hemoglobin. A person with anemia cannot produce hemoglobin efficiently, so red blood cell production drops. The red blood cells that are produced may contain less hemoglobin than those of a person without this condition.

Wrong Choices Explained

(1), (4) *Cell type A would be fewer in number and larger in size* and *cell type A would be larger in size and darker in appearance* are not correct responses. The cell labeled *A* is a white blood cell. White blood cells lack hemoglobin and so would be unaffected by the anemic condition.

(3) *Cell type B would be larger in size and greater in number* is not the correct response. Cell size would be unaffected by anemia. The cells' appearance would be lighter, not darker, due to the lack of sufficient hemoglobin.

24. *The experimental group and control group should be set up identically, except that the control group should receive a placebo rather than a dose of Lowervil.* OR *All parts of the experiment have to be the same except that the control group does not receive the drug.*

    (NOTE: Any correct, complete-sentence answer is acceptable.)

25. *Using a large number in the experiment in order to eliminate bias is important.* OR *If the experiment is run with a small number of subjects, then statistical significance will be difficult to measure.*

    (NOTE: Any correct, complete-sentence answer is acceptable.)

26. *The researcher should carefully measure the blood pressure of both experimental and control subjects before and after the administration of the drug.* OR *The researcher should establish a baseline blood pressure for each participant in the study then monitor blood pressure changes after administration of the drug or placebo.*

    (NOTE: Any correct, complete-sentence answer is acceptable.)

27. *Veins and arteries may be blocked and tissue damage may result.* OR *The body may reject the new organ.*

    (NOTE: Any correct, complete-sentence answer is acceptable.)

28. *A transplant patient might take an immunosuppressant drug in order to prevent the rejection of the new organ or tissue.*

    (NOTE: Any correct, complete-sentence answer is acceptable.)

29. *The drug may weaken the patient's ability to fight diseases.* OR *The drug may leave the patient less able to fight infection.*

    (NOTE: Any correct, complete-sentence answer is acceptable.)

30. *Tissues or organs grown from stem cells of the patient would not be rejected by the patient's immune system.* OR *Organs produced by this process would not be foreign material and would not be attacked by the patient's immune system.*

(NOTE: Any correct, complete-sentence answer is acceptable.)

Question Set 2.9 (p. 217)

1. **4** From the information given in the diagram, the statement that best describes some organisms in the food web illustrated is *raccoons, fish, and ducks are secondary consumers*. Each of these organisms is shown consuming another organism that is either itself a primary or secondary consumer.

Wrong Choices Explained
(1) The statement *minnows and fish are primary consumers* is not supported by the information in the diagram. Fish are shown consuming aquatic crustaceans, which are primary consumers.

(2) The statement *algae and floating plants are decomposers* is not supported by information given in the diagram. Algae and floating plants are producers.

(3) The statement *aquatic crustaceans are omnivores* is unsupported by the information in the diagram. Aquatic crustaceans are shown consuming only plant matter.

2. **2** Of the statements given, *population B competed more successfully for food than population A did* best describes the data shown in the graph. We are led to believe that populations *A* and *B* inhabit the same environmental niche and compete for the same food supply in their range. Over time, these two stable populations undergo a significant change, perhaps in the face of a sudden reduction in available food supply. As a result of this change, population *A* is virtually eliminated, while population *B* increases in number.

Wrong Choices Explained
(1) The statement *all of the plant populations in this habitat decreased* is not supported by the data. While some of the plant populations likely decreased, assuming that all plant populations decreased is unreasonable.

(3) The statement *population A produced more offspring than population B did* is false. The data shows population *B* increasing in number, indicating that many offspring are being produced. At the same time, we see population *A* decreasing in number, indicating a reduced rate of reproductive activity.

(4) The statement *population A consumed the members of population B* is false. The question states that these populations are herbivores, so *A* would not consume members of *B*. Also, if *A* had consumed *B*, then we would expect that *B* would decrease in relation to *A*, but this is not the case.

3. **1** The *tundra* is a biome characterized by very cold average temperatures such that the *subsoil is permanently frozen*. Because of the lack of soil moisture, only surface plants such as *lichens and mosses* have a chance for survival.

Wrong Choices Explained
(2) The *taiga* is a biome characterized by cool average temperatures. Coniferous forests are common, as are caribou and moose.

(3) The *temperate deciduous forest* biome is characterized by moderate average temperatures, good rainfall, broadleaf (deciduous) forests, and index animals such as deer and squirrels.

(4) The *grassland* biome is characterized by moderate average temperatures, low rainfall, grassy prairies, and index animals including bison and prairie dogs.

(5) The *desert* biome is characterized by wide variations in daily temperature, little rainfall, sagebrush and cactus, and index animals such as roadrunners, coyotes, and rattlesnakes.

(6) The *tropical forest* biome is characterized by high average temperature, abundant rainfall, dense tropical forests, and index animals such as the parrot and monkey.

4. **6** The *tropical forest* biome is characterized by high average temperature, abundant rainfall, dense tropical forests, and index animals such as the parrot and monkey.

Wrong Choices Explained
(1), (2), (3), (4), (5) See question 3.

5. **5** The *desert* biome is characterized by wide variations in daily temperature, little rainfall, sagebrush and cactus, and index animals such as roadrunners, coyotes, and rattlesnakes.

Wrong Choices Explained
(1), (2), (3), (4), (6) See question 3.

6. **2** Because a number of different types of organisms are present, both plant and animal species, this diagram illustrates the level of ecological organization known as a *community*.

Wrong Choices Explained

(1) The term *biosphere* refers to the entire inhabited portion of the planet, a much broader level of organization than that illustrated in the diagram.

(3) The *habitat* is the natural environment for a species, including both biotic and abiotic factors. This diagram illustrates only biotic factors.

(4) The *biome* level of ecological organization is a major grouping of large, but similar ecosystems, a much broader level of organization than is illustrated in the diagram.

7. **1** As the *owl population increases*, interspecies and intraspecies competition for what is presumably a fixed food supply will increase dramatically. This food supply may be taxed beyond its ability to support the owl and hawk populations, and eventually many members of these populations would die of starvation.

Wrong Choices Explained

(2) An *increase in the mouse population* will benefit the hawks and owls since additional food resources will be made available. This should allow a slight increase in the owl and hawk populations over time.

(3) A *decrease in the hawk population* will also benefit the food supply since hawks compete with owls for a limited food supply.

(4) A *decrease in the owl population* will benefit the food supply since owls compete with hawks for a limited food supply.

8. **2** *All the* Homo sapiens *living in New York State* represent a population. The term "population" is used to refer to all the members of a single species inhabiting a given geographic area at a given time. The species *Homo sapiens* (humans) inhabiting the geographic area of New York State constitutes such a population.

Wrong Choices Explained

(1), (3), (4) All the *vertebrates*, *plant and animal species*, or *flowering plants* inhabiting New York State are much too broad groupings to represent a population since each of these represents multiple species.

9. **4** *Temperature and oxygen content* are examples of abiotic factors in the environment.

Wrong Choices Explained

(1), (2), (3) Each of these distracters references the measurement of living things, representing biotic factors. These biotic factors include *scavengers*, *green plants and snails*, and *fish*.

326

10. **3** When green plants are eaten by primary consumers that are in turn eaten by secondary consumers, energy is passed from level to level in the *food chain*.

Wrong Choices Explained
(1) *The process of photosynthesis* is involved in the food chain only to the extent that the process is responsible for capturing light energy and using it to manufacture the substances used as food by the consumers.
(2) *Natural selection* describes the mechanism by which favorable variations within a species are thought to be selected by natural forces to increase in frequency (and by which unfavorable traits are gradually reduced in frequency) in a species population. The term does not relate to this question.
(4) *Ecological succession* refers to the changes that occur in ecological communities over long periods of time. The term does not relate to · this question.

11. **2** The *nitrogen cycle* depends primarily on various chemical and biological processes associated with the metabolism of nitrogen compounds. Plant and animal nitrogenous wastes are processed by a variety of soil bacteria specialized for this task. None of the processes mentioned deal primarily with the metabolism of nitrogen compounds.

Wrong Choices Explained
(1), (4) The *carbon-hydrogen-oxygen cycle* is very much associated with the processes of photosynthesis and respiration, which function to recycle these elements through the environment as complementary biological processes.
(3) The *water cycle* depends on the physical processes of transpiration, evaporation, and condensation to operate in nature.

12. **3** The tapeworm's association with the human is harmful to the human because nutrients are removed by the tapeworm before they can be absorbed by the intestine. Therefore the relationship is one of *parasitism* (+, –).

Wrong Choices Explained
(1) *Commensalism* is a symbiotic relationship in which the host organism is not harmed but the other organism is benefited (0, +).
(2) *Mutualism* is a symbiotic relationship in which both organisms in the relationship are benefited (+, +).

13. **2** The bacteria use the legume as a site from which to carry on their metabolic activities; the legume utilizes nutrients from the bacteria's activities. This relationship can be characterized as *mutualism* since both organisms in the relationship are benefited (+, +).

Wrong Choices Explained
(1), (3) See question 12.

14. **3** The flea's association with the dog is harmful to the dog because blood is removed from the dog by the flea, making it unavailable for the dog to use. Therefore, the relationship is one of *parasitism* (+, –).

Wrong Choices Explained
(1), (2) See question 12.

15. **4** *Lichens* → *grasses* → *shrubs* → *trees* represents a typical ecological succession from bare rock. Bare rock succession requires a specialized type of pioneer organism. Lichens are known for their ability to survive on bare rock and to begin the conversion of rock into soil. Lichens are succeeded by grasses that help soil building by adding organic matter to the soil. Shrubs with shallow root structures are next, followed by trees with deeper root structures.

Wrong Choices Explained
(1), (2), (3) These distracters all contain incorrect sequences of succession communities.

16. **4** This ecosystem will be self-sustaining if *materials cycle between the organisms labeled A and the organisms labeled B*. Organisms labeled *A* include the animals illustrated in the diagram; organisms labeled *B* include the plants. In any balanced ecosystem, materials such as carbon, hydrogen, and oxygen cycle between these two types of organisms.

Wrong Choices Explained
(1), (2) If *the organisms labeled A outnumber the organisms labeled B* or *the organisms labeled A are equal in number to the organisms labeled B* the entire ecosystem would be disrupted. In any stable ecosystem, the number and biomass of producer organisms must be greater than that of consumers.
(3) If *the type of organisms represented by B are eliminated*, this would represent an extreme condition that would lead to the total collapse of the ecosystem.

17. **2** *Abiotic factors* (nonliving factors) in the environment are necessary for the survival of living things. The abiotic materials and energy sources mentioned in the question are all essential to the survival of the plant as it carries out its life functions.

Wrong Choices Explained

(1) *Biotic factors* important to plants might include other living plant and animal species in the plant's environment. No biotic factors are mentioned in the statement.

(3), (4) *Symbiotic relationships* and *carnivore-herbivore relationships* always involve other living things and, as such, are biotic relationships.

18. **3** The statement *decomposers release a material that is acted upon by other organisms* is best represented in the diagram. The diagram shows various types of bacteria taking in nitrogen-containing compounds and releasing others after having acted upon them internally. Some of these nitrogen-containing materials are in turn acted upon by other types of organisms, such as plants and animals.

Wrong Choices Explained

(1) The statement *respiration and photosynthesis are interrelated* is not supported by information given in the diagram. No reference to these life processes is made in the diagram nor are their chemical reactants or products shown.

(2) The statement *transpiration and condensation are related to the water cycle* is not supported by information given in the diagram. No reference to these abiotic processes is made in the diagram.

(4) The statement *predators and their prey are involved in many interactions* is not supported by information given in the diagram. No reference to these nutritional relationships is made in the diagram.

19. **2** The statement *stored energy decreases from consumer 2 to consumer 3* is correct. In any environment, the total available energy decreases with each successive step of the food chain, energy being lost as heat and motion at each exchange of food matter between levels. The biomass pyramid represents this phenomenon.

Wrong Choices Explained

(1) The statement *the producer organisms contain the least amount of stored energy* is not correct. In fact, the producer level of the biomass pyramid contains the most energy of any level.

(3) The statement *consumer 3 contains the greatest amount of stored energy* is not correct. In fact, the top consumer level of the biomass pyramid contains the least energy of any level.

(4) The statement *stored energy increases from the producer to consumer 1* is not correct. In fact, the total amount of stored energy decreases in the biomass pyramid from producer to any consumer level.

20. **2** The statement *it persists until the environment changes* is correct. The climax stage is the the final stage of succession. It is characterized by a

self-perpetuating community of plant and animal species. When the environment changes drastically and the climax community is disturbed, the process of succession may begin again.

Wrong Choices Explained
(1) The statement *it changes rapidly* is incorrect. In fact, the climax community displays a marked stability over long periods of time.
(3) The statement *it is the first community to inhabit an area* is incorrect. The first community to inhabit an area is known as the pioneer community.
(4) The statement *it consists entirely of plants* is not correct. Stable climax communities are characterized by the diversity of plant and animal life that inhabit them.

21. **4** Of the choices given, *an increase in sediment, fallen leaves, and tree limbs accumulating in the bottom of the pond* would most likely lead to terrestrial succession. As these materials build up on the pond bottom, pond water is displaced and becomes gradually shallower. Eventually, terrestrial plants will gain a foothold on this moist organic mat and terrestrial succession will begin.

Wrong Choices Explained
(1) *A decrease in the number of suspended particles in the pond water* would not tend to speed the rate of terrestrial succession of a pond. The sedimentation rate would decrease and water displacement would slow.
(2) *An increase in current velocity of the pond water* would not tend to speed the rate of sediment buildup in the pond. Suspended particles and accumulated sediment could be more readily washed away.
(3) *A decrease in the number of diverse organisms in the shallow water of the pond* would signal a decline in biological activity in the pond. This might lead to a decrease in the rate of organic matter deposition in the pond, thereby slowing the rate of filling and the rate of terrestrial succession.

22. **2** *Competition between the squirrels* is the most likely result when resources are limited. This may result in a reduction in the squirrel population over time to bring the natural community back into proper balance.

Wrong Choices Explained
(1) *An increase in the number of squirrels* is not a likely result of this situation. In fact, a decrease in the population is likely to occur.
(3) *Increased habitats for the squirrels* is not a likely result of this situation. The habitat is limited to the areas capable of supporting the population. These areas will be unlikely to increase in the short term.

(4) *A greater diversity of food for the squirrels* is not a likely result of this situation. The squirrels' food requirements are most likely limited to relatively few choices. It is unlikely that the squirrels would be capable of changing them at will.

23. **4** The *mouse, deer, and cricket* are the primary consumers shown in the diagram. Each of these organisms is shown drawing its nutrition from plant matter, which is the main characteristic defining a primary consumer.

Wrong Choices Explained
 (1), (2), (3) The *mouse, snake, and hawk*, the *snake, hawk, and frog*, and the *cricket, frog, and deer* are all combinations that contain organisms that are not primary consumers. The diagram shows that snakes consume mice, hawks consume mice and frogs, and frogs consume crickets.

24. **1** *X—biotic factors; Y—abiotic factors* is the information that belongs in areas *X* and *Y*. Biotic factors are those relating to the presence and activities of living things in the environment. Abiotic factors are those relating to the nonliving portion of the environment.

Wrong Choices Explained
 (2) *X—ecological relationships; Y—biotic relationships* is not a correct response. The subcategories of area *Y* are clearly abiotic, not biotic, factors.
 (3) *X—abiotic factors; Y—interacting populations* is not a correct response. The subcategories of area *X* are clearly biotic, not abiotic, factors.
 (4) *X—energy flow; Y—biotic factors* is not a correct response. The sub-categories of area *Y* are clearly abiotic, not biotic, factors.

25. **3** *Parasitism* best describes the relationship between the dogwood and the fungus. In this relationship, the parasite (fungus) is benefited (+) and the host organism (dogwood) is harmed (–).

Wrong Choices Explained
 (1) *Commensalism* is a type of symbiosis in which one organism is helped (+) and the host organism is neither helped nor harmed (0).
 (2) *Mutualism* is a type of symbiosis in which one organism is helped (+) and the host organism is helped as well (+).
 (4) *Saprophytism* refers to plants and fungi that depend on decaying organic matter for their food supply.

26. Sample essay: *"The organisms in this food web interact on many different nutritional levels. Producers absorb the Sun's energy and convert it to food energy. The plants are consumed by herbivores, which are then consumed by carnivores, including the wolf (or worms or fish). The energy is absorbed by the grass, is passed to the deer when the deer eats grass, and eventually reaches the wolf when the wolf consumes the deer. Decomposers are important in the food web because they recycle the materials that make up the bodies of dead organisms so that new organisms can use them."*

"If the wolf population decreases, then the deer population will increase because no predator is present to limit it. As the deer population increases, its members will consume the grass at a much faster rate. After a year, the grass population will decrease as a result."

"When the farmer sprays pesticides near the pond, the pesticides will wash into the pond. Their presence there will pollute the water, killing beneficial insect and other species. The pesticide-contaminated insects will be consumed by the fish. As this occurs, the pesticide will build up in the bodies of the fish until they die or become inedible to humans."

(NOTE: Any essay that correctly addresses the points required in the question is acceptable.)

Question Set 2.10 (p. 242)

1. **1** The use of *biological controls of insect pests* is one of the positive effects humans have had on the environment in the past several years. In this example, the ladybugs and praying mantises represent nonharmful insects (biological controls) that prey on harmful ones.

Wrong Choices Explained

(2) This choice is a nonsense distracter since scientists do not refer to insect pests as being *exploited*.

(3) The term *abiotic control* has no explicit meaning in the context of controlling insect pests.

(4) *Biocides* are chemical substances toxic to living things. Their use in the environment to control insect pests can have unintended, severe negative consequences.

2. **2** *Preservation of species* is a major positive outcome of the measures listed in this question. Wildlife refuges enable wild animals to live in natural surroundings and to thrive without interference by humans. Game laws help to ensure that now common species will not be overhunted in the future.

Wrong Choices Explained

(1), (4) The *use of biocides* and *exploitation of species* are considered major negative aspects of human involvement in the environment. They would not result from wildlife refuges and game laws.

(3) The *use of biological controls* is a positive aspect of human involvement in the environment. However, it would not result from wildlife refuges and game laws.

3. **1** *Protection of natural habitat* is a positive effect that humans have had on their environment in the past several years. This practice has brought several species, including the American bald eagle, back from the brink of extinction.

Wrong Choices Explained

(2) *Importation of food into their nesting sites* is an activity that would have little positive value for the preservation of a wild species. In fact, the less interference by humans in the lives of wild species, the better for the species' ultimate survival.

(3) *Preservation of other eagle species that occupy the same niche* would have a negative impact on the survival of this species since inter-species competition would be made much more significant by this change.

(4) *Increased use of pesticides* would have a negative impact since many pesticides can interfere with the reproductive cycles of eagles and other bird species.

4. **2** *Using insecticides to kill insects that compete with humans for food* is a human activity that would be more likely to have a negative impact on the environment than the other three. Insecticides contain chemicals that may be toxic to beneficial insects, amphibians, fish, and birds, among other animal species. If these chemicals enter the food chain, they may build up in concentration and eventually pose a threat to top-level feeders, including humans.

Wrong Choices Explained

(1) *Using reforestation and cover cropping to control soil erosion* is generally considered a positive human activity since its goal is the stabilization of soils that might otherwise erode into streams and be lost as a resource.

(3) *Developing research aimed toward the preservation of endangered species* is generally considered a positive human activity since its goal is the maintenance of biodiversity in the environment. Biodiversity is thought to be of vital importance in maintaining the overall health of the environment.

(4) *Investigating the use of biological controls for pests* is generally considered a positive human activity since its goal is using the natural enemies of these pests instead of a chemical biocide to control the pests' numbers.

5. **4** *Habitat destruction* is most likely to endanger species. Some organisms depend on the conditions available in relatively limited habitat zones. When these zones are disturbed or destroyed, the organisms cannot find suitable conditions elsewhere, cannot breed, and may be eliminated altogether in a very short time.

Wrong Choices Explained

(1) *Cover cropping* is a positive human effect on the environment. It involves planting crops on cultivated land in order to control soil erosion.

(2) *Use of pollution controls* is a positive human effect on the environment. It involves using techniques to limit the amount of pollution that enters our air, water, and soil resources.

(3) *Use of erosion controls* is a positive human effect on the environment. It involves using a variety of techniques to limit the amount of soil lost to erosion.

6. **1** *Retinoid by-products may cause fetal deformities* is the reason pregnant women are advised not to use medicines containing retinoids. This information is found in the third paragraph of the passage.

Wrong Choices Explained

(2) *Retinoid by-products cause parasites to invade developing frogs* is not supported by the passage. Parasites are mentioned in the fourth paragraph of the passage as being a possible alternate cause for the frog deformities referenced in the first and second paragraphs of the passage. The passages makes no reference to retinoids causing parasites to invade developing frogs.

(3) *Retinoid by-products mimic the effects of pesticides on fetal tissue* is not supported by the passage. The key word "mimic" is found in the second and third paragraphs of the passage. In the third paragraph, the passage notes that certain pesticides can mimic retinoids, but no direct reference to the effects of pesticides on fetal tissue is made.

(4) *Retinoid by-products reduce abnormalities in maternal tissue* is not supported by the passage. Nowhere in the passage is this statement made. In fact, the third paragraph states that pregnant women are warned not to use retinoids because of their possible deforming effect on fetal tissues.

7. **3** *Other animals in the ponds containing deformed frogs did not have abnormalities* is the reason some scientists argue that pesticides may not be the cause of the frog deformities. Instead, some scientists believe that the deformities may be caused by parasites of the frogs. This information is found in the fourth paragraph of the passage.

Wrong Choices Explained
   (1) *Pesticide use has decreased over the last four years* is not supported by the passage. No statement to this effect is made in the passage.
   (2) *New pesticides are used in skin-care products* is not supported by the passage. No statement to this effect is made in the passage.
   (4) *Laboratory experiments have determined that a pesticide can mimic retinoids* is supported by the passage in the third paragraph. However, this fact is not used in the passage as supporting evidence that pesticides may not be the cause of the frog deformities.

8. **2** *Parasites that affect frogs usually do not affect fish* is a possible reason for the absence of deformed fish in the ponds that contained deformed frogs. While this is not stated explicitly in the passage, an inference to this effect may be made from the information supplied in the fourth paragraph of the passage.

Wrong Choices Explained
   (1) *Fish can swim away from chemicals introduced into the pond* is not a possible reason for the absence of deformed fish in the ponds that contained deformed frogs. Any water-soluble chemical introduced into an aquatic environment will diffuse through the water until it reaches equal concentration everywhere in the system. Therefore, fish cannot swim away from it.
   (3) *Fish cannot develop deformities* is not a possible reason for the absence of deformed fish in the ponds that contained deformed frogs. Any organism whose embryonic cells are assaulted by a mutagenic chemical or parasite can develop deformities.
   (4) *Frogs and fish are not found in the same habitat* is not a possible reason for the absence of deformed fish in the ponds that contained deformed frogs. In fact, fish and frogs are frequently found inhabiting the same aquatic habitats such as ponds, streams, and marshes.

9. **4** *Factors that affect frogs may also affect other organisms* is an inference that can be made from the information in the passage. The fact that pregnant women are warned against using retinoids is evidence that scientists understand that chemical substances can be harmful to all living things, especially in their developmental stages.

Wrong Choices Explained

(1) *Only a few isolated incidents of frog deformities have been observed* is not an inference that should be made from the information in this passage. In fact, the passage leaves the impression that many such deformities have been noted and studied over the past 4 years.

(2) *If frog parasites are controlled, all frog deformities will stop* is not an inference that should be made from the information in this passage. Parasites as a possible cause of frog deformities has not been firmly established by scientific research. Several possible causes for these deformities are being studied.

(3) *Deformities in frogs are of little significance* is not an inference that should be made from the information in this passage. In fact, scientists are sufficiently concerned about this phenomenon that significant resources are being put into researching the possible causes of the frog deformities and their relationship to human birth defects. This research has led scientists to warn pregnant women not to use substances containing retinoids or their by-products.

10. *Some pesticides mimic retinoids, which can cause deformities inside growing animals such as frogs.*

    (NOTE: Any correct, complete-sentence response is acceptable.)

11. Sample essay: *"The Brazilian Amazon River rain forest is being destroyed by logging and agriculture. Loggers come in and strip the forest of trees, which are sold for timber. Farmers then follow and plant crops in the cleared land, selling the crops they produce. The soil in these areas is too thin and unproductive to serve as farmland for long, so new areas are continually being cleared."*

    "When this happens to an area of the rain forest, the environment is changed drastically. Trees are removed, which exposes the soil to drying. Understory plants cannot survive the changed conditions and die out rapidly. Animals that depend on the deep forest habitat for food and shelter are forced to migrate to other forest habitats or die from exposure or lack of food. Humans are affected because the soils in these areas quickly erode and cause silting of the rivers."

    "Steps can be taken to preserve these areas from further destruction. Laws can be passed to limit the amount of forest that can be destroyed in this way in any one year. Parks and preserves can be established to ensure that sizable tracts of this habitat are never destroyed."

    (NOTE: Any essay that correctly addresses the points required in the question is acceptable.)

12. *Chemical herbicides may cause air or water pollution that could kill species other than the loosestrife.* OR *The beetle is a natural enemy of the loosestrife and so less likely to cause habitat disruption than herbicides.*

    (NOTE: Any correct, complete-sentence response is acceptable.)

13. *The beetles might multiply and compete with native insect species.* OR *It is possible that the beetles could also consume native species of plants such as the cattail.*

    (NOTE: Any correct, complete-sentence response is acceptable.)

14. *The spray may reach other areas and harm people, pets, or other animals in that area.* OR *The spray could kill beneficial insects.*

    (NOTE: Any correct, complete-sentence response is acceptable.)

15. *Predators or parasites of the mosquitoes could be released into the area to control them.* OR *Swamp areas could be drained to reduce the breeding areas.* OR *Sterilized male mosquitoes could be released to mate with females to reduce the rate of reproduction.*

    (NOTE: Any correct, complete-sentence response is acceptable.)

16. Positive: *Biological control (or draining or sterilization) means that no pesticide is released that could harm other species.*

    Negative: *Habitat modifications could be detrimental to other species.* OR *Predators or parasites released could harm other species or spread disease.*

    (NOTE: Any correct, complete-sentence response is acceptable.)

GLOSSARY OF
BIOLOGICAL TERMS

abiotic factor Any of several nonliving, physical conditions that affect the survival of an organism in its environment.

absorption The process by which water and dissolved solids, liquids, and gases are taken in by the cell through the cell membrane.

accessory organ In human beings, any organ that has a digestive function but is not part of the food tube. (See **liver**; **gallbladder**; **pancreas**.)

acid A chemical that releases hydrogen ion (H^+) in solution with water.

acid precipitation A phenomenon in which there is thought to be an interaction between atmospheric moisture and the oxides of sulfur and nitrogen that results in rainfall with low pH values.

active immunity The immunity that develops when the body's immune system is stimulated by a disease organism or a vaccination.

active site The specific area of an enzyme molecule that links to the substrate molecule and catalyzes its metabolism.

active transport A process by which materials are absorbed or released by cells against the concentration gradient (from low to high concentration) with the expenditure of cell energy.

adaptation Any structural, biochemical, or behavioral characteristic of an organism that helps it to survive potentially harsh environmental conditions.

addition A type of chromosome mutation in which a section of a chromosome is transferred to a homologous chromosome.

adenine A nitrogenous base found in DNA and RNA molecules.

adenosine triphosphate (ATP) An organic compound that stores respiratory energy in the form of chemical bond energy for transport from one part of the cell to another.

adrenal cortex A portion of the adrenal gland that secretes steroid hormones that regulate various aspects of blood composition.

adrenal gland An endocrine gland that produces several hormones, including adrenaline. (See **adrenal cortex**; **adrenal medulla**.)

adrenal medulla A portion of the adrenal gland that secretes the hormone adrenaline, which regulates various aspects of the body's metabolic rate.

adrenaline A hormone of the adrenal medulla that regulates general metabolic rate, the rates of heartbeat and breathing, and the conversion of glycogen to glucose.

aerobic phase of respiration The reactions of aerobic respiration in which two pyruvic acid molecules are converted to six molecules of water and six molecules of carbon dioxide.

aerobic respiration A type of respiration in which energy is released from organic molecules with the aid of oxygen.

aging A stage of postnatal development that involves differentiation, maturation, and eventual deterioration of the body's tissues.

air pollution The addition, due to technological oversight, of some unwanted factor (for example, chemical oxides, hydrocarbons, particulates) to our air resources.

albinism A condition, controlled by a single mutant gene, in which the skin lacks the ability to produce skin pigments.

alcoholic fermentation A type of anaerobic respiration in which glucose is converted to ethyl alcohol and carbon dioxide.

allantois   A membrane that serves as a reservoir for wastes and as a respiratory surface for the embryos of many animal species.

allele   One of a pair of genes that exist at the same location on a pair of homologous chromosomes and exert parallel control over the same genetic trait.

allergy   A reaction of the body's immune system to the chemical composition of various substances.

alveolus   One of many air sacs within the lung that function to absorb atmospheric gases and pass them on to the bloodstream.

amino acid   An organic compound that is the component unit of proteins.

amino group   A chemical group having the formula $-NH_2$ that is found as a part of all amino acid molecules.

ammonia   A type of nitrogenous waste with high solubility and high toxicity.

amniocentesis   A technique for detecting genetic disorders in unborn human beings in which a small amount of amniotic fluid is removed and the chromosome content of its cells analyzed. (See **karyotyping**.)

amnion   A membrane that surrounds the embryo in many animal species and contains a fluid to protect the developing embryo from mechanical shock.

amniotic fluid   The fluid within the amnion membrane that bathes the developing embryo.

amylase   An enzyme specific for the hydrolysis of starch.

anaerobic phase of respiration   The reactions of aerobic respiration in which glucose is converted to two pyruvic acid molecules.

anaerobic respiration   A type of respiration in which energy is released from organic molecules without the aid of oxygen.

anal pore   The egestive organ of the paramecium.

anemia   A disorder of the human transport system in which the ability of the blood to carry oxygen is impaired usually because of reduced numbers of red blood cells.

angina pectoris   A disorder of the human transport system in which chest pain signals potential damage to the heart muscle due to narrowing of the opening of the coronary artery.

animal   One of the five biological kingdoms; it includes multicellular organisms whose cells are not bounded by cell walls and that are incapable of photosynthesis (for example, human being).

Annelida   A phylum of the animal kingdom whose members (annelids) include the segmented worms (for example, earthworm).

antenna   A receptor organ found in many arthropods (for example, grasshopper), which is specialized for detecting chemical stimuli.

anther   The portion of the stamen that produces pollen.

antibody   A chemical substance produced in response to the presence of a specific antigen that neutralizes that antigen in the immune response.

antigen   A chemical substance, usually a protein, recognized by the immune system as a foreign invader and that is neutralized by a specific antibody.

anus   The organ of egestion of the digestive tract.

aorta   The principal artery carrying blood from the heart to the body tissues.

aortic arches   A specialized part of the earthworm's transport system that serves as a pumping mechanism for the blood fluid.

apical meristem   A plant growth region located at the tip of the root or tip of the stem.

appendicitis   A disorder of the human digestive tract in which the appendix becomes inflamed as a result of bacterial infection.

aquatic biome   An ecological biome composed of many different water environments.

artery   A thick-walled blood vessel that carries blood away from the heart under pressure.

arthritis   A disorder of the human locomotor system in which skeletal joints become inflamed, swollen, and painful.

Arthropoda   A phylum of the animal kingdom whose members (arthropods) have bodies with chitinous exoskeletons and jointed appendages (for example, grasshopper).

artificial selection   A technique of plant/animal breeding in which individual organisms displaying desirable characteristics are chosen for breeding purposes.

asexual reproduction   A type of reproduction in which new organisms are formed from a single parent organism.

asthma   A disorder of the human respiratory system in which the respiratory tube becomes constricted by swelling brought on by some irritant.

atrium   In human beings, one of the two thin-walled upper chambers of the heart that receive blood.

autonomic nervous system   A subdivision of the peripheral nervous system consisting of nerves associated with automatic functions (for example, heartbeat, breathing).

autosome   One of several chromosomes present in the cell that carry genes controlling "body" traits not associated with primary and secondary sex characteristics.

autotroph   An organism capable of carrying on autotrophic nutrition. Self-feeder.

autotrophic nutrition   A type of nutrition in which organisms manufacture their own organic foods from inorganic raw materials.

auxin   A biochemical substance, a plant hormone, produced by plants that regulates growth patterns.

axon   An elongated portion of a neuron that conducts nerve impulses, usually away from the cell body of the neuron.

base   A chemical that releases hydroxyl ion (OH^-) in solution with water.

bicarbonate ion   The chemical formed in the blood plasma when carbon dioxide is absorbed from body tissues.

bile   In human beings, a secretion of the liver that is stored in the gallbladder and that emulsifies fats.

binary fission   A type of cell division in which mitosis is followed by equal cytoplasmic division.

binomial nomenclature   A system of naming used in biological classification that consists of the genus and species names (for example, *Homo sapiens*).

biocide use   The use of pesticides that eliminate one undesirable organism but that have, due to technological oversight, unanticipated effects on beneficial species as well.

biological controls   The use of natural enemies of various agricultural pests for pest control, thereby eliminating the need for biocide use—a positive aspect of human involvement with the environment.

biomass   The total mass of living material present at the various trophic levels in a food chain.

biome   A major geographical grouping of similar ecosystems, usually named for the climax flora in the region (for example, Northeast Deciduous Forest).

biosphere   The portion of the earth in which living things exist, including all land and water environments.

biotic factor   Any of several conditions associated with life and living things that affect the survival of living things in the environment.

birth   In placental mammals, a stage of embryonic development in which the baby passes through the vaginal canal to outside of the mother's body.

blastula   In certain animals, a stage of embryonic development in which the embryo resembles a hollow ball of undifferentiated cells.

341

blood The complex fluid tissue that functions to transport nutrients and respiratory gases to all parts of the body.

blood typing An application of the study of immunity in which the blood of a person is characterized by its antigen composition.

bone A tissue that provides mechanical support and protection for bodily organs and levers for the body's locomotive activities.

Bowman's capsule A cup-shaped portion of the nephron responsible for filtering of soluble blood components.

brain An organ of the central nervous system responsible for regulating conscious and much unconscious activity in the body.

breathing A mechanical process by which air is forced into the lung by means of muscular contraction of the diaphragm and rib muscles.

bronchiole One of several subdivisions of the bronchi that penetrate the lung interior and terminate in alveoli.

bronchitis A disorder of the human respiratory system in which the bronchi become inflamed.

bronchus One of the two major subdivisions of the breathing tube; the bronchi are ringed with cartilage and conduct air from the trachea to the lung interior.

Bryophyta A phylum of the plant kingdom that consists of organisms lacking vascular tissues (for example, moss).

budding A type of asexual reproduction in which mitosis is followed by unequal cytoplasmic division.

bulb A type of vegetative propagation in which a plant bulb produces new bulbs that may be established as independent organisms with identical characteristics.

cambium The lateral meristem tissue in woody plants responsible for annual growth in stem diameter.

cancer Any of a number of conditions characterized by rapid, abnormal, and uncontrolled division of affected cells.

capillary A very small, thin-walled blood vessel that connects an artery to a vein and through which all absorption into the blood fluid occurs.

carbohydrate An organic compound composed of carbon, hydrogen, and oxygen in a 1:2:1 ratio (for example, $C_6H_{12}O_6$).

carbon-fixation reactions A set of biochemical reactions in photosynthesis in which hydrogen atoms are combined with carbon and oxygen atoms to form PGAL and glucose.

carbon 14 A radioactive isotope of carbon used to trace the movement of carbon in various biochemical reactions, and also used in the carbon dating of fossils.

carbon-hydrogen-oxygen cycle A process by which these three elements are made available for use by other organisms through the chemical reactions of respiration and photosynthesis.

carboxyl group A chemical group having the formula —COOH and found as part of all amino acid and fatty acid molecules.

cardiac muscle A type of muscle tissue in the heart and arteries associated with the rhythmic nature of the pulse and heartbeat.

cardiovascular disease In human beings, any disease of the circulatory organs.

carnivore A heterotrophic organism that consumes animal tissue as its primary source of nutrition. (See **secondary consumer**.)

carrier An individual who, though not expressing a particular recessive trait, carries this gene as part of his/her heterozygous genotype.

carrier protein A specialized molecule embedded in the cell membrane that aids the movement of materials across the membrane.

cartilage A flexible connective tissue found in many flexible parts of the body (for example, knee); common in the embryonic stages of development.

catalyst Any substance that speeds up or slows down the rate of a chemical reaction. (See **enzyme**.)

cell plate A structure that forms during cytoplasmic division in plant cells and serves to separate the cytoplasm into two roughly equal parts.

cell theory A scientific theory that states, "All cells arise from previously existing cells" and "Cells are the unit of structure and function of living things."

cell wall A cell organelle that surrounds and gives structural support to plant cells; cell walls are composed of cellulose.

central nervous system The portion of the vertebrate nervous system that consists of the brain and the spinal cord.

centriole A cell organelle found in animal cells that functions in the process of cell division.

centromere The area of attachment of two chromatids in a double-stranded chromosome.

cerebellum The portion of the human brain responsible for the coordination of muscular activity.

cerebral hemorrhage A disorder of the human regulatory system in which a broken blood vessel in the brain may result in severe dysfunction or death.

cerebral palsy A disorder of the human regulatory system in which the motor and speech centers of the brain are impaired.

cerebrum The portion of the human brain responsible for thought, reasoning, sense interpretation, learning, and other conscious activities.

cervix A structure that bounds the lower end of the uterus and through which sperm must pass in order to fertilize the egg.

chemical digestion The process by which nutrient molecules are converted by chemical means into a form usable by the cells.

chemosynthesis A type of autotrophic nutrition in which certain bacteria use the energy of chemical oxidation to convert inorganic raw materials to organic food molecules.

chitin A polysaccharide substance that forms the exoskeleton of the grasshopper and other arthropods.

chlorophyll A green pigment in plant cells that absorbs sunlight and makes possible certain aspects of the photosynthetic process.

chloroplast A cell organelle found in plant cells that contains chlorophyll and functions in photosynthesis.

Chordata A phylum of the animal kingdom whose members (chordates) have internal skeletons made of cartilage and/or bone (for example, human being).

chorion A membrane that surrounds all other embryonic membranes in many animal species, protecting them from mechanical damage.

chromatid One strand of a double-stranded chromosome.

chromosome mutation An alteration in the structure of a chromosome involving many genes. (See **nondisjunction; translocation; addition; deletion**.)

cilia Small, hairlike structures in paramecia and other unicellular organisms that aid in nutrition and locomotion.

classification A technique by which scientists sort, group, and name organisms for easier study.

cleavage A series of rapid mitotic divisions that increase cell number in a developing embryo without a corresponding increase in cell size.

climax community A stable, self-perpetuating community that results from an ecological succession.

cloning A technique of genetic investigation in which undifferentiated cells of an organism are used to produce new organisms with the same set of traits as the original cells.

closed-transport system A type of circulatory system in which the transport fluid is always enclosed within blood vessels (for example, earthworm, human).

clot A structure that forms as a result of enzyme-controlled reactions following the rupturing of a blood vessel and serves as a plug to prevent blood loss.

codominance A type of intermediate inheritance that results from the simultaneous expression of two dominant alleles with contrasting effects.

codon See **triplet codon**.

Coelenterata A phylum of the animal kingdom whose members (coelenterates) have bodies that resemble a sack (for example, hydra, jellyfish).

coenzyme A chemical substance or chemical subunit that functions to aid the action of a particular enzyme. (See **vitamin**.)

cohesion A force binding water molecules together that aids in the upward conduction of materials in the xylem.

commensalism A type of symbiosis in which one organism in the relationship benefits and the other is neither helped nor harmed.

common ancestry A concept central to the theory of evolution that postulates that all organisms share a common ancestry whose closeness varies with the degree of shared similarity.

community A level of biological organization that includes all of the species populations inhabiting a particular geographic area.

comparative anatomy The study of similarities in the anatomic structures of organisms, and their use as an indicator of common ancestry and as evidence of organic evolution.

comparative biochemistry The study of similarities in the biochemical makeups of organisms, and their use as an indicator of common ancestry and as evidence of organic evolution.

comparative cytology The study of similarities in the cell structures of organisms, and their use as an indicator of

common ancestry and as evidence of organic evolution.

comparative embryology The study of similarities in the patterns of embryological development of organisms, and their use as an indicator of common ancestry and as evidence of organic evolution.

competition A condition that arises when different species in the same habitat attempt to use the same limited resources.

complete protein A protein that contains all eight essential amino acids.

compound A substance composed of two or more different kinds of atom (for example, water: H_2O).

compound light microscope A tool of biological study capable of producing a magnified image of a biological specimen by using a focused beam of light.

conditioned behavior A type of response that is learned but that becomes automatic with repetition.

conservation of resources The development and application of practices to protect valuable and irreplaceable soil and mineral resources—a positive aspect of human involvement with the environment.

constipation A disorder of the human digestive tract in which fecal matter solidifies and becomes difficult to egest.

consumer Any heterotrophic animal organism (for example, human being).

coronary artery An artery that branches off the aorta to feed the heart muscle.

coronary thrombosis A disorder of the human transport system in which the heart muscle becomes damaged as a result of blockage of the coronary artery.

corpus luteum A structure resulting from the hormone-controlled transformation of the ovarian follicle that produces the hormone progesterone.

corpus luteum stage A stage of the menstrual cycle in which the cells of the follicle are transformed into the corpus luteum under the influence of luteinizing hormone (LH).

cotyledon A portion of the plant embryo that serves as a source of nutrition for the young plant before photosynthesis begins.

cover cropping A proper agricultural practice in which a temporary planting (cover crop) is used to limit soil erosion between seasonal plantings of main crops.

crop A portion of the digestive tract of certain animals that stores food temporarily before digestion.

cross-pollination A type of pollination in which pollen from one flower pollinates flowers of a different plant of the same species.

crossing-over A pattern of inheritance in which linked genes may be separated during synapsis in the first meiotic division, when sections of homologous chromosomes may be exchanged.

cuticle A waxy coating that covers the upper epidermis of most leaves and acts to help the leaf retain water.

cutting A technique of plant propagation in which vegetative parts of the parent plant are cut and rooted to establish new plant organisms with identical characteristics.

cyclosis The circulation of the cell fluid (cytoplasm) within the cell interior.

cyton The cell body of the neuron, which generates the nerve impulse.

cytoplasm The watery fluid that provides a medium for the suspension of organelles within the cell.

cytoplasmic division The separation of daughter nuclei into two new daughter cells.

cytosine A nitrogenous base found in both DNA and RNA molecules.

daughter cell A cell that results from mitotic cell division.

daughter nucleus One of two nuclei that form as a result of mitosis.

deamination A process by which amino acids are broken down into their component parts for conversion into urea.

death The irreversible cessation of bodily functions and cellular activities.

deciduous A term relating to broad-leaved trees that shed their leaves in the fall.

decomposer Any saprophytic organism that derives its energy from the decay of plant and animal tissues (for example, bacteria of decay, fungus); the final stage of a food chain.

decomposition bacteria In the nitrogen cycle, bacteria that break down plant and animal protein and produce ammonia as a by-product.

dehydration synthesis A chemical process in which two organic molecules may be joined after removing the atoms needed to form a molecule of water as a by-product.

deletion A type of chromosome mutation in which a section of a chromosome is separated and lost.

dendrite A cytoplasmic extension of a neuron that serves to detect an environmental stimulus and carry an impulse to the cell body of the neuron.

denitrifying bacteria In the nitrogen cycle, bacteria that convert excess nitrate salts into gaseous nitrogen.

deoxygenated blood Blood that has released its transported oxygen to the body tissues.

deoxyribonucleic acid (DNA) A nucleic acid molecule known to be the chemically active agent of the gene; the fundamental hereditary material of living organisms.

deoxyribose A five-carbon sugar that is a component part of the nucleotide unit in DNA only.

desert A terrestrial biome characterized by sparse rainfall, extreme temperature variation, and a climax flora that includes cactus.

diabetes A disorder of the human regulatory system in which insufficient insulin production leads to elevated blood sugar concentrations.

diarrhea A disorder of the human digestive tract in which the large intestine fails to absorb water from the waste matter, resulting in watery feces.

diastole The lower pressure registered during blood pressure testing. (See **systole**.)

differentiation The process by which embryonic cells become specialized to perform the various tasks of particular tissues throughout the body.

diffusion A form of passive transport in which soluble substances are absorbed or released by cells.

digestion The process in which complex foods are broken down by mechanical or chemical means for use by the body.

dipeptide A chemical unit composed of two amino acid units linked by a peptide bond.

diploid chromosome number The number of chromosomes found characteristically in the cells (except gametes) of sexually reproducing species.

disaccharidase Any disaccharide-hydrolyzing enzyme.

disaccharide A type of carbohydrate known also as a double sugar; all disaccharides have the molecular formula $C_{12}H_{22}O_{11}$.

disjunction The separation of homologous chromosome pairs at the end of the first meiotic division.

disposal problems Problems, due to technological oversight, that result when commercial and technological activities produce solid and/or chemical wastes that must be disposed of.

dissecting microscope A tool of biological study that magnifies the image of a biological specimen up to 20 times normal size for purposes of gross dissection.

dominance A pattern of genetic inheritance in which the effects of a dominant allele mask those of a recessive allele.

dominant allele (gene) An allele (gene) whose effect masks that of its recessive allele.

double-stranded chromosome The two-stranded structure that results from chromosomal replication.

Down's syndrome In human beings, a condition characterized by mental and physical retardation that may be caused by the nondisjunction of chromosome number 21.

Drosophila The common fruit fly, an organism that has served as an object of genetic research in the development of the gene-chromosome theory.

ductless gland See **endocrine gland**.

ecology The science that studies the interactions of living things with each other and with the nonliving environment.

ecosystem The basic unit of study in ecology, including the plant and animal community in interaction with the nonliving environment.

ectoderm An embryonic tissue that differentiates into skin and nerve tissue in the adult animal.

effector An organ specialized to produce a response to an environmental stimulus; effectors may be muscles or glands.

egestion The process by which undigested food materials are eliminated from the body.

electron microscope A tool of biological study that uses a focused beam of electrons to produce an image of a biological specimen magnified up to 25,000 times its normal size.

element The simplest form of matter; an element is a substance (for example, nitrogen) made up of a single type of atom.

embryo An organism in the early stages of development following fertilization.

embryonic development A series of complex processes by which animal and plant embryos develop into adult organisms.

emphysema A disorder of the human respiratory system in which lung tissue deteriorates, leaving the lung with diminished capacity and efficiency.

emulsification A process by which fat globules are surrounded by bile to form fat droplets.

endocrine (ductless) gland A gland (for example, thyroid, pituitary) specialized for producing and secreting hormones directly into the bloodstream; such glands lack ducts.

endoderm An embryonic tissue that differentiates into the digestive and respiratory tract lining in the adult animal.

endoplasmic reticulum (ER) A cell organelle known to function in the transport of cell products from place to place within the cell.

environmental laws Federal, state, and local legislation enacted in an attempt to protect environmental resources—a positive aspect of human involvement with the environment.

enzymatic hydrolysis An enzyme-controlled reaction by which complex food molecules are broken down chemically into simpler subunits.

enzyme An organic catalyst that controls the rate of metabolism of a single type of substrate; enzymes are protein in nature.

enzyme-substrate complex A physical association between an enzyme molecule and its substrate within which the substrate is metabolized.

epicotyl A portion of the plant embryo that specializes to become the upper stem, leaves, and flowers of the adult plant.

epidermis The outermost cell layer in a plant or animal.

epiglottis In a human being, a flap of tissue that covers the upper end of the trachea during swallowing and prevents inhalation of food.

esophagus A structure in the upper portion of the digestive tract that conducts the food from the pharynx to the midgut.

essential amino acid An amino acid that cannot be synthesized by the human body but must be obtained by means of the diet.

estrogen A hormone, secreted by the ovary that regulates the production of female secondary sex characteristics.

evolution Any process of gradual change through time.

excretion The life function by which living things eliminate metabolic wastes from their cells.

exoskeleton A chitinous material that covers the outside of the bodies of most arthropods and provides protection for internal organs and anchorage for muscles.

exploitation of organisms Systematic removal of animals and plants with commercial value from their environments to sell them—a negative aspect of human involvement with the environment.

extensor A skeletal muscle that extends (opens) a joint.

external development Embryonic development that occurs outside the body of the female parent (for example, birds).

external fertilization Fertilization that occurs outside the body of the female parent (for example, fish).

extracellular digestion Digestion that occurs outside the cell.

fallopian tube See **oviduct**.

fatty acid An organic molecule that is a component of certain lipids.

fauna The animal species comprising an ecological community.

feces The semisolid material that results from the solidification of undigested foods in the large intestine.

fertilization The fusion of gametic nuclei in the process of sexual reproduction.

filament The portion of the stamen that supports the anther.

flagella Microscopic, whiplike structures found on certain cells that aid in locomotion and circulation.

flexor   A skeletal muscle that flexes (closes) a joint.

flora   The plant species comprising an ecological community.

flower   The portion of a flowering plant specialized for sexual reproduction.

fluid-mosaic model   A model of the structure of the cell membrane in which large protein molecules are thought to be embedded in a bilipid layer.

follicle   One of many areas within the ovary that serve as sites for the periodic maturation of ova.

follicle stage   The stage of the menstrual cycle in which an ovum reaches its final maturity under the influence of the hormone FSH.

follicle-stimulating hormone (FSH)   A pituitary hormone that regulates the maturation of and the secretion of estrogen by the ovarian follicle.

food chain   A series of nutritional relationships in which food energy is passed from producer to herbivore to carnivore to decomposer; a segment of a food web.

food web   A construct showing a series of interrelated food chains and illustrating the complex nutritional interrelationships that exist in an ecosystem.

fossil   The preserved direct or indirect remains of an organism that lived in the past, as found in the geologic record.

fraternal twins   In human beings, twin offspring that result from the simultaneous fertilization of two ova by two sperm; such twins are not genetically identical.

freshwater biome   An aquatic biome made up of many separate freshwater systems that vary in size and stability and may be closely associated with terrestrial biomes.

fruit   Any plant structure that contains seeds; a mechanism of seed dispersal.

Fungi   One of the five biological kingdoms; it includes organisms unable to manufacture their own organic foods (for example, mushroom).

gallbladder   An accessory organ that stores bile.

gallstones   A disorder of the human digestive tract in which deposits of hardened cholesterol lodge in the gallbladder.

gamete   A specialized reproductive cell produced by organisms of sexually reproducing species. (See **sperm; ovum; pollen; ovule**.)

gametogenesis   The process of cell division by which gametes are produced. (See **meiosis; spermatogenesis; oogenesis**.)

ganglion   An area of bunched nerve cells that acts as a switching point for nerve impulses traveling from receptors and to effectors.

garden pea   The research organism used by Mendel in his early scientific work in genetic inheritance.

gastric cecum   A gland in the grasshopper that secretes digestive enzymes.

gastrula   A stage of embryonic development in animals in which the embryo assumes a tube-within-a-tube structure and distinct embryonic tissues (ectoderm, mesoderm, endoderm) begin to differentiate.

gastrulation   The process by which a blastula becomes progressively more indented, forming a gastrula.

gene   A unit of heredity; a discrete portion of a chromosome thought to be responsible for the production of a single type of polypeptide; the factor responsible for the inheritance of a genetic trait.

gene frequency   The proportion (percentage) of each allele for a particular trait that is present in the gene pool of a population.

gene linkage   A pattern of inheritance in which genes located along the same chromosome are prevented from assorting independently but are linked together in their inheritance.

gene mutation   An alteration of the chemical nature of a gene that changes its ability to control the production of a polypeptide chain.

gene pool The sum total of all the inheritable genes for the traits in a given sexually reproducing population.

gene-chromosome theory A theory of genetic inheritance that is based on current understanding of the relationships between the biochemical control of traits and the process of cell division.

genetic counseling Clinical discussions concerning inheritance patterns that are designed to inform prospective parents of the potential for expression of a genetic disorder in their offspring.

genetic engineering The use of various techniques to move genes from one organism to another.

genetic screening A technique for the detection of human genetic disorders in which bodily fluids are analyzed for the presence of certain marker chemicals.

genome The total genetic makeup (DNA) of an organism.

genotype The particular combination of genes in an allele pair.

genus A level of biological classification that represents a subdivision of the phylum level; having fewer organisms with great similarity (for example, *Drosophila, Paramecium*).

geographic isolation The separation of species populations by geographic barriers, facilitating the evolutionary process.

geologic record A supporting item of evidence of organic evolution, supplied within the earth's rock and other geologic deposits.

germination The growth of the pollen tube from a pollen grain; the growth of the embryonic root and stem from a seed.

gestation The period of prenatal development of a placental mammal; human gestation requires approximately nine months.

gizzard A portion of the digestive tract of certain organisms, including the earthworm and the grasshopper in which food is ground into smaller fragments.

global warming A phenomenon characterized by a gradual increase in the average temperature of Earth that is thought to be the result of the addition of greenhouse gases to the atmosphere through industrialization and other human technologies.

glomerulus A capillary network lying within Bowman's capsule of the nephron.

glucagon A hormone, secreted by the islets of Langerhans, that regulates the release of blood sugar from stored glycogen.

glucose A monosaccharide produced commonly in photosynthesis and used by both plants and animals as a fuel in the process of respiration.

glycerol An organic compound that is a component of certain lipids.

glycogen A polysaccharide synthesized in animals as a means of storing glucose; glycogen is stored in the liver and in the muscles.

goiter A disorder of the human regulatory system in which the thyroid gland enlarges because of a deficiency of dietary iodine.

Golgi complex Cell organelles that package cell products and move them to the plasma membrane for secretion.

gonad An endocrine gland that produces the hormones responsible for the production of various secondary sex characteristics. (See **ovary**; **testis**.)

gout A disorder of the human excretory system in which uric acid accumulates in the joints, causing severe pain.

gradualism A theory of the time frame required for organic evolution that assumes that evolutionary change is slow, gradual, and continuous.

grafting A technique of plant propagation in which the stems of desirable plants are attached (grafted) to rootstocks of related varieties to produce new plants for commercial purposes.

grana The portion of the chloroplast within which chlorophyll molecules are concentrated.

grassland A terrestrial biome characterized by wide variation in temperature and a climax flora that includes grasses.

greenhouse gases Any of a number of gases in Earth's atmosphere that have the effect of trapping solar radiation in the form of heat (e.g., carbon dioxide, methane).

growth A process by which cells increase in number and size, resulting in an increase in size of the organism.

growth-stimulating hormone (GSH) A pituitary hormone regulating the elongation of the long bones of the body.

guanine A nitrogenous base found in both DNA and RNA molecules.

guard cell One of a pair of cells that surround the leaf stomate and regulate its size.

habitat The environment or set of ecological conditions within which an organism lives.

haploid chromosome number The number of chromosomes commonly found in the gametes of sexually reproducing species.

Hardy-Weinberg principle A hypothesis, advanced by G. H. Hardy and W. Weinberg that states that the gene pool of a population should remain stable as long as a set of ideal conditions is met.

heart In human beings, a four-chambered muscular pump that facilitates the movement of blood throughout the body.

helix Literally a spiral; a term used to describe the twisted ladder shape of the DNA molecule.

hemoglobin A type of protein specialized for the transport of respiratory oxygen in certain organisms, including earthworms and human beings.

herbivore A heterotrophic organism that consumes plant matter as its primary source of nutrition. (See **primary consumer**.)

hermaphrodite An animal organism that produces both male and female gametes.

heterotroph An organism that typically carries on heterotrophic nutrition.

heterotroph hypothesis A scientific hypothesis devised to explain the probable origin and early evolution of life on earth.

heterotrophic nutrition A type of nutrition in which organisms must obtain their foods from outside sources of organic nutrients.

heterozygous A term used to refer to an allele pair in which the alleles have different, contrasting effects (for example, *Aa, RW*).

high blood pressure A disorder of the human transport system in which systolic and diastolic pressures register higher than normal because of narrowing of the artery opening.

histamine A chemical product of the body that causes irritation and swelling of the mucous membranes.

homeostasis The condition of balance and dynamic stability that characterizes living systems under normal conditions.

homologous chromosomes A pair of chromosomes that carry corresponding genes for the same traits.

homologous structures Structures present within different species that can be shown to have had a common origin but that may or may not share a common function.

homozygous A term used to refer to an allele pair in which the alleles are identical in terms of effect (for example, *AA, aa*).

hormone A chemical product of an endocrine gland that has a regulatory effect on the cell's metabolism.

host The organism harmed in a parasitic relationship.

hybrid A term used to describe a heterozygous genotype. (See **heterozygous**.)

hybridization A technique of plant/animal breeding in which two varieties of the same species are crossbred in the hope of producing offspring with the favorable traits of both varieties.

hydrogen bond   A weak electrostatic bond that holds together the twisted strands of DNA and RNA molecules.

hydrolysis   The chemical process by which a complex food molecule is split into simpler components through the addition of a molecule of water to the bonds holding it together.

hypocotyl   A portion of the plant embryo that specializes to become the root and lower stem of the adult plant.

hypothalamus   An endocrine gland whose secretions affect the pituitary gland.

identical twins   In human beings, twin offspring resulting from the separation of the embryonic cell mass of a single fertilization into two separate masses; such twins are genetically identical.

importation of organisms   The introduction of nonactive plants and animals into new areas where they compete strongly with native species—a negative aspect of human involvement with the environment.

in vitro fertilization   A laboratory technique in which fertilization is accomplished outside the mother's body using mature ova and sperm extracted from the parents' bodies.

inbreeding   A technique of plant/animal breeding in which a purebred variety is bred only with its own members so as to maintain a set of desired characteristics.

independent assortment   A pattern of inheritance in which genes on different, nonhomologous chromosomes are free to be inherited randomly and regardless of the inheritance of the others.

ingestion   The mechanism by which an organism takes in food from its environment.

inorganic compound   A chemical compound that lacks the element carbon or hydrogen (for example, table salt: NaCl).

insulin   A hormone, secreted by the islets of Langerhans, that regulates the storage of blood sugar as glycogen.

intercellular fluid (ICF)   The fluid that bathes cells and fills intercellular spaces.

interferon   A substance, important in the fight against human cancer, that may now be produced in large quantities with techniques of genetic engineering.

intermediate inheritance   Any pattern of inheritance in which the offspring expresses a phenotype different from the phenotypes of its parents and usually representing a form intermediate between them.

internal development   Embryonic development that occurs within the body of the female parent.

internal fertilization   Fertilization that occurs inside the body of the female parent.

interneuron   A type of neuron, located in the central nervous system, responsible for the interpretation of impulses received from sensory neurons.

intestine   A portion of the digestive tract in which chemical digestion and absorption of digestive end products occur.

intracellular digestion   A type of chemical digestion carried out within the cell.

iodine   A chemical stain used in cell study; an indicator used to detect the presence of starch. (See **staining**.)

islets of Langerhans   An endocrine gland, located within the pancreas, that produces the hormones insulin and glucagon.

karyotype   An enlarged photograph of the paired homologous chromosomes of an individual cell that is used in the detection of certain genetic disorders involving chromosome mutation.

karyotyping   A technique for the detection of human genetic disorders in which a karyotype is analyzed for abnormalities in chromosome structure or number.

kidney The excretory organ responsible for maintaining the chemical composition of the blood. (See **nephron**.)

kidney failure A disorder of the human excretory system in which there is a general breakdown of the kidney's ability to filter blood components.

kingdom A level of biological classification that includes a broad grouping of organisms displaying general structural similarity; five kingdoms have been named by scientists.

lacteal A small extension of the lymphatic system, found inside the villus, that absorbs fatty acids and glycerol resulting from lipid hydrolysis.

lactic acid fermentation A type of anaerobic respiration in which glucose is converted to two lactic acid molecules.

large intestine A portion of the digestive tract in which undigested foods are solidified by means of water absorption to form feces.

lateral meristem A plant growth region located under the epidermis or bark of a stem. (See **cambium**.)

Latin The language used in biological classification for naming organisms by means of binomial nomenclature.

lenticel A small pore in the stem surface that permits the absorption and release of respiratory gases within stem tissues.

leukemia A disorder of the human transport system in which the bone marrow produces large numbers of abnormal white blood cells. (See **cancer**.)

lichen A symbiosis of alga and fungus that frequently acts as a pioneer species on bare rock.

limiting factor Any abiotic or biotic condition that places limits on the survival of organisms and on the growth of species populations in the environment.

lipase Any lipid-hydrolyzing enzyme.

lipid An organic compound composed of carbon, hydrogen, and oxygen in which hydrogen and oxygen are *not* in a 2:1 ratio (for example, a wax, plant oil); many lipids are constructed of a glycerol and three fatty acids.

liver An accessory organ that stores glycogen, produces bile, destroys old red blood cells, deaminates amino acids, and produces urea.

lock-and-key model A theoretical model of enzyme action that attempts to explain the concept of enzyme specificity.

lung The major organ of respiratory gas exchange.

luteinizing hormone (LH) A pituitary hormone that regulates the conversion of the ovarian follicle into the corpus luteum.

lymph Intercellular fluid (ICF) that has passed into the lymph vessels.

lymph node One of a series of structures in the body that act as reservoirs of lymph and also contain white blood cells as part of the body's immune system.

lymph vessel One of a branching series of tubes that collect ICF from the tissues and redistribute it as lymph.

lymphatic circulation The movement of lymph throughout the body.

lymphocyte Λ type of white blood cell that produces antibodies.

lysosome A cell organelle that houses hydrolytic enzymes used by the cell in the process of chemical digestion.

Malpighian tubules In arthropods (for example, grasshopper), an organ specialized for the removal of metabolic wastes.

maltase A specific enzyme that catalyzes the hydrolysis (and dehydration synthesis) of maltose.

maltose A type of disaccharide; a maltose molecule is composed of two units of glucose joined together by dehydration synthesis.

marine biome An aquatic biome characterized by relatively stable conditions of moisture, salinity, and temperature.

marsupial mammal See **nonplacental mammal**.

mechanical digestion Any of the processes by which foods are broken apart physically into smaller particles.

medulla The portion of the human brain responsible for regulating the automatic processes of the body.

meiosis The process by which four monoploid nuclei are formed from a single diploid nucleus.

meningitis A disorder of the human regulatory system in which the membranes of the brain or spinal cord become inflamed.

menstrual cycle A hormone-controlled process responsible for the monthly release of mature ova.

menstruation The stage of the menstrual cycle in which the lining of the uterus breaks down and is expelled from the body via the vaginal canal.

meristem A plant tissue specialized for embryonic development. (See **apical meristem; lateral meristem; cambium**.)

mesoderm An embryonic tissue that differentiates into muscle, bone, the excretory system, and most of the reproductive system in the adult animal.

messenger RNA (mRNA) A type of RNA that carries the genetic code from the nuclear DNA to the ribosome for transcription.

metabolism All of the chemical processes of life considered together; the sum total of all the cell's chemical activity.

methylene blue A chemical stain used in cell study. (See **staining**.)

microdissection instruments Tools of biological study used to remove certain cell organelles from within cells for examination.

micrometer (μm) A unit of linear measurement equal in length to 0.001 millimeter (0.000001 meter), used for expressing the dimensions of cells and cell organelles.

mitochondrion A cell organelle that contains the enzymes necessary for aerobic respiration.

mitosis A precise duplication of the contents of a parent cell nucleus followed by an orderly separation of these contents into two new, identical daughter nuclei.

mitotic cell division A type of cell division that results in the production of two daughter cells identical to each other and to the parent cell.

Monera One of the five biological kingdoms; it includes simple unicellular forms lacking nuclear membranes (for example, bacteria).

monohybrid cross A genetic cross between two organisms both heterozygous for a trait controlled by a single allele pair. The phenotypic ratio resulting is 3:1; the genotypic ratio is 1:2:1.

monosaccharide A type of carbohydrate known also as a simple sugar; all monosaccharides have the molecular formula $C_6H_{12}O_6$.

motor neuron A type of neuron that carries command impulses from the central nervous system to an effector organ.

mucus A protein-rich mixture that bathes and moistens the respiratory surfaces.

multicellular Having a body that consists of large groupings of specialized cells (for example, human being).

multiple alleles A pattern of inheritance in which the existence of more than two alleles is hypothesized, only two of which are present in the genotype of any one individual.

muscle A type of tissue specialized to produce movement of body parts.

mutagenic agent Any environmental condition that initiates or accelerates genetic mutation.

mutation Any alteration of the genetic material, either a chromosome or a gene, in an organism.

mutualism A type of symbiosis beneficial to both organisms in the relationship.

nasal cavity   A series of channels through which outside air is admitted to the body interior and is warmed and moistened before entering the lung.

natural selection   A concept, central to Darwin's theory of evolution, to the effect that the individuals best adapted to their environment tend to survive and to pass their favorable traits on to the next generation

negative feedback   A type of endocrine regulation in which the effects of one gland may inhibit its own secretory activity while stimulating the secretory activity of another gland.

nephridium   An organ found in certain organisms, including the earthworm, specialized for the removal of metabolic wastes.

nephron   The functional unit of the kidney. (See **glomerulus**; **Bowman's capsule**.)

nerve   A structure formed from the bundling of neurons carrying sensory or motor impulses.

nerve impulse   An electrochemical change in the surface of the nerve cell.

nerve net   A network of nervelike cells in coelenterates such as the hydra.

neuron   A cell specialized for the transmission of nerve impulses.

neurotransmitter   A chemical substance secreted by a neuron that aids in the transmission of the nerve impulse to an adjacent neuron.

niche   The role that an organism plays in its environment.

nitrifying bacteria   In the nitrogen cycle, bacteria that absorb ammonia and convert it into nitrate salts.

nitrogen cycle   The process by which nitrogen is recycled and made available for use by other organisms.

nitrogen-fixing bacteria   A type of bacteria responsible for absorbing atmospheric nitrogen and converting it to nitrate salts in the soil.

nitrogenous base   A chemical unit composed of carbon, hydrogen, and nitrogen that is a component part of the nucleotide unit.

nitrogenous waste   Any of a number of nitrogen-rich compounds that result from the metabolism of proteins and amino acids in the cell. (See **ammonia**; **urea**; **uric acid**.)

nondisjunction   A type of chromosome mutation in which the members of one or more pairs of homologous chromosomes fail to separate during the disjunction phase of the first meiotic division.

nonplacental mammal   A type of mammal (marsupial) in which internal development is accomplished without the aid of a placental connection.

nucleic acid   An organic compound composed of repeating units of nucleotide.

nucleolus   A cell organelle located within the nucleus that functions in protein synthesis.

nucleotide   The repeating unit making up the nucleic acid polymer (for example, DNA, RNA).

nucleus   A cell organelle that contains the cell's genetic information in the form of chromosomes.

nutrition   The life function by which living things obtain food and process it for their use.

omnivore   A heterotrophic organism that consumes both plant and animal matter as sources of nutrition.

one gene–one polypeptide   A scientific hypothesis concerning the role of the individual gene in protein synthesis.

oogenesis   A type of meiotic cell division in which one ovum and three polar bodies are produced from each primary sex cell.

open transport system   A type of circulatory system in which the transport fluid is *not* always enclosed within blood vessels (for example, grasshopper).

oral cavity   In human beings, the organ used for the ingestion of foods.

oral groove The ingestive organ of the paramecium.

organ transplant An application of the study of immunity in which an organ or tissue of a donor is transplanted into a compatible recipient.

organelle A small, functional part of a cell specialized to perform a specific life function (for example, nucleus, mitochondrion).

organic compound a chemical compound that contains the elements carbon and hydrogen (for example, carbohydrate, protein).

organic evolution The mechanism thought to govern the changes in living species over geologic time.

osmosis A form of passive transport by which water is absorbed or released by cells.

ovary A female gonad that secretes the hormone estrogen, which regulates female secondary sex characteristics; the ovary also produces ova, which are used in reproduction.

overcropping A negative aspect of human involvement with the environment in which soil is overused for the production of crops, leading to exhaustion of soil nutrients.

overgrazing The exposure of soil to erosion due to the loss of stabilizing grasses when it is overused by domestic animals—a negative aspect of human involvement with the environment.

overhunting A negative aspect of human involvement with the environment in which certain species have been greatly reduced or made extinct by uncontrolled hunting practices.

oviduct A tube that serves as a channel for conducting mature ova from the ovary to the uterus; the site of fertilization and the earliest stages of embryonic development.

ovulation The stage of the menstrual cycle in which the mature ovum is released from the follicle into the oviduct.

ovule A structure located within the flower ovary that contains a mono-

ploid egg nucleus and serves as the site of fertilization.

ovum A type of gamete produced as a result of oogenesis in female animals; the egg, the female sex cell.

oxygen I8 A radioactive isotope of oxygen that is used to trace the movement of this element in biochemical reaction sequences.

oxygenated blood Blood that contains a high percentage of oxyhemoglobin.

oxyhemoglobin Hemoglobin that is loosely bound to oxygen for purposes of oxygen transport.

pH A chemical unit used to express the concentration of hydrogen ion (H^+), or the acidity, of a solution.

palisade layer A cell layer found in most leaves that contains high concentrations of chloroplasts.

pancreas An accessory organ that produces enzymes that complete the hydrolysis of foods to soluble end products; also the site of insulin and glucagon production.

parasitism A type of symbiosis from which one organism in the relationship benefits, while the other (the host) is harmed, but not ordinarily killed.

parathormone A hormone of the parathyroid gland that regulates the metabolism of calcium in the body.

parathyroid gland An endocrine gland whose secretion, parathormone, regulates the metabolism of calcium in the body.

passive immunity A temporary immunity produced as a result of the injection of preformed antibodies.

passive transport Any process by which materials are absorbed into the cell interior from an area of high concentration to an area of low concentration without the expenditure of cell energy (for example, osmosis, diffusion).

penis A structure that permits internal fertilization through direct implantation of sperm into the female reproductive tract.

peptide bond   A type of chemical bond that links the nitrogen atom of one amino acid with the terminal carbon atom of a second amino acid in the formation of a dipeptide.

peripheral nerves   Nerves in the earthworm and grasshopper that branch from the ventral nerve cord to other parts of the body.

peripheral nervous system   A major subdivision of the nervous system that consists of all the nerves of all types branching through the body. (See **autonomic nervous system; somatic nervous system.**)

peristalsis   A wave of contractions of the smooth muscle lining; the digestive tract that causes ingested food to pass along the food tube.

petal   An accessory part of the flower that is thought to attract pollinating insects.

phagocyte   A type of white blood cell that engulfs and destroys bacteria.

phagocytosis   The process by which an ameba surrounds and ingests large food particles for intracellular digestion.

pharynx   The upper part of the digestive tube that temporarily stores food before digestion.

phenotype   The observable trait that results from the action of an allele pair.

phenylketonuria (PKU)   A genetically related human disorder in which the homozygous combination of a particular mutant gene prevents the normal metabolism of the amino acid phenylalanine.

phloem   A type of vascular tissue through which water and dissolved sugars are transported in plants from the leaf downward to the roots for storage.

phosphate group   A chemical group made up of phosphorus and oxygen and that is a component part of the nucleotide unit.

phosphoglyceraldehyde (PGAL)   An intermediate product formed during photosynthesis that acts as the precursor of glucose formation.

photochemical reactions   A set of biochemical reactions in photosynthesis in which light is absorbed and water molecules are split. (See **photolysis.**)

photolysis   The portion of the photochemical reactions in which water molecules are split into hydrogen atoms and made available to the carbon-fixation reactions.

photosynthesis   A type of autotrophic nutrition in which green plants use the energy of sunlight to convert carbon dioxide and water into glucose.

phylum   A level of biological classification that is a major subdivision of the kingdom level, containing fewer organisms with greater similarity (for example, Chordata).

pinocytosis   A special type of absorption by which liquids and particles too large to diffuse through the cell membrane may be taken in by vacuoles formed at the cell surface.

pioneer autotrophs   The organisms supposed by the heterotroph hypothesis to have been the first to evolve the ability to carry on autotrophic nutrition.

pioneer species   In an ecological succession, the first organisms to inhabit a barren environment.

pistil   The female sex organ of the flower. (See **stigma; style; ovary.**)

pituitary gland   An endocrine gland that produces hormones regulating the secretions of other endocrine glands; the master gland.

placenta   In placental mammals, a structure composed of both embryonic and maternal tissues that permits the diffusion of soluble substances to and from the fetus for nourishment and the elimination of fetal waste.

placental mammal   A mammal species in which embryonic development occurs internally with the aid of a placental connection to the female parent's body.

plant   One of the five biological kingdoms; it includes multicellular organisms whose cells are bounded by cell walls and that are capable of photosynthesis (for example, maple tree).

plasma　The liquid fraction of blood, containing water and dissolved proteins.

plasma membrane　A cell organelle that encloses the cytoplasm and other cell organelles and regulates the passage of materials into and out of the cell.

platelet　A cell-like component of the blood that is important in clot formation.

polar body　One of three nonfunctional cells produced during oogenesis that contain monoploid nuclei and disintegrate soon after completion of the process.

polio　A disorder of the human regulatory system in which viral infection of the central nervous system may result in severe paralysis.

pollen　The male gamete of the flowering plant.

pollen tube　A structure produced by the germinating pollen grain that grows through the style to the ovary and carries the sperm nucleus to the ovule for fertilization.

pollination　The transfer of pollen grains from anther to stigma.

pollution control　The development of new procedures to reduce the incidence of air, water, and soil pollution—a positive aspect of human involvement with the environment.

polyploidy　A type of chromosome mutation in which an entire set of homologous chromosomes fail to separate during the disjunction phase of the first meiotic division.

polysaccharide　A type of carbohydrate composed of repeating units of monosaccharides that form a polymeric chain.

polyunsaturated fat　A type of fat in which many bonding sites are unavailable for the addition of hydrogen atoms.

population　All the members of a particular species in a given geographical location at a given time.

population control　The use of various practices to slow the rapid growth in the human population—a positive aspect of human interaction with the environment.

population genetics　A science that studies the genetic characteristics of a sexually reproducing species and the factors that affect its gene frequencies.

postnatal development　The growth and maturation of an individual from birth, through aging, to death.

prenatal development　The embryonic development that occurs before birth within the uterus. (See **gestation**.)

primary consumer　Any herbivorous organism that receives food energy from the producer level (for example, mouse); the second stage of a food chain.

primary sex cell　The diploid cell that undergoes meiotic cell division to produce monoploid gametes.

producer　Any autotrophic organism capable of trapping light energy and converting it to the chemical bond energy of food (for example, green plants); the organisms forming the basis of the food chain.

progesterone　A hormone produced by the corpus luteum and/or placenta that has the effect of maintaining the uterine lining and suppressing ovulation during gestation.

protease　Any protein-hydrolyzing enzyme.

protein　A complex organic compound composed of repeating units of amino acid.

Protista　One of the five biological kingdoms; it includes simple unicellular forms whose nuclei are surrounded by nuclear membranes (for example, ameba, paramecium).

pseudopod　A temporary, flowing extension of the cytoplasm of an ameba that is used in nutrition and locomotion.

pulmonary artery　One of two arteries that carry blood from the heart to the lungs for reoxygenation.

pulmonary circulation Circulation of blood from the heart through the lungs and back to the heart.

pulmonary vein One of four veins that carry oxygenated blood from the lungs to the heart.

pulse Rhythmic contractions of the artery walls that help to push the blood fluid through the capillary networks of the body.

punctuated equilibrium A theory of the time frame required for evolution that assumes that evolutionary change occurs in bursts with long periods of relative stability intervening.

pyramid of biomass A construct used to illustrate the fact that the total biomass available in each stage of a food chain diminishes from producer level to consumer level.

pyramid of energy A construct used to illustrate the fact that energy is lost at each trophic level in a food chain and is most abundant at the producer level.

pyruvic acid An intermediate product in the aerobic or anaerobic respiration of glucose.

receptor An organ specialized to receive a particular type of environmental stimulus.

recessive allele (gene) An allele (gene) whose effect is masked by that of its dominant allele.

recombinant DNA DNA molecules that have been moved from one cell to another in order to give the recipient cell a genetic characteristic of the donor cell.

recombination The process by which the members of segregated allele pairs are randomly recombined in the zygote as a result of fertilization.

rectum The portion of the digestive tract in which digestive wastes are stored until they can be released to the environment.

red blood cell Small, nonnucleated cells in the blood that contain hemoglobin and carry oxygen to body tissues.

reduction division See **meiosis.**

reflex A simple, inborn, involuntary response to an environmental stimulus.

reflex arc The complete path involving a series of three neurons (sensory, interneuron, and motor) working together in a reflex action.

regeneration A type of asexual reproduction in which new organisms are produced from the severed parts of a single parent organism; the replacement of loss or damaged tissues.

regulation The life process by which living things respond to changes within and around them, and by which all life processes are coordinated.

replication An exact self-duplication of the chromosome during the early stages of cell division; the exact self-duplication of a molecule of DNA.

reproduction The life process by which new cells arise from preexisting cells by cell division.

reproductive isolation The inability of species varieties to interbreed and produce fertile offspring because of variations in behavior or chromosome structure.

respiration The life function by which living things convert the energy of organic foods into a form more easily used by the cell.

response The reaction of an organism to an environmental stimulus.

restriction enzyme Any of a number of specific enzymes capable of "snipping" DNA molecules at particular base sequences.

rhizoid A rootlike fiber produced by fungi that secrete hydrolytic enzymes and absorb digested nutrients.

ribonucleic acid (RNA) A type of nucleic acid that operates in various ways to facilitate protein synthesis.

ribose A five-carbon sugar found as a component part of the nucleotides of RNA molecules only.

ribosomal RNA (rRNA) The type of RNA that makes up the ribosome.

ribosome A cell organelle that serves as the site of protein synthesis in the cell.

root   A plant organ specialized to absorb water and dissolved substances from the soil as well as to anchor the plant to the soil.

root hair   A small projection of the growing root that serves to increase the surface area of the root for absorption.

roughage   A variety of undigestible carbohydrates that add bulk to the diet and facilitate the movement of foods through the intestine.

runner   A type of vegetative propagation in which an above-ground stem (runner) produces roots and leaves and establishes new organisms with identical characteristics.

saliva   A fluid secreted by salivary glands that contains hydrolytic enzymes specific to the digestion of starches.

salivary gland   The gland that secretes saliva, which is important in the chemical digestion of certain foods.

salt   A chemical composed of a metal and a nonmetal joined by means of an ionic bond (for example, sodium chloride).

saprophyte   A heterotrophic organism that obtains its nutrition from the decomposing remains of dead plant and animal tissues (for example, fungi, bacteria).

saturated fat   A type of fat molecule in which all available bonding sites on the hydrocarbon chains are taken up with hydrogen atoms.

scrotum   A pouch extending from the wall of the lower abdomen that houses the testes at a temperature optimum for sperm production.

secondary consumer   Any carnivorous animal that derives its food energy from the primary consumer level (for example, a snake); the third level of a food chain.

secondary sex characteristics   The physical features, different in males and females, that appear with the onset of sexual maturity.

seed   A structure that develops from the fertilized ovule of the flower and germinates to produce a new plant.

seed dispersal   Any mechanism by which seeds are distributed in the environment so as to widen the range of a plant species. (See **fruit**.)

segregation   The random separation of the members of allele pairs that occurs during meiotic cell division.

self-pollination   A type of pollination in which the pollen of a flower pollinates another flower located on the same plant organism.

sensory neuron   A type of neuron specialized for receiving environmental stimuli, which are detected by receptor organs.

sepal   An accessory part of the flower that functions to protect the bud during development.

sessile   A term that relates to the unmoving state of certain organisms, including the hydra.

seta   One of several small, chitinous structures (setae) that aid the earthworm in its locomotor function.

sex chromosomes   A pair of homologous chromosomes carrying genes that determine the sex of an individual; these chromosomes are designated as X and Y.

sex determination   A pattern of inheritance in which the conditions of maleness and femaleness are determined by the inheritance of a pair of sex chromosomes (XX = female; XY = male).

sex linkage   A pattern of inheritance in which certain nonsex genes are located on the X sex chromosome but have no corresponding alleles on the Y sex chromosome.

sex-linked trait   A genetic trait whose inheritance is controlled by the genetic pattern of sex linkage (for example, color blindness).

sexual reproduction   A type of reproduction in which new organisms form as a result of the fusion of gametes from two parent organisms.

shell An adaptation for embryonic development in many terrestrial, externally developing species that protects the developing embryo from drying and physical damage (for example, birds).

sickle-cell anemia A genetically related human disorder in which the homozygous combination of a mutant gene leads to the production of abnormal hemoglobin and crescent-shaped red blood cells.

skeletal muscle A type of muscle tissue associated with the voluntary movements of skeletal levers in locomotion.

small intestine In human beings, the longest portion of the food tube, in which final digestion and absorption of soluble end products occur.

smooth muscle See **visceral muscle**.

somatic nervous system A subdivision of the peripheral nervous system that is made up of nerves associated with voluntary actions.

speciation The process by which new species are thought to arise from previously existing species.

species A biological grouping of organisms so closely related that they are capable of interbreeding and producing fertile offspring (for example, human being).

species presentation The establishment of game lands and wildlife refuges that have permitted the recovery of certain endangered species—a positive aspect of human involvement with the environment.

sperm A type of gamete produced as a result of spermatogenesis in male animals; the male reproductive cell.

spermatogenesis A type of meiotic cell division in which four sperm cells are produced from each primary sex cell.

spinal cord The part of the central nervous system responsible for reflex action as well as impulse conduction between the peripheral nervous system and the brain.

spindle apparatus A network of fibers that form during cell division and to which centromeres attach during the separation of chromosomes.

spiracle One of several small pores in arthropods, including the grasshopper, that serve as points of entry of respiratory gases from the atmosphere to the tracheal tubes.

spongy layer A cell layer found in most leaves that is loosely. packed and contains many air spaces to aid in gas exchange.

spore A specialized asexual reproductive cell produced by certain plants.

sporulation A type of asexual reproduction in which spores released from special spore cases on the parent plant germinate and grow into new adult organisms of the species.

staining A technique of cell study in which chemical stains are used to make cell parts more visible for microscopic study.

stamen The male reproductive structure in a flower. (See **anther**; **filament**.)

starch A type of polysaccharide produced and stored by plants.

stem A plant organ specialized to support the leaves and flowers of a plant as well as to conduct materials between the roots and the leaves.

stem cell An undifferentiated cell that resides in the tissues of the body and can give rise to new, specialized cells of any kind.

stigma The sticky upper portion of the pistil, which serves to receive pollen.

stimulus Any change in the environment to which an organism responds.

stomach A muscular organ that acts to liquefy food and that produces gastric protease for the hydrolysis of protein.

stomate A small opening that penetrates the lower epidermis of a leaf and through which respiratory and photosynthetic gases diffuse.

strata The layers of sedimentary rock that contain fossils, whose ages may be determined by studying the patterns of sedimentation.

stroke A disorder of the human regulatory system in which brain function is impaired because of oxygen starvation of brain centers.

stroma An area of the chloroplast within which the carbon-fixation reactions occur; each stroma lies between pairs of grana.

style The portion of the pistil that connects the stigma to the ovary.

substrate A chemical metabolized by the action of a specific enzyme.

succession A situation in which an established ecological community is gradually replaced by another until a climax community is established.

survival of the fittest The concept, frequently associated with Darwin's theory of evolution, that in the intraspecies competition among naturally occurring species, the organisms best adapted to the particular environment will survive.

sweat glands In human beings, the glands responsible for the production of perspiration.

symbiosis A term that refers to a variety of biotic relationships in which organisms of different species live together in close physical association.

synapse The gap that separates the terminal branches of one neuron from the dendrites of an adjacent neuron.

synapsis The intimate, highly specific pairing of homologous chromosomes that occurs in the first meiotic division, forming tetrads.

synthesis The life function by which living things manufacture the complex compounds required to sustain life.

systemic circulation The circulation of blood from the heart through the body tissues (except the lungs) and back to the heart.

systole The higher pressure registered during blood pressure testing. (See **diastole**.)

taiga A terrestrial biome characterized by long, severe winters and climax flora that includes coniferous trees.

Tay-Sachs disease A genetically related human disorder in which fatty deposits in the cells, particularly of the brain, inhibit proper functioning of the nervous system.

technological oversight A term relating to human activities that adversely affect environmental quality due to failure to assess the environmental impact of a technological development adequately.

teeth Structures located in the mouth that are specialized to aid in the mechanical digestion of foods.

temperate deciduous forest A terrestrial biome characterized by moderate climatic conditions and climax flora that includes deciduous trees.

template A pattern or design provided by the DNA molecule for the synthesis of protein molecules.

tendon A type of connective tissue that attaches a skeletal muscle to a bone.

tendonitis A disorder of the human locomotor system in which the junction between a tendon and a bone becomes irritated and inflamed.

tentacle A grasping structure in certain organisms, including the hydra, that contains stinging cells and is used for capturing prey.

terminal branch A cytoplasmic extension of the neuron that transmits a nerve impulse to adjacent neurons via the secretion of neurotransmitters.

terrestrial biome A biome that comprises primarily land ecosystems, the characteristics of which are determined by the major climate zone of the earth.

testcross A genetic cross to determine the genotype of an organism expressing a dominant phenotype; the unknown is crossed with a homozygous recessive.

testis A gonad in human males that secretes the hormone testosterone, which regulates male secondary sex characteristics; the testis also produces sperm cells for reproduction.

testosterone A hormone secreted by the testis that regulates the production of male secondary sex characteristics.

tetrad A grouping of four chromatids that results from synapsis.

thymine A nitrogenous base found only in DNA.

thyroid gland An endocrine gland that regulates the body's general rate of metabolism through secretion of the hormone thyroxin.

thyroid-stimulating hormone (TSH) A pituitary hormone that regulates the secretions of the thyroid gland.

thyroxin A thyroid hormone that regulates the body's general metabolic rate.

tongue A structure that aids in the mechanical digestion of foods.

trachea A cartilage-ringed tube that conducts air from the mouth to the bronchi.

tracheal tube An adaptation in arthropods (for example, grasshopper) that functions to conduct respiratory gases from the environment to the moist internal tissues.

Tracheophyta A phylum of the plant kingdom whose members (tracheophytes) contain vascular tissues and true roots, stems, and leaves (for example, geranium, fern, bean, maple tree, corn).

transfer RNA (tRNA) A type of RNA that transports specific amino acids from the cytoplasm to the ribosome for protein synthesis.

translocation A type of chromosome mutation in which a section of a chromosome is transferred to a nonhomologous chromosome.

transpiration The evaporation of water from leaf stomates.

transpiration pull A force that aids the upward conduction of materials in the xylem by means of the evaporation of water (transpiration) from leaf surfaces.

transport The life function by which substances are absorbed, circulated, and released by living things.

triplet codon A group of three nitrogenous bases that provide information for the placement of amino acids in the synthesis of proteins.

tropical forest A terrestrial biome characterized by a warm, moist climate and a climax flora that includes many species of broad-leaved trees.

tropism A plant growth response to an environmental stimulus.

tuber A type of vegetative propagation in which an underground stem (tuber) produces new tubers, each of which is capable of producing new organisms with identical characteristics.

tundra A terrestrial biome characterized by permanently frozen soil and climax flora that includes lichens and mosses.

tympanum A receptor organ in arthropods (for example, grasshopper) that is specialized to detect vibrational stimuli.

ulcer A disorder of the human digestive tract in which a portion of its lining erodes and becomes irritated.

ultracentrifuge A tool of biological study that uses very high speeds of centrifugation to separate cell parts for examination.

umbilical cord In placental mammals, a structure containing blood vessels that connects the placenta to the embryo.

unicellular Having a body that consists of a single cell (for example, paramecium).

uracil A nitrogenous base that is a component part of the nucleotides of RNA molecules only.

urea A type of nitrogenous waste with moderate solubility and moderate toxicity.

ureter In human beings, a tube that conducts urine from the kidney to the urinary bladder.

urethra In human beings, a tube that conducts urine from the urinary bladder to the exterior of the body. The urethra also conducts semen to the outside of the body.

uric acid A type of nitrogenous waste with low solubility and low toxicity.

urinary bladder An organ responsible for the temporary storage of urine.

urine A mixture of water, salts, and urea excreted from the kidney.

use and disuse A term associated with the evolutionary theory of Lamarck, since proved incorrect.

uterus In female placental mammals, the organ within which embryonic development occurs.

vaccination An inoculation of dead or weakened disease organisms that stimulates the body's immune system to produce active immunity.

vacuole A cell organelle that contains storage materials (for example, starch, water) housed inside the cell.

vagina In female placental mammals, the portion of the reproductive tract into which sperm are implanted during sexual intercourse and through which the baby passes during birth.

variation A concept, central to Darwin's theory of evolution, that refers to the range of adaptation that can be observed in all species.

vascular tissues Tubelike plant tissues specialized for the conduction of water and dissolved materials within the plant. (See **xylem**; **phloem**.)

vegetative propagation A type of asexual reproduction in which new plant organisms are produced from the vegetative (nonfloral) parts of the parent plant.

vein (human) A relatively thin-walled blood vessel that carries blood from capillary networks back toward the heart.

vein (plant) An area of vascular tissues located in the leaf that aid the upward transport of water and minerals through the leaf and the transport of dissolved sugars to the stem and roots.

vena cava One of two major arteries that return blood to the heart from the body tissues.

ventral nerve cord The main pathway for nerve impulses between the brain and peripheral nerves of the grasshopper and earthworm.

ventricle One of two thick-walled, muscular chambers of the heart that pump blood out to the lungs and body.

villi Microscopic projections of the lining of the small intestine that absorb the soluble end products of digestion. (See **lacteal**.)

visceral muscle A type of muscle tissue associated with the involuntary movements of internal organs (for example, peristalsis in the small intestine).

vitamin a type of nutrient that acts as a coenzyme in various enzyme-controlled reactions.

water cycle The mechanism by which water is made available to living things in the environment through the processes of precipitation, evaporation, runoff, and percolation.

water pollution A type of technological oversight that involves the addition of some unwanted factor (for example, sewage, heavy metals, heat, toxic chemicals) to our water resources.

Watson-Crick model A model of DNA structure devised by J. Watson and F. Crick that hypothesizes a twisted ladder arrangement for the DNA molecule also known as a "double helix."

white blood cell A type of blood cell that functions in disease control. (See **phagocyte**; **lymphocyte**.)

xylem A type of vascular tissue through which water and dissolved minerals are transported upward through a plant from the root to the stems and leaves.

yolk A food substance, rich in protein and lipid, found in the eggs of many animal species.

yolk sac The membrane that surrounds the yolk food supply of the embryos of many animal species.

zygote The single diploid cell that results from the fusion of gametes in sexual reproduction; a fertilized egg.

Examination
June 2015
Living Environment

PART A

Answer all questions in this part. [30]

Directions (1–30): For *each* statement or question, record in the space provided the *number* of the word or expression that, of those given, best completes the statement or answers the question.

1 Materials are transported within a single-celled organism by the

1 nucleus   3 mitochondrion
2 cytoplasm   4 ribosome   1_____

2 Which row in the chart below correctly pairs a food molecule with its building block?

Row	Food Molecule	Building Block
(1)	starch	amino acid
(2)	sugar	starch
(3)	protein	amino acid
(4)	amino acid	sugar

2_____

3 The flow of energy in an ecosystem is best described as energy moving in

   1 one direction from the Sun to the producers and then to the consumers

   2 one direction from a consumer to a producer and then to the Sun as heat and light

   3 two directions between the producers that are present

   4 two directions, back and forth, between the producers and the consumers   3 _____

4 Occasionally, during pregnancy, the placenta can separate from the uterus. This causes a disruption in development and sometimes death of the fetus. Harm to the developing fetus might occur because the placenta

   1 transfers oxygen and nutrients to the fetal blood

   2 sends maternal blood into the fetus

   3 supplies milk for the fetus

   4 breaks down wastes of the fetus   4 _____

5 Which process produces only identical offspring?

   1 meiotic cell division

   2 selective breeding

   3 cloning

   4 fertilization   5 _____

6 A photograph of human cells as seen with a compound light microscope is shown below. A cell structure is labeled A.

Structure A is most likely a

1 mitochondrion that synthesizes food for the cell
2 nucleus that is the site of food storage
3 mitochondrion that absorbs energy from the Sun
4 nucleus that is responsible for the storage of information

6____

7 A land-dwelling organism, A, and an aquatic single-celled organism, B, are represented below.

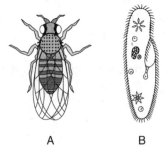

A   B

Which statement best explains how A and B are able to survive in their environments?

1 The organelles in B perform similar functions to the organ systems in A.
2 The transport system in B is more complex than the transport system in A.
3 Both A and B take in oxygen from the water.
4 Only A can pass on traits to offspring.

7____

8 A man is exposed to large amounts of ultraviolet radiation while sunbathing at the beach. This exposure causes a genetic change in the DNA of a skin cell. In the future, this change can be passed on to

1 his male and female children
2 his male children, only
3 all cells in his body
4 his skin cells, only 8 _____

9 Palm oil, produced from palm trees, is not only a biofuel, but is also used in food additives, cosmetics, and lubricants. Palm tree plantations are now cultivated in areas that were formerly natural forests. One ecological concern raised by this expansion is that

1 the natural forest ecosystem may harm the palm trees
2 the use of the land for agriculture will increase the biodiversity of the area
3 humans are changing the basic processes of the palm trees
4 planting large expanses of one crop reduces the biodiversity of the area 9 _____

10 Fishermen have harvested certain fish to the point where the population of that fish is decreasing. This level of direct harvesting could cause

1 ecosystems to be improved for future generations
2 ecosystems to be severely damaged
3 the restoration of environmental stability
4 all other fish species to increase in number 10 _____

11 Which phrase best describes a gene?

1 a segment of a DNA molecule found only in the body cells of an organism

2 a segment of a DNA molecule found only in the gametes of an organism

3 a segment of a DNA molecule that contains the instructions for producing a trait in an organism

4 a segment of a DNA molecule that contains the instructions for producing all the characteristics of an organism   11____

12 The molecule DNA contains the four bases listed below.

A – adenine
C – cytosine
G – guanine
T – thymine

Which base pairings normally occur during DNA replication?

1 Guanine pairs with cytosine. Thymine pairs with thymine.

2 Adenine pairs with thymine. Cytosine pairs with guanine.

3 Thymine pairs with guanine. Cytosine pairs with adenine.

4 Cytosine pairs with cytosine. Thymine pairs with thymine.   12____

13 Evolution of a species could occur as a result of changes in the

1 DNA in muscle cells
2 base sequences in liver cells
3 genes in an egg cell
4 number of chromosomes in a fetal bone cell   13____

14 One positive impact that industrialization has had is that

1 industrialization produces waste gases that pollute the air
2 fossil fuels used by industries help reduce finite resources
3 industrialization has been a source of many jobs for people
4 new technologies have increased acid rain   14____

15 When receiving x rays, individuals wear a lead shield over major organs in order to limit the body's exposure to radiation. One reason for this procedure is to

1 protect the patient against broken bones
2 prevent mutations in gametes
3 improve circulation in the patient
4 increase the chance of a change in DNA   15____

16 When an ant in a colony dies, the live ants will throw the dead ant out of the anthill. If a live ant from the colony, ant X, is sprayed with a chemical characteristic of dead ants, the live ants will repeatedly throw this ant out of the anthill until they can no longer detect the chemical on ant X. What is the best explanation for this behavior?

1 The ants are responding to a chromosomal mutation in ant X.
2 The chemical is exhibiting a feedback mechanism.
3 The live ants must continue this behavior until they have eliminated ant X.
4 The chemical acts as a stimulus for a particular behavior.   16____

17 Rabbits produce large numbers of offspring during each reproductive season, yet the number of rabbits within a given population changes very little from year to year. The stability of the population size is most likely the result of

1 the development of mutations in young rabbits
2 environmental factors that keep the population in check
3 rabbits continuing to reproduce when the population is large
4 the survival of more female rabbits than male rabbits   17 _____

18 Genetic engineering has the potential to correct human genetic disorders. In gene therapy, a defective gene is replaced by using a virus to insert a normal gene into the cells of an individual. This treatment will be most successful if the virus is inserted into cells that

1 lack a nucleus
2 are recycled after death, rather than removed from the body
3 carry out one specific function, rather than multiple functions
4 continue to divide during the life of the patient   18 _____

19 In one town, some people support a proposal to build a shopping mall on a large, undeveloped lot, because it would increase business and create new jobs. As a trade-off, the shopping mall would cause a decrease in the

1 amount of air pollution
2 volume of garbage and litter
3 amount of wastewater entering the local sewage system
4 variety of wildlife populations in the area   19 _____

20 The human female reproductive system is represented below.

Which structure produces chemicals that regulate the reproductive cycle?

1 A	3 C	
2 B	4 D	20 _____

21 The diagram below represents a cell structure involved in converting energy stored in organic molecules into a form used by animal cells.

The arrows represent the movement of which substances?

1 carbon dioxide and sugar
2 oxygen and ATP
3 ATP and carbon dioxide
4 oxygen and sugar   21 _____

22 The diagram below shows a concept map.

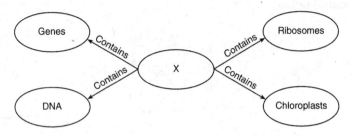

Which label correctly identifies what X represents in the concept map?

1 nucleus   3 autotrophic cell

2 chromosome   4 heterotrophic cell   22 _____

23 The diagrams below represent two molecules that are involved in metabolic activities in some living cells.

The shape of each of the molecules is important because

1 molecules having different shapes are always found in different organisms

2 the shape of a molecule determines how it functions in chemical reactions

3 the shape of a molecule determines the age of an organism

4 if the shape of any molecule in an organism changes, the DNA in that organism will also change   23 _____

24 In the early 1900s, experiments were conducted on two caterpillar species. The members of the two species were each divided into two groups. One group of each species was placed under red light, while the other group of each species was kept in the dark. When the caterpillars developed into butterflies, their wings showed extreme color differences. Exposure to red light resulted in intensely colored wings, while those kept in the dark had paler wing colors. The color differences were most likely due to

1 mutations in the color-producing genes
2 the caterpillars in the red light producing more DNA
3 gene expression being affected by the environment
4 the caterpillars in the dark evolving less than those in the light

24 _____

25 A student used a microscope to observe a single-celled organism. As he watched, it looked as if the organism split into two cells. He made drawings, shown below, of the organism over a short period of time.

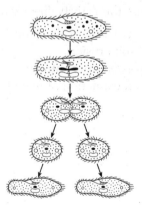

Which process did the student record in his drawings?

1 genetic engineering   3 selective breeding
2 asexual reproduction   4 gamete formation

25 _____

26 Medical professionals are concerned with the increase in the number of bacterial species that are resistant to antibiotics. Once resistance appears in a bacterial population, it spreads rapidly. This is most likely because

1 populations of resistant bacteria are small
2 exposure to antibiotics increases the rate of repro-duction in bacteria
3 resistant bacteria are small when compared to non-resistant bacteria.
4 resistant bacteria survive in greater numbers and pass the trait to their offspring

26 _____

27 When getting a vaccination, which substance is injected into the body?

1 bacteria to combat a pathogen
2 white blood cells to engulf a pathogen
3 a weakened form of a virus
4 antibiotics to kill a virus

27 _____

28 Many beverage companies are required to recycle bottles and cans because this activity directly reduces

1 air pollution and destruction of the ozone shield
2 overpopulation and soil erosion
3 solid waste and depletion of resources
4 thermal pollution and extinction of wildlife

28 _____

29 The diagram below shows some of the DNA in a bacterium into which a human gene, X, has been successfully inserted.

The bacteria that result from reproduction of this cell will most likely have the ability to

1 replicate all of the genetic instructions found in humans
2 produce vaccines to be used to immunize humans
3 produce a human blood cell according to instructions in gene X
4 produce the human protein coded for by gene X   29 _____

30 The Eurasian water milfoil is a nonnative species, which was once commonly sold as an aquarium plant, and is now found growing in many lakes in New York State. It has few natural enemies, and grows rapidly, crowding out many native species. This plant ruins fishing areas and interferes with boating and other water sports. This is an example of

1 human consumption of finite resources
2 an unintended consequence of adding an organism to an ecosystem
3 an abiotic factor having a negative effect on an ecosystem
4 the introduction of a species that has increased the long-term biodiversity of an ecosystem   30 _____

PART B-1

Answer all questions in this part. [13]

Directions (31–43): For *each* statement or question, record in the space provided the *number* of the word or expression that, of those given, best completes the statement or answers the question.

31 The graph below shows the size of a population of foxes over a period of years.

If the line did not stay around the carrying capacity, but continued to rise, which concept would this graph best illustrate?

1 environmental stability
2 genetic variety
3 behavioral change
4 overproduction   31 _____

32 A food web is represented below.

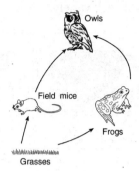

Which organism would receive the *least* amount of transferred solar energy?

1 grasses   3 frogs
2 owls   4 field mice   32 _____

33 Birch bolete is a fungus that normally grows on the roots of birch trees in New York State. During the life of the fungus and the birch, each organism receives nutrients from the various biochemical processes of the other. According to this information, it can be inferred that these two species

1 are both predators
2 require the same amount of sunlight
3 require a similar soil pH
4 recycle the remains of dead organisms   33 _____

34 The photographs below show different varieties of cattle and characteristics of each variety.

A
Good resistance to heat but poor beef

B
Good beef but poor resistance to heat

C
Good resistance to heat and good beef

Which statement best explains the development of variety *C*?

1 Nuclei from body cells taken from variety *A* were inserted into egg cells lacking nuclei taken from variety *B*.

2 Selective breeding was used to combine desirable traits from both varieties *A* and *B*.

3 The need to adapt to changes in the environment led to the selection of advantageous characteristics in the offspring of variety *B*.

4 Mutations that occurred in the body cells of variety *A* were passed on to the offspring generation after generation.

34 _____

35 The diagram below represents a remora fish attached to a shark.

A remora fish has an adhesive disk or sucker on its head, which it uses to attach itself to larger fishes, such as sharks. This attachment causes the shark no harm. The remora fish eat scraps of food that the sharks drop as they feed. This is an example of

1 an adaptation to a specialized niche
2 an adaptation of a successful parasite
3 competition between two fish species for food
4 competition for abiotic resources

35 _____

36 Each row in the chart below represents a different population of the same species of insect. Which row shows the population with the greatest chance of survival in a changing environment?

36 _____

37 The development of nerve, muscle, and skin cells is represented in the diagram below.

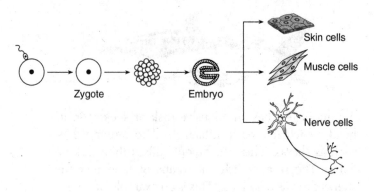

Which statement best explains how each of the different cell types can develop from the same embryo?

1 The cells have identical genetic instructions, but different parts of these instructions are being expressed in each cell.

2 The cells have identical genetic instructions, and all parts of these instructions are being expressed in each cell.

3 The cells are produced by asexual reproduction and contain identical genetic instructions.

4 The cells contain genetic instructions from two different parents and will express the instructions from one parent, only.   37 _____

38 The graph below represents the effect of pH on three different enzymes at normal body temperature.

The graph illustrates that enzymes 1, 2, and 3

1 are not affected by pH
2 work best at different pH levels
3 work best in an acidic environment
4 work best in a basic environment   38 _____

39 The human male reproductive system is represented below.

Which structure produces cells that have the potential to become gametes?

1 A   3 C
2 B   4 D   39 _____

40 Some scientists have collected and stored seeds for many types of food-producing plants. The purpose of this is to

1 increase the destruction of environments
2 continue the deforestation of world ecosystems
3 decrease the dependence on plants for food
4 preserve the diversity of plant species   40 _____

41 Which diagram best illustrates the relationship between the number of cells, tissues, and organs in a complex multicellular organism?

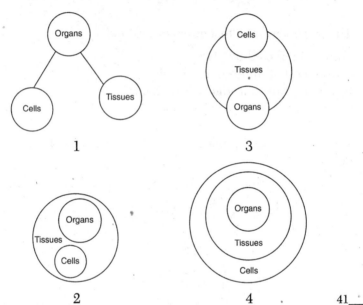

41 _____

Base your answers to questions 42 and 43 on the diagram below, which represents an ameba engulfing bacteria, and on your knowledge of biology.

Bacterium   Bacteria are in   Bacteria are
a cell structure   broken down

Ameba

42 This ameba would most likely be classified as a

1 decomposer
2 producer
3 consumer
4 pathogen   42 _____

43 The activity taking place is

1 photosynthesis
2 differentiation
3 autotrophic nutrition
4 heterotrophic nutrition   43 _____

PART B–2

Answer all questions in this part. [12]

Directions (44–55): For those questions that are multiple choice, record your answers in the space provided. For all other questions in this part, record your answers in accordance with the directions.

Base your answers to questions 44 through 48 on the information and data table below and on your knowledge of biology.

The Enzyme Catalase

Catalase is an enzyme found in nearly all living organisms that breathe or are exposed to oxygen. According to recent scientific studies, low levels of catalase may play a role in the graying process of human hair. The body naturally produces hydrogen peroxide, and catalase breaks it down into water and oxygen. If there is a dip in catalase levels, hydrogen peroxide cannot be broken down. This causes hydrogen peroxide to bleach hair from the inside out. Scientists believe this finding may someday be used in anti-graying treatments for hair.

A pharmaceutical company, investigating ways to prevent hair from turning gray, took tissue samples from two different individuals. Both individuals were the same age. Each of the samples was placed in a solution of hydrogen peroxide. The volume of oxygen gas produced was measured every 5 minutes for 25 minutes. The data the company collected are shown below.

**Oxygen Production in the Breakdown of
Hydrogen Peroxide by Catalase**

Time (min)	Sample from Person A (mL oxygen)	Sample from Person B (mL oxygen)
5	2.0	4.5
10	3.5	8.5
15	5.0	12.0
20	7.5	15.5
25	9.5	20.0

Directions (44–46): Using the information in the data table, construct a line graph on the grid on the next page, following the directions below.

44 Mark an appropriate scale, without any breaks in the data, on each labeled axis.   [1]

45 Plot the data from the data table for the sample from person A on the grid. Connect the points and surround each point with a small circle.   [1]

Example: ⊙—⊙⌒⊙

46 Plot the data from the data table for the sample from person B on the grid. Connect the points and surround each point with a small triangle.   [1]

Example: △—△⌒△

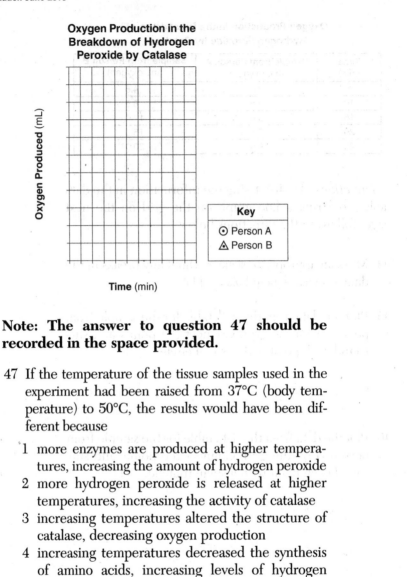

Oxygen Production in the Breakdown of Hydrogen Peroxide by Catalase

Oxygen Produced (mL)

Time (min)

Key
⊙ Person A
△ Person B

Note: The answer to question 47 should be recorded in the space provided.

47 If the temperature of the tissue samples used in the experiment had been raised from 37°C (body temperature) to 50°C, the results would have been different because

1 more enzymes are produced at higher temperatures, increasing the amount of hydrogen peroxide
2 more hydrogen peroxide is released at higher temperatures, increasing the activity of catalase
3 increasing temperatures altered the structure of catalase, decreasing oxygen production
4 increasing temperatures decreased the synthesis of amino acids, increasing levels of hydrogen peroxide

47 _____

48 According to the data provided, which person, *A* or *B*, is more likely to be the first to have gray hair? Support your answer. [1]

Person: _____

_____

_____

Base your answers to questions 49 through 52 on the information and graph below and on your knowledge of biology.

An investigation was carried out to determine the effect of drinking an excessive amount of water on urine flow. A subject drank 1 liter of water in 5 minutes, and then urine output was measured. The graph shows how the human adult kidneys responded to regulate water balance in the body. Urine output was measured every 10 minutes for a little over 3 hours. Normal output for an average adult is approximately 0.5–1 mL/min.

Urine Production in an Adult with Normal Kidney Function

Note: The answer to question 49 should be recorded in the space provided.

49 One half-hour after the liter of water was consumed, the urine produced by the kidneys was

1 between 2 and 3 mL/min
2 between 4 and 5 mL/min
3 eight times greater than normal
4 below the normal range   49 _____

Note: The answer to question 50 should be recorded in the space provided.

50 The change in urine production during this 3-hour period was most likely the result of

1 antibody production
2 homeostatic feedback
3 enzymatic breakdown of the water consumed
4 nerve cell malfunctions of the kidneys   50 _____

51 Identify a structure, in organisms that do not have kidneys, that is adapted to regulate water balance. [1]

_____

52 Approximately how long did it take, in minutes, for the body to return to normal after the intake of water? [1]

_____ **minutes**

Base your answers to questions 53 through 55 on the information and data table below and on your knowledge of biology.

The data table summarizes the changes that occurred to farmland in the years immediately following its abandonment. The land is located in a very stable ecosystem. It was abandoned after years of overuse and weathering, which resulted in the depletion of soil nutrients.

Common Types of Vegetation Present

Years Since Abandoned	Grasses and Weeds	Shrubs	Pine Forest	Hardwood Forest
1	X			
18	X	X	X	
30			X	
70			X	X
100				X
118 (present)				X

53 Which type of vegetation appears to have the lowest soil nutrient requirements? Support your answer with information from the data table. [1]

Lowest soil nutrient requirement vegetation: _____

_____

_____

_____

54 Assuming the ecosystem remains undisturbed, which type of vegetation would you expect to be most common in this area 200 years after it was first abandoned? Support your answer. [1]

Most common vegetation: _____

_____

_____

_____

55 Describe how the types of vegetation present on this farmland would change if a fire burned down all the trees 120 years after the land was abandoned. [1]

_____

_____

PART C

Answer all questions in this part. [17]

Directions (56–72): Record your answers in the spaces provided.

Base your answers to questions 56–58 on the information and photograph below and on your knowledge of biology.

The photograph below is part of an advertisement used by a company selling solar panels. The company claims that their panels, like plants, provide clean, renewable energy. They also claim that using solar panels will have a positive effect on the biosphere by reducing global warming.

Source:http://www.stockwatch.in/files/Energy.jpg

56–58 Explain why these claims are valid. In your answer, be sure to:

- explain why both plants and solar panels provide renewable energy, rather than nonrenewable energy [1]
- state how the widespread use of solar panels to generate electricity can help to reduce global warming [1]
- state how the energy-capturing process used by plants worldwide can help to reduce global warming [1]

_____

_____

_____

_____

_____

_____

_____

_____

Base your answers to questions 59 and 60 on the information below and on your knowledge of biology.

Fungi are interesting organisms that interact with humans in many ways. Yeasts are fungi used in the food industry to produce products such as bread and certain beverages. Some fungi are valuable in medicine. For example, the drug cyclosporine, which is capable of suppressing the response of the immune system to foreign antigens, and the antibiotic penicillin are both products from fungi. Other fungi are less welcomed by humans. The irritation of athlete's foot is caused by a fungus, and a number of allergies are caused by reproductive spores released by fungi.

59 Describe the role of a drug like cyclosporine when transplanting organs from one person to another person. [1]

_____

_____

60 Explain the difference between an infection caused by a fungus and an allergy caused by a fungus. [1]

_____

_____

_____

Base your answers to questions 61 and 62 on the information below and on your knowledge of biology.

Female mosquitoes need a meal of blood from a person or other animal in order to produce eggs. It has been discovered that mosquitoes have cells on their antennae that can detect the insect repellent known as DEET. The repellent is not harmful to mosquitoes, but when mosquitoes detect DEET, they will not land on the surface where the DEET has been applied. This protects people from being bitten by mosquitoes.

Recently, scientists found some mosquitoes that are resistant to DEET because they do not detect its presence. They bred these mosquitoes and eventually produced a population consisting of about 50% DEET-resistant insects.

61 Identify the process most likely responsible for a mosquito initially becoming resistant to DEET. [1]

_____

62 Mosquitoes with DEET resistance have been found in natural environments. Explain how the continued use of this repellent may cause the percentage of these resistant mosquitoes to increase in the future. [1]

_____

_____

Base your answers to questions 63 through 67 on the passage below and on your knowledge of biology.

Ocean-dwelling (marine) iguanas and land iguanas inhabit the Galapagos Islands. Some scientists believe that both types of iguanas diverged from a common ancestor. Marine iguanas eat algae. Land iguanas feed on cacti. Algae are more abundant in the ocean than cacti are on the islands. Both species lay their eggs in the sand.

Rats, cats, and goats have been introduced to the islands by humans. Rats feed on iguana eggs, cats eat baby iguanas, and goats eat cacti.

63 Identify the process by which ancestral iguanas developed into the present-day marine iguanas and land iguanas of the Galapagos Islands. [1]

Process: _____

64 Identify *one* organism in the Galapagos Islands that directly limits the population of both the marine iguanas and land iguanas. [1]

Organism: _____

65 Which population of iguanas, marine or land, would you expect to be larger? Support your answer. [1]

Population of iguana: _____

_____

_____

66 Would the introduction of goats have a greater effect on the population of the marine iguanas or the land iguanas? Support your answer. [1]

Population of iguana: _____

_____

_____

67 Identify *one* technique that can be used to support the conclusion that these two species of iguana developed from a common ancestor. [1]

Technique: _____

Base your answers to questions 68–72 on the information and diagram below and on your knowledge of biology.

The presence of air is believed to be important for root growth in bean plants. The apparatus available to conduct an investigation is shown below. There are enough bottles and other materials to have multiple setups. Air (for aeration) can be bubbled into the bottle through the rubber tube.

Bean seedling
Plastic modeling clay
Rubber tube from aerating pump
Plant nutrient solution

Source: Biology Handbook, SED 1960

68–72 Design an experiment to test the effect of aeration on the growth of roots of bean seedlings. In your answer, be sure to:

- state *one* hypothesis the experiment would test [1]
- describe how the control group will be treated differently from the experimental group [1]
- identify the dependent variable in the experiment [1]
- state *one* reason why many setups should be used in both the experimental and control groups [1]
- state *one* reason why several different kinds of seedlings were *not* tested in this experiment [1]

_____

_____

_____

_____

_____

_____

_____

_____

_____

PART D

Answer all questions in this part. [13]

Directions (73–85): **For those questions that are multiple choice, record your answers in the space provided. For all other questions in this part, record your answers in accordance with the directions.**

Note: The answer to question 73 should be recorded in the space provided.

73 The buildup of waste products in muscle cells that are active might cause

1 digestion 3 increased fatigue
2 cellular respiration 4 decreased heart rate 73 _____

Note: The answer to question 74 should be recorded in the space provided.

74 The diagram below shows the evolutionary relationships among several types of mammals.

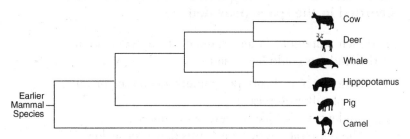

Which mammal would be most closely related to a hippopotamus?

1 deer 3 pig
2 whale 4 cow 74 _____

Base your answers to questions 75 and 76 on the information and data table below and on your knowledge of biology.

A group of students obtained the following data while trying to determine the effect of exercise on pulse rate.

Effect of Exercise on Pulse Rate

Student	Resting Pulse Rate (beats per minute)	Pulse Rate After Exercising (beats per minute)
A	66	92
B	82	107
C	65	97
D	74	124
E	79	118
F	68	98
G	89	122

Note: The answer to question 75 should be recorded in the space provided.

75 Which statement is an example of an observation the students could have made?

1 Pulse rates in beats per minute decrease for all people after exercise.
2 Student A most likely exercises regularly.
3 The pulse rate of student C was dangerously low.
4 The pulse rate of student F increased by 30 beats per minute.

75 _____

Note: The answer to question 76 should be recorded in the space provided.

76 Which two body systems were most actively involved in this experiment?

1 respiratory and immune
2 digestive and endocrine
3 respiratory and circulatory
4 immune and circulatory

76 _____

77 The diagram below represents a green plant cell viewed with the high power of a compound light microscope before and after a particular substance was added.

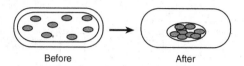

Before   After

Identify a substance that could have been added to the slide to bring about the change shown. [1]

_____

78 Using the DNA base sequences below, identify which *two* species are more closely related. Support your answer. [1]

　　Species A: CAC GTG GAC AGA GGA CAC CTC

　　Species B: CAT GTG GAC AGA GGA CAC CTC

　　Species C: CAC GTA GAC TGA GGA CTT CTC

Species: _____ and _____

_____

_____

79 A student observing onion cells using a microscope was having difficulty seeing any detail in the cells. State *one* action the student could take to improve the detail. [1]

_____

Base your answers to questions 80 and 81 on the diagram below and on your knowledge of biology. The diagram represents the results of paper chromatography performed on extracts from five organisms.

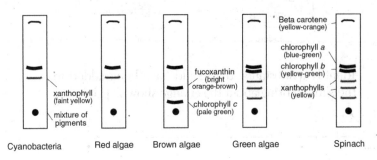

Cyanobacteria   Red algae   Brown algae   Green algae   Spinach

80 Identify *one* pigment molecule common to all five organisms. [1]

_____

Note: The answer to question 81 should be recorded in the space provided.

81 Which two organisms are most closely related?

1 cyanobacteria and green algae
2 red algae and spinach
3 brown algae and red algae
4 red algae and cyanobacteria

81 _____

Base your answers to questions 82 through 84 on the diagram below and on your knowledge of biology. The diagram shows variations in the beaks of finches in the Galapagos Islands.

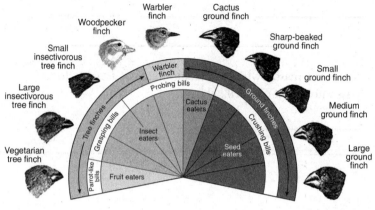

Source: www.pbs.org

Note: The answer to question 82 should be recorded in the space provided.

82 In this diagram, the variety of beak sizes and shapes are adaptations directly related to successful

1 feeding   3 defense
2 camouflage   4 singing   82 _____

83 State *one* reason why the large ground finch and the woodpecker finch can live successfully on the same island. [1]

_____

_____

84 Identify *one* finch in the diagram that is *least* likely to compete with any of the other finches. Support your answer. [1]

_____

_____

_____

85 State *one* reason why a molecule may *not* be able to pass into or out of a cell. [1]

_____

_____

ANSWERS TO REGENTS EXAMINATION

LIVING ENVIRONMENT

JUNE 2015

PART A

1. 2	6. 4	11. 3	16. 4	21. 3	26. 4
2. 3	7. 1	12. 2	17. 2	22. 3	27. 3
3. 1	8. 4	13. 3	18. 4	23. 2	28. 3
4. 1	9. 4	14. 3	19. 4	24. 3	29. 4
5. 3	10. 2	15. 2	20. 1	25. 2	30. 2

PART B–1

31. 4	34. 2	36. 1	38. 2	40. 4	42. 3
32. 2	35. 1	37. 1	39. 1	41. 4	43. 4
33. 3					

PART B–2

47. 3
49. 1
50. 2

PART D

73. 3
74. 2
75. 4
76. 3
81. 4
82. 1

In Parts B-2, C, and D, for any answers not listed here, you are required to show your work and explain your answers.

Examination
June 2016
Living Environment

PART A

Answer all questions in this part. [30]

Directions (1–30): For *each* statement or question, record in the space provided the *number* of the word or expression that, of those given, best completes the statement or answers the question.

1 Which statement is an example of the interdependence of organisms?

  1 Owls hunt at night.
  2 Ants get food from insects and protect insects from predators.
  3 Ticks feed on the blood of animals and the ticks grow larger.
  4 Crows feed on dead mice.

1 _____

2 Residents of a town are concerned that a recently built factory could pose health risks. Scientists were asked to investigate the effects of the factory on the health of local residents. The most relevant information they reported was that

  1 in a survey, residents felt that the air in town looks dirtier now
  2 there have been reports that other types of factories have been linked with health issues
  3 residents have occasionally seen smoke coming from the factory
  4 local medical facilities have recently reported a 15% increase in the number of patients treated for asthma

2 _____

3 Farmers may someday clone their best milk-producing cow into a whole herd. What potential disadvantage might be important to consider in having such a large group of clones on one farm?

1 It may be difficult to tell the animals apart.
2 Lack of variation may limit survival in the herd.
3 The cows could be fertilized by only one type of bull.
4 The cows could be mated only with each other.   3_____

4 DNA replication occurs in preparation for

1 mitosis, only
2 meiosis, only
3 both mitosis and meiosis
4 neither mitosis nor meiosis   4_____

5 An individual eats a hamburger. Which two systems must interact to transfer the nutrients in the hamburger to human muscle tissue?

1 respiratory and excretory
2 digestive and immune
3 digestive and circulatory
4 circulatory and respiratory   5_____

6 The diagram below shows cell A completing a life process.

Cell A

Cell A performs functions similar to the tissues and systems in complex, multicellular organisms. This process results in

1 increased genetic variation
2 the maintenance of homeostasis
3 a reduction in competition
4 increased autotrophic nutrition   6_____

7 The shape of a protein molecule directly determines its

  1 movements through the cytoplasm
  2 functions inside and outside of cells
  3 roles in building water molecules
  4 circulation throughout the body   7_____

8 The diagram below represents a cell and some molecules in its environment.

Which molecule would require the use of energy in order to be brought into the cell?

  1 1   3 3
  2 2   4 4   8_____

9 Many domestic plants that are currently used for food by humans share a wild plant ancestor. The changes that have occurred in four common plants and the results are shown in the chart below.

Wild Plant Ancestor	Change That Occurred	Resulting Modern Plant
wild mustard	reduced flower development	broccoli
wild mustard	sterile flowers	cauliflower
wild mustard	enlargement of leaves	kale
wild mustard	shortened stem length	cabbage

What event most likely produced the changes that occurred in the wild plant ancestor?

  1 Mutations in wild mustard sex cells were passed on to offspring.
  2 Humans did not like to eat wild mustard.
  3 Competition for survival occurred in all ecosystems of the world.
  4 Ancient herbivores overgrazed wild mustard.   9_____

10 A strand of DNA in a skin cell contains the bases:

A-T-G-C-C-A-T-C-G-G-T-A

After the cell is exposed to ultraviolet light, the strand contains the bases:

A-T-G-G-C-C-A-T-C-G-G-T-A

Which statement describes the result of this exposure?

1 A new base has been inserted.
2 A base has been deleted.
3 One base has been substituted for another.
4 There have been no changes in the bases.   10____

11 An ameba, a one-celled organism, can move, ingest, and transport materials within the cell, because it has

1 organs   3 tissues
2 organelles   4 systems   11____

12 In humans, the placenta is essential to the embryo for

1 nutrition, excretion, and reproduction
2 respiration, nutrition, and excretion
3 movement, reproduction, and nutrition
4 coordination, movement, and growth   12____

13 A student infected by a common cold virus ran a low-grade fever. After a few days, the student's temperature returned to normal and the student was free of cold symptoms. The fever served as

1 an antigen in the circulatory system
2 an immune response to a pathogen
3 a biological catalyst
4 a weakened pathogen   13____

14 Many animals have developed courtship behaviors. Males will often dance, swim, or sing in a particular way to attract a female. Males who are more successful at the courtship behavior will have a greater chance of having more offspring. This behavior is a result of

1 natural selection
2 genetic engineering
3 asexual reproduction
4 gene manipulation   14 _____

15 After a zygote is formed, specialization of cells occurs. Through which process do the cells of a zygote become specialized?

1 sexual reproduction   3 fertilization
2 meiosis   4 differentiation   15 _____

16 A farmer wanted to rid his apple trees of a particular leaf-eating insect. He sprayed his trees with an insecticide that killed 98% of the insects. The survival of 2% of this population of insects is most likely due to

1 genes obtained from another species
2 certain chemicals that stimulated overproduction
3 variations that resulted from sexual reproduction
4 their ability to produce food from the pesticide   16 _____

17 Which occurrence represents an example of evolution?

  1 Exposure to radiation reduces the rate of mutation in leaf cells.
  2 A mutation in a liver cell causes a person to produce an enzyme that is less efficient.
  3 Cells in a zygote eventually change into bone cells or skin cells.
  4 Some antibiotics are almost useless, because pathogens have developed a resistance to these antibiotics.

17 _____

18 Populations of aspen trees in the western United States are being destroyed by an unexplained illness. The altered landscape is affecting the animals that live there. Populations of deer mice are increasing greatly in these areas. Unfortunately, these mice often carry a virus that is deadly to humans. This scenario best illustrates that

  1 a change in the environment always results in disease
  2 humans are the cause of the breakdown of this ecosystem
  3 the stability of this ecosystem is limited by the amount of water available
  4 every population in an ecosystem is linked with other populations

18 _____

19 The paramecium is a single-celled organism that reproduces asexually. The offspring of a paramecium usually contain

  1 only half of the genes of the parent cells
  2 more DNA than the parent cell
  3 genetic material identical to that of the parent cell
  4 fewer mutations than the parent cell

19 _____

20 A dead or weakened pathogen used to establish immunity would most likely be found in

  1 a pesticide   3 a vaccine
  2 an antibiotic   4 a toxin

20 _____

21 Which statement is true for all of the organisms in the ecosystem represented in the diagram below?

1 They use energy to combine the inorganic molecules carbon dioxide and water into energy-rich organic compounds.
2 Stored energy cannot be used by these organisms as a source of energy for life processes.
3 Energy stored in inorganic molecules is released during cellular respiration in these organisms.
4 Energy is used by the organisms to obtain and transport materials, and to eliminate wastes.   21 _____

22 When a natural disaster destroys a stable ecosystem, the area is temporarily less stable than before. This is most likely due to

1 a decrease in biodiversity
2 an increase in the number of food chains
3 an increase in the number of species
4 a decrease in the rate of mutation   22 _____

23 An individual walks out of his air-conditioned (75°F) home into the hot outside environment (85°F). His ability to adjust to this changing environment involves a mechanism similar to

1 the regulation of water loss by guard cells in plant leaves
2 the digestion of carbohydrates by enzymes
3 using ATP for the diffusion of water
4 glucose production in the pancreas   23 _____

24 Nonrenewable resources are

1 not finite and are not depleted over time
2 not finite and are depleted over time
3 finite and are not depleted over time
4 finite and are depleted over time   24 _____

25 Dodder, a plant with no chlorophyll, grows on a living plant of a different species from which it obtains nutrients. Which pair of terms describes this relationship?

1 parasite and host
2 predator and prey
3 producer and decomposer
4 consumer and scavenger   25 _____

26 Three human hormones most directly involved in sexual reproduction are

1 estrogen, insulin, and progesterone
2 testosterone, estrogen, and insulin
3 progesterone, ATP, and testosterone
4 estrogen, progesterone, and testosterone   26 _____

27 At one point, scientists observed that the ozone shield was getting thinner. They warned that the loss of the effectiveness of this shield may lead to an increase in

1 allergies to ozone
2 mutations that lead to cancer
3 viral diseases, such as AIDS
4 ice formation at the poles   27 _____

28 As it grows from a seed to a mature plant, a plant will grow taller and thicker. Which are abiotic factors most responsible for the increase in the mass of the plant?

1 water, minerals, bacteria
2 sunlight, oxygen, plant receptors
3 minerals, water, plant enzymes
4 water, sunlight, carbon dioxide   28 _____

411

29 Structures in an animal cell are represented in the diagram below.

Which row in the chart correctly identifies the functions of structures A, B, and C?   29 ____

Row	Structure A	Structure B	Structure C
(1)	waste removal	extract energy from nutrients	protein synthesis
(2)	information storage	transport of materials	storage of liquids
(3)	protein synthesis	storage of wastes	reproduction
(4)	cell communication	transport of materials	waste removal

30 The diagram below represents how air pollution may move across the eastern United States.

Movement of Air Pollution

In order to reduce the amount of air pollution in Pennsylvania, which change is necessary?

1 Laws must be passed to protect endangered species.
2 The use of natural resources must be increased.
3 More coal-burning power plants must be built.
4 The cooperation between the different states must be improved.   30 ____

PART B–1

Answer all questions in this part. [13]

Directions (31–43): For *each* statement or question, record in the space provided the *number* of the word or expression that, of those given, best completes the statement or answers the question.

Base your answers to questions 31 and 32 on the diagram below and on your knowledge of biology. The diagram represents part of a food web.

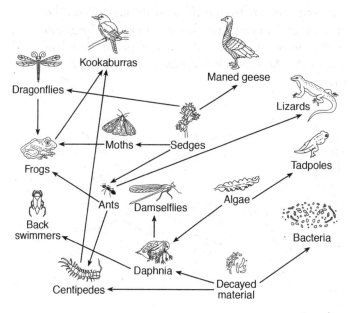

31 Which sequence of organisms represents a food chain within this food web?

1 tadpoles → algae → daphnia → back swimmers
2 sedges → ants → frogs → kookaburras
3 algae → daphnia → decayed material → bacteria
4 dragonflies → sedges → ants → centipedes   31____

32 Which population would be most immediately affected by the removal of the lizard population

1 sedges   3 ants
2 algae   4 centipedes   32____

33 During an investigation, a student measures out 15 grams of salt. Then, he measures 15 milliliters of water and adds the salt to it. Next, he measures a 1 centimeter wide by 4 centimeters long section of plant leaf. Which list of tools is arranged in the order that the student used them?

1 graduated cylinder, ruler, balance
2 balance, ruler, graduated cylinder
3 graduated cylinder, balance, ruler
4 balance, graduated cylinder, ruler   33 _____

Base your answers to questions 34 through 36 on the graph below and on your knowledge of biology. The graph shows the masses of different types of plants found in an area of the Adirondack Mountains after a forest fire occurred.

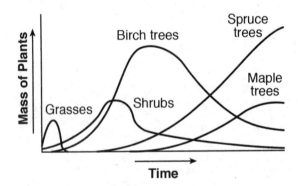

34 Based on the information provided in the graph, the process that is occurring is

1 ecological succession   3 selective breeding
2 biological evolution   4 genetic engineering   34 _____

35 The time shown in the graph is most likely measured in

1 days   3 months
2 weeks   4 years   35 _____

414

36 The mass of plants shown in the graph refers to the
mass of a number of

1 populations   3 ecosystems
2 decomposers   4 communities   36 ____

Base your answers to questions 37 and 38 on the
diagram below and on your knowledge of biology. The
diagram represents a portion of a starch molecule.

37 The building blocks for this molecule are

1 amino acids   3 fats
2 simple sugars   4 molecular bases   37 ____

38 The energy in this molecule is stored

1 in the bonds between atoms
2 in the oxygen found in the molecule
3 when the carbon atoms break off
4 when water breaks this molecule apart   38 ____

Base your answer to question 39 on the information below and on your knowledge of biology.

A genetically modified pig, nicknamed the "enviropig," has the ability to produce a bacterial enzyme in its saliva that helps reduce the amount of phosphorus in its wastes. Phosphorus pollution is a serious environmental concern. Enviropigs are expensive, but the cost is balanced against the benefit to the environment. There is also a concern that the US Department of Agriculture still has not cleared enviropig meat for human consumption.

39 Government agencies and citizens should propose the use of enviropig in the future only after

1 developing ways to remove the bacterial enzyme
2 assessing risks, costs, and benefits
3 people have eaten lots of enviropig meat and determined the effects
4 a different, cheaper pig can be produced regardless of the output of phosphorus   39 ____

Base your answers to questions 40 and 41 on the information below and on your knowledge of biology.

Researchers have produced rice plants that can withstand being completely submerged for up to two weeks. This is good news for farmers in the flood regions of Southeast Asia. The farmers in this region rely heavily on this crop. The diagram below illustrates the process used to genetically modify plants, such as rice.

40 The molecules used to cut, copy, and connect the DNA segments used in this process are

1 sugars   3 indicators

2 enzymes   4 antigens   40 ____

41 The best explanation for these modified rice plants being flood resistant is that

1 the gene for flood resistance was inserted into plant cells, which grew into plants whose cells are expressing this gene

2 they were produced by fertilization, using gametes from two flood-resistant bacterial cells

3 there was a mutation in the bacterial DNA after it was inserted into the plant that caused it to be flood resistant

4 the researchers used selective breeding for the flood-resistance trait

41 _____

42 The diagram below represents two cells viewed using the same magnification with the same microscope.

Mitochondria

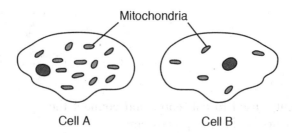

Cell A   Cell B

One possible conclusion that can be drawn about the activity of these two cells is that

1 more active transport occurs in cell *B* than in cell *A*

2 more active transport occurs in cell *A* than in cell *B*

3 cell *B* uses some of the extra mitochondria to make food

4 cell *A* is a plant cell since it has a cell wall

42 _____

43 The diagram below represents the processes leading to the formation of a human embryo.

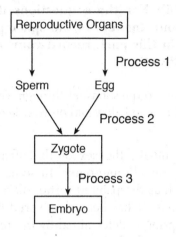

The correct sequence for processes 1, 2, and 3 represented in the diagram is   43 ____

Row	Process 1	Process 2	Process 3
(1)	gamete formation	cell division	fertilization
(2)	cell division	gamete formation	fertilization
(3)	gamete formation	fertilization	cell division
(4)	fertilization	gamete formation	cell division

PART B–2

Answer all questions in this part. [12]

Directions (44–55): **For those questions that are multiple choice, record your answers in the space provided. For all other questions in this part, record your answers in accordance with the directions.**

Base your answers to questions 44 through 47 on the information and data table below and on your knowledge of biology.

For most animals, the sex of the offspring is determined by sex chromosomes. In some species of reptiles, such as the painted turtle, there are no sex chromosomes. It has been discovered that the sex of the offspring is determined by the temperature of the nest in which the egg develops.

**Sex of Painted Turtle Offspring
at Various Nest Temperatures**

Temperature (°C)	Sex of Offspring	
	Males (%)	Females (%)
19	0	100
20	5	95
21	20	80
22	25	75
23	0	100
24	0	100
25	0	100

Directions (44–46): Using the information in the data table, construct a line graph on the grid below, following the directions below.

44 Mark an appropriate scale, without any breaks in the data, on each axis. [1]

45 Plot the data for percent males on the grid. Connect the points and surround each point with a small circle. [1]

Example:

46 Plot the data for percent females on the grid. Connect the points and surround each point with a small triangle. [1]

Example:

Sex of Painted Turtles at Various Nest Temperatures

Offspring (%)

Key
⊙ Males
△ Females

Temperature (°C)

Note: The answer to question 47 should be recorded in the space provided.

47 The fact that the sex of the painted turtle offspring is controlled by the temperature of the nest is an example of .

1 Manipulation of genes caused the movement of embryonic skin cells.

2 Exposure to light stimulated the development of cells containing ozone.

3 Uncontrolled mitotic division occurred as a result of gene mutations.

4 An immune reaction triggered the formation of excess blood cells.

47 _____

Base your answers to questions 48 through 50 on the information and graphs below and on your knowledge of biology. The graphs represent the results of two investigations using leaf disks from spinach plants.

Small disks were cut from spinach leaves that had been treated to remove any air from inside the leaf. The disks were placed in a solution that allowed them to carry out photosynthesis. At first, all the disks sank to the bottom of the container. These disks were then used for two different investigations.

Investigation 1

The disks were divided into five groups. Each group was exposed to light of a different intensity, measured in watts per meter squared (W/m^2). Some of the disks began to float. The results of the first investigation are shown in the graph below.

48 State the relationship between increasing light intensity and the percentage of disks floating at the conclusion of Investigation 1. [1]

_____

_____

Note: The answer to question 49 should be recorded in the space provided.

49 The substance produced inside the leaf disks that caused them to float to the surface of the solution is

1 ozone   3 water

2 oxygen   4 nitrogen   49 _____

Investigation 2

A number of freshly prepared disks were placed in five containers. These containers were then each exposed to light of a different color. The results of the second investigation are shown in the graph below.

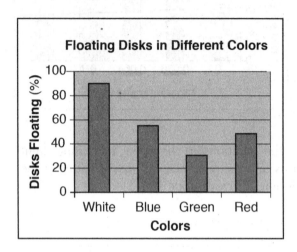

Note: The answer to question 50 should be recorded in the space provided.

50 Which color of light appears to be *least* effective for photosynthesis in spinach leaves?

1 white	3 green
2 blue	4 red

50 ____

Base your answer to question 51 on the information and photograph below and on your knowledge of biology. The photograph shows an oriental hornet.

Oriental hornets are unique insects. A yellow pigment in the body of the insect converts solar energy to electrical energy. Plants also convert energy from the Sun.

Yellow pigment

51 Identify the organelle present in plants where this conversion takes place.   [1]

_____

Base your answer to question 52 on the information below and on your knowledge of biology.

The SUNY Solar Car Model Racing Team's Sunhawk: Car of the Future?

The Sunhawk, a car built by students at SUNY New Paltz, prompted Forbes Magazine to ask "Is The $250,000 Sunhawk the Solar Car of the Future?" These cars show the most advanced solar technology and vehicle construction.

52 There are trade-offs involved in the use of solar-powered cars. Provide *one* advantage and *one* disadvantage of owning a solar car. [1]

Advantage: _____

Disadvantage: _____

Base your answer to question 53 on the information below and on your knowledge of biology.

Ten years ago, scientists discovered a well-preserved set of dinosaur remains in China. This dinosaur, which walked on Earth about 125 million years ago, had feathers and was about the same size as a turkey—but don't be fooled. This dino's bite was a lot worse than a turkey's gobble. After a close (and careful!) examination of the dino's teeth, scientists recently concluded that this dinosaur was probably poisonous. The study was led by David Burnham, who works and teaches at the University of Kansas in Lawrence.

Source: *www.sciencenewsforkids.org*

53 State *one* inference that could be made based on the fact that this dinosaur had feathers. [1]

_____

_____

Base your answer to questions 54 and 55 on the information below and on your knowledge of biology.

A captive New Caledonian crow forages for food using a stick tool. (Credit: Dr. Simon Walker)

A Great Larvae Meal

New Caledonian crows consume a wide range of foods. These crows require tools to extract the larvae of wood boring beetles from their burrows. A bird pokes a larva with a stick until the larva is disturbed enough to bite the stick and hang on to it. The bird is then able to pull the larva out of its burrow. These larvae, with their unusual diet, have a distinct chemical that can be found in the feathers and blood of crows—allowing scientists to determine the percentage of the crows' diet that is made up of beetle larvae. Scientists found that the beetle larvae are so energy-rich that just a few could satisfy the daily energy requirement for a crow. The crows with the greatest skill in using a twig as a tool benefit most in terms of nutrition.

54 State *one* reason why the offspring of crows skilled at using twigs as tools would have the greatest chance of survival. [1]

_____

_____

55 State *one* reason why some members of a population of crows equally skilled in the use of twigs have different rates of survival. [1]

_____

_____

PART C

Answer all questions in this part. [17]

Directions (56–72): **Record your answers in the spaces provided.**

Base your answers to question 56–57 on the information below and on your knowledge of biology.

African violet plants are grown for their delicate, colorful flowers and furry, soft leaves. People often want to touch the leaves and brush the hairy leaves with their fingers. Growers and plant owners were concerned that this could negatively affect the plant. Of particular concern was the presence of body lotion or other skin products on the hands of persons touching the leaves.

A student thought this might be the basis of a science project. He selected two African violet plants. Ten leaves on each of the two plants were brushed with a gloved hand for 30 seconds, once a day, for a period of five days. The difference was that leaves of the second plant were brushed with a gloved hand that had hand lotion applied to the glove.

56–57 As part of the peer review process, evaluate the student's experiment. As part of your evaluation, be sure to:

- state *one* possible hypothesis for the experiment proposed by the student [1]
- describe the type of data that should be collected to determine if the brushing with lotion was having a *negative* effect on the African violet plant [1]

_____

_____

_____

_____

_____

_____

Base your answers to questions 58 and 59 on the information and chart below and on your knowledge of biology.

Scientists studied the distribution of a species of pocket mouse that lived in the sandy desert regions of the southwestern United States. They are eaten by a variety of predators. Pocket mice are active at night, and feed on seeds and grasses. A single female mouse can reproduce several times each year, producing a litter of 3 to 13 offspring each time. Each new litter is considered a generation.

A volcanic eruption that resulted in lava flows changed the color of the area that the mice inhabit from light brown to black. Data from the scientists' research of the population are shown in the chart below.

Changes in Pocket Mouse Fur Color after a Volcanic Eruption

Number of Generations	Percentage of Pocket Mice with Light Brown Fur	Percentage of Pocket Mice with Black Fur
10	95%	5%
25	90%	10%
50	75%	25%
100	5%	95%

58 State the role of mutation *or* recombination in the appearance of the trait for black fur color in the pocket mouse population. [1]

_____

_____

59 Explain why the percentage of black pocket mice changed so much after the volcanic eruption. [1]

_____

_____

Base your answers to questions 60 and 61 on the information below and on your knowledge of biology.

In 2003, as a result of the Human Genome Project, the complete sequence of all the bases in human DNA was released to the public. Although knowing the entire sequence of bases has proven valuable, scientists are currently working to map genes. Mapping genes involves determining the exact location of each gene. Since much of human DNA does not code for a protein, it is challenging to figure out which segments are actual genes. Often, scientists look at the percent composition of bases in a segment of DNA. If the segment of DNA has a large percentage of C and G bases (together over 50%), it is likely that it is a gene and codes for a protein.

60 A scientist analyzes the bases in a segment of DNA from a human skin cell to determine if it codes for a protein. The base A is 12% of the bases in this segment of DNA. Calculate the percentage of bases that would be C. [1]

_____ %

61 Is it likely this segment of DNA codes for a protein? Circle yes *or* no and support your answer. [1]

Circle one: Yes   *or*   No

_____

_____

Base your answers to questions 62 through 64 on the illustration and information below and on your knowledge of biology. The illustration is of a Tasmanian devil.

Source: http://www.statelibrary.tas.gov.au

The Tasmanian devil is the largest surviving carnivorous marsupial in Australia. It is in danger of extinction due to an unusual type of cancer called Devil Facial Tumor Disease (DFTD). It can be passed from one individual to another through wounds that occur when they fight over food. Tumor cells in the mouth of an infected animal break off and enter the wound on an uninfected animal. The tumor cells multiply in the body of the newly infected devil, forming new tumors that eventually kill the animal.

Recent research has shown that the immune system of a Tasmanian devil accepts tumor cells from another devil as if they were cells from its own body. The tumor cells are ignored by the immune system. No immune response develops against them, and the cancerous cells multiply. Scientists predict that DFTD could wipe out all the remaining Tasmanian devils in 25 years, unless a treatment is developed.

62 Using the terms antigens and antibodies, explain why the tumor cells are ignored by the immune system in Tasmanian devils.   [1]

_____

_____

63 Explain how cancer cells differ from normal cells.   [1]

_____

_____

64 Describe *one* possible way to maintain a population of healthy, uninfected Tasmanian devils until a treatment or cure can be found.   [1]

_____

_____

Base your answers to questions 65 through 67 on the information below and on your knowledge of biology. The photograph below shows a Canada lynx, a mammal native to North America.

Source: http://www.allposters.com

Lynx are found in areas where there is deep, soft snow cover during the winter months. The body design of the Canada lynx helps keep the animal on top of the soft snow. Several unique characteristics, such as the design of its feet and its weight, enable the cat to successfully chase and catch snowshoe hares, its primary source of food. Snowshoe hares are also able to remain on top of the snow.

Increased winter recreation has created packed snow trails in lynx habitat. This allows coyotes and cougars to compete with lynx.

65 Explain why coyotes were *not* in competition with the lynx prior to the presence of packed snow trails. [1]

_____

_____

66 Describe the niche that the lynx, coyote, and cougar are competing to fill. [1]

_____

_____

67 Explain how the carrying capacity affects the number of predators in an area. [1]

_____

_____

Base your answers to questions 68 and 69 on the information below and on your knowledge of biology.

Cowherds Discovering Ticks Are for the Birds

South African cowherds [cowboys] are discovering that when it comes to debugging their cattle, nature knows best. Generations of cattle owners who dipped their livestock in pesticides ended up killing not only the ticks that feast on them, but also the red-billed oxpeckers [birds] that eat the ticks. Now environmentalists want to cut out the pesticides, hand the job back to the birds and in the process save them from extinction....

The bird is famous for its bright red bill, yellow ringed eyes and voracious appetite for ticks. An oxpecker can eat 13,000 of them [ticks] in a day, and meals are everywhere— on antelope, horses, cattle, buffalo, rhino, lion, elephant and leopard. The ticks carry a host of illnesses, including red-water disease, a common killer of cattle, but [ticks] are harmless to oxpeckers....

Source: Eric Naki, Associated Press writer

68 State *one* way that the use of pesticides to kill ticks could lead to the decline of the oxpecker population. [1]

_____

_____

69 State *one* ecological advantage of using oxpeckers to solve the problem with these ticks. [1]

_____

_____

Base your answers to questions 70 through 72 on the information below and on your knowledge of biology.

Over the past few decades, researchers have observed declining numbers in two species of penguins native to the West Antarctic peninsula. New evidence is pointing to a decline in their food supply as the primary cause for the recent drops in their numbers. These penguins feed on krill, small animals that grow and develop under ice masses. The graphs below show data related to two factors: atmospheric carbon dioxide (CO_2) levels and Antarctic ice mass. The diagram of a generalized Antarctic food web illustrates the role of the penguins.

Direct CO_2 Measurements: 2005-2011

Antarctica Mass Variation Since 2002

Source: http://www.nasa.gov

Antarctic Food Web

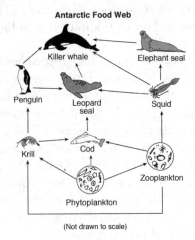

(Not drawn to scale)

Source: School Improvement in Maryland, www.mdk12.org

70 State *one* possible relationship between CO_2 levels and the change in Antarctic ice mass.   [1]

_____

_____

71 Explain why the change in ice mass is resulting in a decline in the penguin populations.   [1]

_____

_____

72 State *one* specific way in which humans might have caused the changes in atmospheric CO_2 levels.   [1]

_____

_____

PART D

Answer all questions in this part. [13]

Directions (73–85): **For those questions that are multiple choice, record your answers in the space provided. For all other questions in this part, record your answers in accordance with the directions.**

Base your answers to questions 73 through 75 on the diagram below and on your knowledge of biology. The diagram represents an experimental setup.

Starch indicator and water
Dialysis tube
Starch solution

Note: The answer to question 73 should be recorded in the space provided.

73 Which statement best describes what would most likely be observed after 20 minutes?

  1 The contents of the dialysis tube would turn blue-black.

  2 The liquid in the beaker would turn blue-black.

  3 The dialysis tube would burst.

  4 There would be no change visible.   73____

Note: The answer to question 74 should be recorded in the space provided.

74 Which term correctly identifies the process by which molecules move through the dialysis tube membrane?

1 paper chromatography   3 diffusion
2 active transport   4 digestion   74 _____

Note: The answer to question 75 should be recorded in the space provided.

75 A student filled a dialysis tube with 97% water solution and sealed the ends. The tube and its contents had a mass of 55 grams. The student placed the tube in a solution, and the mass of the tube and its contents increased to 60 grams. Into which solution was the dialysis tube placed?

1 0% water   3 97% water
2 95% water   4 99% water   75 _____

Base your answers to questions 76 and 77 on the information and diagram below and on your knowledge of biology.

Scientists attempted to determine the evolutionary relationships between three different finch species, A, B, and C. In order to do this, they examined the physical characteristics and DNA of these species. DNA was extracted from all three species and analyzed using gel electrophoresis. The results are shown in the diagram.

Species A   Species B   Species C

Note: The answer to question 76 should be recorded in the space provided.

76 Which statement best describes the method used above to determine the evolutionary relationships between three species of finches?

1 Examine the structure of the beaks and compare them.
2 Observe behavioral and physical characteristics of all the finches and group them by similarities.
3 Obtain molecular evidence from all three species and identify similarities.
4 Compare common ancestors of all three of the species to see if they are the same.

76 _____

77 Based on the data they collected using gel electrophoresis, label the branching tree diagram below. Write the letters A, B, and C, to represent the possible evolutionary relationships between species A, B, and C. [1]

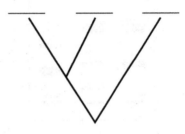

Base your answers to questions 78 through 80 on the information and diagram below and on your knowledge of biology.

Two species of finches found on a particular Galapagos island eat the seeds of a certain variety of plant. The relative strength of their beaks is shown in the graph below.

442

78 One of the finch species has a slightly smaller, weaker beak. Is this species 1 or species 2? Support your answer with information from the graph.   [1]

Species: _____

_____

_____

79 Select the point on the graph where beak strength of the two bird species is equal. Support your answer.   [1]

Point: _____

_____

_____

80 If the environment on the island changed and the seeds of more of the plants became harder to crack open, describe what the graph might look like after many years have passed.   [1]

_____

_____

Base your answers to questions 81 and 82 on the information and diagram below and on your knowledge of biology.

Using a microscope and a wet-mount slide, a student observed a pond water sample containing paramecia, which are single-celled freshwater organisms. He noticed that there was a structure within each living paramecium that contracted regularly—about four times each minute. He researched the organism in his science textbook and found that the structure was a contractile vacuole and its function was to remove excess water from the paramecium.

In the diagram below, a paramecium is represented as seen through a microscope. The function of the contractile vacuole is described.

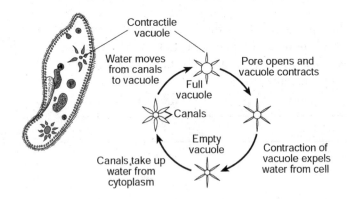

He decided to determine if the concentration of salt in the environment of the paramecium would affect the rate at which the contractile vacuole would contract.

Note: The answer to question 81 should be recorded in the space provided.

81 The process used to remove excess water from the paramecium by the contractile vacuole is

1 synthesis   3 active transport
2 digestion   4 passive transport   81 _____

Note: The answer to question 82 should be recorded in the space provided.

82 The student predicted that the contractile vacuole would contract fewer times in one minute in a solution that had a higher concentration of salt than that found in typical pond water. This prediction is most likely

1 correct, because a high concentration of salt in the environment will force water into the cell, causing the contractile vacuole to pump
2 incorrect, because salt would be entering the cell, and the contractile vacuole would have to pump it out
3 incorrect, because the concentration of salt in the environment should not affect a cell
4 correct, because water would be moving out of the cell into the salt solution   82 _____

Base your answers to questions 83 through 85 on the information below and on your knowledge of biology.

Five individuals had their pulses taken in beats per minute (bpm) before and after exercise. The data are shown in the chart below.

Pulse Rates

Individual	Pulse before Exercise (bpm)	Pulse after Exercise (bpm)
A	68	100
B	70	120
C	54	130
D	64	122
E	75	115

83 State *one* reason why an individual's pulse rate increased during exercise.   [1]

_____

_____

84 Calculate the average pulse rate *before* exercise for this group, to the *nearest tenth*.   [1]

_____**bpm**

85 State why the individuals in this group have different pulse rates before exercise.   [1]

_____

_____

ANSWERS TO REGENTS EXAMINATION

LIVING ENVIRONMENT

JUNE 2016

PART A

1. 2	6. 2	11. 2	16. 3	21. 4	26. 4
2. 4	7. 2	12. 2	17. 4	22. 1	27. 2
3. 2	8. 4	13. 2	18. 4	23. 1	28. 4
4. 3	9. 1	14. 1	19. 3	24. 4	29. 1
5. 3	10. 1	15. 4	20. 3	25. 1	30. 4

PART B–1

31. 2	34. 1	36. 1	38. 1	40. 2	42. 2
32. 3	35. 4	37. 2	39. 2	41. 1	43. 3
33. 4					

PART B–2

47. 4
49. 2
50. 3

PART D

73. 1
74. 3
75. 4
76. 3
81. 3
82. 4

In Parts B-2, C, and D, for any answers not listed here, you are required to show your work and explain your answers.

Index

449

NOTES

NOTES

NOTES

Barron's Regents Exams and Answers

Biology
The Living Environment

GABRIELLE I. EDWARDS
Former Science Consultant
Board of Education
City of New York

Former Assistant Principal Supervision
Science Department
Franklin D. Roosevelt High School
Brooklyn, New York

MARION CIMMINO
Former Teacher, Biology and Laboratory Techniques
Franklin D. Roosevelt High School
Brooklyn, New York

FRANK J. FODER
Teacher, Advanced Placement Biology
Franklin D. Roosevelt High School
Brooklyn, New York

G. SCOTT HUNTER
Former Teacher, Biology and Advanced Biology
Former Consultant, State Education Department Bureaus of Science and Testing
Former School Business Administrator, Schodack, NY
Former Superintendent of Schools
Mexico and Chatham, New York

Barron's Educational Series, Inc.

All inquiries should be addressed to:
Barron's Educational Series, Inc.
250 Wireless Boulevard
Hauppauge, New York 11788
www.barronseduc.com

ISBN: 978-0-8120-3197-3
ISSN 1069-2940

10%
**POST-CONSUMER
WASTE**
Paper contains a minimum
of 10% post-consumer
waste (PCW). Paper used
in this book was derived
from certified, sustainable
forestlands.

PRINTED IN THE UNITED STATES OF AMERICA
9 8 7 6 5 4 3 2 1

Contents

Glossary   126

Regents Examinations, Answers, and Student Self-Appraisal Guides   161

How to Use This Book

ORGANIZATION OF THE BOOK

Study Questions

Section 1 of the book consists of 173 questions taken from past Regents examinations in biology. Most of the questions require that you select a correct response from four choices given. A few questions provide you with a list of words or phrases from which to select the one that best matches a given description. Others are constructed-response, graphical analysis, or reading comprehension questions. You should become familiar with the formats of the questions that appear on the Living Environment Regents examination.

Each question in Section 1 of this book has a well-developed answer. Each answer provides the number of the correct response, the reason why the response is correct, and explanations of why the other choices are incorrect.

A useful feature of Section 1 is the student Self-Appraisal Guide. This device allows you to determine where your learning strengths and weaknesses lie in the major topics of each unit. For specific topics within the units, the numbers of related questions are given. As you attempt to answer each question in Section 1, you may wish to circle on the appraisal form the numbers of the questions that you are unable to answer. The circled items then help you to identify at a glance subject matter areas in which you need additional study.

New York State Learning Standards

COMMENCEMENT STANDARDS

There are several commencement standards required of students in New York State public schools regarding their performance in math, science, and technology. The Core Curriculum for The Living Environment addresses two of these standards:

Standard 1: **Students will use mathematical analysis, scientific inquiry, and engineering design, as appropriate, to pose questions, seek answers, and develop solutions.**

Standard 4: **Students will understand and apply scientific concepts, principles, and theories pertaining to the physical setting and living environment and recognize the historical development of ideas in science.**

The Core Curriculum for The Living Environment was built from these two commencement standards. It is important to recognize that the Core Curriculum is not a syllabus. It does not prescribe what will be taught and learned in any particular classroom. Instead, it defines the skills and understandings that you must master in order to achieve the commencement standards for life science.

Instead of memorizing a large number of details at the commencement level, then, you are expected to develop the skills needed to deal with science on the investigatory level, generating new knowledge from experimentation and sharpening your abilities in data analysis. You are also expected to read and understand scientific literature, taking from it the facts and concepts necessary for a real understanding of issues in science. You are required to pull many facts together from different sources to develop your own opinions about the moral and ethical problems facing modern society concerning technological advances. These are thinking skills that do not respond to simple memorization of facts and scientific vocabulary.

KEY IDEAS, PERFORMANCE INDICATORS, AND MAJOR UNDERSTANDINGS

Each commencement standard is subdivided into a number of Key Ideas. Key Ideas are broad, unifying statements about what you need to know. Within Standard 1, three Key Ideas are concerned with laboratory investigation and data analysis. Together, these unifying principles develop your ability to deal with data and understand how professional science is carried out in biology. Within Standard 4, seven Key Ideas present a set of concepts that are central to the science of biology. These Key Ideas develop your understanding of the essential characteristics of living things that allow them to be successful in diverse habitats.

Within each Key Idea, Performance Indicators are presented that indicate which skills you should be able to demonstrate through your mastery of the Key Idea. These Performance Indicators give guidance to both you and your teacher about what is expected of you as a student of biology.

Performance Indicators are further subdivided into Major Understandings. Major Understandings give specific concepts that you must master in order to achieve each Performance Indicator. It is from these Major Understandings that the Regents assessment material will be drawn.

LABORATORY COMPONENT

A meaningful laboratory experience is essential to the success of this or any other science course. You are expected to develop a good sense of how scientific inquiry is carried out by the professional scientist and how these same techniques can assist in the full understanding of concepts in science. The Regents requirement of 1,200 minutes of successful laboratory experience, coupled with satisfactory written reports of your findings, should be considered a minimum.

Students are required to complete four laboratory experiences required by the New York State Education Department and tested on the Regents Examination. See Barron's *Let's Review: Biology—The Living Environment* for a complete treatment of this requirement.

Regents Examinations

Section 2 of the book consists of actual Biology Regents examinations and answers. These Regents examinations are based on the New York State Core Curriculum for the Living Environment.

Assessments: Format and Scoring

The format of the Regents assessment for The Living Environment is as follows, based on actual Regents examinations.

Part A: Variable number of multiple-choice questions that test the student's knowledge of specific factual information. **All** questions must be answered on Part A.

Part B: Variable number of questions, representing a mixture of multiple-choice and constructed-response items. Questions may be based on the student's direct knowledge of biology, interpretation of experimental data, analysis of readings in science, and ability to deal with representations of biological phenomena. **All** questions must be answered on Part B.

Part C: Variable number of multiple-choice and constructed-response questions. Questions may be based on the student's direct knowledge of biology, interpretation of experimental data, analysis of readings in science, and ability to deal with representations of biological phenomena. **All** questions must be answered on Part C.

Part D—Laboratory component: This component of the examination aims to assess student knowledge of and skills on any of four required laboratory experiences supplied to schools by the New York State Education Department. Questions in Part D can be a combination of multiple-choice and constructed-response questions similar to those found in Parts A, B, and C of the Living Environment Regents Examination. The content of these questions will reflect specific laboratory experiences. You are strongly encouraged to include review of these laboratory experiences as part of your year-end Regents preparation activity.

The following chart summarizes the current laboratory requirement for New York State public schools:

LABORATORY REQUIREMENTS

Laboratory Title	Description
The Beaks of Finches	Explores the adaptive advantages of beaks with different physical characteristics
Relationships and Biodiversity	Explores the relationship between DNA structure and the biochemistry of inheritance
Making Connections	Explores the effects of physical activity on human metabolic activities
Diffusion Through a Membrane	Explores the nature of cross-membrane transport in living cells
Adaptations for Reproductive Success in Flowering Plants*	Not yet available
DNA Technology*	Not yet available
Environmental Conditions and Seed Germination*	Not yet available

*Not yet available as of this writing.

Studying questions from past Regents examinations is an invaluable aid in developing a mind set that will enable you to approach questions with understanding. Although exact questions are not repeated, question types are repeated. If you practice questions that require interpretation, problem solving, and graph construction, you will do well on the entire exam. During the school year, the 30 required laboratory lessons teach you certain manipulative skills. Questions involving identification, measurement, and other laboratory procedures are based on the laboratory exercises. Review of past materials gives you insight as to the types of questions that you may be asked to answer. Study the questions in the Regents exams in this book diligently.

HOW TO STUDY

General Suggestions for Study

You've spent all year learning many different facts and concepts about biology—far more than you could ever hope to remember the "first time around." Your teacher has drilled you on these facts and concepts; you've done homework, taken quizzes and tests, performed laboratory experiments, and reviewed the material at intervals throughout the year.

Now it's time to put everything together. The Regents exam is only a few weeks away. If you and your teacher have planned properly, you will have finished all new information about 3 weeks before the exam. Now you have to make efficient use of the days and weeks ahead to review all that you've learned in order to score high on the Regents.

The task ahead probably seems impossible, but it doesn't have to be! You've actually retained much more of the year's material than you realize! The review process should be one that helps you to recall the many facts and concepts you've stored away in your memory. Your Barron's resources, including Barron's *Let's Review: The Living Environment,* will help you to review this material efficiently.

You also have to get yourself into the right frame of mind. It won't help to be nervous and stressed during the review process. The best way to avoid being stressed during any exam is to be well rested, prepared, and confident. We're here to help you prepare and to build your confidence. So let's get started on the road to a successful exam experience!

To begin, carefully read and follow the steps outlined below.

1. Start your review early; don't wait until the last minute. Allow at least 2 weeks to prepare for the Regents exam. Set aside an hour or two a day over the next few weeks for your review. Less than an hour a day is insufficient time for you to concentrate on the material meaningfully; more than 2 hours daily will yield diminishing returns on your investment of time.

2. Find a quiet, comfortable place to study. You should seat yourself at a well-lighted work surface, free of clutter, in a room without distractions of any kind. You may enjoy watching TV or listening to music curled up in a soft chair, but these and other diversions should be avoided when doing intense studying.

3. Make sure you have the tools you need, including this book, a pen and pencil, and some scratch paper for taking notes and doing calculations. Keep your class notebook at hand for looking up information between test-taking sessions. It will also help to have a good review text, such as

Barron's *Let's Review: The Living Environment,* available for reviewing important concepts quickly and efficiently.

4. Concentrate on the material in the "Study Questions and Answers" section of this book. Read carefully and thoughtfully. Think about the questions that you review, and try to make sense out of them. Choose the answers carefully. (See the following section, "Using This Book for Study," for additional tips on question-answering techniques.)

5. Use available resources, including a dictionary and the glossary in this book, to look up the meanings of unfamiliar words in the practice questions. Remember that the same terms can appear on the Regents exam you will take, so take the opportunity to learn them now.

6. Remember: study requires time and effort. Your investment in study now will pay off when you take the Regents exam.

Using This Book for Study

This book is an invaluable tool if used properly. Read carefully and try to answer *all* the questions in Section 1 and on the practice exams. The more you study and practice, the more you will increase your knowledge of biology and the likelihood that you will earn a high grade on the Regents exam. To maximize your chances, use this book in the following way:

1. Answer all of the questions in the section entitled "Study Questions and Answers." Check your responses by using "Answers to Topic Questions," including "Wrong Choices Explained," following each question set. Record the number of correct responses on each topic in the "Self-Appraisal Guide" at the end of the section to identify your areas of strength and weakness. Use a good review text, such as Barron's *Let's Review: The Living Environment,* to study each area on which you did poorly. Finally, go back to the questions you missed on the first round and be sure that you fully understand what each question asks and why the correct answer is what it is.

2. When you have completed the questions in "Study Questions and Answers," go on to the examination section. Select the first complete examination and take it under test conditions.

3. Interpret the term *test conditions* as follows:
   - Be well rested; get a good night's sleep before attempting *any* exam.
   - Find a quiet, comfortable room in which to work.
   - Allow no distractions of any kind.
   - Select a well-lighted work surface free of clutter.
   - Have your copy of this book with you.
   - Bring to the room a pen, a pencil, some scratch paper, and a watch or alarm clock set for the 3-hour exam limit.

4. Take a deep breath, close your eyes for a moment, and RELAX! Tell yourself, "I know this stuff!" You have lots of time to take the Regents exam; use it to your advantage by reducing your stress level. Forget about your plans for later. For the present, your number 1 priority is to do your best, whether you're taking a practice exam in this book, or the real thing.

5. Read all test directions carefully. Note how many questions you must answer to complete each part of the exam. If test questions relate to a reading passage, diagram, chart, or graph, be sure you fully understand the given information before you attempt to answer the questions that relate to it.

6. When answering multiple-choice questions on the Regents exam, TAKE YOUR TIME! Pay careful attention to the "stem" of the question; read it over several times. These questions are painstakingly written by the test preparers, and every word is chosen to convey a specific meaning. If you read the question carelessly, you may answer a question that was never asked! Then read each of the four multiple-choice answers carefully, using a *pencil* to mark in the test booklet the answer you think is correct.

7. Remember that three of the multiple-choice answers are *incorrect;* these incorrect choices are called "distracters" because they seem like plausible answers to poorly prepared or careless students. To avoid being fooled by these distracters, you must think clearly, using everything you have learned about biology since the beginning of the year. This elimination process is just as important to your success on the Living Environment Regents exam as knowing the correct answer! If more than one answer seems to be correct, reread the question to find the words that will help you to distinguish between the correct answer and the distracters. When you have made your best judgment about the correct answer, circle the number in *pencil* in your test booklet.

8. Constructed-response questions appear in a number of different forms. You may be asked to select a term from a list, write the term on the answer sheet, and define the term. You may be asked to describe some biological phenomenon or state a biological fact using a complete sentence. You may be asked to read a value from a diagram of a measuring instrument and write that value in a blank on the answer paper. When answering this type of question, care should be exercised to follow directions precisely. If a complete sentence is called for, it must contain a subject and a verb, must be punctuated, and must be written understandably in addition to answering the scientific part of the question accurately. Values must be written clearly and accurately and include a unit of measure, if appropriate. Failure to follow the directions for a question may result in a loss of credit for that question.

9. A special type of constructed-response question is the essay or paragraph question. Typically, essay or paragraph questions provide an opportunity to earn multiple credits for answering the question correctly. As in the constructed-response questions described above, it is important that you follow the directions given if you hope to earn the maximum number of credits for the question. Typically, the question outlines exactly what must be included in your essay to gain full credit. Follow these directions step by step, double-checking to be certain that all question components are addressed in your answer. In addition, your essay or paragraph should follow the rules of good grammar and good communication so that it is readable and understandable. And, of course, it should contain correct information that answers all the parts of the question asked.

10. Graphs and charts are a special type of question that requires you to organize and represent data in graphical format. Typically for such questions, you are expected to place unorganized data in ascending order in a data chart or table. You may also be asked to plot organized data on a graph grid, connect the plotted points, and label the graph axes appropriately. Finally, questions regarding data trends and extrapolated projections may be asked, requiring you to analyze the data in the graph and draw inferences from it. As with all examination questions, always follow all directions for the question. Credit can be granted only for correctly following directions and accurately interpreting the data.

11. When you have completed the exam, relax for a moment. Check your time; have you used the entire 3 hours? Probably not. Resist the urge to quit. Go back to the beginning of the exam, and, in the time remaining, *retake the exam in its entirety.* Try to ignore the penciled notations you made the first time. If you come up with a different answer the second time through, read over the question with extreme care before deciding which response is correct. Once you have decided on the correct answer, mark your choice in ink in the answer booklet.

12. Score the exam using the Answer Key at the end of the exam. Review the "Answers Explained" section for each question to aid your understanding of the exam and the material. Remember that it's just as important to understand why the incorrect responses are incorrect as it is to understand why the correct responses are correct!

13. Finally, focus your between-exam study on your areas of weakness in order to improve your performance on the next practice exam. Complete all the practice exams in this book using the techniques outlined above.

14. Be sure to sign the declaration on your answer sheet. Unless this declaration is signed, your paper cannot be scored.

Test-Taking Tips—A Summary

The following pages contain seven tips to help you achieve a good grade on the Living Environment Regents exam.

TIP 1

Be confident and prepared.

SUGGESTIONS

- Review previous tests.
- Use a clock or watch, and take previous exams at home under examination conditions (i.e., don't have the radio or television on).
- Get a review book. (The preferred book is Barron's *Let's Review: The Living Environment*.)
- Talk over the answers to questions on these tests with someone else, such as another student in your class or someone at home.
- Finish all your homework assignments.
- Look over classroom exams that your teacher gave during the term.
- Take class notes carefully.
- Practice good study habits.
- Know that there are answers for every question.
- Be aware that the people who made up the Regents exam want you to pass.
- Remember that thousands of students over the last few years have taken and passed a Biology Regents. You can pass, too!
- Complete your study and review at least one day before taking the examination. Last-minute "cramming" may hurt, rather than enhance, your performance on the exam.

- Visit *www.barronseduc.com* or *www.barronsregents.com* for the latest information on the Regents exams.
- Use Barron's website to communicate directly with subject specialists.
- Be well rested when you enter the exam room. A good night's sleep is essential preparation for any examination.
- On the night before the exam day: lay out all the things you will need, such as clothing, pens, and admission cards.
- Bring with you two pens, two pencils, an eraser, and, if your school requires it, an identification card. Decide before you enter the room that you will remain for the entire 3-hour examination period, and either bring a wristwatch or sit where you can see a clock.
- Once you are in the exam room, arrange things, get comfortable, and attend to personal needs (the bathroom).
- Before beginning the exam, take a deep breath, close your eyes for a moment, and RELAX! Repeat this technique any time you feel yourself "tensing up" during the exam.
- Keep your eyes on your own paper; do not let them wander over to anyone else's paper.
- Be polite in making any reasonable demands of the exam room proctor, such as changing your seat or having window shades raised or lowered.

TIP 2

Read test instructions carefully.

SUGGESTIONS
- Be familiar with the format of the examination.
- Know how the test will be graded.
- Read all directions carefully. Be sure you fully understand supplemental information (reading passages, charts, diagrams, graphs) before you attempt to answer the questions that relate to it.
- Underline important words and phrases.
- Ask for assistance from the exam room proctor if you do not understand the directions.

TIP 3

Read each question carefully and read each choice before recording your answer.

SUGGESTIONS

- When answering the questions, TAKE YOUR TIME! Be sure to read the "stem" of the question and each of the four multiple-choice answers very carefully.
- If you are momentarily "stumped" by a question, put a check mark next to it and go on; come back to the question later if you have time.
- Remember that three of the multiple-choice answers (known as "distracters") are incorrect. If more than one answer seems to be correct, reread the question to find the words that will help you to distinguish between the correct answer and the distractors.
- When you have made your best judgment about the correct answer, circle the appropriate number in pencil on your answer sheet.

TIP 4

Budget your test time (3 hours).

SUGGESTIONS

- Bring a watch or clock to the test.
- The Regents examination is designed to be completed in 1½ to 2 hours.
- If you are absolutely uncertain of the answer to a question, mark your question booklet and move on to the next question.
- If you persist in trying to answer every difficult question *immediately*, you may find yourself rushing or unable to finish the remainder of the examination.
- When you have completed the exam, relax for a moment. Then go back to the beginning, and, in the time remaining, *retake the exam in its entirety*. Pay particular attention to questions you skipped the first time. Once you have decided on a correct response for multiple-choice questions, mark an "X" in ink through the penciled circle on the answer sheet.

- Plan to stay in the room for the entire three hours. If you finish early, read over your work—there may be some things that you omitted or that you may wish to add. You also may wish to refine your grammar, spelling, and penmanship.

TIP 5

Use your reasoning skills.

SUGGESTIONS
- Answer *all* questions.
- Relate (connect) the question to anything that you studied, wrote in your notebook, or heard your teacher say in class.
- Relate (connect) the question to any film, demonstration, or experiment you saw in class, any project you did, or to anything you may have learned from newspapers, magazines, or television.
- Look over the entire test to see whether one part of it can help you answer another part.

TIP 6

Don't be afraid to guess.

SUGGESTIONS
- In general, go with your first answer choice.
- Eliminate obvious incorrect choices.
- If still unsure of an answer, make an educated guess.
- There is no penalty for guessing; therefore, answer ALL questions. An omitted answer gets no credit.

TIP 7

Sign the Declaration.

SUGGESTIONS
- Be sure to sign the declaration on your answer sheet.
- Unless this declaration is signed, your paper cannot be scored.

New York State Regents Biology—The Living Environment: Core Curriculum and Assessment Map

The chart on pages 17–22 lists the Key Ideas and Performance Indicators from the New York State Regents Commencement Standard 1 and Standard 4 for *The Living Environment* (1999). For each Regents examination (assessment) listed across the top of the chart, item (question) numbers have been matched to the Performance Indicator(s) most closely associated with their content.

The information compiled in this chart represents a "map" showing how closely these assessments have mirrored the curriculum required by the state of New York for this course of study. It provides guidance to the teacher and student concerning the emphasis placed on each Performance Indicator by the New York State Education Department assessment development process on four recent Regents examinations.

By following this map, you will be sure to hit every major topic that is required by the state of New York for achievement of the Standards for *The Living Environment*. At the same time, you will be better able to focus your efforts on areas of needed study that will maximize your performance on the Regents examination without wasting valuable time on areas of the curriculum that are less likely to be represented on the examination.

Although some areas of the Core Curriculum have been less-well emphasized than others on Regents examinations to date, both teachers and students are cautioned that *all* parts of the Core Curriculum are required and are subject to testing. The Regents examination may be changed at any time and without notice to include items testing Performance Indicators that may not have been well emphasized in the past.

Students, you can find in Barron's *Let's Review: Biology—The Living Environment* (Hunter) an appropriate body of content, easy-to-read explanation, and practice exercises (with Answers Explained) developed especially for the New York State high school biology student. This resource will ease your way through *The Living Environment* course of study from start to finish and will prepare you well for the culminating Regents examination.

Teachers, you are encouraged to develop Standards-Based Learning Units (SBLUs) and Essential Questions that will provide appropriate local content to illustrate these required Performance Indicators. Barron's *Let's Review: Biology—The Living Environment* (Hunter) provides an excellent source of material for the development of SBLUs and Essential Questions that is consistent with the New York State Core Curriculum for this course. In addition, it includes many concepts that have traditionally been taught in the New York State Regents biology program since 1989 and relates this content to the new Regents Core Curriculum for *The Living Environment*.

LIVING ENVIRONMENT CURRICULUM MAP AND QUESTION INDEX

STANDARD 1, KEY IDEA 1: THE CENTRAL PURPOSE OF SCIENTIFIC INQUIRY IS TO DEVELOP EXPLANATIONS OF NATURAL PHENOMENA IN A CONTINUING AND CREATIVE PROCESS.

Performance Indicator	Description	June 2015	August 2015	June 2016	August 2016
1.1	The student should be able to elaborate on basic scientific and personal explanations of natural phenomena and develop extended visual models and mathematical formulations to represent one's thinking.				
1.2	The student should be able to hone ideas through reasoning, library research, and discussion with others, including experts.		21		39
1.3	The student should be able to work toward reconciling competing explanations and clarify points of agreement and disagreement.				
1.4	The student should be able to coordinate explanations at different levels of scale, points of focus, and degrees of complexity and specificity, and recognize the need for such alternative representations of the natural world.				47

NOTE TO TEACHERS: Periodically, errors are made in assigning Regents exam questions to the appropriate New York State Curriculum Standards (a.k.a.: attribution). For June 2016, you will see discrepancies in attribution of the following questions: 7, 18, 21, 25, 28, 49, 50, 64, 68, and 69. Every effort has been made to assure that each exam question is attributed to the appropriate Standard.

STANDARD 1, KEY IDEA 2: BEYOND THE USE OF REASONING AND CONSENSUS, SCIENTIFIC INQUIRY INVOLVES THE TESTING OF PROPOSED EXPLANATIONS INVOLVING THE USE OF CONVENTIONAL TECHNIQUES AND PROCEDURES AND USUALLY REQUIRING CONSIDERABLE INGENUITY.

Performance Indicator	Description	June 2015	August 2015	June 2016	August 2016
2.1	The student should be able to devise ways of making observations to test proposed explanations.	67		57	
2.2	The student should be able to refine research ideas through library investigations, including electronic information retrieval and reviews of literature, and through peer feedback obtained from review and discussion.				
2.3	The student should be able to develop and present proposals including formal hypotheses to test explanations (i.e. predict what should be observed under specific conditions if the experiment is true).	68	70	56	62
2.4	The student should be able to carry out research for testing explanations, including selecting and developing techniques, acquiring and building apparatus, and recording observations as necessary.	69, 70, 71, 72 67, 68	60	33	32, 56, 57

STANDARD 1, KEY IDEA 3: THE OBSERVATIONS MADE WHILE TESTING PROPOSED EXPLANATIONS, WHEN ANALYZED USING CONVENTIONAL AND INVENTED METHODS, PROVIDE NEW INSIGHTS INTO NATURAL PHENOMENA.

Performance Indicator	Description	June 2015	August 2015	June 2016	August 2016
3.1	The student should be able to use various methods of representing and organizing observations (e.g. diagrams, tables, charts, graphs, equations, matrices) and insightfully interpret the organized data.	44, 45, 46, 48	44, 45, 46	44, 45, 46, 48, 50	44, 45
3.2	The student should be able to apply statistical analysis techniques when appropriate to test if chance alone explains the results.				
3.3	The student should be able to assess correspondence between the predicted result contained in the hypothesis and the actual result, and reach a conclusion as to whether the explanation on which the prediction was based is supported.			53	
3.4	The student should be able to, based on the results of the test and through public discussion, revise the explanation and contemplate additional research.				33, 36
3.5	The student should be able to develop a written report for public scrutiny that describes the proposed explanation, including a literature review, the research carried out, its result, and suggestions for further research.				

STANDARD 4, KEY IDEA 1: LIVING THINGS ARE BOTH SIMILAR TO AND DIFFERENT FROM EACH OTHER AND FROM NONLIVING THINGS.

Performance Indicator	Description	June 2015	August 2015	June 2016	August 2016
1.1	The student should be able to explain how diversity of populations within ecosystems relates to the stability of ecosystems.	3	1, 34, 55	1, 3, 18, 21, 25, 31, 32, 36	14, 52, 53, 54, 55, 69
1.2	The student should be able to describe and explain the structures and functions of the human body at different organizational levels (e.g. systems, tissues, cells, organelles).	6, 41, 47	2, 10, 30, 35, 36, 52, 53, 54, 56, 57	5	1, 31, 40, 48
1.3	The student should be able to explain how a one-celled organism is able to function despite lacking the levels of organization present in more complex organisms.	1, 2, 7, 21, 22, 42, 43	9	6, 8, 11, 29, 42	38, 46

STANDARD 4, KEY IDEA 2: ORGANISMS INHERIT GENETIC INFORMATION IN A VARIETY OF WAYS THAT RESULT IN CONTINUITY OF STRUCTURE AND FUNCTION BETWEEN PARENTS AND OFFSPRING.

Performance Indicator	Description	June 2015	August 2015	June 2016	August 2016
2.1	The student should be able to explain how the structure and replication of genetic material result in offspring that resemble their parents.	11, 12, 24, 25, 34, 37	3, 6, 8, 24	10, 19, 40, 47, 60, 61	5, 11, 15, 35
2.2	The student should be able to describe and explain how the technology of genetic engineering allows humans to alter genetic makeup of organisms.	18, 29	5, 25	41	4, 8, 49, 50, 51

STANDARD 4, KEY IDEA 3: INDIVIDUAL ORGANISMS AND SPECIES CHANGE OVER TIME.

Performance Indicator	Description	June 2015	August 2015	June 2016	August 2016
3.1	The student should be able to explain the major patterns of evolution.	8, 13, 15, 26, 31, 36, 61, 62, 63	11, 12, 13, 14, 16, 31, 32, 33, 37	9, 14, 16, 17, 54, 55, 58, 59	9, 13, 16, 17, 18, 37

STANDARD 4, KEY IDEA 4: THE CONTINUITY OF LIFE IS SUSTAINED THROUGH REPRODUCTION AND DEVELOPMENT.

Performance Indicator	Description	June 2015	August 2015	June 2016	August 2016
4.1	The student should be able to explain how organisms, including humans, reproduce their own kind.	4, 5, 17, 20, 39	15, 17, 18, 19, 20, 38	4, 12, 15, 26, 43	12, 19, 20, 21, 22

STANDARD 4, KEY IDEA 5: ORGANISMS MAINTAIN A DYNAMIC EQUILIBRIUM THAT SUSTAINS LIFE.

Performance Indicator	Description	June 2015	August 2015	June 2016	August 2016
5.1	The student should be able to explain the basic biochemical processes in living organisms and their importance in maintaining dynamic equilibrium.	23, 38, 58	22, 23, 41, 48, 49	7, 37, 38, 49, 51	2, 23
5.2	The student should be able to explain disease as a failure of homeostasis.	27	47, 58, 59	13, 20, 62, 63	24, 42, 63, 64
5.3	The student should be able to relate processes at the system level to the cellular level in order to explain dynamic equilibrium.	16, 49, 50, 51, 52, 59, 60	7, 28, 61, 62, 63	23	7, 10, 25, 41, 43

STANDARD 4, KEY IDEA 6: PLANTS AND ANIMALS DEPEND ON EACH OTHER AND THEIR PHYSICAL ENVIRONMENT.

Performance Indicator	Description	June 2015	August 2015	June 2016	August 2016
6.1	The student should be able to explain factors that limit growth of individuals and populations.	64, 65	39, 40	28, 67	65, 66, 67, 70
6.2	The student should be able to explain the importance of preserving diversity of species and habitats.	30, 32, 33, 35	4, 64, 65, 66	65, 66	26, 27, 71, 72
6.3	The student should be able to explain how the living and nonliving environments change over time and respond to disturbances.	53, 54, 55	67, 68, 69	22, 34, 35	28

STANDARD 4, KEY IDEA 7: HUMAN DECISIONS AND ACTIVITIES HAVE A PROFOUND IMPACT ON THE PHYSICAL AND LIVING ENVIRONMENT.

Performance Indicator	Description	June 2015	August 2015	June 2016	August 2016
7.1	The student should be able to describe the range of interrelationships of humans with the living and nonliving environment.		50, 51	64	
7.2	The student should be able to explain the impact of technological development and growth in the human population on the living and nonliving environment.	9, 10, 14, 56, 57, 58, 66	26, 42, 43, 72	2, 24, 27, 30, 68, 70, 71, 72	3, 29, 34, 58, 60, 61
7.3	The student should be able to explain how individual choices and societal actions can contribute to improving the environment.	19, 28, 40	27, 29, 71	39, 52, 69	6, 30, 59, 68

Tips for Teachers

CLASSROOM USE

All teachers will be able to use this book with their students as a companion to their regular textbooks and will find that their students gain considerable self-confidence and ability in test taking through its consistent use.

The Living Environment Core Curriculum (1999) defines the skills and abilities students should have at the point of commencement at the upper-secondary level. It is assumed that science concepts have been taught and assessed at an age-appropriate level throughout their career, so that little additional detail needs to be presented at the upper-secondary level.

An excellent companion to this book (and any comprehensive biology text) is Barron's *Let's Review: Biology—The Living Environment* (Hunter). The factual material and organization of this book lend themselves well to the development of Standards-Based Learning Units (SBLUs) and Essential Questions. The level of detail is consistent with what students really have to know in order to do well on the New York State Regents Examination on the Living Environment.

APPLICATION-BASED CURRICULUM

The curriculum focus can be characterized as application-based—one that is less concerned about content and more concerned about thinking. It is less about *how much* students know and more about *what they can do* with what they know. The latter, after all, is what real learning is all about; these are the abilities that will last a lifetime, not facts and scientific terminology.

This being said, it is acknowledged that students will have a difficult time expressing their views and making moral and ethical judgments about science if they lack a working knowledge of scientific principles and do not have at least a passing understanding of the terms used by biologists. For this reason, teachers and administrators will need to develop local curricula that complement the Core Curriculum. It is up to the teacher or administrator to decide what examples and factual knowledge best illustrate the concepts presented in the Core Curriculum, what concepts need to be reinforced and enhanced, what experiences will add measurably to students' understanding of science, and what examples of local interest should be included.

The teacher will immediately recognize the need to go beyond this level in the classroom, with examples, specific content, and laboratory experiences that complement and illuminate these Major Understandings. It is at this level that the locally developed curriculum is essential. Each school system is challenged to develop an articulated K-12 curriculum in mathematics, science, and technology that will position students to achieve a passing standard at the elementary and intermediate levels, such that success is maximized at the commencement level.

The addition of factual content must be accomplished without contradicting the central philosophy of the learning standards. If local curricula merely revert to the fact-filled syllabi of the past, then little will have been accomplished in the standards movement other than to add yet another layer of content and requirements on the heads of students. A balance must be struck between the desire to build students' ability to think and analyze and the desire to add to the content they are expected to master.

LABORATORY EXPERIENCE

The reduction of factual detail in the Core Curriculum (1982–1999) should allow a more in-depth treatment of laboratory investigations to be planned and carried out than was possible under the previous syllabus. Laboratory experiences should be designed to address Standard 1 (inquiry techniques) but should also take into account Standards 2 (information systems), 6 (interconnectedness of content), and 7 (problem-solving approaches). They should also address the laboratory skills listed in Appendix A of the Core Curriculum.

Part D of the examination assesses student knowledge of and skills on any of four required laboratory experiences supplied to schools by the New York State Education Department. The specific laboratory experiences required in any year will vary according to a preset schedule (see chart on next page).

Questions on this section can be a combination of multiple-choice and constructed-response questions similar to those found in Parts A, B, and C of the Living Environment Regents Examination. The content of these questions reflect the four specific laboratory experiences required for a particular year. Teachers are strongly encouraged to include review of these laboratory experiences as part of their year-end Regents preparation activity.

The following chart summarizes the current laboratory requirement for New York State public schools:

Laboratory Title	Description
The Beaks of Finches	Explores the adaptive advantages of beaks with different physical characteristics
Relationships and Biodiversity	Explores the relationship between DNA structure and the biochemistry of inheritance
Making Connections	Explores the effects of physical activity on human metabolic activities
Diffusion Through a Membrane	Explores the nature of cross-membrane transport in living cells
Adaptations for Reproductive Success in Flowering Plants*	Not yet available
DNA Technology*	Not yet available
Environmental Conditions and Seed Germination*	Not yet available

*Not yet available as of this writing

Study Questions and Answers

QUESTIONS ON STANDARD 1—Scientific Inquiry: Students will use mathematical analysis, scientific inquiry, and engineering design, as appropriate, to pose questions, seek answers, and develop solutions.

Key Idea 1—Purpose of Scientific Inquiry: The central purpose of scientific inquiry is to develop explanations of natural phenomena in a continuing and creative process.

Performance Indicator	Description
1.1	The student should be able to elaborate on basic scientific and personal explanations of natural phenomena and develop extended visual models and mathematical formulations to represent one's thinking.
1.2	The student should be able to hone ideas through reasoning, library research, and discussion with others, including experts.
1.3	The student should be able to work toward reconciling competing explanations and clarify points of agreement and disagreement.
1.4	The student should be able to coordinate explanations at different levels of scale, points of focus, and degrees of complexity and specificity, and recognize the need for such alternative representations of the natural world.

Base your answers to questions 1 through 4 on the passage below and on your knowledge of biology.

To Tan or Not to Tan

Around 1870, scientists discovered that sunshine could kill bacteria. In 1903, Niels Finsen, an Icelandic researcher, won the Nobel Prize for his use of sunlight therapy against infectious diseases. Sunbathing then came into wide use as a treatment for tuberculosis, Hodgkin's disease (a form of cancer), and common wounds. The discovery of vitamin D, the "sunshine vitamin," reinforced the healthful image of the Sun. People learned that it was better to live in a sun-filled home than a dark dwelling. At that time, the relationship between skin cancer and exposure to the Sun was not known.

In the early twentieth century, many light-skinned people believed that a deep tan was a sign of good health. However, in the 1940s, the rate of skin cancer began to increase and reached significant proportions by the 1970s. At this time, scientists began to realize how damaging deep tans could really be.

Tanning occurs when ultraviolet radiation is absorbed by the skin, causing an increase in the activity of melanocytes, cells that produce the pigment melanin. As melanin is produced, it is absorbed by cells in the upper region of the skin, resulting in the formation of a tan. In reality, the skin is building up protection against damage caused by the ultraviolet radiation. It is interesting to note that people with naturally dark skin also produce additional melanin when their skin is exposed to sunlight.

Exposure to more sunlight means more damage to the cells of the skin. Research has shown that, although people usually do not get skin cancer as children, each time a child is exposed to the Sun without protection, the chance of that child getting skin cancer as an adult increases.

Knowledge connecting the Sun to skin cancer has greatly increased since the late 1800s. Currently, it is estimated that ultraviolet radiation is responsible for more than 90% of skin cancers. Yet, even with this knowledge, about 2 million Americans use tanning parlors that expose patrons to high doses of ultraviolet radiation. A recent survey showed that at least 10% of these people would continue to do so even if they knew for certain that it would give them skin cancer.

Many of the deaths due to this type of cancer can be prevented. The cure rate for skin cancer is almost 100% when it is treated early. Reducing exposure to harmful ultraviolet radiation helps to prevent it. During the past 15 years, scientists have tried to undo the tanning myth. If the word "healthy" is separated from the word "tan," maybe the occurrence of skin cancer will be reduced.

1. State *one* known benefit of daily exposure to the Sun. [1]

_____

2. Explain what is meant by the phrase "the tanning myth." [1]

_____

3. Which statement concerning tanning is correct?
   (1) Tanning causes a decrease in the ability of the skin to regulate body temperature.
   (2) Radiation from the Sun is the only radiation that causes tanning.
   (3) The production of melanin, which causes tanning, increases when skin cells are exposed to the Sun.
   (4) Melanocytes decrease their activity as exposure to the Sun increases, causing a protective coloration on the skin.   3_____

4. Which statement concerning ultraviolet radiation is *not* correct?
   (1) It may damage the skin.
   (2) It stimulates the skin to produce antibodies.
   (3) It is absorbed by the skin.
   (4) It may stimulate the skin to produce excess pigment.   4_____

5. Current knowledge concerning cells is a result of the investigations and observations of many scientists. The work of these scientists forms a well-accepted body of knowledge about cells. This body of knowledge is an example of a
   (1) hypothesis   (3) theory
   (2) controlled experiment (4) research plan   5_____

6. In his theory of evolution, Lamarck suggested that organisms will develop and pass on to offspring variations that they need in order to survive in a particular environment. In a later theory of evolution, Darwin proposed that changing environmental conditions favor certain variations that promote the survival of organisms. Which statement is best illustrated by this information?
   (1) Scientific theories that have been changed are the only ones supported by scientists.
   (2) All scientific theories are subject to change and improvement.
   (3) Most scientific theories are the outcome of a single hypothesis.
   (4) Scientific theories are not subject to change.   6_____

Base your answers to questions 7 and 8 on the passage below and on your knowledge of biology.

The number in the parentheses ( ) at the end of a sentence is used to identify that sentence.

They Sure Do Look Like Dinosaurs

When making movies about dinosaurs, film producers often use ordinary lizards and enlarge their images many times (1). We all know, however, that although they look like dinosaurs and are related to dinosaurs, lizards are not actually dinosaurs (2).

Recently, some scientists have developed a hypothesis that challenges this view (3). These scientists believe that some dinosaurs were actually the same species as some modern lizard that had grown to unbelievable sizes (4). They think that such growth might be due to a special type of DNA called repetitive DNA, often referred to as "junk" DNA because scientists do not understand its functions (5). These scientists studied pumpkins that can reach sizes of nearly 1,000 pounds and found them to contain large amounts of repetitive DNA (6). Other pumpkins that grow to only a few ounces in weight have very little of this kind of DNA (7). In addition, cells that reproduce uncontrollably have almost always been found to contain large amounts of this DNA (8).

7. State *one* reason why scientists formerly thought of repetitive DNA as "junk." [1]

_____

_____

8. Write the number of the sentence that provides evidence supporting the hypothesis that increasing amounts of repetitive DNA are responsible for increased sizes of organisms. [1]

_____

ANSWERS EXPLAINED Questions 1–8

Standard 1, Key Idea 1, Performance Indicators 1–4

1. One response is required. Acceptable responses include:
- *Kills bacteria*
- *Produces vitamin D*
- *Treats diseases and/or wounds*

2. One response is required. Acceptable responses include:
- *The "tanning myth" involves people believing that a tan is a sign of good health.*
- *The "tanning myth" says that a good tan is good for people.*

3. **3** *The production of melanin, which causes tanning, increases when skin cells are exposed to the Sun* is the correct statement concerning tanning. Melanin is a dark pigment that is produced in specialized skin cells in response to ultraviolet radiation in sunlight or an artificial source. This information is found in the third paragraph of the passage.

WRONG CHOICES EXPLAINED:

(1) *Tanning causes a decrease in the ability of the skin to regulate body temperature* is not a correct statement concerning tanning. There is no known relationship between tanning and body temperature regulation.

(2) *Radiation from the Sun is the only radiation that causes tanning* is not a correct statement concerning tanning. Tanning can also occur when the skin is exposed to artificial sources of ultraviolet radiation. A reference is made in the fourth paragraph of the passage to "tanning parlors" where people can be exposed to artificial doses of ultraviolet radiation.

(4) *Melanocytes decrease their activity as exposure to the Sun increases, causing a protective coloration on the skin* is not a correct statement concerning tanning. Melanin is produced as a protective pigment that helps prevent deep penetration of ultraviolet radiation into the deep layers of the skin. When ultraviolet radiation is absorbed by melanocytes, their activity increases, not decreases.

4. **2** *It stimulates the skin to produce antibodies* is not a correct statement concerning ultraviolet radiation. There is no information in the passage relating to the production of antibodies as a result of absorption of ultraviolet radiation, and no known research indicates this type of relationship.

WRONG CHOICES EXPLAINED:

(1), (3), (4) *It may damage the skin, it is absorbed by the skin,* and *it may stimulate the skin to produce excess pigment* are all correct statements concerning ultraviolet radiation. Ultraviolet radiation is an invisible but extremely powerful form of electromagnetic radiation. It can penetrate unshielded living tissues and alter the genetic makeup of the cells it encounters. In humans, this radiation can cause the production of melanin from melanocytes; in extreme cases, it can stimulate the growth of skin cancer.

5. **3** The body of knowledge described in this question is an example of a *theory*. When scientists begin to study a phenomenon in nature, their first step is normally to investigate it through repeated observation and experimentation. As a result of the analysis of the large quantity of data gathered during this process, the scientists then formulate a theory ("well-accepted body of knowledge") that describes the phenomenon in a way that is consistent with the data.

WRONG CHOICES EXPLAINED:

(1) A *hypothesis* is not the body of knowledge described in this question. Scientists develop a hypothesis ("educated guess") around their preliminary observations concerning a natural phenomenon. The hypothesis may be proven accurate or inaccurate as a result of the experimentation used to test it. For this reason, a hypothesis cannot be considered a "well-accepted body of knowledge."

(2) A *controlled experiment* is not the body of knowledge described in this question. A controlled experiment is a scientific method used to test an experimental hypothesis. The data that results from a controlled experiment can be used to support the development of a "well-accepted body of knowledge," but it does not constitute that body of knowledge.

(4) A *research plan* is not the body of knowledge described in this question. A research plan may be a series of controlled experiments designed to test various aspects of a natural phenomenon. The data that results from the research plan can be used to support the development of a "well-accepted body of knowledge," but it does not constitute that body of knowledge.

6. **2** *All scientific theories are subject to change and improvement* is the statement best illustrated by the information given. Both Lamarck and Darwin developed their theories of evolution based on observations made and inferences drawn before there was a good understanding of the genetic basis of variation. Lamarck's earlier theory of "use and disuse" was disproven by later experiments of other scientists. Darwin's later theory of "natural selection," though much closer to the currently accepted scientific theory of evolution, has been modified and improved on by the work of later scientists who have had the benefit of modern-day research in genetics, paleontology, and other sciences.

WRONG CHOICES EXPLAINED:

(1) *Scientific theories that have been changed are the only ones supported by scientists* is not the statement best illustrated by the information given. Scientists generally support theories that have stood the test of good scientific research. A theory that has not changed, as long as it is still supported by such research, is generally supported by most scientists.

(3) *Most scientific theories are the outcome of a single hypothesis* is not the statement best illustrated by the information given. In fact, scientific theories are based on the results of many experiments that each contain their own independent hypotheses.

(4) *Scientific theories are not subject to change* is not the statement best illustrated by the information given. Scientists are constantly questioning and reevaluating scientific theories. It is likely that a vast majority of all scientific theories undergo at least some modification.

7. One response is required that indicates a reason why scientists formerly thought of repetitive DNA as "junk." Acceptable responses include: [1]
   • *Scientists did not understand the function of repetitive DNA.*
   • *They didn't know what it did, and so they thought it was junk.*

8. One credit is allowed for indicating that either sentence 6 or sentence 7 provides evidence supporting the hypothesis that increased amounts of repetitive DNA are responsible for increased sizes of organisms. These sentences give information about the results of scientific investigations that measured the amount of repetitive DNA in the cells of giant pumpkins and miniature pumpkins and found that giant pumpkins contain more of this kind of DNA than miniature pumpkins.

Key Idea 2—Methods of Scientific Inquiry: Beyond the use of reasoning and consensus, scientific inquiry involves the testing of proposed explanations involving the use of conventional techniques and procedures and usually requiring considerable ingenuity.

Performance Indicator	Description
2.1	The student should be able to devise ways of making observations to test proposed explanations.
2.2	The student should be able to refine research ideas through library investigations, including electronic information retrieval and reviews of literature, and through peer feedback obtained from review and discussion.
2.3	The student should be able to develop and present proposals including formal hypotheses to test explanations (i.e., predict what should be observed under specific conditions if the experiment is true).
2.4	The student should be able to carry out research for testing explanations, including selecting and developing techniques, acquiring and building apparatus, and recording observations as necessary.

Base your answers to questions 9 and 10 on the diagram below of the field of view of a light compound microscope and on your knowledge of microscopes.

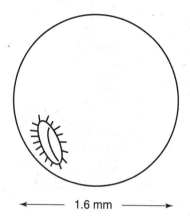

←——— 1.6 mm ———→

9. In order to center the organism in the field of view, the slide should be moved
    (1) down and to the right   (3) up and to the right
    (2) down and to the left   (4) up and to the left   9____

10. The approximate length of the organism is
    (1) 500 μm   (3) 50 μm
    (2) 1,600 μm   (4) 1.6 μm   10____

11. After viewing an organism under low power, a student switches to high power. The student should first
    (1) adjust the mirror
    (2) center the organism
    (3) raise the objective and switch to high power
    (4) close the diaphragm   11____

12. Using one or more complete sentences, explain why a specimen viewed under the high-power objective of a microscope appears darker than when it is viewed under low power.

_____

_____

Base your answers to questions 13 and 14 on the diagram below of a compound light microscope.

13. The letter *C* represents
   (1) the mirror
   (2) the diaphragm
   (3) the eyepiece
   (4) the high-power objective   13____

14. Select and name *one* of the labeled parts, and in one or more complete sentences describe its function.

_____

_____

15. The letter "p" as it normally appears in print is placed on the stage of a compound light microscope. Which best represents the image observed when a student looks through the microscope?
   (1) p   (2) q   (3) b   (4) d   15 ____

16. To separate the parts of a cell by differences in density, a biologist would probably use
   (1) a microdissection instrument
   (2) an ultracentrifuge
   (3) a compound light microscope
   (4) an electron microscope   16____

17. The diagram below represents the field of view of a microscope. What is the approximate diameter, in micrometers, of the cell shown in the field?

(1) 50 μm   (3) 1,000 μm
(2) 500 μm   (4) 2,000 μm   17____

18. Base your answer to the question on the diagram below.

How many millimeters long is the organism resting on the metric ruler?

Questions 19 through 21 are based on the experiment described below.

A test tube was filled with a molasses solution, sealed with a membrane, and inverted into a beaker containing 200 mL of distilled water. A second test tube was filled with a starch solution, sealed with a membrane, and inverted into a beaker containing 200 mL of distilled water. After several hours, the water in each beaker was tested for the presence of molasses and starch.

The diagrams show the setup of the experiment.

At the Start of the Experiment

After Several Hours

Answer each question related to the experiment in one or more complete sentences.

19. What principle was being tested in the experiment?

20.   What reagents were used in the experiment to test for the presence of molasses and starch?

_____

21.   Draw one conclusion from this experiment.

_____

Questions 22 and 23 are based on the experiment described below.

An opaque disk was placed on several leaves of a geranium plant. The remaining leaves of the plant were untreated. After the plant had been exposed to sunlight, a leaf on which a disk had been placed was removed and tested as shown in parts *B* and *C* of the diagram below.

A. Leaf in Sunlight   *B.* Leaf after   *C.* Leaf after
                             Boiling in Alcohol   Testing with Iodine

Answer each question related to the experiment in one or more complete sentences.

22.   What conclusion can be drawn from the result of the experiment?

_____

23.   What process was being investigated by the experiment?

_____

Questions 24 and 25 are based on the experiment described below.

A student added 15 mL of water to each of three test tubes, labeled A, B, and C. A 1-cc piece of raw potato was added to tube B. A 1-cc piece of cooked potato was added to tube C. Five drops of hydrogen peroxide (H_2O_2) were added to each test tube. The results are shown in the following diagram.

24. What conclusion can be drawn from the experiment?

_____

25. Which test tube is the control? Explain the reason for your choice.

_____

Base your answers to questions 26 through 28 on the diagram of the measuring device shown below.

26. What is the name of this measuring device?

_____

27.   In one complete sentence describe the procedure that you would follow to read the meniscus.

---

28.   What must a student do to obtain a volume of 85 milliliters of liquid in this measuring device?

    (1) Add 2.0 mL.   (3) Add 2.5 mL.

    (2) Remove 2.0 mL.   (4) Remove 8.7 mL.   28_____

ANSWERS EXPLAINED Questions 9–28

Standard 1, Key Idea 2, Performance Indicators 1–4

9.   **2**   Specimens viewed under the microscope appear upside-down, backward, and reversed.

WRONG CHOICES EXPLAINED:

(1), (3), (4)   With any of these choices, the specimen would be moved out of the field of view.

10.   **1**   The field of view is given as 1.6 mm. 1 mm = 1000 μm. 1.6 mm × 1000 μm = 1600 μm. The diagram shows that three specimens would fit across the field of view. One-third of 1600 μm = 533 μm. Of the choices given, *500 μm* (choice 1) is closest to this value.

WRONG CHOICES EXPLAINED:

(2), (3), (4)   Each of these choices is mathematically incorrect.

11.   **2**   The student should first *center the organism*. The field of view is smaller under high power; therefore, less of a specimen can be seen. If the organism is not centered, it may fall out of the field of view under high power.

WRONG CHOICES EXPLAINED:

(1)   The *mirror is adjusted* for maximum light under low power. Because the diameter of the high-power objective is very small, it is impossible to adjust the light under high power.

(3)   A compound light microscope is parfocal; that is, it is not necessary to *lift the high-power objective* to focus under high power. The specimen remains in focus when switching from low power to high power.

(4)   *Closing the diaphragm* reduces the amount of light entering the objective. Therefore, the specimen would appear very dark and would be difficult to see.

12.   *The diameter of the high-power objective is smaller than the diameter of the low-power objective. Less light enters through the high-power objective, and therefore the specimen appears darker.*

13.   **2**   The letter *C* represents *the diaphragm.*

WRONG CHOICES EXPLAINED:
   (1) *The mirror* is represented by *D*.
   (3), (4) *The eyepiece* and *the high-power objective* are not labeled on the diagram.

14.   *Coarse adjustment (A)—used to focus a specimen under the low-power objective.*

or

Low-power objective (B)—along with the standard eyepiece, magnifies a specimen 100×.

or

Diaphragm (C)—regulates the amount of light entering the objectives.

or

Mirror (D)—provides a source of light that illuminates the specimen.

15.   **4**   The image of a specimen as seen under a microscope is upside-down (*d*). The right side is on the left side, and the top is on the bottom.

WRONG CHOICES EXPLAINED:
   (1) In this choice (*p*) there is no change in the way the image of the letter appears.
   (2) In this choice (*q*) the image of the letter is reversed in only one direction: The right and left sides are reversed.
   (3) In this choice (*b*) the image of the letter is reversed in only one direction: The top and bottom are reversed.

16.   **2**   The *ultracentrifuge* is a machine that spins at a very high speed. A test tube of a liquid containing the parts of ruptured cells is placed in the machine. Each cell part has its own density (mass per unit volume). When the machine rotates, the cell parts fall to different levels in the test tube depending on their density.

WRONG CHOICES EXPLAINED:
   (1) A *microdissection instrument* enables a biologist to remove a cell part from a single living cell. A micromanipulator is an example of such an instrument.
   (3) A cell is transparent under a light microscope. Its structures cannot be seen unless the cell is stained. A *compound light microscope* can be used to view, but not to separate, cell parts.

(4) *An electron microscope* uses beams of electrons to view freeze-dried specimens; it cannot be used to separate cell parts for study.

17. **2** Study the information given in the diagram. Notice that the diameter of the circle is 2 mm. Since 1 mm is equal to 1,000 µm, 2 mm are equal to 2,000 µm. In relation to the entire circle, how large is the cell? Is the cell one-half as large or one-fourth as large? Dividing the circle into four parts shows us that the diameter of the cell is about one-quarter the diameter of the circle. Dividing 4 into 2,000 results in *500 µm*.

WRONG CHOICES EXPLAINED:
(1) *50 µm* is too small. The cell is ten times larger than 50.
(3) *1,000 µm* is too large. The cell is not one-half the diameter.
(4) *2,000 µm* is the diameter of the circle. The cell is only one-fourth as large.

18. The organism is *26 millimeters* long.

19. The principle of *diffusion* was being tested in the experiment.

20. *Benedict's solution* was used to test for the presence of molasses in the beaker. *Iodine* was used to test for the presence of starch in the beaker.

21. *Molasses can diffuse through a membrane.*
or .
Starch cannot diffuse through a membrane.

22. *No starch was produced in the area covered by the disk.*

23. *The process of photosynthesis was being investigated.*

24. *Raw potato contains an enzyme that breaks down hydrogen peroxide.*
or
Cooking a potato destroys the enzyme that breaks down hydrogen peroxide.

25. *Test tube A is the control.* A control is the part of the experiment that provides the basis of comparison for the variable being tested.

26. The device is known as a *graduated cylinder*.

27. *The meniscus should be read at eye level.*

28. **2** To obtain a volume of 85 mL, *2.0 mL* must be removed. The graduated cylinder contains 87 mL of liquid.

Key Idea 3—Analysis in Scientific Inquiry: The observations made while testing proposed explanations, when analyzed using conventional and invented methods, provide new insights into natural phenomena.

Performance Indicator	Description
3.1	The student should be able to use various methods of representing and organizing observations (e.g., diagrams, tables, charts, graphs, equations, matrices) and insightfully interpret the organized data.
3.2	The student should be able to apply statistical analysis techniques, when appropriate, to test if chance alone explains the results.
3.3	The student should be able to assess correspondence between the predicted result contained in the hypothesis and actual result, and reach a conclusion as to whether the explanation on which the prediction was based is supported.
3.4	The student should be able to, based on the results of the test and through public discussion, revise the explanation and contemplate additional research.
3.5	The student should be able to develop a written report for public scrutiny that describes the proposed explanation, including a literature review, the research carried out, its result, and suggestions for further research.

29. If curve *A* in the diagram represents a population of hawks in a community, what would most likely be represented by curve *B*?

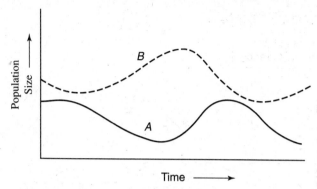

(1) the dominant trees in that community
(2) a population with which the hawks have a mutualistic relationship
(3) variations in the numbers of producers in that community
(4) a population on which the hawks prey

29 _____

Base your answers to questions 30 and 31 on this graph and on your knowledge of biology. The graph below depicts changes in the population growth rate of Kaibab deer.

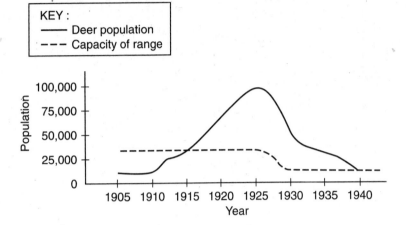

KEY :
—— Deer population
– – – Capacity of range

30. About how many deer could the range have supported in 1930 without some of them starving to death?
(1) 12,000   (3) 50,000
(2) 35,000   (4) 100,000   30____

31. In which year were the natural predators of the deer most likely being killed off faster than they could reproduce?
(1) 1905   (2) 1920   (3) 1930   (4) 1940   31____

32. Which process is illustrated by the diagram?

(1) migration   (3) speciation
(2) adaptive radiation   (4) isolation   32____

Base your answers to questions 33 through 36 on the information and data table below and on your knowledge of biology. The table shows the average systolic and diastolic blood pressure measured in millimeters of mercury (mm Hg) for humans between the ages of 2 and 14 years.

Data Table

Age	Average Blood Pressure (mm Hg)	
	Systolic	Diastolic
2	100	60
6	101	64
10	110	72
14	119	76

DIRECTIONS (33–36): Using the information in the data table, construct a line graph on the grid provided, following the directions below.

33. Mark an appropriate scale on each labeled axis.

34. Plot the data for systolic blood pressure on your graph. Surround each point with a small triangle and connect the points.

35. Plot the data for diastolic blood pressure on your graph. Surround each point with a small circle and connect the points.

The Effect of Age on Human Blood Pressure

Average Blood Pressure (mm of Hg)

Age (yr)

Key

⚠ Systolic blood pressure

⊙ Diastolic blood pressure

36. Using one or more complete sentences, state one conclusion that compares systolic blood pressure to diastolic blood pressure in humans between the ages of 2 and 14 years.

37. The graph below shows the results of an experiment.

At 16°C, what percentage of the brook trout reached maximum size?
(1) 30%
(2) 55%
(3) 75%
(4) 95%

37 _____

38. An experiment is represented in the diagram below.

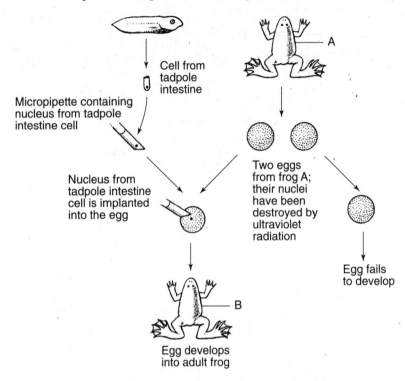

Micropipette containing nucleus from tadpole intestine cell

Cell from tadpole intestine

Nucleus from tadpole intestine cell is implanted into the egg

Two eggs from frog A; their nuclei have been destroyed by ultraviolet radiation

Egg fails to develop

Egg develops into adult frog

An inference that can be made from this experiment is that
(1) adult frog B will have the same genetic traits as the tadpole
(2) adult frog A can develop only from an egg and a sperm
(3) fertilization must occur in order for frog eggs to develop into adult frogs
(4) the nucleus of a body cell fails to function when transferred to other cell types

38 _____

39. The charts below show the relationship of recommended weight to height in men and women age 25–29.

Height-Weight Charts

MEN Age 25–29 Weight (lb)					WOMEN Age 25–29 Weight (lb)				
Height Feet	Inches	Small Frame	Medium Frame	Large Frame	Height Feet	Inches	Small Frame	Medium Frame	Large Frame
5	2	128–134	131–141	138–150	4	10	102–111	109–121	118–131
5	3	130–136	133–143	140–153	4	11	103–113	111–123	120–134
5	4	132–138	135–145	142–156	5	0	104–115	113–126	122–137
5	5	134–140	137–148	144–160	5	1	106–118	115–129	125–140
5	6	136–142	139–151	146–164	5	2	108–121	118–132	128–143
5	7	138–145	142–154	149–168	5	3	111–124	121–135	131–147
5	8	140–148	145–157	152–172	5	4	114–127	124–138	134–151
5	9	142–151	148–160	155–176	5	5	117–130	127–141	137–155
5	10	144–154	151–163	158–180	5	6	120–133	130–144	140–159
5	11	146–157	154–166	161–184	5	7	123–136	133–147	143–163
6	0	149–160	157–170	164–188	5	8	126–139	136–150	146–167
6	1	152–164	160–174	168–192	5	9	129–142	139–153	149–170
6	2	155–168	164–178	172–197	5	10	132–145	142–156	152–173
6	3	158–172	167–182	176–202	5	11	135–148	145–159	155–176
6	4	162–176	171–187	181–207	6	0	138–151	148–162	158–179

The recommended weight for a 6'0" tall man with a small frame is closest to that of a

    (1) 5'10" man with a medium frame
    (2) 5'9" woman with a large frame
    (3) 6'0" man with a medium frame
    (4) 6'0" woman with a medium frame   39 _____

Base your answers to questions 40 through 43 on the information below and on your knowledge of biology.

A group of biology students extracted the photosynthetic pigments from spinach leaves using the solvent acetone. A spectrophotometer was used to measure the percent absorption of six different wavelengths of light by the extracted pigments. The wavelengths of light were measured in units known as nanometers (nm). One nanometer is equal to one-billionth of a meter. The following data were collected:

yellow light (585 nm)—25.8% absorption
blue light (457 nm)—49.8% absorption
orange light (616 nm)—32.1% absorption
violet light (412 nm)—49.8% absorption
red light (674 nm)—41.0% absorption
green light (533 nm)—17.8% absorption

40. Complete all three columns in the data table below so that the wavelength of light either increases or decreases from the top to the bottom of the data table.

Color of Light	Wavelength of Light (nm)	Percent Absorption by Spinach Extract

DIRECTIONS (41–42): Using the information in the data table, construct a line graph on the grid provided, following the directions below.

41. Mark an appropriate scale on the axis labeled "Percent Absorption."

42. Plot the data from the data table on your graph. Surround each point with a small circle and connect the points.

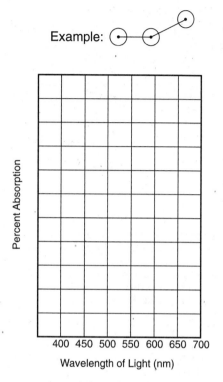

43. Which statement is a valid conclusion that can be drawn from the data obtained in this investigation?

    (1) Photosynthetic pigments in spinach plants absorb blue light and violet light more efficiently than red light.

    (2) The data would be the same for all pigments in spinach plants.

    (3) Green light and yellow light are not absorbed by spinach plants.

    (4) All plants are efficient at absorbing violet light and red light.   43_____

44. The graph below represents the results of an investigation of the growth of three identical bacterial cultures incubated at different temperatures.

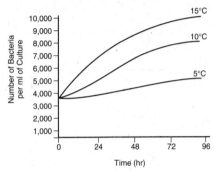

BIO/F126

Which inference can be made from this graph?
(1) Temperature is unrelated to the reproductive rate of bacteria.
(2) Bacteria cannot grow at a temperature of 5°C.
(3) Life activities in bacteria slow down at high temperatures.
(4) Refrigeration will most likely slow the growth of these bacteria.

44 _____

45. A study was conducted using two groups of ten plants of the same species. During the study, the plants were kept under identical environmental conditions. The plants in one group were given a growth solution every 3 days. The heights of the plants in both groups were recorded at the beginning of the study and at the end of a 3-week period. The data showed that the plants given the growth solution grew faster than those not given the solution.

When other researchers conduct this study to test the accuracy of the results, they should
(1) give growth solution to both groups
(2) make sure that the conditions are identical to those in the first study
(3) give an increased amount of light to both groups of plants
(4) double the amount of growth solution given to the first group

45 _____

46. Worker bees acting as scouts are able to communicate the distance of a food supply from the hive by performing a "waggle dance." The graph below shows the relationship between the distance of a food supply from the hive and the number of turns in the waggle dance every 15 seconds.

Using one or more complete sentences, state the relationship between the distance of the food supply from the hive and the number of turns a bee performs in the waggle dance every 15 seconds.

47. Based on experimental results, a biologist in a laboratory reports a new discovery. If the experimental results are valid, biologists in other laboratories should be able to perform
  (1) an experiment with a different variable and obtain the same results
  (2) the same experiment and obtain different results
  (3) the same experiment and obtain the same results
  (4) an experiment under different conditions and obtain the same results   47_____

ANSWERS EXPLAINED Questions 29–47

Standard 1, Key Idea 3, Performance Indicators 1–5

29. **4** The diagram shows the population growth cycle for two organisms. An examination of the graph shows that the population growth cycle of the hawks closely follows the cycle of population B. There is a slight lag in the cycles. This type of graph is used to show a predator-prey relationship. *The hawks prey on population B.*

WRONG CHOICES EXPLAINED:
(1) Hawks are *not* herbivores. They do not live off the *dominant trees* in the community.
(2) If the two populations benefited equally from the relationship (were *mutualistic*), the peaks of the two graphs would coincide.
(3) Hawks are carnivores. They do not depend directly on the *producers* in the community.

30. **1** According to the graph, the range could support *12,000* deer in 1930.

WRONG CHOICES EXPLAINED:
(2) The carrying capacity of the range was *35,000*. The *carrying capacity* is the maximum number of individuals that can be supported by the area. The number is usually constant unless severe environmental changes occur.
(3) The actual number of deer occupying the range in 1930 was *50,000*.
(4) A population of *100,000* deer was reached in 1925.

31. **2** In nature, a predator-prey relationship keeps the prey population in check. In *1920*, the population of deer increased. It was at this time that the predators were removed by human hunting.

WRONG CHOICES EXPLAINED:
(1) The deer probably entered the region in *1905*. It takes time for an organism to adjust to a new environment.
(3) In *1930*, the deer population, which had exceeded the carrying capacity, was declining. The decline was caused by starvation. The deer had consumed almost all of the available vegetation in the area.
(4) By *1940*, the deer population reached the new carrying capacity of the range. The carrying capacity had been greatly reduced by overgrazing by the previously unchecked deer population.

32.   **4**   As the result of *isolation*, the members of populations *A* and *B* are separated and are prevented from interbreeding. Variations that occur in one area are not transmitted to the individuals in the other area. Consequently, over a long period of time, the genetic differences become accentuated, and the new variations are maintained.

WRONG CHOICES EXPLAINED:

(1) *Migration* is not possible when populations are separated by a geographical barrier such as water.

(2) *Adaptive radiation* refers to a branching evolution and is not depicted in the diagram.

(3) *Speciation* indicates formation of new species from a parent population. This is not indicated in the diagram.

33–35.

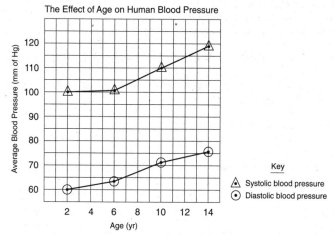

36.   *Systolic pressure is higher than diastolic pressure.* or *Both systolic pressure and diastolic pressure increase between the ages of 2 and 14.*

[*Note:* Other correct complete-sentence responses are acceptable.]

37.   **4** Brook trout growth is represented by the lighter solid line on the graph. Tracing along the horizontal axis to 16°C and then tracing up to the point at which the brook trout line is encountered, we see that the growth rate is between 80% and 100% but closer to 100%; we estimate it is about 95%.

WRONG CHOICES EXPLAINED:
(1) Brook trout growth rate is at *30%* when the water temperature is about 18°C. Largemouth bass growth rate is about 30% at 16°C.
(2) None of the species indicated have a growth rate of 55% at 16°C. Brook trout growth rate is at 55% at about 17.5°C.
(3) Brook trout growth rate falls to 75% at about 17°C. The growth rate of northern pike is at 75% at about 16°C.

38.   **1** The diagram represents an experiment in cloning. Because the nucleus (which contains the genetic material) used in this experiment comes from a tadpole, the egg will produce new cells that have *genetic characteristics identical to those of the tadpole.*

WRONG CHOICES EXPLAINED:
(2) It is unclear from this diagram how frog A came into being.
(3) This conclusion is refuted by the results of this experiment; a new frog was created without the use of fertilization.
(4) This conclusion is refuted by the results of this experiment; the tadpole nucleus was taken from an intestinal cell, which is a "body" cell.

39.   **4** The chart on the left shows that a man 6'0" tall with a small frame has an ideal weight range of 149–160 pounds. Of the choices given, the closest comparison can be made to the ideal weight range of a *6'0" woman with a medium frame* at 148–162 pounds.

WRONG CHOICES EXPLAINED:
(1) The weight range shown for a *5'10" man with a medium frame* is 151–163 pounds.
(2) The weight range shown for a *5'9" woman with a large frame* is 149–170 pounds.
(3) The weight range shown for a *6'0" man with a medium frame* is 157–170 pounds.

40.

Color of Light	Wavelength of Light (nm)	Percent Absorption by Spinach Extract
violet	412	49.8
blue	457	49.8
green	533	17.8
yellow	585	25.8
orange	616	32.1
red	674	41.0

41–42.

Wavelength of Light (nm)

43. **1** The statement *photosynthetic pigments in spinach plants absorb blue light and violet light more efficiently than red light* is a valid conclusion that can be drawn from the data. The high point of the chart/graph data is clearly shown to be above the blue and violet wavelengths of light.

WRONG CHOICES EXPLAINED:
(2) The statement *the data would be the same for all pigments in spinach plants* is not supported by the results of this experiment. The chart/graph data show considerable variation in the experimental results as the wavelength of light varies.
(3) The statement *green light and yellow light are not absorbed by spinach plants* is not supported by the results of this experiment. Although the chart/graph data show a lower absorption rate at these wavelengths, there is still some absorption in this range.

(4) The statement *all plants are efficient at absorbing violet light and red light* is not supported by the results of this experiment. The experimental data are limited to the absorption of light by pigments found in one type of plant. These data cannot be extended to all plants unless all other types of plants are tested under the same experimental conditions and the results are found to be similar.

44. **4** Of those given, *refrigeration will most likely slow the growth of these bacteria* is the most reasonable inference that can be made from the graph data. The graph clearly shows a slower rate of growth (reproduction) at 5°C than at 10°C or 15°C.

WRONG CHOICES EXPLAINED:

(1) The inference that *temperature is unrelated to the reproductive rate of bacteria* is not supported by the data. Temperature is the independent (experimental) variable in this study. It clearly has an influence on the bacterial reproductive rate.

(2) The inference that *bacteria cannot grow at a temperature of 5°C* is not supported by the data. The graph clearly shows that growth at this temperature, while slow, occurs at a steady pace.

(3) The inference that *life activities in bacteria slow down at high temperatures* is not supported by the data. The data indicate that, if anything, bacterial activity increases with increasing temperature. No data are shown for bacterial growth at temperatures above 15°C, and so we cannot draw any inference about what happens to the rate of bacterial growth at these extremes.

45. **2** These researchers should *make sure that the conditions are identical to those in the first study.* The validity of any scientific experiment can be verified only if the same results are obtained under the same experimental conditions. Any change in these conditions invalidates the results of the verification study.

WRONG CHOICES EXPLAINED:

(1) If the researchers *give growth solution to both groups,* there will be no control group against which to compare the experimental group. The results of the verification study will be invalid because the experimental conditions will have been changed.

(3) If the researchers *give an increased amount of light to both groups of plants,* the original experimental method will be altered. The results of the verification study will be invalid because the experimental conditions will have been changed.

(4) If the researchers *double the amount of growth solution given to the first group,* the original experimental method will not be followed. The results of the verification study will be invalid because the experimental conditions will have been changed.

46. *The number of turns in the waggle dance decreases as the distance of the food supply from the hive increases.* Or *The closer to the hive the food source is located, the more turns there are in the waggle dance.* [*Note:* Any correct, complete-sentence answer is acceptable.]

47. **3** Other biologists in other laboratories should be able to perform *the same experiment and obtain the same results* if the experimental results are valid. Any experimental results obtained by one scientist must be validated through independent research by other scientists following the same procedures.

WRONG CHOICES EXPLAINED:
(1) If different scientists perform *an experiment with a different variable and obtain the same results,* the original experimental results will be invalidated.

(2), (4) If different scientists perform *the same experiment and obtain different results,* or *an experiment under different conditions and obtain the same results,* they will neither validate nor invalidate the results of the original experiment. All variables and conditions must be kept the same if the experimental results are to be properly tested.

QUESTIONS ON STANDARD 4—Biological Concepts: Students will understand and apply scientific concepts, principles, and theories pertaining to the physical setting and living environment and recognize the historical development of ideas in science.

Key Idea 1—Application of Scientific Principles: Living things are both similar to and different from each other and from nonliving things.

Performance Indicator	Description
1.1	The student should be able to explain how diversity of populations within ecosystems relates to the stability of ecosystems.
1.2	The student should be able to describe and explain the structures and functions of the human body at different organizational levels (e.g., systems, tissues, cells, organelles).
1.3	The student should be able to explain how a one-celled organism is able to function despite lacking the levels of organization present in more complex organisms.

48. In which life function is the potential energy of organic compounds converted to a form of stored energy that can be used by the cell?

    (1) transport   (3) excretion

    (2) respiration   (4) regulation   48_____

49. Which life activity is *not* required for the survival of an individual organism?

    (1) nutrition   (3) reproduction

    (2) respiration   (4) synthesis   49_____

50. Which function of human blood includes the other three?

    (1) transporting nutrients   (3) maintaining homeostasis

    (2) transporting oxygen   (4) collecting wastes   50_____

51. In the human body, the blood with the greatest concentration of oxygen is found in the

(1) left atrium of the heart   (3) nephrons of the kidney

(2) cerebrum of the brain   (4) lining of the intestine   51 _____

52. Which type of vessel normally contains valves that prevent the backward flow of materials?

(1) artery   (3) capillary

(2) arteriole   (4) vein   52 _____

DIRECTIONS (53–55): For each of questions 53 through 55, select the excretory structure, chosen from the list below, that best answers the question. Then record its number in the space provided at the right.

Excretory Structures

(1) Alveolus

(2) Nephron

(3) Sweat gland

(4) Liver

53. Which structure forms urine from water, urea, and salts?

_____

54. Which structure removes carbon dioxide and water from the blood?

_____

55. Which structure is involved in the breakdown of red blood cells?

_____

56. The bones of the lower arm are connected to the muscles of the upper arm by

(1) ligaments   (3) cartilage

(2) tendons   (4) skin   56 _____

57. The diagram below shows the same type of molecules in area A and area B. With the passage of time, some molecules move from area A to area B.

Molecules

Selectively Permeable Membrane

The movement is the result of the process of
(1) phagocytosis   (3) diffusion
(2) pinocytosis   (4) cyclosis   57_____

58. Which is the principal inorganic compound found in cytoplasm?
(1) lipid   (3) water
(2) carbohydrate   (4) nucleic acid   58_____

59. A specific organic compound contains only the elements carbon, hydrogen, and oxygen in a ratio of 1:2:1. This compound is most probably a
(1) nucleic acid   (3) protein
(2) carbohydrate   (4) lipid   59_____

60. Compared to ingested food molecules, end-product molecules of digestion are usually
(1) smaller and more soluble   (3) smaller and less soluble
(2) larger and more soluble   (4) larger and less soluble   60_____

61. The cellular function of the endoplasmic reticulum is to
(1) provide channels for the transport of materials
(2) convert urea to a form usable by the cell
(3) regulate all cell activities
(4) change light energy into chemical bond energy   61_____

62. In which organelles are polypeptide chains synthesized?
(1) nuclei   (3) ribosomes
(2) vacuoles   (4) cilia

62_____

63. Which organelle contains hereditary factors and controls most cell activities?
(1) nucleus   (3) vacuole
(2) cell membrane   (4) endoplasmic reticulum

63_____

64. Centrioles are cell structures involved primarily in
(1) cell division   (3) enzyme production
(2) storage of fats   (4) cellular respiration

64_____

65. Which cell structure contains respiratory enzymes?
(1) cell wall   (3) mitochondrion
(2) nucleolus   (4) vacuole

65_____

66. Which process is represented below?

enzymes
simple organic molecules → complex organic molecules + H_2O

(1) hydrolysis   (3) digestion
(2) synthesis   (4) respiration

66_____

67. Amino acids derived from the digestion of a piece of meat are transported to living cells of an animal. In the cell they are
(1) converted to cellulose
(2) used to attack invading bacteria
(3) synthesized into specific proteins
(4) incorporated into glycogen molecules

67_____

68. Which of the following variables has the *least* direct effect on the rate of a hydrolytic reaction regulated by enzymes?
(1) temperature
(2) pH
(3) carbon dioxide concentration
(4) enzyme concentration

68_____

69. Which term refers to the chemical substance that aids in the transmission of the impulse through the area indicated by X?

(1) neurotransmitter   (3) neuron

(2) synapse   (4) nerve   69_____

70. Which lists human nervous-system structures in order of increasing size?

(1) neuron, nerve, ganglion, receptor

(2) nerve, ganglion, neuron, receptor

(3) neuron, receptor, ganglion, nerve

(4) ganglion, receptor, nerve, neuron   70_____

71. Glands located within the digestive tube include

(1) gastric glands and thyroid glands

(2) gastric glands and intestinal glands

(3) thyroid glands and intestinal glands

(4) adrenal glands and intestinal glands   71_____

72. In humans, which substance is directly responsible for controlling the calcium levels of the blood?

(1) adrenalin   (3) parathormone

(2) insulin   (4) thyroxin   72_____

Base your answer to question 73 on the word equation below.

glucose → 2 pyruvic acid → 2 ethyl alcohol + 2 carbon dioxide + energy

73. The process represented by the word equation is known as

(1) aerobic respiration   (3) chemosynthesis

(2) fermentation   (4) dehydration synthesis   73_____

74. The excretory organelles of some unicellular organisms are contractile vacuoles and

(1) cell membranes   (3) ribosomes

(2) cell walls   (4) centrioles   74_____

75. Which is a type of asexual reproduction that commonly occurs in many species of unicellular protists?
(1) external fertilization
(2) tissue regeneration
(3) binary fission
(4) vegetative propagation
75_____

ANSWERS EXPLAINED Questions 48–75

Standard 4, Key Idea 1, Performance Indicators 1–3

48. **2** *Respiration* is the life function by which ATP is made available to cells. Carbohydrate molecules are organic compounds. The breakdown of the carbohydrate molecules releases the energy stored in the bonds of the compounds. Potential energy is stored energy. The released potential energy is used to produce ATP.

WRONG CHOICES EXPLAINED:
(1) *Transport* is the life function by which materials are distributed throughout an organism.
(3) *Excretion* is the life function by which the wastes of metabolism are removed from an organism. Carbon dioxide, water, ammonia, and urea are metabolic wastes.
(4) *Regulation* is the life activity by which an organism responds to changes in its environment. The responses are controlled by the nervous system and the endocrine system.

49. **3** *Reproduction*, the life function through which a parent organism gives rise to offspring, is not necessary for the survival of the parent. Although reproduction is not required for the survival of an individual, it is necessary for the survival of a species. If a given species loses its potential for reproduction, it will become extinct.

WRONG CHOICES EXPLAINED:
(1) *Nutrition* is a collective term that refers to the biochemical processes by which cells extract nutrient molecules from food substances. The nutrients are used to build tissues, provide energy, and regulate the many biochemical activities that occur in cells. Without nutrition, cells die and, consequently, so do organisms. Each organism is dependent on adequate nutrition for survival.
(2) *Respiration* refers to the series of chemical changes that fuel molecules undergo to release chemical energy for cells. Respiration is necessary for the survival of the individual. Tissue cells cannot live without a means of obtaining chemical energy to power cellular activities such as active transport and metabolism. Of course, death of tissue cells means death of the individual.

(4) *Synthesis* occurs when small molecules are joined chemically to form large molecules. Enzymes, hormones, and body tissues are the results of syntheses, without which an individual organism cannot survive. Synthesis is a building-up process in which molecules vital to the life of the organism are produced.

50. **3** *Maintaining homeostasis* is the function of human blood that includes the other three. By transporting nutrients, oxygen, wastes, and other materials around the body, the blood helps to make essential materials available to every living body cell while removing potentially harmful materials from these tissues. Equal distribution of these materials helps to promote a steady state in the tissues essential to homeostatic balance.

WRONG CHOICES EXPLAINED:

(1), (2), (4) *Transporting nutrients, transporting oxygen,* and *collecting wastes* are all functions of the blood that are involved in maintaining homeostasis. Nutrients provide cells with dissolved food molecules. Oxygen is used by cells in the release of energy from these food molecules. Wastes such as urea and carbon dioxide are carried away from the cells for excretion into the environment.

51. **1** Blood that has just returned from the lungs has the greatest concentration of oxygen. The *left atrium of the heart* receives blood directly from the lungs.

WRONG CHOICES EXPLAINED:

(2) Brain tissue is one of the largest consumers of oxygen. The blood circulating in the *cerebellum* gives up most of its oxygen to the nerve cells.

(3) The largest concentration of metabolic wastes is found in the *nephrons*. The nephrons are filtering units in the kidney.

(4) The largest concentration of digested nutrients is found in the *lining of the intestine*. Absorption of nutrients occurs through the villi in the small intestine.

52. **4** *Veins* are blood vessels that carry blood to the heart. They contain valves that prevent the backflow of blood. The blood in veins is usually deoxygenated; the exception is the pulmonary vein in which the blood is rich in oxygen.

WRONG CHOICES EXPLAINED:

(1) *Arteries* are blood vessels that transport blood away from the heart. Arteries are rather thick-walled and pump blood in rhythm with the heart. They have no valves.

(2) Small arteries are *arterioles*. This type of blood vessel functions similarly to arteries. Arterioles lead into capillaries.

(3) *Capillaries* are the smallest blood vessels. They are one cell thick and permit diffusion of water, nutrients, gases, and other substances into and out of the bloodstream. Capillaries have no valves; they are the connecting vessels between arterioles and venules.

53. **2** The *nephron* is the unit of structure of the kidney. Each nephron has a glomerulus, Bowman's capsule, and kidney tubules. The kidney tubules filter out excess water, salts, and the wastes from protein metabolism. Urea and salts dissolved in water form urine.

54. **1** The *alveolus* is an air sac in the lung. It not only permits the diffusion of oxygen from the lungs into the bloodstream but also aids in the diffusion of carbon dioxide and water vapor out of the blood into the lungs.

55. **4** The *liver* is the largest gland in the body. One of its functions is to destroy old red blood cells and change the waste products into bile. The liver also synthesizes the anticoagulant known as heparin.

56. **2** *Tendons* are tough connective tissues made strong by fibers. Tendons connect muscles to bones. The movable joints function when muscles pull on tendons.

WRONG CHOICES EXPLAINED:
(1) *Ligaments* are strong connective tissues that contain elastic muscle fibers. Ligaments connect bone to bone.
(3) *Cartilage* is a supporting tissue that provides strength to body structures without rigidity. Cartilage supports structures such as the ears and nose and covers the ends of bones that form joints. The ground substance, or matrix, of cartilage is made of protein.
(4) *Skin* is composed of epithelial tissue. Skin serves as a body covering and has no function in the movement of bones or muscles.

57. **3** *Diffusion* is the process that results in the movement of molecules from a region of higher concentration (area A) to a region of lower concentration (area B). This net movement occurs until the concentrations of molecules have reached equilibrium between area A and area B.

WRONG CHOICES EXPLAINED:
(1), (2) *Phagocytosis* and *pinocytosis* are processes by which certain protists engulf their food and enclose it within a vacuole for digestion.
(4) *Cyclosis* refers to the streaming of cytoplasm in the cell, a simple form of intracellular transport.

58. **3** Inorganic compounds are compounds that do not contain carbon atoms. *Water*, the universal solvent, is the principal inorganic compound of cytoplasm. Water is the medium through which all chemical reactions take place in the cell.

WRONG CHOICES EXPLAINED:
(1), (2), (4) *Lipids, carbohydrates,* and *nucleic acids* are organic compounds. Organic compounds are carbon-containing compounds.

59. **2**   Glucose is the building block of *carbohydrate* molecules. The ratio of carbon to hydrogen to oxygen is 1:2:1 in glucose and all reducing sugars. By dehydration synthesis, many glucose molecules form complex carbohydrates. However, the 1:2:1 ratio holds.

WRONG CHOICES EXPLAINED:
(1) A *nucleic acid* is composed of a phosphate group, a protein base, and a five-carbon sugar. DNA and RNA are nucleic acids. The CHO 1:2:1 ratio is not applicable.
(3) *Proteins* are built from amino acids, which, in addition to carbon, hydrogen, and oxygen, contain nitrogen. Some protein molecules also contain sulfur. The 1:2:1 ratio of elements does not apply to proteins because proteins are tissue builders whereas carbohydrates are fuel molecules.
(4) *Lipids* are fats and are composed of three fatty acid molecules and one glycerol molecule. The 1:2:1 ratio of carbon to hydrogen to oxygen does not apply to fats.

60.   **1**   The end products of digestion are usually *smaller and more soluble* than the ingested food molecules. Digestion makes available nutrient molecules that can diffuse across cell membranes and enter the cytoplasm of cells. Carbohydrates are broken down into glucose molecules. Fats are hydrolyzed into fatty acids and glycerol. Proteins are digested into their component amino acid molecules. Each of these end products of digestion is able to diffuse across cell membranes and enter into the biochemical activities of cells.

WRONG CHOICES EXPLAINED:
(2) Synthesis produces *larger molecules*. Larger molecules are more complex and are usually less soluble than smaller, simpler ones. Digestion results in smaller nutrient molecules.
(3) *Smaller molecules* are usually more soluble than larger ones. Digestion produces molecules that are more soluble than the complex nutrient molecules that were ingested.
(4) *Molecules derived from digestion* of ingested food are not larger than the molecules from which they came. Molecules produced by the digestion of complex carbohydrates, proteins, and fats are more soluble and are able to dissolve in water. Thus, these molecules can cross cell membranes.

61.   **1**   The endoplasmic reticulum is a network of membranes that extends throughout the cell. The membranes form channels that *provide for the movement of materials through the cell.*

WRONG CHOICES EXPLAINED:

(2) *Urea* is a metabolic waste. It is a poisonous nitrogen compound. Urea must be removed from the cells if an organism is to survive.

(3) The *nucleus* is the organelle in the cell that regulates all cellular activities.

(4) The *chloroplasts* are organelles in plant cells that contain the green pigment chlorophyll. Chloroplasts are necessary for the process of photosynthesis.

62. **3** Proteins are polypeptide chains. Proteins are synthesized in the *ribosomes*.

WRONG CHOICES EXPLAINED:

(1) The *nuclei* contain the genetic material carried in the chromosomes.

(2) *Vacuoles* are saclike organelles in the cytoplasm. Food vacuoles and contractile vacuoles are two common types of vacuoles.

(4) *Cilia* are microscopic hairs used for locomotion by some protozoans.

63. **1** The *nucleus* contains the hereditary factors. Nuclei of plant and animal cells house the chromosomes, which are composed of deoxyribonucleic acid. Molecules of DNA function as genes. Points on the chromosomes are genes. Genetic information is passed from parent to offspring by way of the genes. Chromosomes are part of the fine structure of the nucleus. DNA molecules contribute to the chemical structure. Genes are sites or points that dot the length of the chromosome. Genes, DNA molecules, and chromosomes function in passing along hereditary factors.

WRONG CHOICES EXPLAINED:

(2) The *cell membrane* encloses the contents of the cell and directs the flow of materials into and out of the cell. The cell membrane does not contribute to the passing of genetic material from one generation to the next. The function of the membrane is to control cellular transport.

(3) A *vacuole* is a fluid-filled space in the cytoplasm. Vacuoles help to regulate the internal pressure of the cell. The vacuoles in fat cells are filled with oil.

(4) The membranes that line the cytoplasmic canals within cells are known collectively as the *endoplasmic reticulum.* This cytoplasmic fine structure aids in the transport of molecules from the cell membrane to various sites within the cell. Neither the endoplasmic reticulum nor the vacuoles of the cell membrane contain hereditary structures.

64. **1** Centrioles are cell structures involved primarily in *cell division.* Centrioles are organelles that lie in the cytoplasm outside the nucleus; they are also found near the base of each flagellum and cilium. The centrioles of nonflagellated animal cells move to the spindle poles during cell division and seem to send out spindle fibers. The spindle fibers are attached to chromosomes and appear to pull the chromosomes from the center of the cell to the spindle poles.

WRONG CHOICES EXPLAINED: ·

(2) *Fats are stored* in cells. Fat in which the energy is channeled into heat production is stored in brown fat cells of hibernating mammals. At times, fat can be stored in arteries or accumulate around the heart. Fat cells are not involved in cell division.

(3) *Enzyme production* is controlled by the ribosomes that dot the membranes of the endoplasmic reticulum. Molecules of tRNA and mRNA regulate enzyme production.

(4) *Cellular respiration* is the process by which energy is released from glucose molecules. This process takes place in the mitochondria where oxygen is used as the final hydrogen carrier.

65. **3** Cellular respiration occurs in the mitochondria (plural of *mitochondrion*). Each step in the process of cellular respiration is regulated by enzymes. Respiratory enzymes are located in the mitochondria.

WRONG CHOICES EXPLAINED:

(1) The *cell wall* is composed of cellulose. Cell walls give shape and protection to plant cells.

(2) The *nucleolus* contains the materials needed for the synthesis of RNA. It is located in the nucleus of the cell.

(4) A *vacuole* is a rounded sac that serves as a storage place for food and waste products. Some vacuoles, such as contractile vacuoles, maintain a stable internal environment.

66. **2** *Synthesis* is the formation of complex molecules by combining simpler molecules. Water is removed from the simple molecules in this process.

WRONG CHOICES EXPLAINED:

(1) *Hydrolysis* is the addition of water to split complex molecules into simpler molecules. It is the opposite of synthesis.

(3) *Digestion* is another name for hydrolysis.

(4) *Respiration* is the process by which cells obtain energy. Glucose is converted to smaller molecules.

67. **3** Amino acids are the building blocks of proteins. The dehydration synthesis of amino acids *produces protein molecules.*

WRONG CHOICES EXPLAINED:

(1) *Cellulose* is a polysaccharide composed of hundreds of simple sugar molecules. The sugars were joined together by dehydration synthesis.

(2) Antibodies attack *invading bacteria*. Antibodies are protein molecules produced by special white blood cells.

(4) *Glycogen*, a polysaccharide, is a product of the dehydration synthesis of many glucose units.

68. **3** A *hydrolytic reaction* is a reaction in which a molecule is split. Enzymes are needed to speed up such a reaction. Any factor that affects the operation of the enzyme affects the speed at which the reaction takes place. The *concentration of carbon dioxide* has the least effect on enzyme activity.

WRONG CHOICES EXPLAINED:

(1) As the *temperature* is increased up to a point, the rate of the reaction increases. The increase in temperature increases the speed at which the enzyme and the substrate make contact with each other. The substrate is the molecule on which the enzyme acts. A very high temperature destroys the enzyme, and the reaction stops.

(2) Every enzyme works best at a particular *pH*. The enzymes in the stomach work in an acid environment, whereas the enzymes in the intestine work best in a basic medium.

(4) One molecule of an enzyme reacts with one molecule of a substrate. Increasing the *concentration of an enzyme* means that more substrate molecules will be acted on. The rate of the reaction will increase.

69. **1** A *neurotransmitter* is a chemical substance that is released by an impulse arriving at the terminal end of a neuron. The neurotransmitter diffuses across the synapse and stimulates the second nerve cell. Acetylcholine is an example of a neurotransmitter.

WRONG CHOICES EXPLAINED:

(2) A *synapse* is the space between the terminal end of one nerve cell and the dendrites of a second nerve cell. The area marked by an X in the diagram is a synapse.

(3) A *neuron* is a nerve cell that is specially adapted for the conduction of impulses.

(4) A *nerve* is made up of many neurons.

70. **1** A *neuron* is a single microscopic nerve cell. A *nerve* is composed of many nerve cells. A *ganglion* is a large mass of cell bodies of nerve cells; a ganglion functions as a coordinating center for impulses. A *receptor* is an organ specialized to receive environmental stimuli. The eye is an example of a receptor.

WRONG CHOICES EXPLAINED:

(2), (3), (4) In these three choices, either one or several structures are not arranged according to increasing size.

71. **2** *Gastric glands* are embedded in the walls of the stomach. They are duct glands that secrete gastric juice, a mixture of water, hydrochloric acid, rennin, and pepsin. Gastric juice begins the digestion of protein in the stomach. *Intestinal glands* are duct glands that line the walls of the small intestine. They secrete intestinal juice, a mixture of water, proteases, amylases, and lipases. Both types of glands lie within the digestive tube.

WRONG CHOICES EXPLAINED:

(1) Gastric glands are described above. *Thyroid glands* lie outside the digestive tract at the base of the neck, straddled across the larynx. The thyroid is a ductless gland that secretes the hormone known as thyroxin. Thyroid glands do not function in the biochemical process of digestion.

(3) *Thyroid glands* and intestinal glands are described above. Thyroxin controls the metabolism of cells. The explanation above shows why this choice is wrong.

(4) *Adrenal glands* are dual endocrine glands that lie on top of each kidney. They are not within the digestive tract. The adrenal medulla, the inner gland, secretes the hormone adrenaline, also known as epinephrine. This hormone enables the body to function in emergencies. The adrenal cortex secretes about six active hormones, including cortisone, the antiarthritis hormone.

72. **3** *Parathormone* is the hormone secreted by the parathyroid glands. The parathyroids are buried in the thyroids. Parathormone controls the level of calcium in the blood. Lack of blood calcium causes muscles to go into tetany. Tetany, or cramping, of the heart muscle causes death.

WRONG CHOICES EXPLAINED:

(1) *Adrenalin* is the hormone of the adrenal medulla, a ductless gland called the "gland of combat." Adrenaline stimulates the heart to beat faster, increases the rate of breathing, and controls the constriction and dilation of the arteriole walls.

(2) *Insulin* is secreted by the beta cells of the islets of Langerhans, which lie in the pancreas. Insulin controls sugar metabolism; specifically, it makes cell walls permeable to glucose and encourages the phosphorylation of fructose.

(4) *Thyroxin* is released by the thyroid gland. The rate of cellular metabolism is controlled by thyroxin. Iodine is used in the synthesis of thyroxin. People whose thyroid glands fail to develop become cretins; they are mentally retarded and physically undersized.

73. **2** Another name for anaerobic respiration is *fermentation.* In the process of fermentation, glucose is converted to energy, alcohol, and carbon dioxide.

WRONG CHOICES EXPLAINED:

(1) *Aerobic respiration* is another name for cellular respiration. This process requires oxygen. The following is an equation for aerobic respiration.

glucose + oxygen $\rightarrow$ pyruvic acid $\rightarrow$ carbon dioxide + water + energy

(3) *Chemosynthesis* is the synthesis of carbohydrates from inorganic compounds without the use of sunlight as a source of energy. Chemosynthesis is a form of autotrophic nutrition. It is carried out only by certain species of bacteria such as nitrifying bacteria.

(4) *Dehydration synthesis* is the method by which simple molecules are converted to complex molecules.

74. **1** *Cell membranes* and contractile vacuoles are excretory organelles of some unicellular organisms. The cell membrane is a selectively permeable membrane. It permits the diffusion of carbon dioxide and ammonia, two metabolic waste gases.

WRONG CHOICES EXPLAINED:
(2) *Cell walls* are composed of nonliving materials. Many canals penetrate through these walls, allowing the unrestricted passage of molecules.
(3) Proteins are synthesized in *ribosomes*.
(4) *Centrioles* are rodlike particles found in the centrosome. They function during the processes of mitosis and meiosis. Centrioles are found only in animal cells.

75. **3** A unicellular protist (e.g., an ameba) is composed of a single cell. When this cell divides by mitosis, the process is known as *binary fission*.

WRONG CHOICES EXPLAINED:
(1) *External fertilization* is an element of sexual reproduction in many aquatic multicellular species. Both the sexual nature of this process and the fact that it is carried out by multicellular animals eliminate this as a correct choice.
(2) *Tissue regeneration* implies a process that occurs in multicellular organisms.
(4) *Vegetative propagation* is a form of asexual reproduction common to certain species of multicellular plants; it cannot be carried out by unicellular protists.

Key Idea 2—Genetic Continuity: Organisms inherit genetic information in a variety of ways that result in continuity of structure and function between parents and offspring.

Performance Indicator	Description
2.1	The student should be able to explain how the structure and replication of genetic material result in offspring that resemble their parents.
2.2	The student should be able to describe and explain how the technology of genetic engineering allows humans to alter genetic makeup of organisms.

76. Corn plants grown in the dark will be white and usually much taller than genetically identical corn plants grown in light, which will be green and shorter. The most probable explanation for this is that the
- (1) corn plants grown in the dark were all mutants for color and height
- (2) expression of a gene may be dependent on the environment
- (3) plants grown in the dark will always be genetically albino
- (4) phenotype of a plant is independent of its genotype

76____

77. In order for a substance to act as a carrier of hereditary information, it must be
- (1) easily destroyed by enzyme action
- (2) exactly the same in all organisms
- (3) present only in the nuclei of cells
- (4) copied during the process of mitosis

77____

78. During synapsis in meiosis, portions of one chromosome may be exchanged for corresponding portions of its homologous chromosome. This process is known as
- (1) nondisjunction
- (2) polyploidy
- (3) crossing-over
- (4) hybridization

78____

79. A DNA nucleotide is composed of three parts. These three parts may be
- (1) phosphate, adenine, and thymine
- (2) phosphate, deoxyribose, and thymine
- (3) phosphate, glucose, and cytosine
- (4) adenine, thymine, and cytosine

79____

80. A double-stranded DNA molecule replicates as it unwinds and "unzips" along weak
- (1) hydrogen bonds
- (2) carbon bonds
- (3) phosphate groups
- (4) ribose groups

80____

Base your answers to questions 81 through 84 on your knowledge of biology and the diagrams below. The diagram on the left represents a portion of a double-stranded DNA molecule. The diagrams at the right represent specific combinations of nitrogenous bases found in compounds transporting specific amino acids.

81. The amino acid whose genetic code is present in strand I is
(1) lysine
(2) serine
(3) asparagine
(4) phenylalanine

81_____

82. The thymine (*T*) of strand I is accidentally replaced by adenine (*A*). This occurrence is called
(1) segregation
(2) disjunction
(3) cytoplasmic inheritance
(4) gene mutation

82_____

83. The number of different amino acids coded by strand I is
(1) 1
(2) 2
(3) 8
(4) 12

83_____

84. Which represents the sequence of nitrogenous bases in the molecule of messenger RNA synthesized by strand I?
(1) -T-T-C-G-U-C-
(2) -A-A-C-G-T-C-
(3) -U-U-C-G-A-C-
(4) -A-A-G-C-U-G-

84_____

85. Molecules that transport amino acids to ribosomes are known as
(1) protein molecules
(2) RNA molecules
(3) mitochondria
(4) chromosomes

85_____

86. A similarity between DNA molecules and RNA molecules is that they
(1) are built from nucleotides
(2) are double-stranded
(3) contain deoxyribose sugar
(4) contain uracil

86_____

87. What is the function of DNA molecules in the synthesis of proteins?
(1) They catalyze the formation of peptide bonds.
(2) They determine the sequence of amino acids in a protein.
(3) They transfer amino acids from the cytoplasm to the nucleus.
(4) They supply energy for protein synthesis.   87____

88. In pea plants, the trait for smooth seeds is dominant over the trait for wrinkled seeds. When two hybrids are crossed, which results are most probable?
(1) 75% smooth and 25% wrinkled seeds
(2) 100% smooth seeds
(3) 50% smooth and 50% wrinkled seeds
(4) 100% wrinkled seeds   88____

89. A person who is homozygous for blood type *A* has a genotype that can be represented as
(1) I^aI^b (2) I^aI^a (3) I^ai (4) ii 89____

90. Animal breeders often cross breed members of the same litter in order to maintain desirable traits. This procedure is known as
(1) hybridization   (3) natural selection
(2) inbreeding   (4) vegetative propagation   90____

ANSWERS EXPLAINED Questions 76–90

Standard 4, Key Idea 2, Performance Indicators 1–2

76. **2** Corn plants grown in the dark will be white. The most probable explanation for this is that expression of the gene for color may *depend on the environment*. The plants have the genetic information for chlorophyll production. This can be assumed because they are genetically identical to the plants grown in the light. Light is needed to activate the chlorophyll gene.

WRONG CHOICES EXPLAINED:
(1) *Mutations* are sudden changes in the genetic material. Mutations are inherited. Because the plants grown in the dark were genetically identical to those grown in the light, neither group lacked the genetic information for chlorophyll production.

(3) *Albinism* is a condition resulting from the absence of a normal gene for color. Both groups of plants had the normal gene for color.

(4) The *phenotype* is the physical appearance of the organism. The phenotype depends on the genotype.

77. **4** The hereditary information is contained in the chromosomes. During mitosis, the *chromosomes duplicate*. The duplication of chromosomes ensures the equal distribution of identical genetic material to the new cells.

WRONG CHOICES EXPLAINED:

(1) If the hereditary information is *destroyed*, the cells cannot function. The chromosomes contain the information necessary for carrying out all cellular activities.

(2) *No two organisms are exactly alike.* No two organisms have the same hereditary material. Identical twins are the only exception to these statements.

(3) Plasmagenes are *genes located outside the nucleus*. Drug resistance in some bacteria is transmitted through plasmagenes.

78. **3** *Crossing-over* is the exchange of chromosomal material between homologous pairs of chromosomes. This process occurs during synapsis in meiosis.

WRONG CHOICES EXPLAINED:

(1) *Nondisjunction* is the failure of homologous chromosomes to separate from each other during meiosis. Cells with extra chromosomes and cells with too few chromosomes result from nondisjunction.

(2) *Polyploidy* is a condition in which the cells have extra sets of chromosomes beyond the normal 2n number.

(4) *Hybridization* is the crossing of two organisms that are distinctly different from each other. The purpose is to bring together new combinations of genes. Usually the individual with the new gene combinations is more sturdy than either parent. A tangelo is a cross between a tangerine and a grapefruit.

79. **2** A DNA nucleotide is composed of a *deoxyribose sugar molecule, a phosphoric acid molecule, and a nitrogen base.*

WRONG CHOICES EXPLAINED:

(1) *Adenine and thymine* are bases.

(3) *Glucose* is not the sugar molecule in DNA.

(4) *Adenine, thymine, and cytosine* are bases.

80. **1** The two strands of DNA are held together by *hydrogen bonds*. The hydrogen bonds form weak links between the base pairs of each strand.

WRONG CHOICES EXPLAINED:

(2) The sugars and bases of the nucleotides of DNA are organic compounds. Each individual compound is made up of *carbon bonds*.

(3) The nucleotides in each strand are joined together by *phosphate groups*.

(4) There are no *ribose groups* in DNA.

81. **1** *Lysine* is the amino acid whose genetic code is present on DNA strand I. Strand I of the DNA molecule contains two triplet codons (a triplet codon is a three-base sequence): AAG-CTG. Each triplet codon represents a specific amino acid. The amino acids are carried by tRNA molecules. The tRNA codon matches the DNA codon (except that U replaces T). There are two possible tRNA codons that could match the DNA strand I sequence; they are AAG and CUG. Of these, only the AAG tRNA appears in the diagram. The AAG tRNA carries the amino acid known as lysine.

DNA strand I	mRNA	tRNA	
A	U	A	
A	U	A	→ serine
G	C	G	
C	G	C	
T	A	U	→ unnamed amino
G	C	G	acid

WRONG CHOICES EXPLAINED:

(2) The DNA triplet code for *serine* is TTC. This codon does not appear on strand I of the DNA molecule in the diagram.

(3) The DNA triplet code for *asparagine* is GAC. This codon does not appear on strand I of the DNA molecule in the diagram.

(4) The DNA triplet code for *phenylalenine* is TTT. This codon does not appear on strand I of the DNA molecule in the diagram.

82. **4** A gene controls the production of a protein. The substitution of one base for another changes the triplet code. One amino acid will be substituted for another. The result is a *mutation*. The replacement of glutamic acid by valine in a hemoglobin molecule causes sickle cell anemia.

WRONG CHOICES EXPLAINED:

(1) *Segregation* is the separation of alleles from each other during the formation of gametes.

(2) *Disjunction* is the separation of homologous chromosomes during the process of meiosis.

(3) *Cytoplasmic inheritance* is the inheritance of genes located in the cytoplasm, not in the nucleus. The cytoplasmic genes are called plasmagenes.

83. **2** *Two* different amino acids are coded by strand I. Strand I has two triplet codes, six bases.

WRONG CHOICES EXPLAINED:

(1) Only three bases would have to be shown in the diagram to code for *1* amino acid.

(3) 24 bases are needed for *8* amino acids.

(4) 48 bases are needed for *12* amino acids.

84. **3** *UUCGAC* is the correct sequence. Base pairing is an important concept in DNA duplication and RNA synthesis. Adenine pairs with thymine; cytosine pairs with guanine. There is no thymine in RNA. Uracil takes its place.

WRONG CHOICES EXPLAINED:

(1) *TTCGUC* is not correct. Because thymine is present in the base sequences, the molecule cannot be RNA.

(2) *AACGTC* is not correct. The base sequences are not complementary to either strand I or strand II.

(4) *AAGCUG* is not correct. The base sequences are complementary to strand II in the diagram not strand I.

85. **2** Amino acids are transported to the ribosomes by *RNA molecules* known as transfer RNA, tRNA.

WRONG CHOICES EXPLAINED:

(1) *Protein molecules* are synthesized in the ribosomes. The code for the synthesis is contained in mRNA.

(3) Cellular respiration takes place in the *mitochondria*.

(4) *Chromosomes* are structures found in the nucleus. They are composed of DNA and protein. The genes are located on the chromosomes.

86. **1** Both DNA and RNA *are built from nucleotides*.

WRONG CHOICES EXPLAINED:

(2) Only DNA is *double-stranded*. RNA is single-stranded.

(3) *Deoxyribose* is the sugar in the DNA nucleotides. Ribose is the sugar in the RNA nucleotides.

(4) The base thymine is replaced by *uracil* in RNA nucleotides.

87. **2** The *sequence of amino acids* in a protein is determined by the triplet codes in DNA. The codes are carried to the ribosomes when mRNA is synthesized. A DNA strand is the template in mRNA synthesis.

WRONG CHOICES EXPLAINED:

(1) Enzymes *catalyze the formation of peptide bonds*. A peptide bond is a C–N bond formed by the dehydration synthesis of amino acids.

(3) *Amino acids are transferred* from the cytoplasm to the ribosomes, not to the nucleus. Transfer RNA is the carrier molecule.

(4) ATP molecules *supply the energy for protein synthesis.*

88.   **1**   A *hybrid* is an individual that has two different alleles for a particular trait. The hybrid may be represented by the symbols Ss. The hybrid is smooth; the smooth trait is dominant over the wrinkled trait. When two hybrids are crossed, *75% of the offspring will have the smooth trait and 25% of the offspring will have the wrinkled trait.*

	S	s
S	SS	Ss
s	Ss	ss

WRONG CHOICES EXPLAINED:

(2), (3), (4) These choices are incorrect based on the information provided by the Punnett square.

89.   **2**   The term *homozygous* means *pure for the trait.* A homozygous individual has two identical alleles for a gene. There are three alleles for blood type: I^a, I^b, and i. The allele I^a produces a protein for blood type A; the allele I^b produces a protein for blood type B; the allele i does not produce either protein. The type of blood is determined by the combination of alleles. The genotype refers to the allelic combination. Because the person in the question is homozygous for type A blood, his genotype is $I^a I^a$.

WRONG CHOICES EXPLAINED:

(1) The I^a and I^b alleles are both dominant over the i allele. When both I^a and I^b alleles occur in the same person, the person has type AB blood.

(3) The type of blood represented by the genotype $I^a i$ is type A.

(4) The genotype of a person with type O blood is ii.

90.   **2**   The mating of members of the same litter to maintain desirable traits is known as *inbreeding.* Because mating pairs come from the same litter, they are genetically similar to each other. Inbreeding is used to maintain pure breeds.

WRONG CHOICES EXPLAINED:

(1) *Hybridization,* or outbreeding, is the mating of organisms with contrasting traits. It is the opposite of inbreeding.

(3) Factors in the environment select the organisms that are best adapted to survive in the environment. This principle is known as *natural selection.* It is an essential feature in the theory of evolution.

(4) *Vegetative propagation* is asexual reproduction in plants.

Key Idea 3—Organic Evolution: Individual organisms and species change over time.

Performance Indicator	Description
3.1	The student should be able to explain the major patterns of evolution.

91. In modern classification, protozoa and algae are known as molds, and bacteria are known as
- (1) bryophytes
- (2) plants
- (3) protists
- (4) animals

91 _____

92. Most modern biologists agree that an ideal classification system should reflect
- (1) nutritional similarities among organisms
- (2) habitat requirements of like groups
- (3) distinctions between organisms based on size
- (4) evolutionary relationships among species

92 _____

93. Which term includes the other three?
- (1) genus
- (2) species
- (3) kingdom
- (4) phylum

93 _____

94. In one modern classification system, organisms are grouped into three
- (1) kingdoms
- (2) phyla
- (3) genera
- (4) species

94 _____

95. Which is one basic assumption of the heterotroph hypothesis?
- (1) More complex organisms appeared before less complex organisms.
- (2) Living organisms did not appear until there was oxygen in the atmosphere.
- (3) Large autotrophic organisms appeared before small photosynthesizing organisms.
- (4) Autotrophic activity added molecular oxygen to the environment.

95 _____

96. According to the heterotroph hypothesis, scientists believe that life arose in
    (1) a desert environment   (3) a vacuum
    (2) a forest environment   (4) an ocean environment   96_____

97. From an evolutionary standpoint, the greatest advantage of sexual reproduction is the
    (1) variety of organisms produced
    (2) appearance of similar traits generation after generation
    (3) continuity within a species
    (4) small number of offspring produced   97_____

98. According to modern theories of evolution, which of the following factors would be *least* effective in bringing about species changes?
    (1) geographic isolation   (3) genetic recombination
    (2) changing environments   (4) asexual reproduction   98_____

99. A factor that tends to cause species to change is a
    (1) stable environment   (3) recombination of genes
    (2) lack of migration   (4) decrease in mutations   99_____

100. If a fossil mammoth were discovered frozen in ice, its cells could be analyzed to determine whether its proteins were similar to those of the modern elephant. This type of investigation is known as comparative
    (1) anatomy   (3) biochemistry
    (2) embryology   (4) ecology   100_____

101. If members of the same species have been geographically isolated from each other for an extended period of time, which will they most likely exhibit?
    (1) mutations identical to each other
    (2) random recombination occurring in the same manner
    (3) evolution of traits of high adaptive value for their particular environments
    (4) evolution into two new species which will have no problem interbreeding   101_____

102. Skeletal similarities between two animals of different species are probably due to the fact that both species
    (1) live in the same environment
    (2) perform the same functions
    (3) are genetically related to a common ancestor
    (4) have survived until the present time   102_____

103. The best means of discovering if there is a close evolutionary relationship between animals is to compare

(1) blood proteins   (3) foods consumed

(2) use of forelimbs   (4) habitats occupied   103_____

104. In the process of evolution, the effect of the environment is to

(1) prevent the occurrence of mutations

(2) act as a selective force on variations in species

(3) provide conditions favorable for the formation of fossils

(4) provide stable conditions favorable to the survival of all species   104_____

105. In a stable population in which the gene frequencies have been constant for a long time, the rate of evolution

(1) increases   (3) remains the same

(2) decreases   (4) increases, then decreases   105_____

106. Certain strains of bacteria that were susceptible to penicillin in the past have now become resistant. The probable explanation for this is that

(1) the mutation rate must have increased naturally

(2) the strains have become resistant because they needed to do so for survival

(3) a mutation was retained and passed on to succeeding generations because it had high survival value

(4) the principal forces influencing the pattern of survival in a population are isolation and mating   106_____

107. The frequency of traits that presently offer high adaptive value to a population may *decrease* markedly in future generations if

(1) conditions remain stable   (3) all organisms with these

(2) the environment changes   traits survive

                                (4) mating remains random   107_____

108. Since the publication of Darwin's theory, evolutionists have developed the concept that

(1) a species produces more offspring than can possibly survive

(2) the individuals that survive are those best fitted to the environment

(3) through time, favorable variations are retained in a species

(4) mutations are partially responsible for the variations within a species   108_____

109.   One factor that Darwin was unable to explain satisfactorily in his theory of evolution was

(1) natural selection
(3) survival of the fittest
(2) overproduction
(4) the source of variations   109_____

ANSWERS EXPLAINED Questions 91–109

Standard 4, Key Idea 3, Performance Indicator 1

91.   **3**   The *protists* include all unicellular organisms and organisms that have both plant and animal features within one cell. Protozoa and algae are protists.

WRONG CHOICES EXPLAINED:

(1) *Bryophytes* are multicellular green plants that do not have vascular tissue. Mosses are examples of bryophytes.

(2) Multicellular photosynthetic organisms make up the *plant* kingdom, which includes both vascular and nonvascular plants.

(4) The *animal* kingdom is composed of multicellular organisms that cannot manufacture their own food. The organisms within this kingdom lack cell walls, and most are capable of some type of locomotion.

92.   **4**   A classification system should reflect *evolutionary relationships among species*. Evolutionary relationships are determined on the basis of the similarities in the anatomy, embryology, and biochemistry among organisms.

WRONG CHOICES EXPLAINED:

(1) All animals from protozoans to humans utilize the same nutrients in a similar manner. The process of photosynthesis is the same in tree cells and unicellular algae. *Nutritional similarities among organisms* are *not* useful in a system of classification.

(2) Both the whale and the fish live in an ocean environment. However, the whale is a mammal. Other than *sharing the same habitat*, the whale has no fish characteristics.

(3) Algae and protozoans are both microscopic organisms. However, algae have plant characteristics and protozoans have animal characteristics. *Size cannot be used* as the basis for a system of classification.

93.   **3**   According to the classification system, the largest grouping of organisms is the *kingdom*. Following this, the other groups, in order, are phylum, class, order, family, genus, species. Depending on its chief characteristics, an organism is placed in one of five kingdoms, Monera, Protist, Fungi, Animal, or Plant.

94. **1** In one modern classification system, organisms are grouped into three *kingdoms:* Animal, Plant, and Protist. Organisms that are not typical plants or animals are classified as protists. Examples are protozoa, slime molds, and bacteria.

WRONG CHOICES EXPLAINED:
(2) A *phylum* (plural phyla) is a large grouping that consists of classes, orders, families, genera, and species.
(3) A *genus* (plural genera) is a classification group composed of species. Members of a genus are more closely related than groups belonging to a given phylum.
(4) The *species* is the unit of classification. All members of a species are so closely related that they can mate and produce viable offspring.

95. **4** A heterotroph is an organism that must get its food from a source outside its own body cells; it cannot synthesize its food from inorganic materials. The heterotroph hypothesis proposes that the first living things on Earth were heterotrophs that obtained their food from the organic materials in the primitive seas. Autotrophs are organisms, such as green plants, that can synthesize their own food. At some stage in Earth's history, *autotrophic activity* used up the carbon dioxide in the air and, as a consequence of photosynthesis, *added molecular oxygen to the atmosphere.*

WRONG CHOICES EXPLAINED:
(1) Coacervates and then relatively *simple cells developed before more complex organisms.*
(2) Living heterotrophs *appeared before molecular oxygen was added to the atmosphere.*
(3) *Small cells that carried on photosynthesis appeared before the more complex vascular and seed plants.* The course of evolution is from the simple to the complex.

96. **4** According to the heterotroph hypothesis, life on earth evolved through a sequence of stages. The gases of the primitive atmosphere, such as methane, ammonia, and hydrogen, were washed down by heavy rains into the *early oceans.* They were acted on by ultraviolet radiation, cosmic rays, the earth's heat, and radioactivity. The bonding together of the molecules resulted in the formation of larger organic molecules.

WRONG CHOICES EXPLAINED:
(1) A *desert* environment could not support "first" life. The intense heat and the rapid evaporation of water are conditions that do not allow for the movement or maintenance of molecules in a fluid medium.
(2) A *forest* environment does not provide the pools of warm water on a continuous basis necessary for aggregate molecules to form.
(3) A *vacuum,* a place without air, cannot support life.

97. **1** Sexual reproduction helps to maximize the number of different allelic combinations that occur in offspring, leading to a greater *variety of organisms produced* within the species' population as a whole. Individuals displaying favorable traits in a changing environment are more likely to survive and to pass these traits on to their offspring, a fact that helps to promote evolutionary change.

WRONG CHOICES EXPLAINED:

(2), (3) *Appearance of similar traits generation after generation* and *continuity within a species* describe conditions that promote stability and uniformity within species, both of which are maximized during asexual reproduction.

(4) Evolutionarily speaking, the larger the number of offspring produced during reproduction, the more successful a particular species variety tends to be in competing with other varieties of the same species. *Small number of offspring produced,* therefore, is not an evolutionary advantage of sexual or any other type of reproduction.

98. **4** Variations among organisms are necessary for speciation. *Asexual reproduction* is least effective in bringing about changes in species. There are no variations among organisms that are reproduced asexually. These organisms are genetically like their parents.

WRONG CHOICES EXPLAINED:

(1) *Geographic isolation* increases the chance that a group of organisms will develop a new gene pool, which will give rise to a new species.

(2) *Changes in the environment* cause shifts in the gene pool. The genes that ensure the survival of organisms increase in the pool. Thus, the environment changes the characteristics of the original population, and a new species is formed.

(3) *Genetic recombination* occurs during meiosis and fertilization. The shuffling of genes results in the appearance of new characteristics in a population.

99. **3** The *recombination of genes* is one factor that tends to cause species to change. Mutations are changes in genes. When like mutations combine in the fertilized egg, the new characteristic will be expressed in the offspring. If this mutation adds to the survival value of the organism, the gene change will be passed on to progeny because individuals having this mutation will live to reproduce.

WRONG CHOICES EXPLAINED:

(1) A *stable environment* will probably not cause species to change. Beneficial mutations become effective in changing environments. For example, the mutations that produced white fur in the polar bear were beneficial. At one time, the polar regions were tropical. A change to a glacial environment was accompanied by a change or changes in the animal species that inhabited the region. Bears with a dark coat color became immediate targets for natural enemies.

(2) *Migration* aids the recombination of genes because organisms have greater opportunities for interbreeding.

(4) A *decrease in mutations* does not aid speciation but slows it.

100. **3** *Biochemistry* is the study of the chemistry of living organisms. Proteins are compounds found only in living organisms.

WRONG CHOICES EXPLAINED:

(1) *Anatomy* is the study of the structure of organisms.

(2) *Embryology* is the study of the development of embryos.

(4) *Ecology* is the study of the relationship of living organisms to each other and to their environment.

101. **3** Geographic isolation involves the separation of organisms by natural barriers. Each group of isolated individuals develops its own gene pool because each group lives under different environmental conditions. In the case of members of the same species who have been geographically isolated from each other, the *selection for individuals with special survival traits* is different in each environment.

WRONG CHOICES EXPLAINED:

(1) *Mutations* are the raw materials for evolution. Although mutations might have been the same in each group, they do not have the same adaptive value in each group.

(2) *Random recombination* occurs in the same manner in each group. However, the recombination process operates on two distinctly different gene pools.

(4) Usually, *different species cannot mate*. If their mating happens to be successful, their offspring will not be fertile.

102. **3** Morphology is the study of the structure and form of living things. When the arm of a human and the wing of a bird are studied, they are seen to have similar bone structure. This indicates that both organisms descended from a *common ancestor*. They have both undergone many changes since then and are now very different from each other. However, they still retain some of the same genes and therefore show a similarity in many parts of their bodies, including the arrangement of the bones in their forelimbs.

103. **1** A close evolutionary relationship between animals can be shown by a study of their *blood proteins*. The precipitin test is used to show such a relationship. A rabbit can be sensitized to human blood by being injected with human serum. When the sensitized rabbit serum is mixed with human serum, a white precipitate forms. If the sensitized rabbit serum is mixed with serum from a chicken, there is no reaction. However, a precipitate does form when the sensitized rabbit serum is mixed with the serum of a chimpanzee. In a like manner, the serum of a dog and a wolf show precipitation with serum sensitized to dog serum.

WRONG CHOICES EXPLAINED:

(2) The *forelimbs* of many unrelated animals may be used for the same purpose.

(3) Many unrelated animals *consume similar food*.

(4) Many unrelated animals *live in the same habitat*.

104. **2**   There are many variations among the organisms of a species. Some variations allow an organism to survive best in a particular environment. Such factors as climate, food supply, and type of predators determine which organisms are *best adapted to that environment*.

WRONG CHOICES EXPLAINED:

(1) *Mutations* occur naturally and randomly. Mutations increase the variations among organisms.

(3) *Fossils* are the remains of organisms that lived in the past. They present evidence that evolution has occurred.

(4) There are many factors in an environment that influence the survival of different species. A *stable environment* preserves the species that have already adapted to that environment.

105. **3**   Stable gene pools are a hallmark of nonevolving populations. The rate of evolution in such populations neither increases nor decreases, but *remain the same*.

WRONG CHOICES EXPLAINED:

(1), (2), and (4) all refer to changing rates of evolution that do not occur.

106. **3**   Bacteria resistant to penicillin developed as a result of mutation. Organisms that did not receive the mutated gene were killed by the antibiotic. Those in which gene mutation occurred survived and *passed the mutation on to succeeding generations*.

WRONG CHOICES EXPLAINED:

(1) *The mutation rate did not increase*. The survivors had the mutated gene that allowed the bacteria to resist the effects of penicillin. These resistant strains reproduced, creating populations that replaced the nonresistant strains.

(2) *Need does not determine mutation*. Mutations are chance occurrences.

(4) Survival of a species depends on the ability of its members to obtain food, carry out respiration, and *reproduce successfully*. *Isolation* does not increase species survival.

107. **2**   As long as environmental conditions remain stable, the alleles controlling traits that promote individual survival in that environment tend to be maintained at a high level in the population's gene pool. When *the environment changes,* however, the factors that promoted survival may no longer be present. Selection pressure may then operate to reduce the frequency of the once-prevailing alleles in favor of alleles controlling other, contrasting traits that increase individuals' chances for survival in the new environment.

WRONG CHOICES EXPLAINED:

(1) As long as *conditions remain stable,* selection pressures on individuals displaying favorable traits remain low, promoting the maintenance of a high frequency of alleles controlling these traits.

(3) If *all organisms with these traits survive,* their genes will be passed on to future generations at a high rate. This will help to maintain a high frequency of alleles controlling these traits in the population.

(4) If *mating remains random,* the probability that alleles will pair in unrestricted combinations will remain high, helping to ensure that the laws of probability will operate freely and that gene frequencies for existing traits will remain stable.

108.   **4**   Darwin proposed his theory of evolution in 1856. His theory did not explain how variation arose in organisms. In 1901, Hugo De Vries discovered the existence of mutations. *Mutations accounted for the rise of variations in organisms.*

WRONG CHOICES EXPLAINED:

(1) One of the principles of Darwin's theory of evolution stated that a species produced *more offspring than could possibly survive.*

(2) Another principle stated that the *individuals that survived were those best suited to the environment.*

(3) *The variations favored by the environment are retained within a species.* It is the environment that determines which variations are favorable.

109.   **4**   Darwin's theory of evolution was completed and published before Mendel completed his study of inheritance in the garden pea. Darwin could not explain *how variations occurred* or how they were passed on from parent to offspring.

WRONG CHOICES EXPLAINED:

(1) Darwin's theory of *natural selection* was divided into five distinct principles that formulated his concept of evolution. These ideas were set forth in his book *The Origin of the Species by Natural Selection.*

(2) *Overproduction* was one of the principles of Darwin's theory. He explained that for a species to continue in existence, it must overproduce in order to maintain the species number. For example, one female codfish lays about 9 million eggs. Not all of these eggs are fertilized and not all codfish lay about 9 million eggs. Not all of these eggs are fertilized and not all codfish fry reach adulthood. If the 9 million eggs per female were fertilized and if these zygotes developed into adult fish, the seas would be overrun with codfish. However, if the number of gametes produced by codfish were greatly reduced, the species would die out. This overproduction of gametes is necessary to maintain codfish survival.

(3) Another of Darwin's principles was *survival of the fittest.* No two organisms are alike; each has variations. These variations may either help or hinder the organism in its struggle for existence. An organism with variations that help it reach food faster is more fit and has a better potential for survival than a slower, less fit member of the species.

Key Idea 4—Reproductive Continuity: The continuity of life is sustained through reproduction and development.

Performance Indicator	Description
4.1	The student should be able to explain how organisms, including humans, reproduce their own kind.

110. Which occurs in a plant cell but *not* in an animal cell during mitotic cell division?
    (1) formation of spindle fibers   (3) formation of a cell plate
    (2) chromosome duplication   (4) cytoplasmic division   110_____

111. A plant cell with 12 chromosomes undergoes normal mitosis. What is the total number of chromosomes in each of the resulting daughter cells?
    (1) 24   (2) 12   (3) 6   (4) 4   111_____

112. Asexual reproduction *differs* from sexual reproduction in that, in asexual reproduction,
    (1) new organisms are usually genetically identical to the parent
    (2) the reproductive cycle involves the production of gametes
    (3) nuclei of sex cells fuse to form a zygote
    (4) offspring show much genetic variation   112_____

113. In most multicellular animals, meiotic cell division occurs in specialized organs known as
    (1) gonads   (3) kidneys
    (2) gametes   (4) cytoplasmic organelles   113_____

114. Which is an important adaptation for reproduction among land animals?
    (1) fertilization of gametes outside the body of the female
    (2) fertilization of gametes within the body of the female
    (3) production of sperm cells with thick cell walls
    (4) production of sperm cells with thin cell walls   114_____

115. In humans, a single primary sex cell may produce four gametes. These gametes are known as
    (1) diploid egg cells   (3) polar bodies
    (2) monoploid egg cells   (4) sperm cells   115_____

116. In sexual reproduction, the $2n$ chromosome number is restored as a direct result of
  (1) fertilization
  (2) gamete formation
  (3) cleavage
  (4) meiosis

116_____

117. In human females, the main function of the follicle-stimulating hormone (FSH) secreted by the pituitary gland is to
  (1) stimulate the adrenal glands to produce cortisone
  (2) stimulate activity in the ovaries
  (3) control the metabolism of calcium
  (4) regulate the rate of oxidation in the body

117_____

118. If the first stage of an uninterrupted human menstrual cycle is the follicle stage, the last stage includes the
  (1) formation of sperm cells in the testis
  (2) release of a mature egg
  (3) buildup of the uterine lining
  (4) shedding of the uterine lining

118_____

119. Which statement best describes internal fertilization?
  (1) It does not require motile gametes.
  (2) It helps to make terrestrial life possible.
  (3) It requires the presence of many eggs.
  (4) It normally occurs in the male.

119_____

120. What are the normal chromosome numbers of a sperm, egg, and zygote, respectively?
  (1) monoploid, monoploid, and monoploid
  (2) monoploid, diploid, and diploid
  (3) diploid, diploid, and diploid
  (4) monoploid, monoploid, and diploid

120_____

121. When compared with the number of gametes produced from a single primary sex cell during oogenesis, the number of gametes produced from a single human primary sex cell during spermatogenesis is usually
  (1) four times as great
  (2) twice as great
  (3) half as great
  (4) the same

121_____

122. In human males, sperm cells are suspended in a fluid medium. The main advantage gained from this adaptation is that the fluid
- (1) removes polar bodies from the surface of the sperm
- (2) activates the egg nucleus so that it begins to divide
- (3) acts as a transport medium for sperm
- (4) provides currents that propel the egg down the oviduct

122 _____

Base your answers to questions 123 and 124 on the diagrams and the information below.

A　　　　**B**　　　　**C**　　　　**D**
BEAN　　CHIMPANZEE　　CHICKEN　　AMEBA

123. Which organisms were produced as a result of fertilization?
- (1) *A*, *B*, and *C*, only
- (2) *B* and *C*, only
- (3) *C* and *D*, only
- (4) *B*, *C*, and *D*, only

123 _____

124. Structures that function in the storage of food to be used by growing embryonic cells are indicated by
- (1) 1 and 3
- (2) 2 and 3
- (3) 2 and 4
- (4) 3 and 4

124 _____

Base your answers to questions 125 through 127 on your knowledge of biology and the information below.

A biologist cut a flap of ectoderm from the top of a developing embryo. He did not remove the piece of ectoderm but just folded it back. Then he cut out the mesoderm underneath and completely removed it. He folded the flap of ectoderm back in place. The ectoderm healed; however, a complete nervous system did not develop.

125. This experiment was most likely performed immediately after
- (1) cleavage
- (2) gestation
- (3) fertilization
- (4) gastrulation

125 _____

126. This experiment interfered with the process of
(1) differentiation (3) cleavage
(2) zygote formation (4) ovulation 126____

127. This experiment demonstrates that the
(1) ectoderm is solely responsible for development of the nervous system
(2) nervous system is destroyed during surgical operations
(3) mesoderm influences the development of the nervous system
(4) digestive enzymes have a major role in the development of embryonic layers 127____

128. In a developing embryo, the mesoderm layer normally gives rise to
(1) epidermal tissue (3) digestive tract lining
(2) skeletal tissue (4) respiratory tract lining 128____

129. What is the function of the placenta in a mammal?
(1) It surrounds the embryo and protects it from shock.
(2) It allows mixing of the maternal and fetal blood.
(3) It permits the passage of nutrients and oxygen from the mother to the fetus.
(4) It replaces the heart of the fetus until the fetus is born. 129____

ANSWERS EXPLAINED Questions 110–129

Standard 4, Key Idea 4, Performance Indicator 1

110. **3** Mitosis is the process by which two identical nuclei are formed. Mitotic cell division is usually followed by cytoplasmic division. A plant cell has a rigid cell wall. Division of the cytoplasm begins with the appearance of a *cell plate* between the two nuclei. The cell plate is composed of membrane fragments from the endoplasmic reticulum.

WRONG CHOICES EXPLAINED:
(1) *Spindle fibers* are elastic-like protein fibers. Chromosome movement is controlled by spindle fibers.
(2) In order for two nuclei to be identical, they must have the same number and kind of chromosomes. The *chromosomes duplicate* before the nucleus divides. The mitotic process is the same in both plant and animal cells.

(4) *Cytoplasmic division*, or cytokinesis, usually follows nuclear division in both plant and animal cells.

111. **2** Chromosomes are structures in the nucleus. During mitosis, two cells with identical chromosomes are formed. Because the cell had 12 chromosomes, the daughter cells must also have *12* chromosomes.

WRONG CHOICES EXPLAINED:
(1) A cell with *24* chromosomes has twice the diploid number. The condition in which there are extra sets of chromosomes is known as polyploidy.
(3) A cell with *6* chromosomes has one-half the diploid number. Monoploid cells arise through meiosis.
(4) A cell with *4* chromosomes can only arise through a complete breakdown of the mitotic or meiotic process.

112. **1** In asexual reproduction, new organisms are produced by a single parent. Asexual reproduction involves the mitotic process. The *genetic material of the offspring is identical to that of the parent*.

WRONG CHOICES EXPLAINED:
(2) *Gametes* are produced by sexually reproducing organisms.
(3) The fusion of sex cells (gametes) to *form a zygote* is characteristic of sexually reproducing organisms.
(4) *Genetic variation* among offspring is characteristic of sexually reproducing organisms. The process of meiosis through synapsis and segregation ensures new combinations of genetic material.

113. **1** *Gonads* are sex glands. In these glands, gametes, or sex cells, are produced from primary sex cells that undergo meiosis, also known as reduction division. Male gonads are called testes, and female gonads, ovaries.

WRONG CHOICES EXPLAINED:
(2) *Gametes* are sex cells and not organs. Sex cells are specialized for fertilization.
(3) *Kidneys* are organs of excretion and are specialized for filtering metabolic wastes out of the blood. The nephron is the unit of structure and function in the kidney. Meiosis does not take place in kidney cells.
(4) *Cytoplasmic organelles* such as mitochondria, ribosomes, lysosomes, and endoplasmic reticula are not organs.

114. **2** A gamete is a reproductive cell that must fuse with another gamete to produce a new individual. Sperm cells and egg cells are gametes. Fertilization is the fusion of an egg cell and a sperm cell. In land animals, *fertilization occurs within the body of the female* and is known as internal fertilization.

WRONG CHOICES EXPLAINED:

(1) External fertilization, the *union of gametes outside the female's body*, occurs in animals that live in a watery environment. Fish and amphibians reproduce by external fertilization.

(3), (4) The question refers to reproduction in animals. Animal cells, including gametes, do not have *cell walls*.

115. **4** Primary sex cells give rise to gametes. Gametes are formed by the process of meiosis. In meiosis, a diploid cell divides twice to form four monoploid cells. In humans the four gametes, which are identical in size, are known as *sperm cells*.

WRONG CHOICES EXPLAINED:

(1) Chromosomes occur in pairs. The diploid number of chromosomes is the full number of chromosomes of all the pairs. Meiosis is cell division in which the nucleus receives one member of each pair of chromosomes. The nucleus of an *egg cell* thus contains half the diploid chromosome number, or the monoploid number.

(2) In formation of the egg cell, the cytoplasm does not divide equally. One large monoploid cell, the egg cell, and three very small cells (polar bodies) are produced from one primary sex cell.

(3) The three small monoploid cells accompanying the egg cell are known as *polar bodies*. Polar bodies degenerate and do *not* function in fertilization.

116. **1** The diploid chromosome number is represented as *2n,* and the monoploid number as *n.* When two gametes in the *n* condition combine, a *2n* cell is produced. *Fertilization* is the union of two gametes.

WRONG CHOICES EXPLAINED:

(2) *Gamete formation* reduces the chromosome number from *2n* to *n.*

(3) *Cleavage* is mitotic cell division without growth. It is the process by which a fertilized egg cell becomes a multicellular embryo.

(4) *Meiosis,* or reduction division, reduces the chromosome number of diploid cells.

117. **2** Follicles contain immature egg cells. The follicles are found in the ovary. FSH *stimulates ovarian* follicle development.

WRONG CHOICES EXPLAINED:

(1) ACTH stimulates the *adrenal glands* to produce cortisone. ACTH is secreted by the pituitary gland.

(3) *Calcium metabolism* is controlled by parathormone. The parathyroid gland secretes parathormone.

(4) Thyroxin secreted by the thyroid gland is the major regulator of the rate of *oxidation.* The hormones from the adrenal glands and the pancreas also play a role in the oxidation of glucose.

118. **4** The menstrual cycle is a series of changes that occur within the female reproductive system. The events of the cycle prepare the uterus to receive an embryo. The lining of the uterus is built up. If the cycle is not interrupted, the egg is not fertilized and no embryo is formed. In the last stage of the cycle, the *lining of the uterus disintegrates and is shed*.

WRONG CHOICES EXPLAINED:
(1) *Sperm cells* are produced by the male.
(2) Ovulation (*release of a mature egg*) occurs midway through the menstrual cycle.
(3) Once a month the *uterus* is prepared to receive an embryo. What happens to the lining of the uterus depends upon presence or absence of an embryo. If an embryo is present, the uterus continues to develop and the menstrual cycle is interrupted.

119. **2** *It helps to make terrestrial life possible* is the correct response. Internal fertilization, as its name implies, occurs within the body of the parent (usually female). The conditions in the female reproductive tract provide an ideal environment for the survival and pairing of gametes, helping to ensure that fertilization occurs successfully. This method of reproduction is especially helpful in the survival of terrestrial animal species, who live where harsh conditions (such as drying, heat, and cold) can easily damage or kill gametes released into the environment for external fertilization.

WRONG CHOICES EXPLAINED:
(1) *Motile gametes* (such as human sperm cells) are common in species employing both external and internal fertilization. Motility (ability to move) enables the sperm cells to swim toward the egg cell in either environment.
(3) Because of the dangers posed to fragile gametes in any environment, the *presence of many eggs* is characteristic of species employing external fertilization. Species using internal fertilization produce relatively few eggs in the reproductive process.
(4) In most species internal *fertilization occurs within the body of the female*, not that of the *male*.

120. **4** *Monoploid, monoploid, and diploid* is the correct combination. Sperm cells and egg cells are monoploid (n) gametes formed during the process of meiotic cell division. A zygote results from the fusion of two monoploid nuclei in fertilization and so must be diploid ($2n$) in chromosome number.

WRONG CHOICES EXPLAINED:
(1), (2), (3) Each of these distracters contains an incorrect combination of choices (see above).

121. **1** In the process of oogenesis, a single primary sex cell gives rise to a single monoploid egg cell and three nonfunctional monoploid polar bodies. The process of spermatogenesis yields four functional monoploid sperm cells for each primary sex cell. Therefore, a comparison of these two processes leads to the conclusion that, per primary sex cell, spermatogenesis yields *four times* as many gametes as oogenesis does.

WRONG CHOICES EXPLAINED:
(2), (3), (4) Each of these distracters contains a mathematical comparison that is not consistent with the explanation above.

122. **3** The fluid surrounding human sperm cells *acts as a transport medium for sperm*. This fluid is known as semen. Its primary function is to provide a protective watery medium for sperm cells as they enter the female reproductive tract.

WRONG CHOICES EXPLAINED:
(1) *Removes polar bodies from the surface of the sperm* is a "nonsense" distracter. Polar bodies are not associated with sperm production.
(2) *Activates the egg nucleus so that it begins to divide* is not a function of semen. The egg is stimulated to divide by the act of fertilization. Semen is not directly involved in this process.
(4) *Provides currents that propel the egg down the oviduct* is not a function of semen. Cilia that line the oviduct are responsible for establishing fluid currents that both carry the egg downward toward the uterus and carry sperm upward toward the ovary. Semen is not directly involved in this process.

123. **1** Fertilization is one of the processes in sexual reproduction. Sexual reproduction is the method of reproduction in the bean plant, chimpanzee, and chicken (*A, B, and C, only*). The ameba reproduces asexually by binary fission.

WRONG CHOICES EXPLAINED:
(2) The bean plant (*B*) was omitted in this choice.
(3) The ameba (*D*) is an incorrect answer. Both the bean (*A*) and the chimpanzee (*B*) were omitted in this choice.
(4) The bean plant (*A*) was omitted and the ameba (*D*) is an incorrect answer.

124. **3** The structures that function in the storage of food for the embryonic cells are labeled *2* and *4*. Structure *2* is the cotyledon of the seed. The yolk *sac* is structure *4*.

WRONG CHOICES EXPLAINED:
(1) Structure *3* refers to the wall of the uterus. The embryo of a chimpanzee is nourished through the placenta, not the uterine wall. Structure *1* is the leaf of the embryo bean plant.
(2) Although structure *2* is a correct answer, structure *3* is incorrect.
(4) Although structure *4* is a correct answer, structure *3* is incorrect.

125.   **4**   The experiment was performed after *gastrulation.* Gastrulation is a stage in embryonic development that gives rise to three germ layers of cells. The three germ layers are the ectoderm, mesoderm, and endoderm.

WRONG CHOICES EXPLAINED:

(1) *Cleavage* is a stage of embryonic development in which the zygote undergoes rapid mitotic divisions. The final result is a ball of cells.

(2) *Gestation* is a prebirth period. It is the time a developing embryo spends in the uterus.

(3) The union of a sperm cell nucleus with an egg cell nucleus is called *fertilization.* The result of the process is a zygote.

126.   **1**   The experiment interfered with the development of a nervous system. The development of special tissues and organisms is known as *differentiation.*

WRONG CHOICES EXPLAINED:

(2) *Zygote formation* must occur before an embryo can develop.

(3) The process of *cleavage* provides the embryo with hundreds of undifferentiated cells.

(4) *Ovulation* is the release of an egg from the ovary.

127.   **3**   The experiment demonstrates that the development of the *nervous system is influenced by the presence of the mesoderm.* The nervous system does not develop when the mesoderm is removed.

WRONG CHOICES EXPLAINED:

(1) If the *ectoderm was solely responsible for the development of the nervous system,* the nervous system would have developed after the mesoderm was removed.

(2) There was no *nervous system* present when the surgery was performed.

(4) The experiment was not concerned with the reasons for the *development of the embryonic layers.*

128.   **2**   Each of the three germ layers of the embryo is responsible for the development of the systems of the body. The *skeletal system* develops from the mesoderm. The muscle system, circulatory system, and excretory system also evolve from the mesoderm.

WRONG CHOICES EXPLAINED:

(1) *Epidermal cells* form the outer covering or skin of the body. The skin and nervous system develop from the ectoderm.

(3), (4) The linings of the *digestive* and *respiratory tracts* develop from the endoderm.

129. **3** The placenta is an area of spongy tissue in the uterus. It is very rich in blood vessels. The placenta functions as a respiratory and excretory organ of the fetus. *Vital materials are exchanged between the capillaries of the fetus and the capillaries of the mother.*

WRONG CHOICES EXPLAINED:

(1) The amnion is a fluid-filled sac surrounding the embryo. The fluid bathes the cells of the fetus and *protects it against shock.*

(2) The circulatory systems of the mother and the fetus are separate from each other. *Blood does not flow from one system into the other.*

(4) The *embryo develops its own heart.* The placenta provides an area for the diffusion of materials into and out of the fetus.

Key Idea 5—Dynamic Equilibrium and Homeostasis: Organisms maintain a dynamic equilibrium that sustains life.

Performance Indicator	Description
5.1	The student should be able to explain the basic biochemical processes in living organisms and their importance in maintaining dynamic equilibrium.
5.2	The student should be able to explain disease as a failure of homeostasis.
5.3	The student should be able to relate processes at the system level to the cellular level in order to explain dynamic equilibrium.

130. Which energy conversion occurs in the process of photosynthesis?
 (1) Light energy is converted to nuclear energy.
 (2) Chemical bond energy is converted to nuclear energy.
 (3) Light energy is converted to chemical bond energy.
 (4) Mechanical energy is converted to light energy   130_____

131. An environmental change that would most likely increase the rate of photosynthesis in a bean plant would be an increase in the

(1) intensity of green light
(2) concentration of nitrogen in the air
(3) concentration of oxygen in the air
(4) concentration of carbon dioxide in the air   131 _____

132. During photosynthesis, molecules of oxygen are released as a result of the "splitting" of water molecules. This is a direct result of the

(1) dark reaction   (3) formation of PGAL
(2) light reaction   (4) formation of CO_2 132 _____

133. An organism that makes its own food without the direct need for any light energy is known as a

(1) chemosynthetic heterotroph (3) photosynthetic het-
(2) chemosynthetic autotroph   erotroph
                                (4) photosynthetic autotroph   133 _____

134. While looking through a microscope at a section of a leaf from a freshwater plant, a student observed some cells in which chloroplasts were moving around with the cytoplasm. This type of movement is known as

(1) pinocytosis   (3) osmosis
(2) synapsis   (4) cyclosis   134 _____

135. By what process does carbon dioxide pass through the stomates into the leaf ?

(1) diffusion   (3) respiration
(2) osmosis   (4) pinocytosis   135 _____

136. Two end products of aerobic respiration are

(1) oxygen and alcohol   (3) carbon dioxide and water
(2) oxygen and water   (4) carbon dioxide and oxy-
                              gen   136 _____

137. Homeostatic regulation of the body is made possible through the coordination of all body systems. This coordination is achieved mainly by

(1) respiratory and   (3) nervous and endocrine
    reproductive systems   systems
(2) skeletal and excretory (4) circulatory and digestive
    systems   systems   137 _____

138. Phenylketonuria (PKU) is an inherited condition charac-terized by mental retardation. The symptoms of the disorder result from an inability to synthesize a single type of

(1) enzyme
(3) blood cell

(2) nutrient
(4) brain cell
138____

139. All the children of a hemophiliac male and a normal fe-male are normal with respect to blood clotting. However, some of their grandsons are hemophiliacs. This is an example of the pattern of hereditary known as

(1) sex determination
(3) incomplete dominance

(2) sex linkage
(4) multiple alleles
139____

140. What is the total number of chromosomes in a typical body cell of a person with Down's syndrome?

(1) 22   (2) 23   (3) 44   (4) 47
140____

ANSWERS EXPLAINED Questions 130–140

Standard 4, Key Idea 5, Performance Indicators 1–3

130. **3** In the process of photosynthesis, carbon dioxide and water molecules are converted to glucose. Light is the energy source for the reaction. *Light energy is transformed into the chemical bond energy* of the glucose molecules.

WRONG CHOICES EXPLAINED:
(1), (2), (4) The Law of Conservation of Energy states that energy cannot be created or destroyed, but it can be changed from one form to another. All the choices refer to this law, but only choice (3) occurs in living organisms.

131. **4** Experiments have shown that the rate of photosynthesis depends on the availability of carbon dioxide. The greater the *concentration of carbon dioxide*, the greater the rate of photosynthesis.

WRONG CHOICES EXPLAINED:
(1) Chlorophyll is a light-absorbing pigment found in chloroplasts. However, it does not absorb much of the green wavelength of light. An increase in the *intensity of green light* has no effect on the photosynthetic rate.
(2) Carbohydrates are the products of photosynthesis. They do not contain atoms of nitrogen. The nitrogen needed by plants comes from nitrates in the soil, not from *the concentration of nitrogen in the air*.

(3) Oxygen is released during photosynthesis. Therefore, *the concentration of oxygen in the air* has no direct effect on photosynthesis.

132.   **2**   Sunlight provides the energy needed to split water molecules into hydrogen and oxygen. Because light is required, this part of photosynthesis is known as the *light reaction*.

WRONG CHOICES EXPLAINED:

(1) The *dark reaction* does not use light energy. In this reaction, the hydrogen released from the light reaction is combined with carbon dioxide.

(3) *PGAL*, phosphoglyceric aldehyde, is the first stable compound formed during the dark reaction. This compound is later converted to glucose.

(4) *Carbon dioxide is not formed* but is used during photosynthesis.

133.   **2**   An autotroph manufactures its own food. A *chemosynthetic autotroph* produces its own food without the use of light energy. It obtains its energy from certain chemical reactions that take place in the cell.

WRONG CHOICES EXPLAINED:

(1) A *chemosynthetic organism* cannot be a heterotroph. Heterotrophs do not have the ability to manufacture their own food. All animals are heterotrophs.

(3) *Photosynthetic organisms* manufacture their own food. They cannot be heterotrophs.

(4) A *photosynthetic autotroph* utilizes light energy. All green plants are photosynthetic autotrophs.

134.   **4**   The movement of chloroplasts in the plant cell was due to the movement of cytoplasm in the cell. *Cyclosis* is the streaming of cytoplasm in a cell.

WRONG CHOICES EXPLAINED:

(1) *Pinocytosis* is the formation of a pocket by an infolding of the cell membrane. Large molecules are brought into the cell by this process.

(2) *Synapsis* is the pairing of homologous chromosomes during meiosis.

(3) *Osmosis* is the movement of water across a selectively permeable membrane. Osmosis is the diffusion of water.

135.   **1**   Carbon dioxide passes through the stomates of a leaf by *diffusion*. Diffusion is passive transport. Molecules move along a concentration gradient from an area of high density to an area of lower density.

WRONG CHOICES EXPLAINED:

(2) *Osmosis* is the diffusion of water.

(3) *Respiration* is an energy-releasing process.

(4) *Pinocytosis* is active transport. Cells use energy to draw in large molecules by the infolding of their cell membranes.

136. **3** Respiration is a process by which cells release energy from glucose molecules. Aerobic respiration requires the presence of oxygen. In the process of aerobic respiration, glucose is oxidized to *carbon dioxide and water*. Both compounds are end products, the results of a chemical reaction.

WRONG CHOICES EXPLAINED:

(1), (2), (4) All three choices are incorrect because oxygen is consumed in aerobic respiration. Oxygen is not the end product of the reaction.

137. **3** Homeostasis refers to the steady state of control of the cell and, in turn, the entire body. The biochemical processes that take place in body cells occur in even and regular sequences. The cells, tissues, and organs in all body systems must function cooperatively so that the organism can carry out its life functions effectively. The coordination of all these biochemical activities is made possible by the work of the *nervous and endocrine systems*. The nervous system carries impulses from sense organs to the brain or spinal cord and then to effector organs such as muscles or glands. The endocrine system secretes hormones that control the functions of certain glands, tissues, and organs. Together the nervous and endocrine systems maintain the homeostasis of the body.

WRONG CHOICES EXPLAINED:

(1) The *respiratory system* is specialized for the intake and distribution of oxygen. It also expels waste gases. The *reproductive system* is specialized for the developing of embryos. Both systems are controlled by the nervous and endocrine systems. In mammals, hormones control gestation and birth.

(2) The *skeletal system* gives support to the body. Hormones control the growth of long bones. Nerve cell fibers help the muscles to function. The *excretory system* coordinates waste removal from the body. Hormones control water loss and reabsorption by the kidney tubules.

(4) Blood and lymph circulate by way of the *circulatory system*. Hydrolysis of food takes place in the *digestive system*. Both systems depend on the nervous and endocrine systems for the coordination of mechanical and biochemical activities.

138. **1** According to the one gene–one enzyme theory, a single gene is responsible for the production of a single enzyme. Because PKU is an inherited defect, the gene for an *enzyme* is absent in the victim.

WRONG CHOICES EXPLAINED:

(2) *Nutrients* are *not* synthesized by animals. Nutrients must be taken in from organic sources.

(3) *Blood cells* are *not* affected nor involved in the PKU disorder.

(4) The development of *brain cells* is affected by the PKU condition. Because of the lack of an enzyme, phenylalanine is converted to phenylpyruvic acid, which accumulates in brain tissue. The result is mental retardation.

139. **2** A trait that appears more often in one sex than in the other sex is said to be *sex-linked*. The gene for the trait is located on a sex chromosome. Hemophilia and color blindness are examples of sex-linked traits.

WRONG CHOICES EXPLAINED:

(1) *Sex determination* is controlled by a pair of sex chromosomes. There are two kinds of sex chromosomes, an X and a Y chromosome. A female has two X chromosomes (XX). A male has one X chromosome and one Y chromosome (XY).

(3) *Incomplete dominance* is a type of inheritance in which neither allele in a hybrid is dominant. The hybrid shows a trait completely different from either parent. The inheritance of color in a Japanese four-o'clock flower is an example of incomplete dominance, or blending.

(4) Alleles are different forms of the same gene. The term *multiple alleles* implies that a gene has more than two forms. Inheritance of human blood type involves multiple alleles.

140. **4** The normal number of chromosomes in human body cells is 46. Down's syndrome results from the presence of an extra chromosome. The extra or 47th chromosome is due to meiotic nondisjunction.

WRONG CHOICES EXPLAINED:

(1), (3) The numbers 22 and 44 do not apply to any known normal human chromosome number.

(2) The number 23 is the monoploid number of chromosomes in human gametes.

Key Idea 6—Interdependence of Living Things: Plants and animals depend on each other and their physical environment.

Performance Indicator	Description
6.1	The student should be able to explain factors that limit growth of individuals and populations.
6.2	The student should be able to explain the importance of preserving diversity of species and habitats.
6.3	The student should be able to explain how the living and nonliving environments change over time and respond to disturbances.

141. Animals *cannot* synthesize nutrients from inorganic raw materials. Therefore, animals obtain their nutrients by
(1) combining carbon dioxide with water
(2) consuming preformed organic compounds
(3) hydrolyzing large quantities of simple sugars
(4) oxidizing inorganic molecules for energy   141____

142. Which organisms carry out heterotrophic nutrition?
(1) ferns   (3) fungi
(2) grasses   (4) mosses   142____

143. Which activity is an example of intracellular digestion?
(1) a grasshopper chewing blades of grass
(2) a maple tree converting starch to sugar in its roots
(3) an earthworm digesting proteins in its intestine
(4) a fungus digesting dead leaves   143____

144. A hydra ingests a daphnia, digests it, and later egests some materials. All these events are most closely associated with the life process known as
(1) transport   (3) growth
(2) synthesis   (4) nutrition   144____

145. Some bacteria are classified as saprophytes because they are organisms that
(1) feed on other living things
(2) feed on dead organic matter
(3) manufacture food by photosynthesis
(4) contain vascular bundles   145____

146. Aerobic organisms are dependent on autotrophs. One reason for this dependency is that most autotrophs provide the aerobic organisms with
(1) oxygen   (3) nitrogen gas
(2) carbon dioxide   (4) hydrogen   146____

147. Of the following, the greatest amount of the Earth's food production is thought to occur in .
(1) coastal ocean waters   (3) taiga forests
(2) desert biomes   (4) tundra biomes   147____

148. Most of the minerals within an ecosystem are recycled and returned to the environment by the direct activities of organisms known as
(1) producers
(2) secondary consumers
(3) decomposers
(4) primary consumers

148_____

149. Which type of organism is *not* shown in the following representation of a food chain?

grass → mouse → snake → hawk

(1) herbivore
(2) decomposer
(3) producer
(4) carnivore

149_____

150. In the food chain shown below, which organism represents a primary consumer?

(1) grasshopper
(2) grass
(3) frog
(4) snake

150_____

151. A lake contains minnows, mosquito larvae, sunfish, algae, and pike. Which of these organisms would probably be present in the largest number?
(1) minnows
(2) larvae
(3) sunfish
(4) algae

151_____

152. An abiotic factor that affects the ability of pioneer organisms such as lichens to survive is the
(1) type of climax vegetation
(2) species of algae
(3) type of substratum
(4) species of bacteria

152_____

153. In order to avoid predators, the clown fish hides unharmed in the stinging tentacles of the sea anemone. The clown fish attracts food to the sea anemone. This is an example of a type of relationship known as
(1) mutualism
(2) commensalism
(3) predator–prey
(4) parasitism

153_____

154. Which world biome has the greatest number of organisms?

(1) tundra
(2) tropical forest

(3) temperate deciduous forest
(4) marine

154_____

155. In a particular area, living organisms and the nonliving environment function together as

(1) a population
(2) a community

(3) an ecosystem
(4) a species

155_____

ANSWERS EXPLAINED Questions 141–155

Standard 4, Key Idea 6, Performance Indicators 1–3

141. **2** The portions of food that are usable to an animal are known as nutrients. Carbohydrates, lipids, and proteins are the organic nutrients needed by all organisms. Animals cannot synthesize their own nutrients. They must eat other organisms that contain the *preformed organic nutrients*.

WRONG CHOICES EXPLAINED:

(1) *Carbon dioxide and water* are inorganic compounds. These compounds are converted to nutrients by plants only. The process is called photosynthesis.

(3) Hydrolysis is the breakdown of compounds to simpler molecules through the action of enzymes in the presence of water. Glucose is a *simple sugar*. The hydrolyzing of glucose results in the release of energy in a cell.

(4) The use of *inorganic molecules* for the production of energy occurs only in certain species of bacteria. Chemosynthesis does not occur in members of the animal kingdom.

142. **3** *Fungi* are nongreen plants. They are heterotrophs, which means that they cannot manufacture their own food. Organisms that carry out heterotrophic nutrition must take in preformed organic molecules.

WRONG CHOICES EXPLAINED:

(1), (2), (4) *Ferns*, *grasses*, and *mosses* are green plants or bryophytes. Green plants are *autotrophs,* which means that they manufacture their own food.

143. **2** Digestion that occurs within a cell is known as intracellular digestion. Plants do not have special digestive systems. Digestion, or the *conversion of starch to sugar*, occurs within the individual cells of a plant, including those of the root.

WRONG CHOICES EXPLAINED:

(1) *Grasshoppers* have a digestive system. Digestion is extracellular and takes place in a digestive tube outside the body cells. Chewing a blade of grass is an example of mechanical digestion taking place in the mouth.

(3) *Earthworms* also have a digestive system. Proteins are digested outside the body cells in a portion of the digestive system known as the intestine.

(4) *Fungi* demonstrate a special form of extracellular digestion. Digestive enzymes are secreted into the external environment. The nutrients from the digested food diffuse into the cells.

144. **4** *Nutrition* is the life process most closely associated with a hydra ingesting, digesting, and egesting a daphnia. Ingestion is the process by which food materials are taken into the body of an organism such as a hydra. Digestion is the process by which the complex food molecules within the daphnia are hydrolyzed to soluble end products. Egestion is the process by which the undigestible materials of the daphnia's body are expelled from the body of the hydra.

WRONG CHOICES EXPLAINED:

(1) *Transport* is the life process by which soluble foods and other materials are circulated through the body of an organism such that they reach all parts of the organism's body.

(2) *Synthesis* is a process by which complex materials are constructed from simpler chemical components. The processes described in the question represent hydrolysis, the opposite of synthesis.

(3) *Growth* is a process that involves an increase in cell number and cell size, leading to an increase in the size of the organism. The end products of digestion can be used to supply raw materials for such growth.

145. **2** *Saprophytes* are organisms that *feed on dead organic matter*. Fungi and the bacteria of decay are examples of saprophytes.

WRONG CHOICES EXPLAINED:

(1) Heterotrophs *live on or off other living organisms*. A dog flea is an example of a heterotroph also known as a parasite.

(3) Autotrophs *manufacture their own food by photosynthesis*. Algae, mosses, and grasses are examples of autotrophs.

(4) Higher plants *contain vascular bundles*. These plants, called *tracheophytes,* include ferns, conifers, and flowering plants.

146. **1** Aerobic organisms need *oxygen* for cellular respiration. Some autotrophs are photosynthetic organisms. Oxygen is released by photosynthesis. Aerobic organisms depend on the autotrophs to release oxygen into the environment.

WRONG CHOICES EXPLAINED:

(2) *Carbon dioxide* is a waste product from the cellular respiration of aerobic organisms.

(3) *Nitrogen gas* makes up 78% of the atmosphere. However, the nitrogen cannot be used in the gaseous form by aerobic organisms and most autotrophs.

(4) *Hydrogen* does *not* exist as a gas on our planet. It is combined with other elements. The hydrogen needed by organisms comes mostly from water and organic compounds.

147. **1** The area of greatest food production is in the region where the greatest rate of photosynthesis occurs. The area must be rich in minerals, water, gases, and light. The *coastal ocean waters* meet these requirements.

WRONG CHOICES EXPLAINED:

(2) There is very little precipitation in the *desert*. Water is the factor that limits plant growth.

(3) The water in the *taiga* is frozen part of the year.

(4) The *tundra* is a frozen plain. Water is frozen almost all year long on the tundra.

148. **3** Bacteria of decay are *decomposers* that release minerals from decaying plant and animal bodies and return them to the environment.

WRONG CHOICES EXPLAINED:

(1) *Producers* are autotrophs, that is, green plants that synthesize food by photosynthesis from carbon dioxide and water in the presence of sunlight.

(2) *Secondary consumers* are animals that eat other animals. For example, a frog feeds on flies. Thus, a frog is a secondary consumer.

(4) *Primary consumers* are organisms that feed on plants only. Herbivores are primary consumers.

149. **2** *Decomposers* are not shown in the food chain. They are organisms that live on dead things. Fungi and bacteria are decomposers.

WRONG CHOICES EXPLAINED:

(1) An *herbivore* is a primary consumer; that is, it eats vegetation. The mouse is the herbivore in the food chain.

(3) A *producer* is a green plant. It depends on sunlight to synthesize its own food. The grass is the producer in the food chain.

(4) A *carnivore* is an animal that eats the flesh of other animals. Both the snake and the hawk are carnivores.

150. **1** The *grasshopper* is a primary consumer because it feeds on vegetation.

WRONG CHOICES EXPLAINED:

(2) *Grass* is an autotroph, that is, a producer or self-feeder. Grass is a green plant that can make its own food.

(3) A *frog* is a secondary consumer; it eats insects that are plant eaters.

(4) A *snake* is a secondary consumer; it eats animals that are primary consumers.

151.   **4**   The organisms in the question make up a food chain. The number of organisms at each level of the food chain decreases as one moves down the chain. The pyramid of energy shown below represents this fact. The producers, which form the base of the pyramid, are the most numerous. *Algae* are the producers in this food chain.

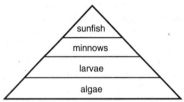

WRONG CHOICES EXPLAINED:

(1) The *minnows* are carnivores. They are the secondary consumers.

(2) The *larvae* are herbivores. They are primary consumers.

(3) The *sunfish* are tertiary consumers and occupy the top level of the pyramid. The organisms at the top of the pyramid are the least numerous.

152.   **3**   A *substratum* is the surface on which organisms grow. Lichens grow on rocks. Rocks are nonliving. The nonliving parts of the environment make up the abiotic factors.

WRONG CHOICES EXPLAINED:

(1) *Climax vegetation* is the type of vegetation that occupies an area in its final stage of succession. Plants make up the biotic, or living, environment.

(2) *Algae* are living organisms. They are part of the biotic (living) environment.

(4) *Bacteria* also make up the biotic environment.

153.   **1**   A relationship between two organisms in which both benefit from the association is known as *mutualism.* The clown fish is protected by the sea anemone. The sea anemone is sessile. The clown fish draws food to the sea anemone.

WRONG CHOICES EXPLAINED:

(2) *Commensalism* is a relationship between two organisms in which one organism is benefited by the association. The second organism is neither harmed nor benefited by the association. Barnacles attached to a whale are an example of a commensal relationship.

(3) A predator is a carnivore that hunts, kills, and eats its prey. The prey is the hunted organism. A *predator-prey* relationship is important in controlling the population of both.

(4) *Parasitism* is a relationship between two organisms in which one organism is benefited by the association whereas the second organism is harmed by the association. The parasitized organism is called the host.

154.   **4**   A biome is a large area dominated by one major type of vegetation and one type of climate. The *marine* biome has the greatest number of organisms.

WRONG CHOICES EXPLAINED:

(1) The *tundra* is a region where the ground is frozen all year long. Mosses and lichens are the dominant vegetation.

(2) The *tropical rain forest* is dominated by broadleaf plants. The region is always warm, and the rainfall is abundant and continuous.

(3) The *temperate deciduous forest* is made up of trees that shed their leaves once a year. The winters are cold, and the summers are warm. The rainfall is distributed throughout the year.

155.   **3**   *An ecosystem* is an area in which communities of living organisms interact with the nonliving environment.

WRONG CHOICES EXPLAINED:

(1) A *population* is all the organisms of a particular species living in a given area.

(2) A *community* is made up of populations of different species that interact with each other.

(3) A *species* is a group of organisms whose members are able to interbreed with each other. The offspring resulting from the matings are fertile and can reproduce.

Key Idea 7—Human Impact on the Environment: Human decisions and activities have a profound impact on the physical and living environment.

Performance Indicator	Description
7.1	The student should be able to describe the range of interrelationships of humans with the living and nonliving environment.
7.2	The student should be able to explain the impact of technological development and growth in the human population on the living and nonliving environment.
7.3	The student should be able to explain how individual choices and societal actions can contribute to improving the environment.

156. Human impact on the environment is most often more dramatic than the impact of most other living things because humans have a greater

    (1) need for water   (3) ability to adapt to change

    (2) need for food   (4) ability to alter the environment   156____

157. Which human activity would have the most direct impact on the oxygen–carbon dioxide cycle?

    (1) reducing the rate of ecological succession

    (2) decreasing the use of water

    (3) destroying large forest areas

    (4) enforcing laws that prevent the use of leaded gasoline   157____

158. Fertilizers used to improve lawns and gardens may interfere with the equilibrium of an ecosystem because they

    (1) cause mutations in all plants

    (2) cannot be absorbed by roots

    (3) can be carried into local water supplies

    (4) cause atmospheric pollution   158____

159. The tall wetland plant purple loosestrife was brought from Europe to the United States in the early 1800s as a garden plant. The plant's growth is now so widespread across the United States that it is crowding out a number of native plants. This situation is an example of

    (1) the results of the use of pesticides

    (2) the recycling of nutrients

    (3) the flow of energy present in all ecosystems

    (4) an unintended effect of adding a species to an ecosystem   159____

160. Choose *one* ecological problem from the list below.

Ecological Problems

Global warming

Destruction of the ozone shield

Loss of biodiversity

Discuss the ecological problem you chose. In your answer be sure to state:

- the problem you selected and *one* human action that may have caused the problem [1]
- *one* way in which the problem can negatively affect humans [1]
- *one* positive action that could be taken to reduce the problem [1]

_____

_____

_____

Base your answers to questions 161 through 163 on the information below and on your knowledge of biology.

The planning board of a community held a public hearing in response to complaints by residents concerning a waste-recycling plant. The residents claimed that the waste-hauling trucks were polluting air, land, and water and that the garbage has brought an increase in rats, mice, and pathogenic bacteria to the area. The residents insisted that the waste-recycling plant be closed permanently.

Other residents recognized the health risks but felt that the benefits of waste recycling outweighed the health issues.

161. Identify two specific health problems that could result from living near the waste-recycling plant [2]

_____

162.   Identify one specific contaminant that might be released into the environment from operation of the waste-recycling plant. [1]

163.   State one ecological benefit of recycling wastes. [1]

164.   Which organism is a near-extinct species?
(1) Japanese beetle   (3) blue whale
(2) dodo bird   (4) passenger pigeon   164_____

165.   Which human activity has probably contributed most to the acidification of lakes in the Adirondack region?
(1) passing environmental protection laws
(2) establishing reforestation projects in lumbered areas
(3) burning fossil fuels that produce air pollutants containing sulfur and nitrogen
(4) using pesticides for the control of insects that feed on trees   165_____

166.   Compared to a natural forest, the wheat field of a farmer lacks
(1) heterotrophs   (3) autotrophs
(2) significant biodiversity   (4) stored energy   166_____

167.   Which factor is not considered by ecologists when they evaluate the impact of human activities on an ecosystem?
(1) amount of energy released from the Sun
(2) quality of the atmosphere
(3) degree of biodiversity
(4) location of power plants   167_____

168.   A new type of fuel gives off excessive amounts of smoke. Before this type of fuel is widely used, an ecologist would most likely want to know
(1) what effect the smoke will have on the environment
(2) how much it will cost to produce the fuel
(3) how long it will take to produce the fuel
(4) if the fuel will be widely accepted by consumers   168_____

169. Which of the following is the most ecologically promising method of insect control?
(1) interference with insect reproductive processes
(2) stronger insecticides designed to kill higher percentages of insects
(3) physical barriers to insect pests
(4) draining marshes and other insect habitats   169____

170. Which is an example of biological control of a pest species?
(1) DDT was used to destroy the red mite.
(2) Most of the predators of a deer population were destroyed by humans.
(3) Gypsy moth larvae (tree defoliators) are destroyed by beetle predators that were cultured and released.
(4) Drugs were used in the control of certain pathogenic bacteria. 170____

171. To ensure environmental quality for the future, each individual should
(1) acquire and apply knowledge of ecological principles
(2) continue to take part in deforestation
(3) use Earth's finite resources
(4) add and take away organisms from ecosystems   171____

172. Ladybugs were introduced as predators into an agricultural area of the United States to reduce the number of aphids (pests that feed on grain crops). Describe the positive and negative effects of this method of pest control. Your response must include at least:

• two advantages of this method of pest control [2]

• two possible dangers of using this method of pest control [2]

_____

_____

_____

_____

_____

173. Some people claim that certain carnivores should be destroyed because they kill beneficial animals. Explain why these carnivores should be protected. Your answer must include information
concerning:

- prey population growth [1]

- extinction [1]

- the importance of carnivores in an ecosystem [1]

ANSWERS EXPLAINED Questions 156–173

Standard 4, Key Idea 7, Performance Indicators 1–3

156.   **4**   The fact that humans have a greater *ability to alter the environment*
means that human impact on the environment is often more dramatic than that of
most other living things. In addition to our ability to make physical changes in the
environment, humans have the unique ability to alter the environment chemically,
introducing many materials that are not found in nature and that cannot be converted to useful products by nature.

WRONG CHOICES EXPLAINED:
(1), (2) On an individual basis, humans' *need for water* and *need for food* are
not significantly greater than those of other living things. However, the fact is that
our large population places incredible demands on the environment to supply these
basic resources. As a result, our tendency to destroy natural habitats to create additional water and agricultural resources is a significant factor affecting the natural
world.
(3) On an individual basis, humans' *ability to adapt to change* is not significantly
greater than that of other living things. However, as a species, we have created artificial environments to protect ourselves from harsh environmental conditions. To
the extent that these artificial environments are dependent on energy and other
natural resources, their construction and maintenance have resulted in significant
alterations of the natural world.

157. **3** *Destroying large forest areas* is the human activity that would have the most direct impact on the oxygen-carbon cycle. Reducing the number of trees over a large area would decrease the forest's ability to absorb carbon dioxide and water and convert them to atmospheric oxygen and glucose. The millions of leaves in a forest are capable of releasing many tons of oxygen gas to the atmosphere. The massive bodies of forest trees can likewise store tons of carbon in the form of complex carbohydrates such as cellulose.

WRONG CHOICES EXPLAINED:
    (1) *Reducing the rate of ecological succession* is not the human activity that would have the most direct impact on the oxygen-carbon cycle. Ecological succession is a process by which one plant-animal community is replaced over time by other plant-animal communities until a stable climax community is established. Reducing its rate would only have the effect of prolonging each successive community longer than might otherwise be expected but would not directly alter the cycling of carbon and oxygen.
    (2) *Decreasing the use of water* is not the human activity that would have the most direct impact on the oxygen-carbon cycle. Water is a precious resource in many parts of the world. Reducing water use so as to conserve it would represent a positive impact of human activity on the environment but would not directly alter the cycling of carbon and oxygen.
    (4) *Enforcing laws that prevent the use of leaded gasoline* is not the human activity that would have the most direct impact on the oxygen-carbon cycle. Lead is a dangerous heavy metal pollutant released when leaded gasoline is burned. Enforcing laws that limit its use would represent a positive impact of human activity on the environment, but would not directly alter the cycling of carbon and oxygen.

158. **3** Fertilizers used to improve lawns and gardens may interfere with the equilibrium of an ecosystem because they *can be carried into local water supplies.* Once dissolved fertilizers enter streams, ponds, wetlands, or lakes, they provide an abundant nutrient source for the growth of algae. As masses of algae die off in the water environment, their decomposition can rob the water of oxygen needed for the survival of fish and other water-dwelling populations, causing their elimination from the habitat. When these species disappear, other species that depend on them for food must migrate or starve. Because the changes caused by the entry of fertilizers into water environments are so significant, it can be said that ecosystem equilibrium is destroyed.

WRONG CHOICES EXPLAINED:
    (1) It is not true that fertilizers *cause mutations in all plants.* Some compounds with chemical structures similar to that of fertilizers are known to stimulate rapid gene mutation in plant cells that may lead to the death of the plant. However, the class of chemical compounds known as fertilizers do not have this effect on all plants.

(2) It is not true that fertilizers *cannot be absorbed by all plants*. When dissolved in water, fertilizers can easily enter plants by being absorbed via simple diffusion into root hairs.

(4) It is not normally true that fertilizers *cause atmospheric pollution*. Most fertilizers are relatively stable chemical compounds that are solids at normal temperatures. For this reason fertilizers are not normally responsible for atmospheric pollution unless they are applied in a gaseous form (such as ammonia) or become airborne (when attached to dry soil particles).

159. **4** The situation described in the question is an example of *an unintended effect of adding a species to an ecosystem*. Although purple loosestrife has adapted well to North American habitats, its rapid growth in wetland environments has stressed or eliminated populations of cattail, pickerelweed, and other native plant species. The introduction of nonnative purple loosestrife to the North American continent has had an unintended negative effect on these native species and on the balance of nature established over many centuries.

WRONG CHOICES EXPLAINED:
(1), (2), (3) The situation described in the question is not an example of *the results of the use of pesticides*, *the recycling of nutrients*, or *the flow of energy present in all ecosystems*. The introduction of a nonnative plant (a living thing) is not the same as the introduction of a chemical pesticide, the recycling of nutrients, or the flow of energy (nonliving things).

160. A three-part response is required that must include the following points:
- One human activity that may have caused the ecological problem selected from the list [1]
- One way the problem may negatively affect humans [1]
- One positive action that could be taken to reduce the problem [1]

Note: No credit is awarded for discussing an ecological problem not on the list.

Acceptable responses include: [3]
- *Global warming is a worldwide ecological problem that may be caused by the release of carbon dioxide and other gases in automobile exhaust. [1] This problem may negatively affect humans if the warming conditions disrupt weather patterns and lead to droughts, floods, or other natural disasters. [1] One positive action that could be taken to help the problem would be to find an energy source for automobiles that would not release carbon dioxide into the atmosphere. [1]*
- *An ecological problem affecting humans is destruction of the ozone layer, which is caused by the use of chemicals known as CFCs as propellants in aerosol sprays. [1] This is a problem for*

humans because the ozone layer protects us from ultraviolet radiation from the sun; without this protection we would have an increased chance of getting skin cancer. [1] A way to help solve this problem would be to ban the use of CFCs in aerosols. [1]

- *Loss of biodiversity is an ecological problem that negatively impacts humans. This problem is caused whenever humans destroy a natural habitat and convert it to other uses. [1] The overall health of our environment depends on the diversity of species that fill different roles in nature. When species diversity and environmental health are reduced, our health is threatened as well. [1] This problem can be reversed only if we use education to learn that protecting natural species is just as important as protecting our own. [1]*

161. Two responses are required. Acceptable responses include:

- *Asthma*
- *Respiratory infections*
- *Allergic reactions*
- *Cancer*
- *Bacterial infections*
- *Viral infections*
- *Disease linked to a pathogen*
- *Poisoning linked to toxic contamination of groundwater*

162. One response is required. Acceptable responses include:

- *Particles in the air*
- *Presence of viruses or bacteria on trucks*
- *Chemicals in air or water*
- *Carcinogens*
- *Mold and fungus spores*

163. One response is required. Acceptable responses include:

- *Conservation of natural resources*
- *Protection of finite resources*
- *Energy conservation*
- *Reduction in pollution*
- *Landfills last longer*
- *Preservation of open space resources*

164.   **3**   The blue whale is near extinction because of uncontrolled hunting by humans.

WRONG CHOICES EXPLAINED:

(1) The Japanese beetle is a plant pest that was accidentally introduced into the United States. Its population is kept in check by the praying mantis, its predator.

(2) The dodo bird became extinct because of hunting by humans.

(4) The passenger pigeon became extinct in the 1900s due to hunting by humans.

165. **3** *Burning fossil fuels that produce air pollutants containing sulfur and nitrogen* is the human activity that has probably contributed the most to the acidification of lakes in the Adirondack region. These pollutants combine with water in the atmosphere to form sulfuric and nitric acids. These acids then enter lakes in rainfall and runoff, adding to the acidic quality of the lake water and killing many susceptible species.

WRONG CHOICES EXPLAINED:

(1) *Passing environmental protection laws* is not an activity that results in the acidification of lakes. In fact, it is a positive human activity that can help to limit the production and release of such gases into the atmosphere.

(2) *Establishing reforestation projects in lumbered areas* is not an activity that results in the acidification of lakes. In fact, it is a positive human activity that can help to replace trees lost because of the acidification of soils by acid rain

(4) *Using pesticides for the control of insects that feed on trees* is not an activity that results in the acidification of lakes. It is a negative human activity carried out to protect commercial crops from destruction and does not normally result in the production of sulfur and nitrogen gases.

166. **2** *Significant biodiversity* is the factor lacking in a wheat field as compared to a natural forest. *Biodiversity* is a term relating to the variety of life forms in an environment. Natural environments, including forests, are typically made up of thousands of species that interact to provide a balanced, ecologically responsive community. By contrast, farm fields are often limited to a small number of different species, and predominantly a single species. Communities lacking in biodiversity are unstable and prone to collapse when environmental conditions change.

WRONG CHOICES EXPLAINED:

(1) *Heterotrophs* are not lacking in a farm field compared to a forest. Heterotrophs are found within a wheat field, although their number and variety are normally limited to those that use wheat or its by-products as food.

(3) *Autotrophs* are not lacking in a farm field compared to a forest. Wheat is a type of autotroph, as are the various weed species that may be interspersed among the wheat plants in the field.

(4) *Stored energy* is not lacking in a farm field compared to a forest. As the wheat grows in the field, it absorbs the Sun's energy and stores it as the chemical bond energy of carbohydrates and other organic compounds.

167.   **1**   The *amount of energy released from the Sun* is not normally considered by an ecologist when evaluating the impact of human activities on an ecosystem. The amount of solar energy emitted by the Sun is generally constant and out of our direct control. Because it is not a variable that can be directly affected by human activities, it is usually not a consideration in decisions of this kind.

WRONG CHOICES EXPLAINED:
   (2) The *quality of the atmosphere* is often a factor considered by ecologists in evaluating the impact of human activities on an ecosystem. Many human activities introduce chemical contaminants into the atmosphere. These chemicals may have a negative impact on the health and survival of humans and other species.
   (3) The *degree of biodiversity* is often a factor considered by ecologists in evaluating the impact of human activities on an ecosystem. Human activities often put pressure on natural species, eliminating those unable to migrate or adapt. As biodiversity in an area declines, so does environmental stability. This situation threatens the health and survival of humans and other species.
   (4) The *location of power plants* is often a factor considered by ecologists in evaluating the impact of human activities on an ecosystem. Fossil fuel plants can pollute the atmosphere and consume valuable petroleum products. Nuclear plants can release radiation and heat into the environment. Hydroelectric, solar, wind, and geothermal plants can destroy natural habitats because of space considerations. Each of these consequences can affect the health and survival of humans and other species.

168.   **1**   An ecologist would want to know *what effect the smoke will have on the environment* before a new type of fuel is widely used. By understanding this effect, the ecologist can make more informed judgments about whether the smoke will harm the environment and human health.

WRONG CHOICES EXPLAINED:
   (2), (3), (4) An ecologist is less likely to want to know *how much it will cost to produce the fuel, how long it will take to produce the fuel,* and *if the fuel will be widely accepted by consumers*. Although these are important questions for the manufacturer, they do not provide critical information for the ecologist, whose main concern is the protection of environmental quality for humans and other organisms.

169.   **1**   Interference with insect reproductive processes is known as biological control. It is the most promising method of controlling insects because it is the least ecologically damaging.

WRONG CHOICES EXPLAINED:
   (2) The use of insecticides is a chemical control of insects. Insecticides kill both harmful and helpful insects. The chemicals accumulate in the bodies of birds, fish, and mammals and interfere with their normal life activities.

(3) It is impossible to set up physical barriers for insects because they are motile and are also carried from place to place by animals and humans.

(4) Draining marshes and other insect habitats has helped to control many insects such as mosquitoes. However, this method interferes with the life cycles of useful organisms living in the area.

170. **3** Insecticides are chemical pest controls. Biological controls are other insect species that feed on or in some way prey on an insect pest species. The example given here is control of the gypsy moth larvae by a certain species of beetle.

WRONG CHOICES EXPLAINED:

(1) DDT is an insecticide and represents chemical control. DDT is no longer used because it destroyed the insect food of birds and other wildlife.

(2) Humans upset the balance of nature (the balance of natural communities) by killing off deer predators. The deer population then increased so dramatically that deer starved to death because there was not enough food to support them.

(4) The use of drugs to cure disease is an example of chemical control of pathogens.

171. **1** Each individual should *acquire and apply knowledge of ecological principles* in order to ensure environmental quality for the future. By understanding how environmental principles operate, we can make more informed judgments about activities that may harm the environment and human health.

WRONG CHOICES EXPLAINED:

(2) If each individual were to *continue to take part in deforestation*, environmental quality would be degraded. Because forests are a natural part of the environment, eliminating them disturbs the balance of nature and can have significant negative consequences for environmental quality.

(3) If each individual were to *use Earth's finite resources*, environmental quality would be threatened. As these resources are used up, fewer remain for future generations. In addition, processing these resources consumes energy, produces pollutants, and adds to the solid waste problem.

(4) If each individual were to *add and take away organisms from ecosystems*, environmental quality would be diminished. Each natural community has established itself based on the particular niches filled by each type of organism. Adding to or taking away from this community upsets the balance of nature and would likely cause negative consequences.

172.   Write one or more paragraphs describing positive and negative effects of this method of pest control. Include the following points:

- Two advantages of this method of pest control [2]
  - ➤ Chemicals are not added to the environment.
  - ➤ Biological controls are more specific than chemical controls.
  - ➤ Ladybugs are less likely to kill beneficial organisms.
  - ➤ Desirable garden plants are protected from aphid attacks.
  - ➤ Birds and other unintended victims of pesticide use are spared.
  - ➤ Human health is protected against the toxic effects of pesticides.
- Two possible dangers of using this method of pest control [2]
  - ➤ The control insects may eat the food of other organisms.
  - ➤ The population of natural predators of the aphids may be eliminated or greatly reduced.
  - ➤ The control organism may become overpopulated.
  - ➤ The control organisms may themselves become pests.

Sample paragraph: *The method of pest control described is known as "biological control." This method of insect control has some distinct advantages over chemical controls: First, biological controls don't release toxic chemicals into the air and water, a fact that helps to protect wildlife and humans from being unintended victims of chemical pesticides. Second, biological controls are usually specific, which means that beneficial insects such as ladybugs and preying mantises aren't harmed. [2] There are also some things we should be careful of in the use of biological controls: First, we should know a lot about the control organism to be sure that it doesn't crowd out our native beneficial organisms. Second, we should remember that the control organism could become a pest, too, if it gets too numerous in the environment. [2]*

173. Write one or more paragraphs explaining why carnivores should be protected. Include the following points:

- Information concerning prey population growth [1]
  - ➢ If predators are destroyed, the prey population will increase.
  - ➢ If unchecked by predation or disease, a natural population will tend to increase in number geometrically.
- Information concerning extinction [1]
  - ➢ If too many carnivores of a particular species are killed, the species may become extinct.
  - ➢ Extinction is a definite possibility when any species has too few members alive to carry out effective breeding.
  - ➢ Complete elimination of any species from its natural range can destabilize the ecosystem.
- Information concerning the importance of carnivores in an environment [1]
  - ➢ By feeding on herbivores, carnivores help keep certain species of plants from being eliminated because of overgrazing in a particular area.
  - ➢ Without predators to limit its number, a prey population could exceed the capacity of its range, resulting in widespread starvation and death of the prey population.
  - ➢ Carnivorous animals are part of the natural scheme that promotes ecological equilibrium.

Sample paragraph: *Carnivores are important in an ecosystem because by reducing the number of prey organisms, the food organisms of the prey are kept from being eliminated from the environment. [1] If the predators were destroyed, the prey population would increase [1], perhaps to the point of consuming so many of the plants that the prey feed on that these plants would become extinct. [1]*

STANDARD/KEY IDEA	QUESTION NUMBERS	NUMBER OF CORRECT RESPONSES	NUMBER OF INCORRECT RESPONSES
1.1 Purpose of Scientific Inquiry	1–8		
1.2 Methods of Scientific Inquiry	9–28		
1.3 Analysis in Scientific Inquiry	29–47		
4.1 Application of Scientific Principles	48–75		
4.2 Genetic Continuity	76–90		
4.3 Organic Evolution	91–109		
4.4 Reproductive Continuity	110–129		
4.5 Dynamic Equilibrium and Homeostasis	130–140		
4.6 Interdependence of Living Things	141–155		
4.7 Human Impact on the Environment	156–173		

Glossary

PROMINENT SCIENTISTS

Crick, Francis A 20th-century British scientist who, with James Watson, developed the first workable model of DNA structure and function.

Darwin, Charles A 19th-century British naturalist whose theory of organic evolution by natural selection forms the basis for the modern scientific theory of evolution.

Fox, Sidney A 20th-century American scientist whose experiments showed that Stanley Miller's simple chemical precursors could be joined to form more complex biochemicals.

Hardy, G. H. A 20th-century British mathematician who, with W. Weinberg, developed the Hardy-Weinberg principle of gene frequencies.

Lamarck, Jean An 18th-century French scientist who devised an early theory of organic evolution based on the concept of "use and disuse."

Linnaeus, Carl An 18th-century Dutch scientist who developed the first scientific system of classification, based on similarity of structure.

Mendel, Gregor A 19th-century Austrian monk and teacher who was the first to describe many of the fundamental concepts of genetic inheritance through his work with garden peas.

Miller, Stanley A 20th-century American scientist whose experiments showed that the simple chemical precursors of life could be produced in the laboratory.

Morgan, Thomas Hunt A 20th-century American geneticist whose pioneering work with Drosophila led to the discovery of several genetic principles, including sex linkage.

Watson, James A 20th-century American scientist who, with Francis Crick, developed the first workable model of DNA structure and function.

Weinberg, W. A 20th-century German physician who, with G. H. Hardy, developed the Hardy-Weinberg principle of gene frequencies.

Weismann, August A 19th-century German biologist who tested Lamarck's theory of use and disuse and found it to be unsupportable by scientific methods.

BIOLOGICAL TERMS

abiotic factor Any of several nonliving, physical conditions that affect the survival of an organism in its environment.

absorption The process by which water and dissolved solids, liquids, and gases are taken in by the cell through the cell membrane.

accessory organ In human beings, any organ that has a digestive function but is not part of the food tube. (See **liver; gallbladder; pancreas.**)

acid A chemical that releases hydrogen ion (H+) in solution with water.

acid precipitation A phenomenon in which there is thought to be an interaction between atmospheric moisture and the oxides of sulfur and nitrogen that results in rainfall with low pH values.

active immunity The immunity that develops when the body's immune system is stimulated by a disease organism or a vaccination.

active site The specific area of an enzyme molecule that links to the substrate molecule and catalyzes its metabolism.

active transport A process by which materials are absorbed or released by cells against the concentration gradient (from low to high concentration) with the expenditure of cell energy.

adaptation Any structural, biochemical, or behavioral characteristic of an organism that helps it to survive potentially harsh environmental conditions.

addition A type of chromosome mutation in which a section of a chromosome is transferred to a homologous chromosome.

adenine A nitrogenous base found in DNA and RNA molecules.

adenosine triphosphate (ATP) An organic compound that stores respiratory energy in the form of chemical-bond energy for transport from one part of the cell to another.

adrenal cortex A portion of the adrenal gland that secretes steroid hormones which regulate various aspects of blood composition.

adrenal gland An endocrine gland that produces several hormones, including **adrenaline.** (See **adrenal cortex; adrenal medulla.**)

adrenal medulla A portion of the adrenal gland that secretes the hormone adrenaline, which regulates various aspects of the body's metabolic rate.

adrenaline A hormone of the adrenal medulla that regulates general metabolic rate, the rates of heartbeat and breathing, and the conversion of glycogen to glucose.

aerobic phase of respiration The reactions of aerobic respiration in which two pyruvic acid molecules are converted to six molecules of water and six molecules of carbon dioxide.

aerobic respiration A type of respiration in which energy is released from organic molecules with the aid of oxygen.

aging A stage of postnatal development that involves differentiation, maturation, and eventual deterioration of the body's tissues.

air pollution The addition, due to technological oversight, of some unwanted factor (e.g., chemical oxides, hydrocarbons, particulates) to our air resources.

albinism A condition, controlled by a single mutant gene, in which the skin lacks the ability to produce skin pigments.

alcoholic fermentation A type of anaerobic respiration in which glucose is converted to ethyl alcohol and carbon dioxide.

allantois A membrane that serves as a reservoir for wastes and as a respiratory surface for the embryos of many animal species.

allele One of a pair of genes that exist at the same location on a pair of homologous chromosomes and exert parallel control over the same genetic trait.

allergy A reaction of the body's immune system to the chemical composition of various substances.

alveolus One of many "air sacs" within the lung that function to absorb atmospheric gases and pass them on to the bloodstream.

amino acid An organic compound that is the component unit of proteins.

amino group A chemical group having the formula $-NH_2$ that is found as a part of all amino acid molecules.

ammonia A type of nitrogenous waste with high solubility and high toxicity.

amniocentesis A technique for the detection of genetic disorders in human beings in which a small amount of amniotic fluid is removed and the chromosome content of its cells analyzed. (See **karyotyping.**)

amnion A membrane that surrounds the embryo in many animal species and contains a fluid to protect the developing embryo from mechanical shock.

amniotic fluid The fluid within the amnion membrane that bathes the developing embryo.

amylase An enzyme specific for the hydrolysis of starch.

anaerobic phase of respiration The reactions of aerobic respiration in which glucose is converted to two pyruvic acid molecules.

anaerobic respiration A type of respiration in which energy is released from organic molecules without the aid of oxygen.

anal pore The egestive organ of the paramecium.

anemia A disorder of the human transport system in which the ability of the blood to carry oxygen is impaired, usually because of reduced numbers of red blood cells.

angina pectoris A disorder of the human transport system in which chest pain signals potential damage to the heart muscle due to narrowing of the opening of the coronary artery.

Animal One of the five biological kingdoms; it includes multicellular organisms whose cells are not bounded by cell walls and which are incapable of photosynthesis (e.g., human being).

Annelida A phylum of the Animal Kingdom whose members (annelids) include the segmented worms (e.g., earthworm).

antenna A receptor organ found in many arthropods (e.g., grasshopper), which is specialized for detecting chemical stimuli.

anther The portion of the stamen that produces pollen.

antibody A chemical substance, produced in response to the presence of a specific antigen, which neutralizes that antigen in the immune response.

antigen A chemical substance, usually a protein, that is recognized by the immune system as a foreign "invader" and is neutralized by a specific antibody.

anus The organ of egestion of the digestive tract.

aorta The principal artery carrying blood from the heart to the body tissues.

aortic arches A specialized part of the earthworm's transport system that serves as a pumping mechanism for the blood fluid.

apical meristem A plant growth region located at the tip of the root or tip of the stem.

appendicitis A disorder of the human digestive tract in which the appendix becomes inflamed as a result of bacterial infection.

aquatic biome An ecological biome composed of many different water environments.

artery A thick-walled blood vessel that carries blood away from the heart under pressure.

arthritis A disorder of the human locomotor system in which skeletal joints become inflamed, swollen, and painful.

Arthropoda A phylum of the Animal Kingdom whose members (arthropods) have bodies with chitinous exoskeletons and jointed appendages (e.g., grasshopper).

artificial selection A technique of plant/animal breeding in which individual organisms displaying desirable characteristics are chosen for breeding purposes.

asexual reproduction A type of reproduction in which new organisms are formed from a single parent organism.

asthma A disorder of the human respiratory system in which the respiratory tube becomes constricted by swelling brought on by some irritant.

atrium In human beings, one of the two thin-walled upper chambers of the heart that receive blood.

autonomic nervous system A subdivision of the peripheral nervous system consisting of nerves associated with automatic functions (e.g., heartbeat, breathing).

autosome One of several chromosomes present in the cell that carry genes controlling "body" traits not associated with primary and secondary sex characteristics.

autotroph An organism capable of carrying on autotrophic nutrition. Self feeder.

autotrophic nutrition A type of nutrition in which organisms manufacture their own organic foods from inorganic raw materials.

auxin A biochemical substance, plant hormone, produced by plants that regulates growth patterns.

axon An elongated portion of a neuron that conducts nerve impulses, usually away from the cell body of the neuron.

base A chemical that releases hydroxyl ion (OH^-) in solution with water.

bicarbonate ion The chemical formed in the blood plasma when carbon dioxide is absorbed from body tissues.

bile In human beings, a secretion of the liver that is stored in the gallbladder and that emulsifies fats.

binary fission A type of cell division in which mitosis is followed by equal cytoplasmic division.

binomial nomenclature A system of naming, used in biological classification, that consists of the genus and species names (e.g., *Homo sapiens*).

biocide use The use of pesticides that eliminate one undesirable organism but that have, due to technological oversight, unanticipated effects on beneficial species as well.

biological controls The use of natural enemies of various agricultural pests for pest control, thereby eliminating the need for biocide use—a positive aspect of human involvement with the environment.

biomass The total mass of living material present at the various trophic levels in a food chain.

biome A major geographical grouping of similar ecosystems, usually named for the climax flora in the region (e.g., Northeast Deciduous Forest).

biosphere The portion of the earth in which living things exist, including all land and water environments.

biotic factor Any of several conditions associated with life and living things that affect the survival of living things in the environment.

birth In placental mammals, a stage of embryonic development in which the baby passes through the vaginal canal to the outside of the mother's body.

blastula In certain animals, a stage of embryonic development in which the embryo resembles a hollow ball of undifferentiated cells.

blood The complex fluid tissue that functions to transport nutrients and respiratory gases to all parts of the body.

blood typing An application of the study of immunity in which the blood of a person is characterized by its antigen composition.

bone A tissue that provides mechanical support and protection for bodily organs, and levers for the body's locomotive activities.

Bowman's capsule A cup-shaped portion of the nephron responsible for the filtration of soluble blood components.

brain An organ of the central nervous system that is responsible for regulating conscious and much unconscious activity in the body.

breathing A mechanical process by which air is forced into the lungs by means of muscular contraction of the diaphragm and rib muscles.

bronchiole One of several subdivisions of the bronchi that penetrate the lung interior and terminate in alveoli.

bronchitis A disorder of the human respiratory system in which the bronchi become inflamed.

bronchus One of the two major subdivisions of the breathing tube; the bronchi are ringed with cartilage and conduct air from the trachea to the lung interior.

Bryophyta A phylum of the Plant Kingdom that consists of organisms lacking vascular tissues (e.g., moss).

budding A type of asexual reproduction in which mitosis is followed by unequal cytoplasmic division.

bulb A type of vegetative propagation in which a plant bulb produces new bulbs that may be established as independent organisms with identical characteristics.

cambium The lateral meristem tissue in woody plants responsible for annual growth in stem diameter.

cancer Any of a number of conditions characterized by rapid, abnormal, and uncontrolled division of affected cells.

capillary A very small, thin-walled blood vessel that connects an artery to a vein and through which all absorption into the blood fluid occurs.

carbohydrate An organic compound composed of carbon, hydrogen, and oxygen in a 1:2:1 ratio (e.g., $C_6H_{12}O_6$).

carbon-14 A radioactive isotope of carbon used to trace the movement of carbon in various biochemical reactions, and also used in the "carbon dating" of fossils.

carbon-fixation reactions A set of biochemical reactions in photosynthesis in which hydrogen atoms are combined with carbon and oxygen atoms to form PGAL and glucose.

carbon-hydrogen-oxygen cycle A process by which these three elements are made available for use by other organisms through the chemical reactions of respiration and photosynthesis.

carboxyl group A chemical group having the formula—COOH and found as part of all amino acid and fatty acid molecules.

cardiac muscle A type of muscle tissue in the heart and arteries that is associated with the rhythmic nature of the pulse and heartbeat.

cardiovascular disease In human beings, any disease of the circulatory organs.

carnivore A heterotrophic organism that consumes animal tissue as its primary source of nutrition. (See **secondary consumer.**)

carrier An individual who, though not expressing a particular recessive trait, carries this gene as part of his/her heterozygous genotype.

carrier protein A specialized molecule embedded in the cell membrane that aids the movement of materials across the membrane.

cartilage A flexible connective tissue found in many flexible parts of the body (e.g., knee); common in the embryonic stages of development.

catalyst Any substance that speeds up or slows down the rate of a chemical reaction. (See **enzyme.**)

cell plate A structure that forms during cytoplasmic division in plant cells and serves to separate the cytoplasm into two roughly equal parts.

cell theory A scientific theory that states, "All cells arise from previously existing cells" and "Cells are the unit of structure and function of living things."

cell wall A cell organelle that surrounds and gives structural support to plant cells; cell walls are composed of cellulose.

central nervous system The portion of the vertebrate nervous system that consists of the brain and the spinal cord.

centriole A cell organelle found in animal cells that functions in the process of cell division.

centromere The area of attachment of two chromatids in a double-stranded chromosome.

cerebellum The portion of the human brain responsible for the coordination of muscular activity.

cerebral hemorrhage A disorder of the human regulatory system in which a broken blood vessel in the brain may result in severe dysfunction or death.

cerebral palsy A disorder of the human regulatory system in which the motor and speech centers of the brain are impaired.

cerebrum The portion of the human brain responsible for thought, reasoning, sense interpretation, learning, and other conscious activities.

cervix A structure that bounds the lower end of the uterus and through which sperm must pass in order to fertilize the egg.

chemical digestion The process by which nutrient molecules are converted by chemical means into a form usable by the cells.

chemosynthesis A type of autotrophic nutrition in which certain bacteria use the energy of chemical oxidation to convert inorganic raw materials to organic food molecules.

chitin A polysaccharide substance that forms the exoskeleton of the grasshopper and other arthropods.

chlorophyll A green pigment in plant cells that absorbs sunlight and makes possible certain aspects of the photosynthetic process.

chloroplast A cell organelle found in plant cells that contains chlorophyll and functions in photosynthesis.

Chordata A phylum of the Animal Kingdom whose members (chordates) have internal skeletons made of cartilage and/or bone (e.g., human being).

chorion A membrane that surrounds all other embryonic membranes in many animal species, protecting them from mechanical damage.

chromatid One strand of a double-stranded chromosome.

chromosome mutation An alteration in the structure of a chromosome involving many genes. (See **nondisjunction; translocation; addition; deletion.**)

cilia Small, hairlike structures in paramecia and other unicellular organisms that aid in nutrition and locomotion.

classification A technique by which scientists sort, group, and name organisms for easier study.

cleavage A series of rapid mitotic divisions that increase cell number in a developing embryo without corresponding increase in cell size.

climax community A stable, self-perpetuating community that results from an ecological succession.

cloning A technique of genetic investigation in which undifferentiated cells of an organism are used to produce new organisms with the same set of traits as the original cells.

closed transport system A type of circulatory system in which the transport fluid is always enclosed within blood vessels (e.g., earthworm, human).

clot A structure that forms as a result of enzyme-controlled reactions following the rupturing of a blood vessel and serves as a plug to prevent blood loss.

codominance A type of intermediate inheritance that results from the simultaneous expression of two dominant alleles with contrasting effects.

codon See **triplet codon.**

Coelenterata A phylum of the Animal Kingdom whose members (coelenterates) have bodies that resemble a sack (e.g., hydra, jellyfish).

coenzyme A chemical substance or chemical subunit that functions to aid the action of a particular enzyme. (See **vitamin.**)

cohesion A force binding water molecules together that aids in the upward conduction of materials in the xylem.

commensalism A type of symbiosis in which one organism in the relationship benefits and the other is neither helped nor harmed.

common ancestry A concept central to the science of evolution which postulates that all organisms share a common ancestry whose closeness varies with the degree of shared similarity.

community A level of biological organization that includes all of the species populations inhabiting a particular geographic area.

comparative anatomy The study of similarities in the anatomical structures of organisms, and their use as an indicator of common ancestry and as evidence of organic evolution.

comparative biochemistry The study of similarities in the biochemical makeups of organisms, and their use as an indicator of common ancestry and as evidence of organic evolution.

comparative cytology The study of similarities in the cell structures of organisms, and their use as an indicator of common ancestry and as evidence of organic evolution.

comparative embryology The study of similarities in the patterns of embryological development of organisms, and their use as an indicator of common ancestry and as evidence of organic evolution.

competition A condition that arises when different species in the same habitat attempt to use the same limited resources.

complete protein A protein that contains all eight essential amino acids.

compound A substance composed of two or more different kinds of atom (e.g., water: H_2O).

compound light microscope A tool of biological study capable of producing a magnified image of a biological specimen by using a focused beam of light.

conditioned behavior A type of response that is learned, but that becomes automatic with repetition.

conservation of resources The development and application of practices to protect valuable and irreplaceable soil and mineral resources—a positive aspect of human involvement with the environment.

constipation A disorder of the human digestive tract in which fecal matter solidifies and becomes difficult to egest.

consumer Any heterotrophic animal organism (e.g., human being).

coronary artery An artery that branches off the aorta to feed the heart muscle.

coronary thrombosis A disorder of the human transport system in which the heart muscle becomes damaged as a result of blockage of the coronary artery.

corpus luteum A structure resulting from the hormone-controlled transformation of the ovarian follicle that produces the hormone progesterone.

corpus luteum stage A stage of the menstrual cycle in which the cells of the follicle are transformed into the corpus luteum under the influence of the hormone LH.

cotyledon A portion of the plant embryo that serves as a source of nutrition for the young plant before photosynthesis begins.

cover-cropping A proper agricultural practice in which a temporary planting (cover crop) is used to limit soil erosion between seasonal plantings of main crops.

crop A portion of the digestive tract of certain animals that stores food temporarily before digestion.

cross-pollination A type of pollination in which pollen from one flower pollinates flowers of a different plant of the same species.

crossing-over A pattern of inheritance in which linked genes may be separated during synapsis in the first meiotic division, when sections of homologous chromosomes may be exchanged.

cuticle A waxy coating that covers the upper epidermis of most leaves and acts to help the leaf retain water.

cutting A technique of plant propagation in which vegetative parts of the parent plant are cut and rooted to establish new plant organisms with identical characteristics.

cyclosis The circulation of the cell fluid (cytoplasm) within the cell interior.

cyton The "cell body" of the neuron, which generates the nerve impulse.

cytoplasm The watery fluid that provides a medium for the suspension of organelles within the cell.

cytoplasmic division The separation of daughter nuclei into two new daughter cells.

cytosine A nitrogenous base found in both DNA and RNA molecules.

daughter cell A cell that results from mitotic cell division.

daughter nucleus One of two nuclei that form as a result of mitosis.

deamination A process by which amino acids are broken down into their component parts for conversion into urea.

death The irreversible cessation of bodily functions and cellular activities.

deciduous A term relating to broadleaf trees which shed their leaves in the fall.

decomposer Any saprophytic organism that derives its energy from the decay of plant and animal tissues (e.g., bacteria of decay, fungus); the final stage of a food chain.

decomposition bacteria In the nitrogen cycle, bacteria that break down plant and animal protein and produce ammonia as a by-product.

dehydration synthesis A chemical process in which two organic molecules may be joined after removing the atoms needed to form a molecule of water as a by-product.

deletion A type of chromosome mutation in which a section of a chromosome is separated and lost.

dendrite A cytoplasmic extension of a neuron that serves to detect an environmental stimulus and carry an impulse to the cell body of the neuron.

denitrifying bacteria In the nitrogen cycle, bacteria that convert excess nitrate salts into gaseous nitrogen.

deoxygenated blood Blood that has released its transported oxygen to the body tissues.

deoxyribonucleic acid (DNA) A nucleic acid molecule known to be the chemically active agent of the gene; the fundamental hereditary material of living organisms.

deoxyribose A five-carbon sugar that is a component part of the nucleotide unit in DNA only.

desert A terrestrial biome characterized by sparse rainfall, extreme temperature variation, and a climax flora that includes cactus.

diabetes A disorder of the human regulatory system in which insufficient insulin production leads to elevated blood sugar concentrations.

diarrhea A disorder of the human digestive tract in which the large intestine fails to absorb water from the waste matter, resulting in watery feces.

diastole The lower pressure registered during blood pressure testing. (See **systole**.)

differentiation The process by which embryonic cells become specialized to perform the various tasks of particular tissues throughout the body.

diffusion A form of passive transport by which soluble substances are absorbed or released by cells.

digestion The process by which complex foods are broken down by mechanical or chemical means for use by the body.

dipeptide A chemical unit composed of two amino acid units linked by a peptide bond.

diploid chromosome number The number of chromosomes found characteristically in the cells (except gametes) of sexually reproducing species.

disaccharidase Any disaccharide-hydrolyzing enzyme.

disaccharide A type of carbohydrate known also as a "double sugar"; all disaccharides have the molecular formula $C_{12}H_{22}O_{11}$.

disjunction The separation of homologous chromosome pairs at the end of the first meiotic division.

disposal problems Problems, due to technological oversight, that result when commercial and technological activities produce solid and/or chemical wastes that must be disposed of.

dissecting microscope A tool of biological study that magnifies the image of a biological specimen up to 20 times normal size for purposes of gross dissection.

dominance A pattern of genetic inheritance in which the effects of a dominant allele mask those of a recessive allele.

dominant allele (gene) An allele (gene) whose effect masks that of its recessive allele.

double-stranded chromosome The two-stranded structure that results from chromosomal replication.

Down's syndrome In human beings, a condition, characterized by mental and physical retardation, that may be caused by the nondisjunction of chromosome number 21.

Drosophila The common fruit fly, an organism that has served as an object of genetic research in the development of the gene-chromosome theory.

ductless gland See **endocrine gland.**

ecology The science that studies the interactions of living things with each other and with the nonliving environment.

ecosystem The basic unit of study in ecology, including the plant and animal community in interaction with the nonliving environment.

ectoderm An embryonic tissue that differentiates into skin and nerve tissue in the adult animal.

effector An organ specialized to produce a response to an environmental stimulus: effectors may be muscles or glands.

egestion The process by which undigested food materials are eliminated from the body.

electron microscope A tool of biological study that uses a focused beam of electrons to produce an image of a biological specimen magnified up to 25,000 times normal size.

element The simplest form of matter; an element is a substance (e.g., nitrogen) made up of a single type of atom.

embryo An organism in the early stages of development following fertilization.

embryonic development A series of complex processes by which animal and plant embryos develop into adult organisms.

emphysema A disorder of the human respiratory system in which lung tissue deteriorates, leaving the lung with diminished capacity and efficiency.

emulsification A process by which fat globules are surrounded by bile to form fat droplets.

endocrine ("ductless") gland A gland (e.g., thyroid, pituitary) specialized for the production of hormones and their secretion directly into the bloodstream; such glands lack ducts.

endoderm An embryonic tissue that differentiates into the digestive and respiratory tract lining in the adult animal.

endoplasmic reticulum (ER) A cell organelle known to function in the transport of cell products from place to place within the cell.

environmental laws Federal, state, and local legislation enacted in an attempt to protect environmental resources—a positive aspect of human involvement with the environment.

enzymatic hydrolysis An enzyme-controlled reaction by which complex food molecules are broken down chemically into simpler subunits.

enzyme An organic catalyst that controls the rate of metabolism of a single type of substrate; enzymes are protein in nature.

enzyme-substrate complex A physical association between an enzyme molecule and its substrate within which the substrate is metabolized.

epicotyl A portion of the plant embryo that specializes to become the upper stem, leaves, and flowers of the adult plant.

epidermis The outermost cell layer in a plant or animal.

epiglottis In a human being, a flap of tissue that covers the upper end of the trachea during swallowing and prevents inhalation of food.

esophagus A structure in the upper portion of the digestive tract that conducts the food from the pharynx to the midgut.

essential amino acid An amino acid that cannot be synthesized by the human body, but must be obtained by means of the diet.

estrogen A hormone, secreted by the ovary, that regulates the production of female secondary sex characteristics.

evolution Any process of gradual change through time.

excretion The life function by which living things eliminate metabolic wastes from their cells.

exoskeleton A chitinous material that covers the outside of the bodies of most arthropods and provides protection for internal organs and anchorage for muscles.

exploitation of organisms Systematic removal of animals and plants with commercial value from their environments, for sale—a negative aspect of human involvement with the environment.

extensor A skeletal muscle that extends (opens) a joint.

external development Embryonic development that occurs outside the body of the female parent (e.g., birds).

external fertilization Fertilization that occurs outside the body of the female parent (e.g., fish).

extracellular digestion Digestion that occurs outside the cell.

fallopian tube See **oviduct**.

fatty acid An organic molecule that is a component of certain lipids.

fauna The animal species comprising an ecological community.

feces The semisolid material that results from the solidification of undigested foods in the large intestine.

fertilization The fusion of gametic nuclei in the process of sexual reproduction.

filament The portion of the stamen that supports the anther.

flagella Microscopic, whiplike structures found on certain cells that aid in locomotion and circulation.

flexor A skeletal muscle that flexes (closes) a joint.

flora The plant species comprising an ecological community.

flower The portion of a flowering plant that is specialized for sexual reproduction.

fluid-mosaic model A model of the structure of the cell membrane in which large protein molecules are thought to be embedded in a bilipid layer.

follicle One of many areas within the ovary that serve as sites for the periodic maturation of ova.

follicle stage The stage of the menstrual cycle in which an ovum reaches its final maturity under the influence of the hormone FSH.

follicle-stimulating hormone (FSH) A pituitary hormone that regulates the maturation of, and the secretion of estrogen by, the ovarian follicle.

food chain A series of nutritional relationships in which food energy is passed from producer to herbivore to carnivore to decomposer; a segment of a food web.

food web A construct showing a series of interrelated food chains and illustrating the complex nutritional interrelationships that exist in an ecosystem.

fossil The preserved direct or indirect remains of an organism that lived in the past, as found in the geologic record.

fraternal twins In human beings, twin offspring that result from the simultaneous fertilization of two ova by two sperm; such twins are not genetically identical.

freshwater biome An aquatic biome made up of many separate freshwater systems that vary in size and stability and may be closely associated with terrestrial biomes.

fruit Any plant structure that contains seeds; a mechanism of seed dispersal.

Fungi One of the five biological kingdoms; it includes organisms unable to manufacture their own organic foods (e.g., mushroom).

gallbladder An accessory organ that stores bile.

gallstones A disorder of the human digestive tract in which deposits of hardened cholesterol lodge in the gallbladder.

gamete A specialized reproductive cell produced by organisms of sexually reproducing species. (See **sperm; ovum; pollen; ovule.**)

gametogenesis The process of cell division by which gametes are produced. (See **meiosis; spermatogenesis; oogenesis.**)

ganglion An area of bunched nerve cells that acts as a switching point for nerve impulses traveling from receptors and to effectors.

garden pea The research organism used by Mendel in his early scientific work in genetic inheritance.

gastric cecum A gland in the grasshopper that secretes digestive enzymes.

gastrula A stage of embryonic development in animals in which the embryo assumes a tube-within-a-tube structure and distinct embryonic tissues (ectoderm, mesoderm, endoderm) begin to differentiate.

gastrulation The process by which a blastula becomes progressively more indented, forming a gastrula.

gene A unit of heredity; a discrete portion of a chromosome thought to be responsible for the production of a single type of polypeptide; the "factor" responsible for the inheritance of a genetic trait.

gene frequency The proportion (percentage) of each allele for a particular trait that is present in the gene pool of a population.

gene linkage A pattern of inheritance in which genes located along the same chromosome are prevented from assorting independently, but are linked together in their inheritance.

gene mutation An alteration of the chemical nature of a gene that changes its ability to control the production of a polypeptide chain.

gene pool The sum total of all the inheritable genes for the traits in a given sexually reproducing population.

gene-chromosome theory A theory of genetic inheritance that is based on current understanding of the relationships between the biochemical control of traits and the process of cell division.

genetic counseling Clinical discussions concerning inheritance patterns that are designed to inform prospective parents of the potential for expression of a genetic disorder in their offspring.

genetic engineering The use of various techniques to move genes from one organism to another.

genetic screening A technique for the detection of human genetic disorders in which bodily fluids are analyzed for the presence of certain marker chemicals.

genotype The particular combination of genes in an allele pair.

genus A level of biological classification that represents a subdivision of the phylum level; having fewer organisms with great similarity (e.g., *Drosophila*, paramecium).

geographic isolation The separation of species populations by geographical barriers, facilitating the evolutionary process.

geologic record A supporting item of evidence of organic evolution, supplied within the earth's rock and other geological deposits.

germination The growth of the pollen tube from a pollen grain; the growth of the embryonic root and stem from a seed.

gestation The period of prenatal development of a placental mammal; human gestation requires approximately 9 months.

gizzard A portion of the digestive tract of certain organisms, including the earthworm and the grasshopper, in which food is ground into smaller fragments.

glomerulus A capillary network lying within Bowman's capsule of the nephron.

glucagon A hormone, secreted by the islets of Langerhans, that regulates the release of blood sugar from stored glycogen.

glucose A monosaccharide produced commonly in photosynthesis and used by both plants and animals as a "fuel" in the process of respiration.

glycerol An organic compound that is a component of certain lipids.

glycogen A polysaccharide synthesized in animals as a means of storing glucose; glycogen is stored in the liver and in the muscles.

goiter A disorder of the human regulatory system in which the thyroid gland enlarges because of a deficiency of dietary iodine.

Golgi complex Cell organelles that package cell products and move them to the plasma membrane for secretion.

gonad An endocrine gland that produces the hormones responsible for the production of various secondary sex characteristics. (See **ovary; testis.**)

gout A disorder of the human excretory system in which uric acid accumulates in the joints, causing severe pain.

gradualism A theory of the time frame required for organic evolution which assumes that evolutionary change is slow, gradual, and continuous.

grafting A technique of plant propagation in which the stems of desirable plants are attached (grafted) to rootstocks of related varieties to produce new plants for commercial purposes.

grana The portion of the chloroplast within which chlorophyll molecules are concentrated.

grassland A terrestrial biome characterized by wide variation in temperature and a climax flora that includes grasses.

growth A process by which cells increase in number and size, resulting in an increase in size of the organism.

growth-stimulating hormone (GSH) A pituitary hormone regulating the elongation of the long bones of the body.

guanine A nitrogenous base found in both DNA and RNA molecules.

guard cell One of a pair of cells that surround the leaf stomate and regulate its size.

habitat The environment or set of ecological conditions within which an organism lives.

Hardy-Weinberg principle A hypothesis, advanced by G. H. Hardy and W. Weinberg, which states that the gene pool of a population should remain stable as long as a set of "ideal" conditions is met.

heart In human beings, a four-chambered muscular pump that facilitates the movement of blood throughout the body.

helix Literally a spiral; a term used to describe the "twisted ladder" shape of the DNA molecule.

hemoglobin A type of protein specialized for the transport of respiratory oxygen in certain organisms, including earthworms and human beings.

herbivore A heterotrophic organism that consumes plant matter as its primary source of nutrition. (See **primary consumer.**)

hermaphrodite An animal organism that produces both male and female gametes.

heterotroph An organism that typically carries on heterotrophic nutrition.

heterotroph hypothesis A scientific hypothesis devised to explain the probable origin and early evolution of life on earth.

heterotrophic nutrition A type of nutrition in which organisms must obtain their foods from outside sources of organic nutrients.

heterozygous A term used to refer to an allele pair in which the alleles have different contrasting effects (e.g., *Aa, RW*).

high blood pressure A disorder of the human transport system in which systolic and diastolic pressures register higher than normal because of narrowing of the artery opening.

histamine A chemical product of the body that causes irritation and swelling of the mucous membranes.

homeostasis The condition of balance and dynamic stability that characterizes living systems under normal conditions.

homologous chromosomes A pair of chromosomes that carry corresponding genes for the same traits.

homologous structures Structures present within different species that can be shown to have had a common origin, but that may or may not share a common function.

homozygous A term used to refer to an allele pair in which the alleles are identical in terms of effect (e.g., *AA, aa*).

hormone A chemical product of an endocrine gland which has a regulatory effect on the cell's metabolism.

host The organism that is harmed in a parasitic relationship.

hybrid A term used to describe a heterozygous genotype. (See **heterozygous.**)

hybridization A technique of plant/animal breeding in which two varieties of the same species are crossbred in the hope of producing offspring with the favorable traits of both varieties.

hydrogen bond A weak electrostatic bond that holds together the twisted strands of DNA and RNA molecules.

hydrolysis The chemical process by which a complex food molecule is split into simpler components through the addition of a molecule of water to the bonds holding it together.

hypocotyl A portion of the plant embryo that specializes to become the root and lower stem of the adult plant.

hypothalamus An endocrine gland whose secretions affect the pituitary gland.

identical twins In human beings, twin offspring resulting from the separation of the embryonic cell mass of a single fertilization into two separate masses; such twins are genetically identical.

importation of organisms The introduction of nonactive plants and animals into new areas where they compete strongly with native species—a negative aspect of human involvement with the environment.

in vitro fertilization A laboratory technique in which fertilization is accomplished outside the mother's body using mature ova and sperm extracted from the parents' bodies.

inbreeding A technique of plant/animal breeding in which a "purebred" variety is bred only with its own members, so as to maintain a set of desired characteristics.

independent assortment A pattern of inheritance in which genes on different, nonhomologous chromosomes are free to be inherited randomly and regardless of the inheritance of the others.

ingestion The mechanism by which an organism takes in food from its environment.

inorganic compound A chemical compound that lacks the element carbon or hydrogen (e.g., table salt: NaCl).

insulin A hormone, secreted by the islets of Langerhans, that regulates the storage of blood sugar as glycogen.

intercellular fluid (ICF) The fluid that bathes cells and fills intercellular spaces.

interferon A substance, important in the fight against human cancer, that may now be produced in large quantities through techniques of genetic engineering.

intermediate inheritance Any pattern of inheritance in which the offspring expresses a phenotype different from the phenotypes of its parents and usually representing a form intermediate between them.

internal development Embryonic development that occurs within the body of the female parent.

internal fertilization Fertilization that occurs inside the body of the female parent.

interneuron A type of neuron, located in the central nervous system, that is responsible for the interpretation of impulses received from sensory neurons.

intestine A portion of the digestive tract in which chemical digestion and absorption of digestive end-products occur.

intracellular digestion A type of chemical digestion carried out within the cell.

iodine A chemical stain used in cell study; an indicator used to detect the presence of starch. (See **staining.**)

islets of Langerhans An endocrine gland, located within the pancreas, that produces the hormones insulin and glucagon.

karyotype An enlarged photograph of the paired homologous chromosomes of an individual cell that is used in the detection of certain genetic disorders involving chromosome mutation.

karyotyping A technique for the detection of human genetic disorders in which a karyotype is analyzed for abnormalities in chromosome structure or number.

kidney The excretory organ responsible for maintaining the chemical composition of the blood. (See **nephron.**)

kidney failure A disorder of the human excretory system in which there is a general breakdown of the kidney's ability to filter blood components.

kingdom A level of biological classification that includes a broad grouping of organisms displaying general structural similarity; five kingdoms have been named by scientists.

lacteal A small extension of the lymphatic system, found inside the villus, that absorbs fatty acids and glycerol resulting from lipid hydrolysis.

lactic acid fermentation A type of anaerobic respiration in which glucose is converted to two lactic acid molecules.

large intestine A portion of the digestive tract in which undigested foods are solidified by means of water absorption to form feces.

lateral meristem A plant growth region located under the epidermis or bark of a stem. (See **cambium.**)

Latin The language used in biological classification for naming organisms by means of binomial nomenclature.

lenticel A small pore in the stem surface that permits the absorption and release of respiratory gases within stem tissues.

leukemia A disorder of the human transport system in which the bone marrow produces large numbers of abnormal white blood cells. (See **cancer.**)

lichen A symbiosis of alga and fungus that frequently acts as a pioneer species on bare rock.

limiting factor Any abiotic or biotic condition that places limits on the survival of organisms and on the growth of species populations in the environment.

lipase Any lipid-hydrolyzing enzyme.

lipid An organic compound composed of carbon, hydrogen, and oxygen in which hydrogen and oxygen are *not* in a 2:1 ratio (e.g., a wax, plant oil); many lipids are constructed of a glycerol and three fatty acids.

liver An accessory organ that stores glycogen, produces bile, destroys old red blood cells, deaminates amino acids, and produces urea.

lock-and-key model A theoretical model of enzyme action that attempts to explain the concept of enzyme specificity.

lung The major organ of respiratory gas exchange.

luteinizing hormone (LH) A pituitary hormone that regulates the conversion of the ovarian follicle into the corpus luteum.

lymph Intercellular fluid (ICF) that has passed into the lymph vessels.

lymph node One of a series of structures in the body that act as reservoirs of lymph and also contain white blood cells as part of the body's immune system.

lymph vessel One of a branching series of tubes that collect ICF from the tissues and redistribute it as lymph.

lymphatic circulation The movement of lymph throughout the body.

lymphocyte A type of white blood cell that produces antibodies.

lysosome A cell organelle that houses hydrolytic enzymes used by the cell in the process of chemical digestion.

Malpighian tubules In arthropods (e.g., grasshopper), an organ specialized for the removal of metabolic wastes.

maltase A specific enzyme that catalyzes the hydrolysis (and dehydration synthesis) of maltose.

maltose A type of disaccharide; a maltose molecule is composed of two units of glucose joined together by dehydration synthesis.

marine biome An aquatic biome characterized by relatively stable conditions of moisture, salinity, and temperature.

marsupial mammal See **nonplacental mammal.**

mechanical digestion Any of the processes by which foods are broken apart physically into smaller particles.

medulla The portion of the human brain responsible for regulating the automatic processes of the body.

meiosis The process by which four monoploid nuclei are formed from a single diploid nucleus.

meningitis A disorder of the human regulatory system in which the membranes of the brain or spinal cord become inflamed.

menstrual cycle A hormone-controlled process responsible for the monthly release of mature ova.

menstruation The stage of the menstrual cycle in which the lining of the uterus breaks down and is expelled from the body via the vaginal canal.

meristem A plant tissue specialized for embryonic development. (See **apical meristem; lateral meristem; cambium.**)

mesoderm An embryonic tissue that differentiates into muscle, bone, the excretory system, and most of the reproductive system in the adult animal.

messenger RNA (m-RNA) A type of RNA that carries the genetic code from the nuclear DNA to the ribosome for transcription.

metabolism All of the chemical processes of life considered together; the sum total of all the cell's chemical activity.

methylene blue A chemical stain used in cell study. (See **staining.**)

microdissection instruments Tools of biological study that are used to remove certain cell organelles from within cells for examination.

micrometer (μm) A unit of linear measurement equal in length to 0.001 millimeter (0.000001 meter), used for expressing the dimensions of cells and cell organelles.

mitochondrion A cell organelle that contains the enzymes necessary for aerobic respiration.

mitosis A precise duplication of the contents of a parent cell nucleus, followed by an orderly separation of these contents into two new, identical daughter nuclei.

mitotic cell division A type of cell division that results in the production of two daughter cells identical to each other and to the parent cell.

Monera One of the five biological kingdoms; it includes simple unicellular forms lacking nuclear membranes (e.g., bacteria).

monohybrid cross A genetic cross between two organisms both heterozygous for a trait controlled by a single allele pair. The phenotypic ratio resulting is 3:1; the genotypic ratio is 1:2:1.

monoploid chromosome number The number of chromosomes commonly found in the gametes of sexually reproducing species.

monosaccharide A type of carbohydrate known also as a "simple sugar"; all monosaccharides have the molecular formula $C_6H_{22}O_6$.

motor neuron A type of neuron that carries "command" impulses from the central nervous system to an effector organ.

mucus A protein-rich mixture that bathes and moistens the respiratory surfaces.

multicellular Having a body that consists of large groupings of specialized cells (e.g., human being).

multiple alleles A pattern of inheritance in which the existence of more than two alleles is hypothesized, only two of which are present in the genotype of any one individual.

muscle A type of tissue specialized to produce movement of body parts.

mutagenic agent Any environmental condition that initiates or accelerates genetic mutation.

mutation Any alteration of the genetic material, either a chromosome or a gene, in an organism.

mutualism A type of symbiosis beneficial to both organisms in the relationship.

nasal cavity A series of channels through which outside air is admitted to the body interior and is warmed and moistened before entering the lung.

natural selection A concept, central to Darwin's theory of evolution, to the effect that the individuals best adapted to their environment tend to survive and to pass their favorable traits on to the next generation.

negative feedback A type of endocrine regulation in which the effects of one gland may inhibit its own secretory activity, while stimulating the secretory activity of another gland.

nephridium An organ found in certain organisms, including the earthworm, specialized for the removal of metabolic wastes.

nephron The functional unit of the kidney. (See **glomerulus; Bowman's capsule.**)

nerve A structure formed from the bundling of neurons carrying sensory or motor impulses.

nerve impulse An electrochemical change in the surface of the nerve cell.

nerve net A network of "nerve" cells in coelenterates such as the hydra.

neuron A cell specialized for the transmission of nerve impulses.

neurotransmitter A chemical substance secreted by a neuron that aids in the transmission of the nerve impulse to an adjacent neuron.

niche The role that an organism plays in its environment.

nitrifying bacteria In the nitrogen cycle, bacteria that absorb ammonia and convert it into nitrate salts.

nitrogen cycle The process by which nitrogen is recycled and made available for use by other organisms.

nitrogen-fixing bacteria A type of bacteria responsible for absorbing atmospheric nitrogen and converting it to nitrate salts in the soil.

nitrogenous base A chemical unit composed of carbon, hydrogen, and nitrogen that is a component part of the nucleotide unit.

nitrogenous waste Any of a number of nitrogen-rich compounds that result from the metabolism of proteins and amino acids in the cell. (See **ammonia; urea; uric acid.**)

nondisjunction A type of chromosome mutation in which the members of one or more pairs of homologous chromosomes fail to separate during the disjunction phase of the first meiotic division.

nonplacental mammal A species of mammal in which internal development is accomplished without the aid of a placental connection (marsupial mammals).

nucleic acid An organic compound composed of repeating units of nucleotide.

nucleolus A cell organelle located within the nucleus that is known to function in protein synthesis.

nucleotide The repeating unit making up the nucleic acid polymer (e.g., DNA, RNA).

nucleus A cell organelle that contains the cell's genetic information in the form of chromosomes.

nutrition The life function by which living things obtain food and process it for their use.

omnivore A heterotrophic organism that consumes both plant and animal matter as sources of nutrition.

one gene-one polypeptide A scientific hypothesis concerning the role of the individual gene in protein synthesis.

oogenesis A type of meiotic cell division in which one ovum and three polar bodies are produced from each primary sex cell.

open transport system A type of circulatory system in which the transport fluid is *not* always enclosed within blood vessels (e.g., grasshopper).

oral cavity In human beings, the organ used for the ingestion of foods.

oral groove The ingestive organ of the paramecium.

organ transplant An application of the study of immunity in which an organ or tissue of a donor is transplanted into a compatible recipient.

organelle A small, functional part of a cell specialized to perform a specific life function (e.g., nucleus, mitochondrion).

organic compound A chemical compound that contains the elements carbon and hydrogen (e.g., carbohydrate, protein).

organic evolution The mechanism thought to govern the changes in living species over geologic time.

osmosis A form of passive transport by which water is absorbed or released by cells.

ovary A female gonad that secretes the hormone estrogen, which regulates female secondary sex characteristics; the ovary also produces ova, which are used in reproduction.

overcropping A negative aspect of human involvement with the environment in which soil is overused for the production of crops, leading to exhaustion of soil nutrients.

overgrazing The exposure of soil to erosion due to the loss of stabilizing grasses when it is overused by domestic animals—a negative aspect of human involvement with the environment.

overhunting A negative aspect of human involvement with the environment in which certain species have been greatly reduced or made extinct by uncontrolled hunting practices.

oviduct A tube that serves as a channel for conducting mature ova from the ovary to the uterus; the site of fertilization and the earliest stages of embryonic development.

ovulation The stage of the menstrual cycle in which the mature ovum is released from the follicle into the oviduct.

ovule A structure located within the flower ovary that contains a monoploid egg nucleus and serves as the site of fertilization.

ovum A type of gamete produced as a result of oogenesis in female animals; the egg, the female sex cell.

oxygen-18 A radioactive isotope of oxygen that is used to trace the movement of this element in biochemical reaction sequences.

oxygenated blood Blood that contains a high percentage of oxyhemoglobin.

oxyhemoglobin Hemoglobin that is loosely bound to oxygen for purposes of oxygen transport.

palisade layer A cell layer found in most leaves that contains high concentrations of chloroplasts.

pancreas An accessory organ which produces enzymes that complete the hydrolysis of foods to soluble end-products; also the site of insulin and glucagon production.

parasitism A type of symbiosis from which one organism in the relationship benefits, while the other (the "host") is harmed, but not ordinarily killed.

parathormone A hormone of the parathyroid gland that regulates the metabolism of calcium in the body.

parathyroid gland An endocrine gland whose secretion, parathormone, regulates the metabolism of calcium in the body.

passive immunity A temporary immunity produced as a result of the injection of preformed antibodies.

passive transport Any process by which materials are absorbed into the cell interior from an area of high concentration to an area of low concentration, without the expenditure of cell energy (e.g., osmosis, diffusion).

penis A structure that permits internal fertilization through direct implantation of sperm into the female reproductive tract.

peptide bond A type of chemical bond that links the nitrogen atom of one amino acid with the terminal carbon atom of a second amino acid in the formation of a dipeptide.

peripheral nerves Nerves in the earthworm and grasshopper that branch from the ventral nerve cord to other parts of the body.

peripheral nervous system A major subdivision of the nervous system that consists of all the nerves of all types branching through the body. (See **autonomic nervous system; somatic nervous system.**)

peristalsis A wave of contraction of the smooth muscle lining; the digestive tract that causes ingested food to pass along the food tube.

petal An accessory part of the flower that is thought to attract pollinating insects.

pH A chemical unit used to express the concentration of hydrogen ion (H^+), or the acidity, of a solution.

phagocyte A type of white blood cell that engulfs and destroys bacteria.

phagocytosis The process by which the ameba surrounds and ingests large food particles for intracellular digestion.

pharynx The upper part of the digestive tube that temporarily stores food before digestion.

phenotype The observable trait that results from the action of an allele pair.

phenylketonuria (PKU) A genetically related human disorder in which the homozygous combination of a particular mutant gene prevents the normal metabolism of the amino acid phenylalanine.

phloem A type of vascular tissue through which water and dissolved sugars are transported in plants from the leaf downward to the roots for storage.

phosphate group A chemical group made up of phosphorus and oxygen that is a component part of the nucleotide unit.

phosphoglyceraldehyde (PGAL) An intermediate product formed during photosynthesis that acts as the precursor of glucose formation.

photochemical reactions A set of biochemical reactions in photosynthesis in which light is absorbed and water molecules are split. (See **photolysis**.)

photolysis The portion of the photochemical reactions in which water molecules are split into hydrogen atoms and made available to the carbon fixation reactions.

photosynthesis A type of autotrophic nutrition in which green plants use the energy of sunlight to convert carbon dioxide and water into glucose.

phylum A level of biological classification that is a major subdivision of the kingdom level, containing fewer organisms with greater similarity (e.g., Chordata).

pinocytosis A special type of absorption by which liquids and particles too large to diffuse through the cell membrane may be taken in by vacuoles formed at the cell surface.

pioneer autotrophs The organisms supposed by the heterotroph hypothesis to have been the first to evolve the ability to carry on autotrophic nutrition.

pioneer species In an ecological succession, the first organisms to inhabit a barren environment.

pistil The female sex organ of the flower. (See **stigma; style; ovary**.)

pituitary gland An endocrine gland that produces hormones regulating the secretions of other endocrine glands; the "master gland."

placenta In placental mammals, a structure composed of both embryonic and maternal tissues that permits the diffusion of soluble substances to and from the fetus for nourishment and the elimination of fetal waste.

placental mammal A mammal species in which embryonic development occurs internally with the aid of a placental connection to the female parent's body.

Plant One of the five biological kingdoms; it includes multicellular organisms whose cells are bounded by cell walls and which are capable of photosynthesis (e.g., maple tree).

plasma The liquid fraction of blood, containing water and dissolved proteins.

plasma membrane A cell organelle that encloses the cytoplasm and other cell organelles and regulates the passage of materials into and out of the cell.

platelet A cell-like component of the blood that is important in clot formation.

polar body One of three nonfunctional cells produced during oogenesis that contain monoploid nuclei and disintegrate soon after completion of the process.

polio A disorder of the human regulatory system in which viral infection of the central nervous system may result in severe paralysis.

pollen The male gamete of the flowering plant.

pollen tube A structure produced by the germinating pollen grain that grows through the style to the ovary and carries the sperm nucleus to the ovule for fertilization.

pollination The transfer of pollen grains from anther to stigma.

pollution control The development of new procedures to reduce the incidence of air, water, and soil pollution—a positive aspect of human involvement with the environment.

polyploidy A type of chromosome mutation in which an entire set of homologous chromosomes fail to separate during the disjunction phase of the first meiotic division.

polysaccharide A type of carbohydrate composed of repeating units of monosaccharide that form a polymeric chain.

polyunsaturated fat A type of fat in which many bonding sites are unavailable for the addition of hydrogen atoms.

population All the members of a particular species in a given geographical location at a given time.

population control The use of various practices to slow the rapid growth in the human population—a positive aspect of human interaction with the environment.

population genetics A science that studies the genetic characteristics of a sexually reproducing species and the factors that affect its gene frequencies.

postnatal development The growth and maturation of an individual from birth, through aging, to death.

prenatal development The embryonic development that occurs within the uterus before birth. (See **gestation.**)

primary consumer Any herbivorous organism that receives food energy from the producer level (e.g., mouse); the second stage of a food chain.

primary sex cell The diploid cell that undergoes meiotic cell division to produce monoploid gametes.

producer Any autotrophic organism capable of trapping light energy and converting it to the chemical bond energy of food (e.g., green plants); the organisms forming the basis of the food chain.

progesterone A hormone produced by the corpus luteum and/or placenta that has the effect of maintaining the uterine lining and suppressing ovulation during gestation.

protease Any protein-hydrolyzing enzyme.

protein A complex organic compound composed of repeating units of amino acid.

Protista One of the five biological kingdoms; it includes simple unicellular forms whose nuclei are surrounded by nuclear membranes (e.g., ameba, paramecium).

pseudopod A temporary, flowing extension of the cytoplasm of an ameba that is used in nutrition and locomotion.

pulmonary artery One of two arteries that carry blood from the heart to the lungs for reoxygenation.

pulmonary circulation Circulation of blood from the heart through the lungs and back to the heart.

pulmonary vein One of four veins that carry oxygenated blood from the lungs to the heart.

pulse Rhythmic contractions of the artery walls that help to push the blood fluid through the capillary networks of the body.

punctuated equilibrium A theory of the time frame required for evolution which assumes that evolutionary change occurs in "bursts" with long periods of relative stability intervening.

pyramid of biomass A construct used to illustrate the fact that the total biomass available in each stage of a food chain diminishes from producer level to consumer level.

pyramid of energy A construct used to illustrate the fact that energy is lost at each trophic level in a food chain, being most abundant at the producer level.

pyruvic acid An intermediate product in the aerobic or anaerobic respiration of glucose.

receptor An organ specialized to receive a particular type of environmental stimulus.

recessive allele (gene) An allele (gene) whose effect is masked by that of its dominant allele.

recombinant DNA DNA molecules that have been moved from one cell to another in order to give the recipient cell a genetic characteristic of the donor cell.

recombination The process by which the members of segregated allele pairs are randomly recombined in the zygote as a result of fertilization.

rectum The portion of the digestive tract in which digestive wastes are stored until they can be released to the environment.

red blood cell Small, nonnucleated cells in the blood that contain hemoglobin and carry oxygen to bodily tissues.

reduction division See **meiosis.**

reflex A simple, inborn, involuntary response to an environmental stimulus.

reflex arc The complete path, involving a series of three neurons (sensory, interneuron, and motor), working together, in a reflex action.

regeneration A type of asexual reproduction in which new organisms are produced from the severed parts of a single parent organism; the replacement of lost or damaged tissues.

regulation The life process by which living things respond to changes within and around them, and by which all life processes are coordinated.

replication An exact self-duplication of the chromosome during the early stages of cell division; the exact self-duplication of a molecule of DNA.

reproduction The life process by which new cells arise from preexisting cells by cell division.

reproductive isolation The inability of species varieties to interbreed and produce fertile offspring, because of variations in behavior or chromosome structure.

respiration The life function by which living things convert the energy of organic foods into a form more easily used by the cell.

response The reaction of an organism to an environmental stimulus.

rhizoid A rootlike fiber produced by fungi that secrete hydrolytic enzymes and absorb digested nutrients.

ribonucleic acid (RNA) A type of nucleic acid that operates in various ways to facilitate protein synthesis.

ribose A five-carbon sugar found as a component part of the nucleotides of RNA molecules only.

ribosomal RNA (r-RNA) The type of RNA that makes up the ribosome.

ribosome A cell organelle that serves as the site of protein synthesis in the cell.

root A plant organ specialized to absorb water and dissolved substances from the soil, as well as to anchor the plant to the soil.

root hair A small projection of the growing root that serves to increase the surface area of the root for absorption.

roughage A variety of undigestible carbohydrates that add bulk to the diet and facilitate the movement of foods through the intestine.

runner A type of vegetative propagation in which an above-ground stem (runner) produces roots and leaves and establishes new organisms with identical characteristics.

saliva A fluid secreted by salivary glands that contains hydrolytic enzymes specific to the digestion of starches.

salivary gland The gland that secretes saliva, important in the chemical digestion of certain foods.

salt A chemical composed of a metal and a nonmetal joined by means of an ionic bond (e.g., sodium chloride).

saprophyte A heterotrophic organism that obtains its nutrition from the decomposing remains of dead plant and animal tissues (e.g., fungus, bacteria).

saturated fat A type of fat molecule in which all available bonding sites on the hydrocarbon chains are taken up with hydrogen atoms.

scrotum A pouch extending from the wall of the lower abdomen that houses the testes at a temperature optimum for sperm production.

secondary consumer Any carnivorous animal that derives its food energy from the primary consumer level (e.g., a snake); the third level of a food chain.

secondary sex characteristics The physical features, different in males and females, that appear with the onset of sexual maturity.

seed A structure that develops from the fertilized ovule of the flower and germinates to produce a new plant.

seed dispersal Any mechanism by which seeds are distributed in the environment so as to widen the range of a plant species. (See **fruit**.)

segregation The random separation of the members of allele pairs that occurs during meiotic cell division.

self-pollination A type of pollination in which the pollen of a flower pollinates another flower located on the same plant organism.

sensory neuron A type of neuron specialized for receiving environmental stimuli, which are detected by receptor organs.

sepal An accessory part of the flower that functions to protect the bud during development.

sessile A term that relates to the "unmoving" state of certain organisms, including the hydra.

seta One of several small, chitinous structures (setae) that aid the earthworm in its locomotor function.

sex chromosomes A pair of homologous chromosomes carrying genes that determine the sex of an individual; these chromosomes are designated as X and Y.

sex determination A pattern of inheritance in which the conditions of maleness and femaleness are determined by the inheritance of a pair of sex chromosomes (XX = female; XY = male).

sex linkage A pattern of inheritance in which certain nonsex genes are located on the X sex chromosome, but have no corresponding alleles on the Y sex chromosome.

sex-linked trait A genetic trait whose inheritance is controlled by the genetic pattern of sex linkage (e.g., color blindness).

sexual reproduction A type of reproduction in which new organisms are formed as a result of the fusion of gametes from two parent organisms.

shell An adaptation for embryonic development in many terrestrial, externally developing species that protects the developing embryo from drying and physical damage (e.g., birds).

sickle-cell anemia A genetically related human disorder in which the homozygous combination of a mutant gene leads to the production of abnormal hemoglobin and crescent-shaped red blood cells.

skeletal muscle A type of muscle tissue associated with the voluntary movements of skeletal levers in locomotion.

small intestine In human beings, the longest portion of the food tube, in which final digestion and absorption of soluble end-products occur.

smooth muscle See **visceral muscle.**

somatic nervous system A subdivision of the peripheral nervous system that is made up of nerves associated with voluntary actions.

speciation The process by which new species are thought to arise from previously existing species.

species A biological grouping of organisms so closely related that they are capable of interbreeding and producing fertile offspring (e.g., human being).

species presentation The establishment of game lands and wildlife refuges that have permitted the recovery of certain endangered species—a positive aspect of human involvement with the environment.

sperm A type of gamete produced as a result of spermatogenesis in male animals; the male reproductive cell.

spermatogenesis A type of meiotic cell division in which four sperm cells are produced for each primary sex cell.

spinal cord The part of the central nervous system responsible for reflex action, as well as impulse conduction between the peripheral nervous system and the brain.

spindle apparatus A network of fibers that form during cell division and to which centromeres attach during the separation of chromosomes.

spiracle One of several small pores in arthropods, including the grasshopper, that serve as points of entry of respiratory gases from the atmosphere to the tracheal tubes.

spongy layer A cell layer found in most leaves that is loosely packed and contains many air spaces to aid in gas exchange.

spore A specialized asexual reproductive cell produced by certain plants.

sporulation A type of asexual reproduction in which spores released from special spore cases on the parent plant germinate and grow into new adult organisms of the species.

staining A technique of cell study in which chemical stains are used to make cell parts more visible for microscopic study.

stamen The male reproductive structure in a flower. (See **anther; filament.**)

starch A type of polysaccharide produced and stored by plants.

stem A plant organ specialized to support the leaves and flowers of a plant, as well as to conduct materials between the roots and the leaves.

stigma The sticky upper portion of the pistil, which serves to receive pollen.

stimulus Any change in the environment to which an organism responds.

stomach A muscular organ that acts to liquefy food and that produces gastric protease for the hydrolysis of protein.

stomate A small opening that penetrates the lower epidermis of a leaf and through which respiratory and photosynthetic gases diffuse.

strata The layers of sedimentary rock that contain fossils, whose ages may be determined by studying the patterns of sedimentation.

stroke A disorder of the human regulatory system in which brain function is impaired because of oxygen starvation of brain centers.

stroma An area of the chloroplast within which the carbon-fixation reactions occur; stroma lie between pairs of grana.

style The portion of the pistil that connects the stigma to the ovary.

substrate A chemical that is metabolized by the action of a specific enzyme.

succession A situation in which an established ecological community is gradually replaced by another until a climax community is established.

survival of the fittest The concept, frequently associated with Darwin's theory of evolution, that in the intraspecies competition among naturally occurring species the organisms best adapted to the particular environment will survive.

sweat glands In human beings, the glands responsible for the production of perspiration.

symbiosis A term which refers to a variety of biotic relationships in which organisms of different species live together in close physical association.

synapse The gap that separates the terminal branches of one neuron from the dendrites of an adjacent neuron.

synapsis The intimate, highly specific pairing of homologous chromosomes that occurs in the first meiotic division, forming tetrads.

synthesis The life function by which living things manufacture the complex compounds required to sustain life.

systemic circulation The circulation of blood from the heart through the body tissues (except the lungs) and back to the heart.

systole The higher pressure registered during blood pressure testing. (See **diastole.**)

taiga A terrestrial biome characterized by long, severe winters and climax flora that includes coniferous trees.

Tay-Sachs A genetically related human disorder in which fatty deposits in the cells, particularly of the brain, inhibit proper functioning of the nervous system.

technological oversight A term relating to human activities that adversely affect environmental quality due to failure to adequately assess the environmental impact of a technological development.

teeth Structures located in the mouth that are specialized to aid in the mechanical digestion of foods.

temperate deciduous forest A terrestrial biome characterized by moderate climatic conditions and climax flora that includes deciduous trees.

template A pattern or design provided by the DNA molecule for the synthesis of protein molecules.

tendon A type of connective tissue that attaches a skeletal muscle to a bone.

tendonitis A disorder of the human locomotor system in which the junction between a tendon and a bone becomes irritated and inflamed.

tentacle A grasping structure in certain organisms, including the hydra, that contains stinging cells and is used for capturing prey.

terminal branch A cytoplasmic extension of the neuron that transmits a nerve impulse to adjacent neurons via the secretion of neurotransmitters.

terrestrial biome A biome that comprises primarily land ecosystems, the characteristics of which are determined by the major climate zone of the earth.

test cross A genetic cross accomplished for the purpose of determining the genotype of an organism expressing a dominant phenotype; the unknown is crossed with a homozygous recessive.

testis A gonad in human males that secretes the hormone testosterone, which regulates male secondary sex characteristics; the testis also produces sperm cells for reproduction.

testosterone A hormone secreted by the testis that regulates the production of male secondary sex characteristics.

tetrad A grouping of four chromatids that results from synapsis.

thymine A nitrogenous base found only in DNA.

thyroid gland An endocrine gland that regulates the body's general rate of metabolism through secretion of the hormone thyroxin.

thyroid-stimulating hormone (TSH) A pituitary hormone that regulates the secretions of the thyroid gland.

thyroxin A thyroid hormone that regulates the body's general metabolic rate.

tongue A structure that aids in the mechanical digestion of foods.

trachea A cartilage-ringed tube that conducts air from the mouth to the bronchi.

tracheal tube An adaptation in arthropods (e.g., grasshopper) which functions to conduct respiratory gases from the environment to the moist internal tissues.

Tracheophyta A phylum of the Plant Kingdom whose members (tracheophytes) contain vascular tissues and true roots, stems, and leaves (e.g., geranium, fern, bean, maple tree, corn).

transfer RNA (t-RNA) A type of RNA that functions to transport specific amino acids from the cytoplasm to the ribosome for protein synthesis.

translocation A type of chromosome mutation in which a section of a chromosome is transferred to a nonhomologous chromosome.

transpiration The evaporation of water from leaf stomates.

transpiration pull A force that aids the upward conduction of materials in the xylem by means of the evaporation of water (transpiration) from leaf surfaces.

transport The life function by which substances are absorbed, circulated, and released by living things.

triplet codon A group of three nitrogenous bases that provide information for the placement of amino acids in the synthesis of proteins.

tropical forest A terrestrial biome characterized by a warm, moist climate and a climax flora that includes many species of broadleaf trees.

tropism A plant growth response to an environmental stimulus.

tuber A type of vegetative propagation in which an underground stem (tuber) produces new tubers, each of which is capable of producing new organisms with identical characteristics.

tundra A terrestrial biome characterized by permanently frozen soil and climax flora that includes lichens and mosses.

tympanum A receptor organ in arthropods (e.g., grasshopper) which is specialized to detect vibrational stimuli.

ulcer A disorder of the human digestive tract in which a portion of its lining erodes and becomes irritated.

ultracentrifuge A tool of biological study that uses very high speeds of centrifugation to separate cell parts for examination.

umbilical cord In placental mammals, a structure containing blood vessels that connects the placenta to the embryo.

unicellular Having a body that consists of a single cell (e.g., paramecium).

uracil A nitrogenous base that is a component part of the nucleotides of RNA molecules only.

urea A type of nitrogenous waste with moderate solubility and moderate toxicity.

ureter In human beings, a tube that conducts urine from the kidney to the urinary bladder.

urethra In human beings, a tube that conducts urine from the urinary bladder to the exterior of the body.

uric acid A type of nitrogenous waste with low solubility and low toxicity.

urinary bladder An organ responsible for the temporary storage of urine.

urine A mixture of water, salts, and urea excreted from the kidney.

use and disuse A term associated with the evolutionary theory of Lamarck, since proved incorrect.

uterus In female placental mammals, the organ within which embryonic development occurs.

vaccination An inoculation of dead or weakened disease organisms that stimulates the body's immune system to produce active immunity.

vacuole A cell organelle that contains storage materials (e.g., starch, water) housed inside the cell.

vagina In female placental mammals, the portion of the reproductive tract into which sperm are implanted during sexual intercourse and through which the baby passes during birth.

variation A concept, central to Darwin's theory of evolution, that refers to the range of adaptation which can be observed in all species.

vascular tissues Tubelike plant tissues specialized for the conduction of water and dissolved materials within the plant. (See **xylem; phloem.**)

vegetative propagation A type of asexual reproduction in which new plant organisms are produced from the vegetative (nonfloral) parts of the parent plant.

vein (human) A relatively thin-walled blood vessel that carries blood from capillary networks back toward the heart.

vein (plant) An area of vascular tissues located in the leaf that aid the upward transport of water and minerals through the leaf and the transport of dissolved sugars to the stem and roots.

vena cava One of two major arteries that return blood to the heart from the body tissues.

ventral nerve cord The main pathway for nerve impulses between the brain and peripheral nerves of the grasshopper and earthworm.

ventricle One of two thick-walled, muscular chambers of the heart that pump blood out to the lungs and body.

villi Microscopic projections of the lining of the small intestine that absorb the soluble end-products of digestion. (See **lacteal.**)

visceral muscle A type of muscle tissue associated with the involuntary movements of internal organs (e.g., peristalsis in the small intestine).

vitamin a type of nutrient that acts as a coenzyme in various enzyme-controlled reactions.

water cycle The mechanism by which water is made available to living things in the environment through the processes of precipitation, evaporation, runoff, and percolation.

water pollution A type of technological oversight that involves the addition of some unwanted factor (e.g., sewage, heavy metals, heat, toxic chemicals) to our water resources.

Watson-Crick model A model of DNA structure devised by J. Watson and F. Crick that hypothesizes a "twisted ladder" arrangement for the DNA molecule.

white blood cell A type of blood cell that functions in disease control. (See **phagocyte; lymphocyte.**)

xylem A type of vascular tissue through which water and dissolved minerals are transported upward through a plant from the root to the stems and leaves.

yolk A food substance, rich in protein and lipid, found in the eggs of many animal species.

yolk sac The membrane that surrounds the yolk food supply of the embryos of many animal species.

zygote The single diploid cell that results from the fusion of gametes in sexual reproduction; a fertilized egg.

Regents Examinations, Answers, and Student Self-Appraisal Guides

Examination
June 2015
Living Environment

PART A

Answer all questions in this part. [30]

Directions **(1–30): For** *each* **statement or question, record in the space provided the** *number* **of the word or expression that, of those given, best completes the statement or answers the question.**

1 Materials are transported within a single-celled organism by the

1 nucleus		3 mitochondrion	
2 cytoplasm		4 ribosome	1 __2__

2 Which row in the chart below correctly pairs a food molecule with its building block?

Row	Food Molecule	Building Block
(1)	starch	amino acid
(2)	sugar	starch
(3)	protein	amino acid
(4)	amino acid	sugar

2 __3__

3 The flow of energy in an ecosystem is best described as energy moving in

  1 one direction from the Sun to the producers and then to the consumers

  2 one direction from a consumer to a producer and then to the Sun as heat and light

  3 two directions between the producers that are present

  4 two directions, back and forth, between the producers and the consumers

3 _1_

4 Occasionally, during pregnancy, the placenta can separate from the uterus. This causes a disruption in development and sometimes death of the fetus. Harm to the developing fetus might occur because the placenta

  1 transfers oxygen and nutrients to the fetal blood

  2 sends maternal blood into the fetus

  3 supplies milk for the fetus

  4 breaks down wastes of the fetus

4 _1_

5 Which process produces only identical offspring?

  1 meiotic cell division

  2 selective breeding

  3 cloning

  4 fertilization

5 _3_

6 A photograph of human cells as seen with a compound light microscope is shown below. A cell structure is labeled A.

Structure A is most likely a

1 mitochondrion that synthesizes food for the cell
2 nucleus that is the site of food storage
3 mitochondrion that absorbs energy from the Sun
4 nucleus that is responsible for the storage of information

6 ___

7 A land-dwelling organism, A, and an aquatic single-celled organism, B, are represented below.

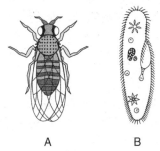

A   B

Which statement best explains how A and B are able to survive in their environments?

1 The organelles in B perform similar functions to the organ systems in A.
2 The transport system in B is more complex than the transport system in A.
3 Both A and B take in oxygen from the water.
4 Only A can pass on traits to offspring.

7 ___

8 A man is exposed to large amounts of ultraviolet radiation while sunbathing at the beach. This exposure causes a genetic change in the DNA of a skin cell. In the future, this change can be passed on to

1 his male and female children
2 his male children, only
3 all cells in his body
4 his skin cells, only   8 __4__

9 Palm oil, produced from palm trees, is not only a biofuel, but is also used in food additives, cosmetics, and lubricants. Palm tree plantations are now cultivated in areas that were formerly natural forests. One ecological concern raised by this expansion is that

1 the natural forest ecosystem may harm the palm trees
2 the use of the land for agriculture will increase the biodiversity of the area
3 humans are changing the basic processes of the palm trees
4 planting large expanses of one crop reduces the biodiversity of the area   9 __4__

10 Fishermen have harvested certain fish to the point where the population of that fish is decreasing. This level of direct harvesting could cause

1 ecosystems to be improved for future generations
2 ecosystems to be severely damaged
3 the restoration of environmental stability
4 all other fish species to increase in number   10 __2__

11 Which phrase best describes a gene?

1 a segment of a DNA molecule found only in the body cells of an organism

· 2 a segment of a DNA molecule found only in the gametes of an organism

3 a segment of a DNA molecule that contains the instructions for producing a trait in an organism

4 a segment of a DNA molecule that contains the instructions for producing all the characteristics of an organism

11 _3_

12 The molecule DNA contains the four bases listed below.

A – adenine
C – cytosine
G – guanine
T – thymine

Which base pairings normally occur during DNA replication?

1 Guanine pairs with cytosine. Thymine pairs with thymine.

2 Adenine pairs with thymine. Cytosine pairs with guanine.

3 Thymine pairs with guanine. Cytosine pairs with adenine.

4 Cytosine pairs with cytosine. Thymine pairs with thymine.

12 _2_

13 Evolution of a species could occur as a result of changes in the

1 DNA in muscle cells

2 base sequences in liver cells

3 genes in an egg cell

4 number of chromosomes in a fetal bone cell

13 _3_

14 One positive impact that industrialization has had is that

1 industrialization produces waste gases that pollute the air
2 fossil fuels used by industries help reduce finite resources
3 industrialization has been a source of many jobs for people
4 new technologies have increased acid rain

14 _3_

15 When receiving x rays, individuals wear a lead shield over major organs in order to limit the body's exposure to radiation. One reason for this procedure is to

1 protect the patient against broken bones
2 prevent mutations in gametes
3 improve circulation in the patient
4 increase the chance of a change in DNA

15 _2_

16 When an ant in a colony dies, the live ants will throw the dead ant out of the anthill. If a live ant from the colony, ant X, is sprayed with a chemical characteristic of dead ants, the live ants will repeatedly throw this ant out of the anthill until they can no longer detect the chemical on ant X. What is the best explanation for this behavior?

1 The ants are responding to a chromosomal mutation in ant X.
2 The chemical is exhibiting a feedback mechanism.
3 The live ants must continue this behavior until they have eliminated ant X.
4 The chemical acts as a stimulus for a particular behavior.

16 _4_

17 Rabbits produce large numbers of offspring during each reproductive season, yet the number of rabbits within a given population changes very little from year to year. The stability of the population size is most likely the result of

   1 the development of mutations in young rabbits
   2 environmental factors that keep the population in check
   3 rabbits continuing to reproduce when the population is large
   4 the survival of more female rabbits than male rabbits   17 _2_

18 Genetic engineering has the potential to correct human genetic disorders. In gene therapy, a defective gene is replaced by using a virus to insert a normal gene into the cells of an individual. This treatment will be most successful if the virus is inserted into cells that

   1 lack a nucleus
   2 are recycled after death, rather than removed from the body
   3 carry out one specific function, rather than multiple functions
   4 continue to divide during the life of the patient   18 _4_

19 In one town, some people support a proposal to build a shopping mall on a large, undeveloped lot, because it would increase business and create new jobs. As a trade-off, the shopping mall would cause a decrease in the

   1 amount of air pollution
   2 volume of garbage and litter
   3 amount of wastewater entering the local sewage system
   4 variety of wildlife populations in the area   19 _4_

20 The human female reproductive system is represented below.

Which structure produces chemicals that regulate the reproductive cycle?

1 A   3 C

2 B   4 D   20 _1_

21 The diagram below represents a cell structure involved in converting energy stored in organic molecules into a form used by animal cells.

The arrows represent the movement of which substances?

1 carbon dioxide and sugar

2 oxygen and ATP

3 ATP and carbon dioxide

4 oxygen and sugar   21 _3_

22 The diagram below shows a concept map.

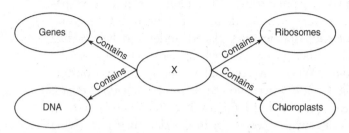

Which label correctly identifies what X represents in the concept map?

1 nucleus   3 autotrophic cell
2 chromosome   4 heterotrophic cell   22 __3__

23 The diagrams below represent two molecules that are involved in metabolic activities in some living cells.

The shape of each of the molecules is important because

1 molecules having different shapes are always found in different organisms
2 the shape of a molecule determines how it functions in chemical reactions
3 the shape of a molecule determines the age of an organism
4 if the shape of any molecule in an organism changes, the DNA in that organism will also change   23 __2__

24 In the early 1900s, experiments were conducted on two caterpillar species. The members of the two species were each divided into two groups. One group of each species was placed under red light, while the other group of each species was kept in the dark. When the caterpillars developed into butterflies, their wings showed extreme color differences. Exposure to red light resulted in intensely colored wings, while those kept in the dark had paler wing colors. The color differences were most likely due to

1 mutations in the color-producing genes
2 the caterpillars in the red light producing more DNA
3 gene expression being affected by the environment
4 the caterpillars in the dark evolving less than those in the light   24 ___

25 A student used a microscope to observe a single-celled organism. As he watched, it looked as if the organism split into two cells. He made drawings, shown below, of the organism over a short period of time.

Which process did the student record in his drawings?
1 genetic engineering   3 selective breeding
2 asexual reproduction   4 gamete formation   25 ___

26 Medical professionals are concerned with the increase in the number of bacterial species that are resistant to antibiotics. Once resistance appears in a bacterial population, it spreads rapidly. This is most likely because

  1 populations of resistant bacteria are small
  2 exposure to antibiotics increases the rate of reproduction in bacteria
  ·3 resistant bacteria are small when compared to non-resistant bacteria.
  4 resistant bacteria survive in greater numbers and pass the trait to their offspring   26 __4__

27 When getting a vaccination, which substance is injected into the body?

  1 bacteria to combat a pathogen
  2 white blood cells to engulf a pathogen
  3 a weakened form of a virus
  4 antibiotics to kill a virus   27 __3__

28 Many beverage companies are required to recycle bottles and cans because this activity directly reduces

  1 air pollution and destruction of the ozone shield
  2 overpopulation and soil erosion
  3 solid waste and depletion of resources
  4 thermal pollution and extinction of wildlife   28 __3__

29 The diagram below shows some of the DNA in a bacterium into which a human gene, *X*, has been successfully inserted.

The bacteria that result from reproduction of this cell will most likely have the ability to

1 replicate all of the genetic instructions found in humans
2 produce vaccines to be used to immunize humans
3 produce a human blood cell according to instructions in gene *X*
4 produce the human protein coded for by gene *X*

29 ___4___

30 The Eurasian water milfoil is a nonnative species, which was once commonly sold as an aquarium plant, and is now found growing in many lakes in New York State. It has few natural enemies, and grows rapidly, crowding out many native species. This plant ruins fishing areas and interferes with boating and other water sports. This is an example of

1 human consumption of finite resources
2 an unintended consequence of adding an organism to an ecosystem
3 an abiotic factor having a negative effect on an ecosystem
4 the introduction of a species that has increased the long-term biodiversity of an ecosystem

30 ___2___

PART B–1

Answer all questions in this part.   [13]

Directions **(31–43): For** *each* **statement or question, record in the space provided the** *number* **of the word or expression that, of those given, best completes the statement or answers the question.**

31 The graph below shows the size of a population of foxes over a period of years.

If the line did not stay around the carrying capacity, but continued to rise, which concept would this graph best illustrate?

1 environmental stability
2 genetic variety
3 behavioral change
4 overproduction

31 __4__

32 A food web is represented below.

Which organism would receive the *least* amount of transferred solar energy?

1 grasses   3 frogs
2 owls   4 field mice   32 _2_

33 Birch bolete is a fungus that normally grows on the roots of birch trees in New York State. During the life of the fungus and the birch, each organism receives nutrients from the various biochemical processes of the other. According to this information, it can be inferred that these two species

1 are both predators
2 require the same amount of sunlight
3 require a similar soil pH
4 recycle the remains of dead organisms   33 _3_

34 The photographs below show different varieties of cattle and characteristics of each variety.

A	B	C
Good resistance to heat but poor beef	Good beef but poor resistance to heat	Good resistance to heat and good beef

Which statement best explains the development of variety *C*?

1 Nuclei from body cells taken from variety *A* were inserted into egg cells lacking nuclei taken from variety *B*.

2 Selective breeding was used to combine desirable traits from both varieties *A* and *B*.

3 The need to adapt to changes in the environment led to the selection of advantageous characteristics in the offspring of variety *B*.

4 Mutations that occurred in the body cells of variety *A* were passed on to the offspring generation after generation.

34 __2__

35 The diagram below represents a remora fish attached to a shark.

A remora fish has an adhesive disk or sucker on its head, which it uses to attach itself to larger fishes, such as sharks. This attachment causes the shark no harm. The remora fish eat scraps of food that the sharks drop as they feed. This is an example of

1 an adaptation to a specialized niche
2 an adaptation of a successful parasite
3 competition between two fish species for food
4 competition for abiotic resources

35 __1__

36 Each row in the chart below represents a different population of the same species of insect. Which row shows the population with the greatest chance of survival in a changing environment?

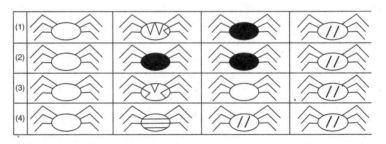

36 __1__

37 The development of nerve, muscle, and skin cells is represented in the diagram below.

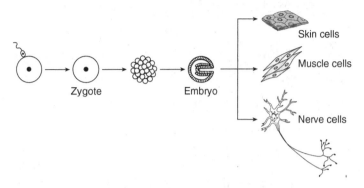

Which statement best explains how each of the different cell types can develop from the same embryo?

1 The cells have identical genetic instructions, but different parts of these instructions are being expressed in each cell.

2 The cells have identical genetic instructions, and all parts of these instructions are being expressed in each cell.

3 The cells are produced by asexual reproduction and contain identical genetic instructions.

4 The cells contain genetic instructions from two different parents and will express the instructions from one parent, only.

37 ___1___

38 The graph below represents the effect of pH on three different enzymes at normal body temperature.

The Effect of pH on Three Different Enzymes at Normal Body Temperature

Key
– – Enzyme 1
...... Enzyme 2
– - – Enzyme 3

The graph illustrates that enzymes 1, 2, and 3

1 are not affected by pH
2 work best at different pH levels
3 work best in an acidic environment
4 work best in a basic environment   38 __2__

39 The human male reproductive system is represented below.

Which structure produces cells that have the potential to become gametes?

1 A   3 C
2 B   4 D   39 __A__

40 Some scientists have collected and stored seeds for many types of food-producing plants. The purpose of this is to

  1 increase the destruction of environments
  2 continue the deforestation of world ecosystems
  3 decrease the dependence on plants for food
  4 preserve the diversity of plant species
                                      40 __4__

41 Which diagram best illustrates the relationship between the number of cells, tissues, and organs in a complex multicellular organism?

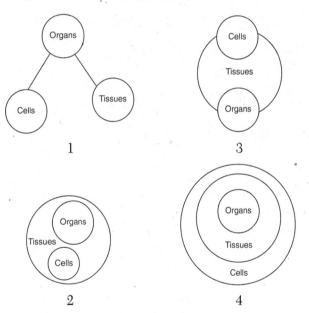

41 __4__

Base your answers to questions 42 and 43 on the diagram below, which represents an ameba engulfing bacteria, and on your knowledge of biology.

42 This ameba would most likely be classified as a

　　1　decomposer
　　2　producer
　　3　consumer
　　4　pathogen   42 _3_

43 The activity taking place is

　　1　photosynthesis
　　2　differentiation
　　3　autotrophic nutrition
　　4　heterotrophic nutrition   43 _4_

PART B-2

Answer all questions in this part. [12]

Directions (44–55): **For those questions that are multiple choice, record in the space provided the *number* of the choice that, of those given, best completes each statement or answers each question. For all other questions in this part, follow the directions given and record your answers in the spaces provided.**

Base your answers to questions 44 through 48 on the information and data table below and on your knowledge of biology.

The Enzyme Catalase

Catalase is an enzyme found in nearly all living organisms that breathe or are exposed to oxygen. According to recent scientific studies, low levels of catalase may play a role in the graying process of human hair. The body naturally produces hydrogen peroxide, and catalase breaks it down into water and oxygen. If there is a dip in catalase levels, hydrogen peroxide cannot be broken down. This causes hydrogen peroxide to bleach hair from the inside out. Scientists believe this finding may someday be used in anti-graying treatments for hair.

A pharmaceutical company, investigating ways to prevent hair from turning gray, took tissue samples from two different individuals. Both individuals were the same age. Each of the samples was placed in a solution of hydrogen peroxide. The volume of oxygen gas produced was measured every 5 minutes for 25 minutes. The data the company collected are shown below.

**Oxygen Production in the Breakdown of
Hydrogen Peroxide by Catalase**

Time (min)	Sample from Person A (mL oxygen)	Sample from Person B (mL oxygen)
5	2.0	4.5
10	3.5	8.5
15	5.0	12.0
20	7.5	15.5
25	9.5	20.0

Directions (44–46): Using the information in the data table, construct a line graph on the grid on the next page, following the directions below.

44 Mark an appropriate scale, without any breaks in the data, on each labeled axis. [1]

45 Plot the data from the data table for the sample from person *A* on the grid. Connect the points and surround each point with a small circle. [1]

Example:

46 Plot the data from the data table for the sample from person *B* on the grid. Connect the points and surround each point with a small triangle. [1]

Example:

Oxygen Production in the Breakdown of Hydrogen Peroxide by Catalase

Time (min)

Key

⊙ Person A
△ Person B

Note: The answer to question 47 should be recorded in the space provided.

47 If the temperature of the tissue samples used in the experiment had been raised from 37°C (body temperature) to 50°C, the results would have been different because

1 more enzymes are produced at higher temperatures, increasing the amount of hydrogen peroxide
2 more hydrogen peroxide is released at higher temperatures, increasing the activity of catalase
3 increasing temperatures altered the structure of catalase, decreasing oxygen production
4 increasing temperatures decreased the synthesis of amino acids, increasing levels of hydrogen peroxide

47 ___3___

48 According to the data provided, which person, *A* or *B*, is more likely to be the first to have gray hair? Support your answer. [1]

Person: ___A___

Person A's catalase does not
work as efficiently as
Person B's.

Base your answers to questions 49 through 52 on the information and graph below and on your knowledge of biology.

An investigation was carried out to determine the effect of drinking an excessive amount of water on urine flow. A subject drank 1 liter of water in 5 minutes, and then urine output was measured. The graph shows how the human adult kidneys responded to regulate water balance in the body. Urine output was measured every 10 minutes for a little over 3 hours. Normal output for an average adult is approximately 0.5–1 mL/min.

**Urine Production in an Adult
with Normal Kidney Function**

Note: The answer to question 49 should be recorded in the space provided.

49 One half-hour after the liter of water was consumed, the urine produced by the kidneys was

   1 between 2 and 3 mL/min
   2 between 4 and 5 mL/min
   3 eight times greater than normal
   4 below the normal range   49 __1__

Note: The answer to question 50 should be recorded in the space provided.

50 The change in urine production during this 3-hour period was most likely the result of

   1 antibody production
   2 homeostatic feedback
   3 enzymatic breakdown of the water consumed
   4 nerve cell malfunctions of the kidneys   50 __2__

51 Identify a structure, in organisms that do not have kidneys, that is adapted to regulate water balance. [1]

   __Cell membrane__

52 Approximately how long did it take, in minutes, for the body to return to normal after the intake of water? [1]

   __160__ **minutes**

Base your answers to questions 53 through 55 on the information and data table below and on your knowledge of biology.

The data table summarizes the changes that occurred to farmland in the years immediately following its abandonment. The land is located in a very stable ecosystem. It was abandoned after years of overuse and weathering, which resulted in the depletion of soil nutrients.

Common Types of Vegetation Present

Years Since Abandoned	Grasses and Weeds	Shrubs	Pine Forest	Hardwood Forest
1	X			
18	X	X	X	
30			X	
70			X	X
100				X
118 (present)				X

53 Which type of vegetation appears to have the lowest soil nutrient requirements? Support your answer with information from the data table. [1]

Lowest soil nutrient requirement vegetation: _____

Weeds - They can grow with very little nutrients.

54 Assuming the ecosystem remains undisturbed, which type of vegetation would you expect to be most common in this area 200 years after it was first abandoned? Support your answer. [1]

Most common vegetation: _Hardwood_
Forest, grows on the
remains of the other
plants, which have already
lived and died.

55 Describe how the types of vegetation present on this farmland would change if a fire burned down all the trees 120 years after the land was abandoned. [1]

The fire would destroy
all vegetation, and
the cycle would restart.

PART C

Answer all questions in this part.   [17]

Directions (56–72): **Record your answers in the spaces provided.**

Base your answer to question 56–58 on the information and photograph below and on your knowledge of biology.

The photograph below is part of an advertisement used by a company selling solar panels. The company claims that their panels, like plants, provide clean, renewable energy. They also claim that using solar panels will have a positive effect on the biosphere by reducing global warming.

Source:http://www.stockwatch.in/files/Energy.jpg

56–58 Explain why these claims are valid. In your answer, be sure to:

- explain why both plants and solar panels provide renewable energy, rather than nonrenewable energy [1]
- state how the widespread use of solar panels to generate electricity can help to reduce global warming [1]
- state how the energy-capturing process used by plants worldwide can help to reduce global warming [1]

plants take carbon dioxide and sun, which are renewable and make energy. Solar panels absorb sunlight, which is renewable to make energy. Using the sun as a source of energy would halt the constant burning of fossil fuels. When plants go through photosynthesis, they remove carbon dioxide from the air.

Base your answers to questions 59 and 60 on the information below and on your knowledge of biology.

Fungi are interesting organisms that interact with humans in many ways. Yeasts are fungi used in the food industry to produce products such as bread and certain beverages. Some fungi are valuable in medicine. For example, the drug cyclosporine, which is capable of suppressing the response of the immune system to foreign antigens, and the antibiotic penicillin are both products from fungi. Other fungi are less welcomed by humans. The irritation of athlete's foot is caused by a fungus, and a number of allergies are caused by reproductive spores released by fungi.

59 Describe the role of a drug like cyclosporine when transplanting organs from one person to another person. [1]

Cyclosporine would prevent the host's body's antibodies from attacking foriegn antigens.

60 Explain the difference between an infection caused by a fungus and an allergy caused by a fungus. [1]

An infection must be treated seriously as the fungus might be harming the body, but allergies are an unecessary immune response to a non-dangerous substance.

Base your answers to questions 61 and 62 on the information below and on your knowledge of biology.

Female mosquitoes need a meal of blood from a person or other animal in order to produce eggs. It has been discovered that mosquitoes have cells on their antennae that can detect the insect repellent known as DEET. The repellent is not harmful to mosquitoes, but when mosquitoes detect DEET, they will not land on the surface where the DEET has been applied. This protects people from being bitten by mosquitoes.

Recently, scientists found some mosquitoes that are resistant to DEET because they do not detect its presence. They bred these mosquitoes and eventually produced a population consisting of about 50% DEET-resistant insects.

61 Identify the process most likely responsible for a mosquito initially becoming resistant to DEET. [1]

Mutated gene (passed
down from
parents).

62 Mosquitoes with DEET resistance have been found in natural environments. Explain how the continued use of this repellent may cause the percentage of these resistant mosquitoes to increase in the future. [1]

_The resistant mosquitos
will live to pass on the
resistance to offspring,
while non-resistant ones
might not live as long._

Base your answers to questions 63 through 67 on the passage below and on your knowledge of biology.

Ocean-dwelling (marine) iguanas and land iguanas inhabit the Galapagos Islands. Some scientists believe that both types of iguanas diverged from a common ancestor. Marine iguanas eat algae. Land iguanas feed on cacti. Algae are more abundant in the ocean than cacti are on the islands. Both species lay their eggs in the sand.

Rats, cats, and goats have been introduced to the islands by humans. Rats feed on iguana eggs, cats eat baby iguanas, and goats eat cacti.

63 Identify the process by which ancestral iguanas developed into the present-day marine iguanas and land iguanas of the Galapagos Islands. [1]

Process: __Natural Selection__

64 Identify *one* organism in the Galapagos Islands that directly limits the population of both the marine iguanas and land iguanas. [1]

Organism: __humans__

65 Which population of iguanas, marine or land, would you expect to be larger? Support your answer. [1]

Population of iguana: __marine__

__More available food.__

66 Would the introduction of goats have a greater effect on the population of the marine iguanas or the land iguanas? Support your answer. [1]

Population of iguana: _____land_____

_____Both live on land._____

_____

67 Identify *one* technique that can be used to support the conclusion that these two species of iguana developed from a common ancestor. [1]

Technique: _____Electrophoresis_____

Base your answer to question 68–72 on the information and diagram below and on your knowledge of biology.

The presence of air is believed to be important for root growth in bean plants. The apparatus available to conduct an investigation is shown below. There are enough bottles and other materials to have multiple setups. Air (for aeration) can be bubbled into the bottle through the rubber tube.

Source: Biology Handbook, SED 1960

68–72 Design an experiment to test the effect of aeration on the growth of roots of bean seedlings. In your answer, be sure to:

- state *one* hypothesis the experiment would test [1]
- describe how the control group will be treated differently from the experimental group [1]
- identify the dependent variable in the experiment [1]
- state *one* reason why many setups should be used in both the experimental and control groups [1]
- state *one* reason why several different kinds of seedlings were *not* tested in this experiment [1]

The more air recieved, the faster the seed will grow. The control group will recieve air, the experimental group will not. The dependent variable is the growth of the roots. Many species would not provide valid results, because different species grow at different rates.

PART D

Answer all questions in this part.   [13]

Directions (73–85): **For those questions that are multiple choice, record in the space provided the *number* of the choice that, of those given, best completes the statement or answers the question. For all other questions in this part, follow the directions given and record your answers in the spaces provided.**

Note: The answer to question 73 should be recorded in the space provided.

73 The buildup of waste products in muscle cells that are active might cause

1 digestion   3 increased fatigue
2 cellular respiration   4 decreased heart rate   73 __3__

Note: The answer to question 74 should be recorded in the space provided.

74 The diagram below shows the evolutionary relationships among several types of mammals.

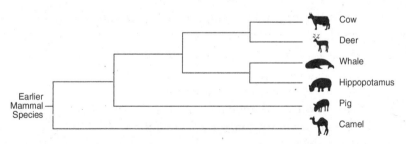

Which mammal would be most closely related to a hippopotamus?

1 deer   3 pig
2 whale   4 cow   74 __2__

Base your answers to questions 75 and 76 on the information and data table below and on your knowledge of biology.

A group of students obtained the following data while trying to determine the effect of exercise on pulse rate.

Effect of Exercise on Pulse Rate

Student	Resting Pulse Rate (beats per minute)	Pulse Rate After Exercising (beats per minute)
A	66	92
B	82	107
C	65	97
D	74	124
E	79	118
F	68	98
G	89	122

Note: The answer to question 75 should be recorded in the space provided.

75 Which statement is an example of an observation the students could have made?

  1 Pulse rates in beats per minute decrease for all people after exercise.

  2 Student *A* most likely exercises regularly.

  3 The pulse rate of student *C* was dangerously low.

  4 The pulse rate of student *F* increased by 30 beats per minute.

75 ___4___

Note: The answer to question 76 should be recorded in the space provided.

76 Which two body systems were most actively involved in this experiment?

  1 respiratory and immune
  2 digestive and endocrine
  3 respiratory and circulatory
  4 immune and circulatory

76 _3_

77 The diagram below represents a green plant cell viewed with the high power of a compound light microscope before and after a particular substance was added.

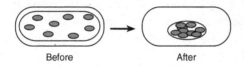

Before   After

Identify a substance that could have been added to the slide to bring about the change shown. [1]

Salt / Saline Solution

78 Using the DNA base sequences below, identify which *two* species are more closely related. Support your answer. [1]

    Species A: CAC GTG GAC AGA GGA CAC CTC

    Species B: CAT GTG GAC AGA GGA CAC CTC

    Species C: CAC GTA GAC TGA GGA CTT CTC

Species: _____A_____ and _____B_____

They have the most
similar sequence.

79 A student observing onion cells using a microscope was having difficulty seeing any detail in the cells. State *one* action the student could take to improve the detail. [1]

Add dye

    Base your answers to questions 80 and 81 on the diagram below and on your knowledge of biology. The diagram represents the results of paper chromatography performed on extracts from five organisms.

Cyanobacteria   Red algae   Brown algae   Green algae   Spinach

80 Identify *one* pigment molecule common to all five organisms. [1]

Chlorophyll A

Note: The answer to question 81 should be recorded in the space provided.

81 Which two organisms are most closely related?

   1 cyanobacteria and green algae
   2 red algae and spinach
   3 brown algae and red algae
   4 red algae and cyanobacteria

81 __4__

Base your answers to questions 82 through 84 on the diagram below and on your knowledge of biology. The diagram shows variations in the beaks of finches in the Galapagos Islands.

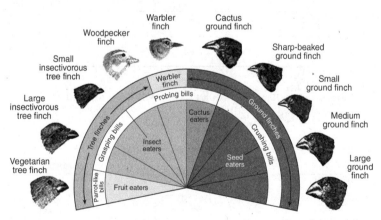

Source: www.pbs.org

Note: The answer to question 82 should be recorded in the space provided.

82 In this diagram, the variety of beak sizes and shapes are adaptations directly related to successful

1 feeding   3 defense
2 camouflage   4 singing   82 ___

83 State *one* reason why the large ground finch and the woodpecker finch can live successfully on the same island. [1]

_____

_____

84 Identify *one* finch in the diagram that is *least* likely to compete with any of the other finches. Support your answer. [1]

_____

_____

_____

85 State *one* reason why a molecule may *not* be able to pass into or out of a cell. [1]

_____

_____

Answers
June 2015

Living Environment

Answer Key

PART A

1. 2	**6.** 4	**11.** 3	**16.** 4	**21.** 3	**26.** 4
2. 3	**7.** 1	**12.** 2	**17.** 2	**22.** 3	**27.** 3
3. 1	**8.** 4	**13.** 3	**18.** 4	**23.** 2	**28.** 3
4. 1	**9.** 4	**14.** 3	**19.** 4	**24.** 3	**29.** 4
5. 3	**10.** 2	**15.** 2	**20.** 1	**25.** 2	**30.** 2

PART B–1

31. 4	**34.** 2	**36.** 1	**38.** 2	**40.** 4	**42.** 3
32. 2	**35.** 1	**37.** 1	**39.** 1	**41.** 4	**43.** 4
33. 3					

PART B–2

44. *See* Answers Explained.
45. *See* Answers Explained.
46. *See* Answers Explained.
47. 3
48. *See* Answers Explained.
49. 1

50. 2
51. *See* Answers Explained.
52. *See* Answers Explained.
53. *See* Answers Explained.
54. *See* Answers Explained.
55. *See* Answers Explained.

PART C. *See* **Answers Explained**.

PART D

73. 3
74. 2
75. 4
76. 3
77. *See* Answers Explained.
78. *See* Answers Explained.
79. *See* Answers Explained.

80. *See* Answers Explained.
81. 4
82. 1
83. *See* Answers Explained.
84. *See* Answers Explained.
85. *See* Answers Explained.

Answers Explained

PART A

1. **2** Materials are transported within a single-celled organism by the *cytoplasm*. The cytoplasm is a watery medium that fills the cell's interior and suspends the cell organelles. Materials that enter the cell are able to diffuse freely within the cytoplasm, reaching target organelles for processing.

WRONG CHOICES EXPLAINED

(1) Materials are *not* transported within a single-celled organism by the *nucleus*. The nucleus is the cell organelle that contains the genetic information necessary to synthesize the proteins necessary for life. The nucleus is not directly involved in transporting material within the cell.

(3) Materials are *not* transported within a single-celled organism by the *mitochondrion*. The mitochondrion is the cell organelle that contains the enzymes that catalyze the biochemical reactions of respiration, a process that provides ATP energy for other cell functions. The mitochondrion is not directly involved in transporting material within the cell.

(4) Materials are *not* transported within a single-celled organism by the *ribosome*. The ribosome is the cell organelle that converts genetic codes into polypeptide chains using amino acids as the building blocks for this process. It is not directly involved in transporting material within the cell.

2. **3** Row *3* of the chart correctly pairs a food molecule with its building block. Proteins are biochemical food molecules that are produced in cells by linking together hundreds or thousands of amino acid molecules into long polypeptide chains.

WRONG CHOICES EXPLAINED

(1) Row *1* of the chart does *not* correctly pair a food molecule with its building block. Starches are biochemical food molecules that are produced in plant cells by linking together simple sugars, not amino acids.

(2) Row *2* of the chart does *not* correctly pair a food molecule with its building block. Sugars are simple biochemical carbohydrate molecules composed of up to two simple sugars, such as glucose, as their building blocks. Starch is not a simple sugar, so it is not the building block of sugar.

(4) Row *4* of the chart does *not* correctly pair a food molecule with its building block. Amino acids are simple biochemical molecules that serve as the building blocks of proteins. Structurally, they are composed of an amine ($-NH_2$) unit, a carboxylic acid ($-COOH$) unit, and a variable radical unit. Amino acids are not composed of sugar.

3. **1** The flow of energy in an ecosystem is best described as energy moving in *one direction from the Sun to the producers and then to the consumers*. Light energy from the Sun is absorbed by green plants (producers) and is used to power the process of photosynthesis. The photosynthetic reactions result in the manufacture of glucose molecules, which contain chemical bond energy. Plants then convert the glucose into other biochemical materials that are consumed by animals and other consumer organisms. Those biochemical materials are used as a source of energy in those consumers. Ultimately, the energy that was absorbed from sunlight and then passed from producers to consumers is dissipated into the environment as waste heat.

WRONG CHOICES EXPLAINED

(2) The flow of energy in an ecosystem is *not* best described as energy moving in *one direction from a consumer to a producer and then to the Sun as heat and light*. This description is the exact opposite of the energy flow pattern that scientists know to be correct.

(3), (4) The flow of energy in an ecosystem is *not* best described as energy moving in *two directions between the producers that are present* or *two directions, back and forth, between the producers and the consumers*. Scientists recognize that energy flows in only one direction in the environment. It always flows from sources of energy, such as the Sun, through producers to consumers until it is dissipated into the environment as heat.

4. **1** Harm to the developing fetus might occur because the placenta *transfers oxygen and nutrients to the fetal blood*. This transfer occurs by diffusion across the cellular boundary that separates the maternal and fetal bloodstreams. If this boundary is significantly damaged due to separation of the connection, infusions of vital oxygen and nutrients will be diminished and damage to or death of fetal tissues may occur as a result.

WRONG CHOICES EXPLAINED

(2) It is *not* true that harm to the developing fetus might occur because the placenta *sends maternal blood into the fetus*. In a normal pregnancy, the bloodstreams of the mother and fetus are always separated by a placental cellular boundary, not intermingled. However, a damaged placental connection may allow such intermingling, a condition that would imperil the life of the fetus.

(3) It is *not* true that harm to the developing fetus might occur because the placenta *supplies milk for the fetus*. The placenta never supplies milk to the fetus. Rather, the placenta supplies dissolved simple nutrients to the fetus via diffusion across the placental cellular boundary that separates the maternal and fetal blood supplies.

(4) It is *not* true that harm to the developing fetus might occur because the placenta *breaks down wastes of the fetus*. Metabolic wastes of the fetus diffuse across the placenta's cellular boundary from the fetal bloodstream to

that of the mother, where they are eliminated from the maternal bloodstream by the mother's excretory processes.

5. **3** *Cloning* is the process that produces only identical offspring. Cloning is a laboratory technique in which offspring that are genetically identical to the parent and to each other are produced using a diploid nucleus of a single "parent" organism. Because all offspring produced by cloning contain identical genetic information, they must also be phenotypically identical to each other.

WRONG CHOICES EXPLAINED
(1) *Meiotic cell division* is *not* the process that produces only identical offspring. Meiotic cell division is a process by which haploid (*n*) gametes (sperm and egg cells) are produced from diploid (2*n*) primary sex cells. This sexual process helps to ensure varied, not identical, offspring.

(2) *Selective breeding* is *not* the process that produces only identical offspring. Selective breeding is a technique used by animal and plant breeders in which organisms with desirable traits are cross-bred with the hope of producing offspring that also display those traits. This is a sexual process that may produce similar, but never identical, offspring.

(4) *Fertilization* is *not* the process that produces only identical offspring. Fertilization is a process in which male and female haploid gametes fuse to produce a diploid zygote with a unique combination of genetic traits. This is a sexual process that may produce similar, but rarely identical, offspring.

6. **4** Structure A is most likely a *nucleus that is responsible for the storage of information*. The organelle shown is the right size and contrast to be identified as a cell nucleus. The nucleus is the cell organelle that contains the genetic information necessary to synthesize the proteins necessary for life.

WRONG CHOICES EXPLAINED
(1), (3) Structure A is *not* most likely a *mitochondrion that synthesizes food for the cell* or a *mitochondrion that absorbs energy from the Sun*. The mitochondrion is the cell organelle that contains the enzymes that catalyze the biochemical reactions of respiration, a process that provides ATP energy for other cell functions. The mitochondrion is not involved in food synthesis or energy absorption.

(2) Structure A is *not* most likely a *nucleus that is the site of food storage*. Cell nuclei do not serve as sites for food storage. This function is carried out in cells by the food vacuole.

7. **1** *The organelles in B perform similar functions to the organ systems in A* is the statement that best explains how A and B are able to survive in their environment. Organism A represents a multicellular organism known as an insect; organism B represents a unicellular organism known as a paramecium.

Organism *B* can perform all the basic life functions necessary to support life through the actions of subcellular structures known as organelles. Organelles are microscopic structures found in living cells that perform specialized life functions for that cell. The term organelle literally means "tiny organ," reflecting the life-enhancing function that these structures perform for the cell.

WRONG CHOICES EXPLAINED
(2) *The transport system in B is more complex than the transport system in A is not* the statement that best explains how *A* and *B* are able to survive in their environment. In fact, organs and organ systems found in organism *A* are more complex than the organelles found in organism *B*.

(3) *Both A and B take in oxygen from the water* is *not* the statement that best explains how *A* and *B* are able to survive in their environment. Organism *B*, an aquatic (freshwater) organism, does absorb dissolved oxygen from its watery environment. However, organism *A* is a terrestrial (land) animal that, in its adult stage, absorbs oxygen directly from the atmosphere after receiving the oxygen through spiracles located along its abdomen, not from a water environment.

(4) *Only A can pass on traits to offspring* is *not* the statement that best explains how *A* and *B* are able to survive in their environment. Both organism *A* and organism *B* contain DNA in their cell nuclei that enable them to produce offspring that carry genetic traits found in the parent.

8. **4** In the future, this change can be passed on to *his skin cells, only*. All skin cells that arise by mitotic cell division from this mutated cell will carry the mutation. Because it did not occur in reproductive tissues, the mutation will not be passed on to future generations.

WRONG CHOICES EXPLAINED
(1), (2) It is *not* true that in the future, this change can be passed on to *his male and female children* or *his male children, only*. Because the DNA mutation did not occur in reproductive tissues, it will not be passed on to future generations.

(3) It is *not* true that in the future, this change can be passed on to *all cells in his body*. Only cells that arise from the mutated skin cell will receive the DNA mutation.

9. **4** One ecological concern raised by this expansion is that *planting large expanses of one crop reduces the biodiversity of the area*. Biodiversity is a term that relates to the variety of plant and animal species that inhabit an area and form a sustainable ecological community. When biodiverse ecosystems such as forests are removed and replaced with a single crop such as palm trees, biodiversity is reduced and ecological sustainability is threatened.

WRONG CHOICES EXPLAINED

(1) It is *not* true that one ecological concern raised by this expansion is that *the natural forest ecosystem may harm the palm trees*. It is true that should the natural forest environment become reestablished by means of ecological succession, nonnative species such as the palm tree would most likely be eliminated from the environment. However, this eventuality would be an agricultural, not an ecological, concern.

(2) It is *not* true that one ecological concern raised by this expansion is that *the use of the land for agriculture will increase the biodiversity of the area*. Converting the area to agricultural activities such as the one described will likely decrease, not increase, biodiversity.

(3) It is *not* true that one ecological concern raised by this expansion is that *humans are changing the basic processes of the palm trees*. Increasingly, agricultural scientists are experimenting with altering the genetic makeups of the plants they wish to grow in former natural environments. In a manner of speaking, these genetic alterations are changing the basic processes that promote the plants' survival. Environmentalists do have concerns about this process. They warn against wholesale genetic manipulation without examining unintended consequences on the environment and on human health.

[NOTE: This is a possible correct answer to the question as written; however, it is not the best answer.]

10. **2** This level of direct harvesting could cause *ecosystems to be severely damaged*. Each species of organism inhabiting a balanced ecosystem fills an environmental niche that promotes its own survival and that of numerous other organisms. Wholesale removal of one species from an ecosystem can cause drastic changes to ecological balance. Wholesale removal can contribute to the total collapse of the ecosystem and the unintended elimination of many other species.

WRONG CHOICES EXPLAINED

(1) It is *not* true that this level of direct harvesting could cause *ecosystems to be improved for future generations*. Wholesale removal of one species from an ecosystem does not improve the ecosystem. It helps to destroy the ecosystem.

(3) It is *not* true that this level of direct harvesting could cause *the restoration of environmental stability*. Wholesale removal of one species from an ecosystem does not restore environmental stability. It disrupts environmental stability.

(4) It is *not* true that this level of direct harvesting could cause *all other fish species to increase in number*. Wholesale removal of one species from an ecosystem does not cause all other fish species to increase. Instead, wholesale removal causes only some species to increase for limited periods of time.

11. **3** *A segment of a DNA molecule that contains the instructions for producing a trait in an organism* is the phrase that best describes a gene. This segment of DNA contains the instructions to code a single polypeptide (enzyme), whose role is to catalyze a single biochemical reaction in the living cell.

WRONG CHOICES EXPLAINED

(1), (2) *A segment of a DNA molecule found only in the body cells of an organism* and *a segment of a DNA molecule found only in the gametes of an organism* are *not* the phrases that best describe a gene. DNA segments (genes) are found in every living cell that contains a nucleus, whether that cell is a somatic (body) cell or a gametic (sex) cell.

(4) *A segment of a DNA molecule that contains the instructions for producing all the characteristics of an organism* is *not* the phrase that best describes a gene. A gene is composed of only one DNA segment. Each body cell contains thousands of DNA segments, or all the genes necessary to code all the enzymes that catalyze every biochemical reaction needed to produce all the traits that define an individual. Gametes contain half of these genes.

12. **2** *Adenine pairs with thymine—cytosine pairs with guanine* are the base pairings that normally occur during DNA replication. Adenine and thymine are chemically complementary to each other but not with cytosine or guanine. Cytosine and guanine are chemically complementary to each other but not with adenine or thymine. This complementarity is responsible for accurate pairings resulting in the production of exact complementary DNA strands during replication.

WRONG CHOICES EXPLAINED

(1), (3), (4) *Guanine pairs with cytosine—thymine pairs with thymine; thymine pairs with guanine—cytosine pairs with adenine*; and *cytosine pairs with cytosine—thymine pairs with thymine* are *not* the base pairings that normally occur during DNA replication. Except for the guanine-cytosine pairing in choice (1), these pairs of bases are not complementary. Therefore these pairs of bases will not form complementary DNA strands during replication.

13. **3** Evolution of a species could occur as a result of changes in the *genes in an egg cell*. If a gamete such as an egg cell experiences changes in one or more genes and if that gamete participates in fertilization, the changed gene may result in an altered phenotype (characteristic) in the resulting offspring. If the altered phenotype provides the offspring with some adaptive advantage, it is possible that the changed gene will be passed on to future generations and will become increasingly common in the species gene pool. If given enough gene changes and enough time, phenotypic changes may accumulate to the point that a new variety of the species is recognized. If

enough phenotypic changes accumulate, the species variety may become reproductively independent of the ancestral species and therefore represent a newly evolved, separate species.

WRONG CHOICES EXPLAINED

(1), (2), (4) It is *not* true that the evolution of a species could occur as a result of changes in the *DNA in muscle cells, base sequences in liver cells*, or *number of chromosomes in a fetal bone cell*. Any gene change that occurs in a somatic (body) cell such as muscle, liver, or bone cells cannot be passed from the individual organism to its offspring. For this reason, such gene changes can never result in evolution. Instead, they are lost upon the death of the organism.

14. **3** One positive impact that industrialization has had is that *industrialization has been a source of many jobs for people*. Industrialization has resulted in monetary profits for industrial companies, some of which have been used to pay salaries to their workers. These salaries, in turn, have been used by the workers to purchase goods and services, an activity that has helped to drive local and national economies. Although industrialization has benefited humans in several economic ways, it has also harmed them in terms of their health and the health of the natural environment.

WRONG CHOICES EXPLAINED

(1), (2), (4) It is *not* true that one positive impact that industrialization has had is that *industrialization produces waste gases that pollute the air, fossil fuels used by industries help reduce finite resources*, or *new technologies have increased acid rain*. Each of these responses describes a negative, not a positive, impact of industrialization on the natural environment and on human health.

15. **2** One reason for this procedure is to *prevent mutations in gametes*. X-rays cannot pass through lead. Placing the lead shield so that it covers the lower abdomen, where gametes are formed and stored prior to fertilization, helps to ensure that the X-rays do not penetrate body tissues and expose gametic DNA to the mutagenic effects of this radiation.

WRONG CHOICES EXPLAINED

(1), (3) It is *not* true that one reason for this procedure is to *protect the patient against broken bones* or to *improve circulation in the patient*. No competent scientific evidence indicates that placing a lead shield over major organs during X-ray procedures has either of these effects.

(4) It is *not* true that one reason for this procedure is to *increase the chance of a change in DNA*. In fact, the reason for placing a lead shield over major organs during X-ray procedures is to decrease, not increase, the chance of a change in DNA.

16. **4** The best explanation for this behavior is that *the chemical acts as a stimulus for a particular behavior*. Such automatic stimulus-response behaviors are common among animal species and represent survival mechanisms that help to protect the health of the entire species.

WRONG CHOICES EXPLAINED

(1) It is *not* true that the best explanation for this behavior is that *the ants are responding to a chromosomal mutation in ant X*. No information is provided that indicates that the chemical caused a mutation in ant X.

(2) It is *not* true that the best explanation for this behavior is that *the chemical is exhibiting a feedback mechanism*. Feedback mechanisms are characteristic of living organisms, not of inanimate chemical substances.

(3) It is *not* true that the best explanation for this behavior is that *the live ants must continue this behavior until they have eliminated ant X*. The ants in the colony are described as responding only to the chemical, not to ant X.

17. **2** The stability of the population size is most likely the result of *environmental factors that keep the population in check*. Such forces are known as limiting factors. They may include abiotic factors such as average temperature, water availability, and soil type. They may also include biotic factors such as food sources, predation, and disease.

WRONG CHOICES EXPLAINED

(1) The stability of the population size is *not* most likely the result of *the development of mutations in young rabbits*. Mutations are random events that result from exposure to mutagenic agents such as radiation and chemicals. Mutations alter DNA and may lead to small phenotypic changes. However, they do not limit the growth of populations.

(3), (4) The stability of the population size is *not* most likely the result of *rabbits continuing to reproduce when the population is large* or of *the survival of more female rabbits than male rabbits*. Either of these events would likely result in an increase, not a decrease, in rabbit population size.

18. **4** This treatment will be most successful if the virus is inserted into cells that *continue to divide during the life of the patient*. By reproducing many cells containing the corrected gene, the phenotypic effects of the gene will be more likely to become widespread in the body and may reverse the genetic disorder in the individual.

WRONG CHOICES EXPLAINED

(1) It is *not* true that this treatment will be most successful if the virus is inserted into cells that *lack a nucleus*. In order to affect the genetic disorder, the corrected gene would have to be inserted into DNA, which is located in the nucleus of the cell.

(2) It is *not* true that this treatment will be most successful if the virus is inserted into cells that *are recycled after death rather than removed from the body*. When a cell dies, the DNA within it becomes inactive. So recycling dead cells will have no effect on correcting the genetic disorder.

(3) It is *not* true that this treatment will be most successful if the virus is inserted into cells that *carry out one specific function rather than multiple functions*. Depending on the nature of the genetic disorder, it may be necessary to insert the virus directly into tissues that carry out one specific function. For example, a corrected gene for insulin production would probably have to be inserted into tissues of the pancreas in order to have any corrective effect on the genetic disorder of diabetes.

[NOTE: This is a possible correct answer to the question as written; however, it is not the *best* answer.]

19. **4** As a trade-off, the shopping mall would cause a decrease in the *variety of wildlife populations in the area*. It is highly likely that as a result of building this project, natural habitat will be reduced and native wildlife populations will be displaced or eliminated. Identifying trade-offs is increasingly employed when conducting a cost-benefit analysis for a particular development. It asks the essential question, "In addition to the dollar costs of constructing a proposed development, are there other, less obvious costs to humans and to the natural environment?" Too often in the past, this question was not asked until after large-scale environmental destruction had occurred.

WRONG CHOICES EXPLAINED

(1), (2), (3) It is *not* true that as a trade-off, the shopping mall would cause a decrease in the *amount of air pollution, volume of garbage and litter*, and *amount of wastewater entering the local sewage system*. It is likely that all of these negative consequences of development will increase, not decrease, as a result of this project. These consequences should also be considered in the cost-benefit analysis of this project.

20. **1** Structure *A* is the structure that produces chemicals that regulate the reproductive cycle. Structure *A* represents the ovary, a reproductive organ that produces gametes (eggs) that may become fertilized and ultimately result in the formation of a new human being. The chemicals produced by structure *A* that regulate the reproductive cycle, known as hormones, include estrogen and progesterone.

WRONG CHOICES EXPLAINED

(2) Structure *B* is *not* the structure that produces chemicals that regulate the reproductive cycle. Structure *B* represents the oviduct (Fallopian tube), an organ that transports the egg from the ovary to the uterus. It also serves as the site of fertilization.

(3) Structure C is *not* the structure that produces chemicals that regulate the reproductive cycle. Structure C represents the uterus, an organ that serves as the site of embryonic development. During pregnancy, the tissues of the placenta produce hormones that help to sustain the development of the fetus and prepare the mother's body to give birth.

(4) Structure D is *not* the structure that produces chemicals that regulate the reproductive cycle. Structure D represents the vagina, an organ that serves as the site of sperm implantation as well as the canal through which the baby passes at birth.

21. **3** The arrows represent the movement of *ATP and carbon dioxide*. The structure illustrated is a mitochondrion, an organelle that houses the reactions of aerobic respiration in the cell. The chemical result of these reactions is the production of energy-rich ATP (adenosine triphosphate), with carbon dioxide being produced as a by-product. Once produced, these substances diffuse out of the mitochondrion and are either used (ATP) or excreted (carbon dioxide) by the body.

WRONG CHOICES EXPLAINED

(1) The arrows do *not* represent the movement of *carbon dioxide and sugar*. Carbon dioxide does diffuse out of the mitochondrion. However, sugar diffuses into the mitochondrion, where it serves as the primary fuel for the respiratory reactions.

(2) The arrows do *not* represent the movement of *oxygen and ATP*. ATP does diffuse out of the mitochondrion. However, oxygen diffuses into the mitochondrion, where it helps to maximize the productivity of the respiratory reactions.

(4) The arrows do *not* represent the movement of *oxygen and sugar*. Oxygen and sugar diffuse into, not out of, the mitochondrion during respiration. Oxygen helps to maximize the efficiency of the respiratory reactions for which sugar serves as the primary fuel.

22. **3** *Autotrophic cell* is the label that correctly identifies what X represents in the concept map. An autotroph is a self-feeder, such as a green plant. Plant cells contain all four of the components shown in the concept map, including chloroplasts that serve as the site of the photosynthetic reactions.

WRONG CHOICES EXPLAINED

(1), (2) *Nucleus* and *chromosome* are *not* the labels that correctly identify what X represents in the concept map. A nucleus and a chromosome contain DNA and genes but not ribosomes or chloroplasts.

(4) *Heterotrophic cell* is *not* the label that correctly identifies what X represents in the concept map. A heterotrophic cell contains DNA, genes, and ribosomes but not chloroplasts.

23. **2** The shape of each of these molecules is important because *the shape of the molecule determines how it functions in chemical reactions*. In an enzyme-substrate reaction, this chemical shape allows the two molecules to interact at their active sites to facilitate a biochemical reaction. If the shape of one or both molecules is altered, this interaction may be slowed or prevented altogether.

WRONG CHOICES EXPLAINED

(1) It is *not* true that the shape of each of these molecules is important because *molecules having different shapes are always found in different organisms*. Each type of biochemical found in living things has its own unique chemical shape. However, this fact does not address the question as asked.

(3), (4) It is *not* true that the shape of each of these molecules is important because *the shape of a molecule determines the age of an organism* or because *if the shape of any molecule in an organism changes, the DNA in that organism will also change*. No competent scientific evidence supports these statements.

24. **3** The color differences were most likely due to *gene expression being affected by the environment*. Although all genes contain the genetic information necessary to produce phenotypic effects (characteristics), some genes require a particular set of environmental conditions in order to be expressed. In this case, the wing color of two particular species of butterfly is affected by the presence or absence of certain wavelengths of light during development.

WRONG CHOICES EXPLAINED

(1) The color differences were *not* most likely due to *mutations in the color-producing genes*. Mutations are random events affecting the base-pair sequences of DNA. It would be extremely unlikely that the same mutation would affect all the caterpillars in this study in the same way at the same time.

(2) The color differences were *not* most likely due to *the caterpillars in the red light producing more DNA*. The amount of DNA in the cells of these species remains constant due to the processes of meiosis and mitosis. The quantity of DNA is not affected by the presence or absence of light.

(4) The color differences were *not* most likely due to *the caterpillars in the dark evolving less than those in the light*. Evolution is a process requiring extremely long periods of time. Evolution cannot occur or be a factor in the time frame of a single scientific experiment.

25. **2** *Asexual reproduction* is the process the student recorded in his drawings. Also known as binary fission, this form of asexual reproduction is one in which a single parent cell undergoes mitotic cell division to produce two offspring cells that are genetically identical to each other and to the parent cell.

WRONG CHOICES EXPLAINED

(1) *Genetic engineering* is *not* the process the student recorded in his drawings. Genetic engineering is a laboratory technique in which a gene for a desired trait (such as the production of a desired biochemical) is snipped from the DNA of a human cell and inserted into the genome of a bacterial cell. This bacterial cell is then cultured to obtain large quantities of the desired biochemical. Genetic engineering is not the process illustrated.

(3) *Selective breeding* is *not* the process the student recorded in his drawings. Selective breeding is a technique used by animal and plant breeders in which organisms with desirable traits are cross-bred with the hope of producing offspring that also display those traits. Selective breeding is accomplished using multicellular animals and plants, not unicellular organisms such as the one illustrated.

(4) *Gamete formation* is *not* the process the student recorded in his drawings. Gamete formation is a biological process carried out by sexually reproducing organisms in which sex cells (eggs or sperm) are produced from primary sex cells located in the gonads (ovary or testis). This process involves two separate cell divisions, not just a single division such as the one illustrated.

26. **4** This is most likely because *resistant bacteria survive in greater numbers and pass the trait to their offspring*. A colony containing hundreds of thousands of bacteria is statistically likely to include a few individual bacteria that have randomly mutated in a manner that gives them a genetic resistance to a particular antibiotic. Use of this antibiotic to help fight an infection will readily eliminate non-resistant individuals, leaving resistant individuals to form new colonies of resistant bacteria. Overuse of antibiotics can have the unintended consequence of selecting for antibiotic-resistant infectious bacteria, creating strains of pathogenic bacteria whose infections cannot be cured by any known antibiotic.

WRONG CHOICES EXPLAINED

(1) It is *not* true that this is most likely because *populations of resistant bacteria are small*. When given a favorable growth medium such as the human body, bacterial colonies can grow at an exponential rate. For this reason, population size is not a factor in determining how rapidly a colony will spread.

(2) It is *not* true that this is most likely because *exposure to antibiotics increases the rate of reproduction in bacteria*. When exposed to antibiotics, the growth of a bacterial colony will normally be significantly decreased, not increased.

(3) It is *not* true that this is most likely because *resistant bacteria are small when compared to non-resistant bacteria*. The gene affecting antibiotic resistance is not known to affect the dimension of a bacterial cell. Resistant and non-resistant bacteria of a given species are normally the same size.

27. **3** When getting a vaccination, *a weakened form of a virus* is the substance injected into the body. A vaccine is normally prepared from weakened, dead, or fragmented virus particles. The body's immune response detects chemical markers on the coat of the virus and manufactures antibodies specific to that virus. In this way, the body develops a natural immunity to that particular viral infection.

WRONG CHOICES EXPLAINED

(1) When getting a vaccination, *bacteria to combat the pathogen* is *not* the substance injected into the body. Bacteria are not effective in combating viral infections. Bacteria are never included as a component of vaccines.

(2) When getting a vaccination, *white blood cells to engulf a pathogen* is *not* the substance injected into the body. White blood cells manufactured by the body can be effective in eliminating foreign invaders such as bacteria and viruses in the bloodstream and surrounding tissues. However, white blood cells are not included as a component of vaccines.

(4) When getting a vaccination, *antibiotics to kill a virus* is *not* the substance injected into the body. Antibiotics produced from molds or fungi are not effective in combating viral infections. Antibiotics are never included as a component of vaccines.

28. **3** Many beverage companies are required to recycle bottles and cans because this activity directly reduces *solid waste and depletion of resources.* Bottles and cans are produced from naturally occurring minerals including silica (sand), bauxite (aluminum ore), and hematite (iron ore). Although sand is a relatively abundant material, supplies of bauxite and hematite are becoming increasingly difficult and costly to extract and refine. Recycling containers made from these substances saves the cost of mining and refining, which preserves remaining supplies of these resources for future generations. In addition, recycling reduces pressure on already-stressed landfills by reducing the quantity of solid waste being discarded.

WRONG CHOICES EXPLAINED

(1), (2), (4) It is *not* true that many beverage companies are required to recycle bottles and cans because this activity directly reduces *air pollution and destruction of the ozone shield, overpopulation and soil erosion,* or *thermal pollution and extinction of wildlife.* Although these factors are not directly reduced by recycling, some of them may be reduced indirectly. For example, air pollution, thermal pollution, and soil erosion may be reduced because the need to mine, transport, and refine these raw materials is reduced through recycling efforts.

29. **4** The bacteria that result from reproduction of this cell will most likely have the ability to *produce the human protein coded for by gene X.* Genetic engineering is a laboratory technique in which a gene for a desired

trait (such as production of a desired protein) is snipped from the DNA of a human cell and inserted into the genome of a bacterial cell. This bacterial cell is then cultured to obtain large quantities of the desired protein.

WRONG CHOICES EXPLAINED

(1) The bacteria that result from reproduction of this cell will *not* most likely have the ability to *replicate all of the genetic instructions found in humans*. The human genome is made up of many thousands of individual instructions, each represented by a unique gene. Gene *X* is only one such gene. So replication of all the genetic instructions in the human genome is not possible.

(2) The bacteria that result from reproduction of this cell will *not* most likely have the ability to *produce vaccines to be used to immunize humans*. Vaccines are mixtures of dead, weakened, or fragmented viruses known to cause certain viral infections. Vaccines are not produced by inserting a human gene into a bacterium.

(3) The bacteria that result from reproduction of this cell will *not* most likely have the ability to *produce a human blood cell according to instructions in gene X*. The individual characteristics of a human blood cell are controlled by dozens, if not hundreds, of genes that are activated within the human genome for that purpose. Gene *X* is only one such gene, so production of a blood cell is not possible.

30. **2** This is an example of *an unintended consequence of adding an organism to an ecosystem*. Eurasian water milfoil is an example of an invasive species that now competes with native water plants for light, soil minerals, growing space, and other limited resources. In addition, its decomposition depletes oxygen supplies needed by fish and other aquatic animal species. These negative effects are unintended consequences of the introduction of this invasive species into the U.S.

WRONG CHOICES EXPLAINED

(1) This is *not* an example of *human consumption of finite resources*. Human consumption of finite resources refers to our depleting natural resources such as minerals, petrochemicals, and other nonrenewable resources for industrial purposes. It does not relate to humans introducing nonnative species into new environments.

(3) This is *not* an example of *an abiotic factor having a negative effect on an ecosystem*. Abiotic (nonliving) factors such as light, temperature, water availability, and soil chemistry are not discussed in this question. Eurasian water milfoil is an example of a biotic (living) factor in the environment.

(4) This is *not* an example of *the introduction of a species that has increased the long-term biodiversity of an ecosystem*. Because Eurasian water milfoil competes with native species, over the long term it will decrease, not increase, the biodiversity of an ecosystem.

PART B–1

31. **4** *Overproduction* is the concept that the graph would best illustrate if the line did not stay around the carrying capacity but continued to rise. Overproduction is the tendency of living populations to increase in number exponentially in the absence of environmental limiting factors such as predation, disease, overcrowding, and starvation.

WRONG CHOICES EXPLAINED

(1) *Environmental stability* is *not* the concept that the graph would best illustrate if the line did not stay around the carrying capacity but continued to rise. If the line continued to rise, the rapid increase in the population would quickly exhaust available resources, causing the ecosystem to become unstable and eventually collapse.

(2) *Genetic variety* is *not* the concept that the graph would best illustrate if the line did not stay around the carrying capacity but continued to rise. Genetic variety is a concept in evolutionary science that describes how genetic variation in a species population can lead to the creation of new species varieties and eventually to new species. The graph does not illustrate genetic variety.

(3) *Behavioral change* is *not* the concept that the graph would best illustrate if the line did not stay around the carrying capacity but continued to rise. Many behaviors in natural animal species are genetically programmed (instinctive) and therefore difficult or impossible to change. The graph does not measure behavioral change.

32. **2** *Owls* are the organisms that would receive the *least* amount of transferred solar energy. This concept is typically illustrated using an energy pyramid, where producer organisms (such as grasses) at the base of the pyramid absorb and store the greatest amount of solar energy. Producers are followed by herbivores (such as field mice) in the second tier of the pyramid and collectively contain only about 90% of the energy held by producers. Finally, carnivores (such as owls) at the top of the pyramid collectively contain only about 90% of the energy held by the herbivores. The energy lost between each level of the pyramid is dissipated into the environment as heat.

[NOTE: Frogs are incorrectly shown in the food web diagram as herbivores. In fact, frogs are primarily insectivorous carnivores.]

WRONG CHOICES EXPLAINED

(1), (4) *Grasses* and *field mice* are *not* the organisms that would receive the *least* amount of transferred solar energy. Grasses are producers, and field mice are herbivores. Both of these groups of organisms contain more transferred solar energy than do the carnivores, represented by owls in the diagram.

(3) *Frogs* are *not* the organisms that would receive the *least* amount of transferred solar energy. As insectivorous carnivores, frogs belong to the same level of the energy pyramid as the owls. In nature, frogs would be part of the group that contains the least amount of transferred solar energy. However, because frogs are preyed upon by owls, it can be inferred that frogs collectively hold a greater amount of transferred solar energy than is collectively held by owls.

33. **3** According to this information, it can be inferred that these two species *require a similar soil pH*. These two species coexist in the same environment. So it can be broadly inferred that both can tolerate the same set of environmental conditions, probably including soil pH.

[NOTE: No information is provided in the question concerning the soil acidity requirements of these two species. Without this information, the only way the question can be answered correctly by the average student is through the process of elimination.]

WRONG CHOICES EXPLAINED
(1) It is *not* true that according to this information, it can be inferred that these two species *are both predators*. Predators are animals that derive their nutritional requirements from the bodies of animals upon which they have preyed. Birch bolete is a fungus, not a predator. A birch tree is a producer, not a predator.

(2) It is *not* true that according to this information it can be inferred that these two species *require the same amount of sunlight*. Because birch bolete is a nonphotosynthetic fungus and the birch tree is a photosynthetic producer, it can be inferred that these two species do not share the same requirements for sunlight.

(4) It is *not* true that according to this information, it can be inferred that these two species *recycle the remains of dead organisms*. The process of recycling the remains of dead organisms is carried on in nature by several species of bacteria and fungi known collectively as saprophytes. Producers such as birch trees do not carry on this process directly, although they do absorb and recycle certain soil nutrients that are derived from saprophytic processes.

34. **2** *Selective breeding was used to combine desirable traits from both varieties A and B* is the statement that best explains the development of variety *C*. Selective breeding is a technique used by animal and plant breeders in which organisms with desirable traits are cross-bred with the hope of producing offspring that also display those traits. In this case, the breeders were hoping to combine the heat resistance of variety *A* and the meat quality of variety *B* into a single new variety, *C*.

WRONG CHOICES EXPLAINED

(1) *Nuclei from body cells taken from variety A were inserted into egg cells lacking nuclei taken from variety B* is *not* the statement that best explains the development of variety *C*. This is a description of the laboratory technique known as cloning. The offspring of cloning would display genetic characteristics identical to the donors of nuclei from variety *A*, not the blended characteristics shown for variety *C*.

(3) *The need to adapt to changes in the environment led to the selection of advantageous characteristics in the offspring of variety B* is *not* the statement that best explains the development of variety *C*. Need is not a factor in biological evolution. Biological evolution is a process by which members of a species displaying favorable characteristics survive harsh environmental conditions to reproduce and pass on their favorable characteristics to offspring. Those members of the species with less favorable characteristics are less successful at passing on their traits to future generations.

(4) *Mutations that occurred in the body cells of variety A were passed on to the offspring generation after generation* is *not* the statement that best explains the development of variety *C*. Mutations received in body cells cannot be passed on to future generations; only those received in sex cells can be passed on.

35. 1 This is an example of *an adaptation to a specialized niche*. The relationship between the shark and the remora is a type of symbiosis known as commensalism. A commensalistic symbiosis is one in which one organism is helped by the relationship without harming or helping the other organism. In this case, the remora is helped by receiving food scraps left by the shark and by conserving energy because it does not have to swim on its own. The shark is neither deprived of food nor slowed down by the attachment of the remora to its skin.

WRONG CHOICES EXPLAINED

(2) This is *not* an example of *an adaptation of a successful parasite*. Parasitism is a type of symbiosis in which one organism (parasite) lives in or on the body of another organism (host) and derives nutritional and other benefits from the relationship while the host is harmed by the relationship. The shark-remora symbiosis is commensalism, not parasitism.

(3) This is *not* an example of *competition between two fish species for food*. The remora is described as eating scraps of food that the shark drops as it feeds, food that the shark would not have consumed even if the remora had not been present. No information is presented that indicates that the remora and shark compete for food resources in their relationship.

(4) This is *not* an example of *competition for abiotic resources*. Abiotic factors are those environmental conditions not linked to the activities of living things. Examples include rock substrata, water salinity, temperature, oxygen

concentration, and solar radiation, among others. No information is presented that indicates that the remora and shark compete for abiotic resources in their relationship.

36. **1** Row *1* is the row that shows the population with the greatest chance of survival in a changing environment. Species with the greatest genetic variation are generally more successful than populations with less genetic variation when these populations are subjected to severe environmental changes. Population 1 has four different varieties, which is more than the other populations illustrated. So population 1 should prove more successful than populations 2, 3, and 4 under changing environmental conditions.

WRONG CHOICES EXPLAINED

(2), (3), (4) Rows *2, 3,* and *4* are *not* the rows that show the population with the greatest chance of survival in a changing environment. Each of these populations has three different varieties, fewer than the four varieties illustrated for population 1. For this reason, these varieties are less likely to be successful than population 1 under changing environmental conditions.

37. **1** *The cells have identical genetic instructions, but different parts of these instructions are being expressed in each cell* is the statement that best explains how each of the different cell types can develop from the same embryo. Every body cell type produced from the zygote by mitotic cell division contains the full diploid set of chromosomes, half of which is contributed by each parent. During the process of differentiation, these cells mature into tissue types with specialized functions. For each tissue type, only those genes needed for the tissue's specialized function are activated, while those not needed are switched off.

WRONG CHOICES EXPLAINED

(2) *The cells have identical genetic instructions, and all parts of these instructions are being expressed in each cell* is *not* the statement that best explains how each of the different cell types can develop from the same embryo. Only those genes needed for the tissue's specialized function are activated, while those not needed are switched off.

(3) *The cells are produced by asexual reproduction and contain identical genetic instructions* is *not* the statement that best explains how each of the different cell types can develop from the same embryo. Although this is a true statement biologically, it does not explain how each of the different cell types shown in the diagram can develop from the same embryo.

(4) *The cells contain genetic instructions from two different parents and will express the instructions from one parent, only* is *not* the statement that best explains how each of the different cell types can develop from the same embryo. The expression of a gene is a function of its relative dominance com-

pared with its allelic counterpart. The gene's expression is not dependent on which parent contributed it during fertilization.

38. **2** The graph illustrates that enzymes 1, 2, and 3 *work best at different pH levels*. An examination of the graphed data reveals that enzyme 1 reaches its optimum rate of reaction at a pH of about 4.5, enzyme 2 at about 8.0, and enzyme 3 at about 10.5.

WRONG CHOICES EXPLAINED
(1) The graph does *not* illustrate that enzymes 1, 2, and 3 *are not affected by pH*. An examination of the graphed data indicates that each of these enzymes is affected differently by the relative acidity (pH) of the system in which they operate. For example, at a pH of 4.5, where enzyme 1 has the fastest activity, enzyme 3 does not work at all.

(3), (4) The graph does *not* illustrate that enzymes 1, 2, and 3 *work best in an acidic environment* or that they *work best in a basic environment*. An examination of the graphed data shows that enzyme 1 works best in an acidic environment (pH of 4.5), while enzymes 2 and 3 work best in a basic environment (pH of 8.0 and 10.5, respectively).

39. **1** Structure *A* is the structure that produces cells that have the potential to become gametes. Structure *A* represents the testis, a reproductive organ that contains primary sex cells that mature into haploid (*n*) sperm cells (gametes) under the influence of male reproductive hormones such as testosterone.

WRONG CHOICES EXPLAINED
(2) Structure *B* is *not* the structure that produces cells that have the potential to become gametes. Structure *B* is the vas deferens, a reproductive organ that transports sperm cells toward the seminal vesicles during sexual intercourse.

(3) Structure *C* is *not* the structure that produces cells that have the potential to become gametes. Structure *C* is the urinary bladder, an excretory organ that stores urine prior to its release into the environment.

(4) Structure *D* is *not* the structure that produces cells that have the potential to become gametes. Structure *D* is the penis, a reproductive organ that implants sperm and seminal fluid into the female reproductive tract during sexual intercourse.

40. **4** The purpose of this is to *preserve the diversity of plant species*. Some scientists are becoming increasingly concerned that as a result of climate change and other human-influenced environmental conditions, the agricultural advances of the past 10,000 years of human experience will be jeopardized or lost to future generations. By preserving the living seeds of food-producing plants, these scientists hope to provide a cache of genetic

diversity that will enable humans to recover from any future agricultural or social collapse.

WRONG CHOICES EXPLAINED

(1), (2), (3) It is *not* true that the purpose of this is to *increase the destruction of environments, continue the deforestation of world ecosystems,* or *decrease the dependence on plants for food.* Each of these purposes has a negative implication relative to the future of the environment. It is extremely unlikely that competent scientists would have these motives for collecting and preserving seeds of food-producing plants.

41. **4** Diagram *4* is the diagram that best illustrates the relationship between the number of cells, tissues, and organs in a complex multicellular organism. The body is made up of millions of cells, dozens of different tissue types made up of specialized cells, and a relatively small number of organs made up of specialized tissues. Diagram 4 most clearly illustrates the relative numbers of cells (the largest circle representing millions of cells), tissues (a medium circle within the universe of cells representing a medium number of tissue types), and organs (a small circle within the universe of tissues representing a relatively small number of organs) in the body of a complex multicellular organism such as a human.

WRONG CHOICES EXPLAINED

(1), (2), (3) Diagrams *1, 2,* and *3* are not the diagrams that best illustrate the relationship between the number of cells, tissues, and organs in a complex multicellular organism. Each of these diagrams illustrates a relationship among these components that does not accurately represent the relative number of the components in the body of a typical multicellular organism.

42. **3** This ameba would most likely be classified as a *consumer*. A consumer organism is one that depends on sources outside its own body for its food supply. The diagram illustrates an ameba consuming bacterial cells for food, indicating that it is a consumer organism.

WRONG CHOICES EXPLAINED

(1) This ameba would *not* most likely be classified as a *decomposer*. Decomposers digest dead plant and animal matter and convert it to soil nutrients usable by plants. Such digestion typically occurs outside the cells of the decomposer, and the digestion products are absorbed into the cells by diffusion. The ameba illustrated is clearly ingesting bacteria prior to digesting them. So the ameba is not considered to be a decomposer.

(2) This ameba would *not* most likely be classified as a *producer*. Producers manufacture their own food from inorganic raw materials in the presence of an energy source such as sunlight. The ameba illustrated is

clearly ingesting bacteria as food, not manufacturing food. So the ameba is not considered to be a producer.

(4) This ameba would *not* most likely be classified as a *pathogen*. Pathogens enter the bodies of host organisms and multiply in large numbers, excreting toxic wastes, disrupting bodily functions, and commandeering healthy cells to produce more copies of the pathogen. The ameba illustrated is clearly ingesting bacteria, not attacking host tissues. So the ameba is not considered to be a pathogen.

[NOTE: Certain species of ameba do cause disease (e.g., amebic dysentery) and so may be considered pathogens.]

43. **4** The activity taking place is *heterotrophic nutrition*. Heterotrophic nutrition is a form of nutritional activity in which an organism consumes the bodies of other organisms for food. The diagram illustrates an ameba consuming bacterial cells for food, indicating that it is carrying on heterotrophic nutrition.

WRONG CHOICES EXPLAINED

(1), (3) It is *not* true that the activity taking place is *photosynthesis* or *autotrophic nutrition*. Autotrophic nutrition is a form of nutritional activity in which an organism manufactures its own food from inorganic raw materials in the presence of an energy source such as sunlight. Photosynthesis is a specific type of autotrophic nutrition carried on by green plants and plantlike organisms. The ameba illustrated is clearly ingesting bacteria as food, not manufacturing food. So the ameba is not considered to be carrying on autotrophic nutrition or photosynthesis.

(2) It is *not* true that the activity taking place is *differentiation*. Differentiation is a process by which embryonic cells mature into tissue types with specialized functions. Nothing in the illustration leads to the conclusion that the ameba is specializing. So the ameba is not considered to be carrying on differentiation.

PART B–2

44. One credit is allowed for correctly marking an appropriate scale, without any breaks in the data, on each labeled axis. [1]

45. One credit is allowed for correctly plotting the data from the data table for the sample from person *A* on the grid, connecting the points, and surrounding each point with a small circle. [1]

46. One credit is allowed for correctly plotting the data from the data table for the sample from person *B* on the grid, connecting the points, and surrounding each point with a small triangle. [1]

47. **3** If the temperature of the tissue samples used in the experiment had been raised from 37°C (body temperature) to 50°C, the results would have been different because *increasing temperatures altered the structure of catalase, decreasing oxygen production.* No information is provided in the question concerning the temperature of the experimental setups or of the range of temperatures tolerated by the enzyme catalase. Based on a general knowledge of biology, it may be broadly inferred that the human

enzyme catalase acts at its optimum rate at or near human body temperature and denatures (deforms and becomes inactive) at or above 40°C. If this is the case, the ability of catalase to catalyze the breakdown of hydrogen peroxide to water and oxygen would be significantly impaired at a temperature of 50°C.

WRONG CHOICES EXPLAINED

(1) It is *not* true that if the temperature of the tissue samples used in the experiment had been raised from 37°C (body temperature) to 50°C, the results would have been different because *more enzymes are produced at higher temperatures, increasing the amount of hydrogen peroxide.* It is likely that catalase is synthesized optimally at temperatures at or near 37°C. A temperature of 50°C would almost certainly stop this synthesis activity altogether.

(2) It is *not* true that if the temperature of the tissue samples used in the experiment had been raised from 37°C (body temperature) to 50°C, the results would have been different because *more hydrogen peroxide is released at higher temperatures, increasing the activity of catalase.* The human enzyme catalase acts at its optimum rate at or near body temperature and denatures (deforms and becomes inactive) at or above 40°C. If this is the case, then the activity rate of catalase would be significantly impaired or stopped altogether at a temperature of 50°C.

(4) It is *not* true that if the temperature of the tissue samples used in the experiment had been raised from 37°C (body temperature) to 50°C, the results would have been different because *increasing temperatures decreased the synthesis of amino acids, increasing levels of hydrogen peroxide.* No information is provided in the question concerning the relationship between the synthesis of amino acids and the production of hydrogen peroxide. This is a nonsense distracter.

48. One credit is allowed for correctly stating which person, A or B, is more likely to be the first to have gray hair and for supporting the answer. Acceptable responses include but are not limited to: [1]

- *Person A—This person's tissue sample produced a lower amount of oxygen, indicating a lower activity rate of catalase.*
- *A—We know that person A has less catalase than person B because less oxygen was released in this experiment.*
- *A—Person A must have more hydrogen peroxide or less catalase present because less oxygen was measured.*
- *Person A—The graphed data indicate that less oxygen was produced over the 25-minute experiment for person A. Person A has either less enzyme (catalase) or less substrate (hydrogen peroxide) present in his/her tissue sample than person B has in his/her tissue sample.*

49. **1** One half-hour after the liter of water was consumed, the urine produced by the kidneys was *between 2 and 3 mL/min*. This value may be found by tracing along the horizontal axis to the 30-minute (one half-hour) mark, then tracing up to the point shown at that time value, then tracing left to the vertical axis to read the value at that point: about 2.2 mL/min or between 2 and 3 mL/min.

WRONG CHOICES EXPLAINED

(2) It is *not* true that one half-hour after the liter of water was consumed, the urine produced by the kidneys was *between 4 and 5 mL/min*. This value cannot be determined using the method described above.

(3), (4) It is *not* true that one half-hour after the liter of water was consumed, the urine produced by the kidneys was *eight times greater than normal* or *below the normal range*. At the average rate of 0.5–1.0 mL/min of urine production, the subject's measured urine output at minute 30 of about 2.2 mL/min was between two and four times the normal output, not eight times greater than normal or below normal.

50. **2** The change in urine production during this 3-hour period was most likely the result of *homeostatic feedback*. Homeostasis is a term used to describe the state of dynamic equilibrium under which living things operate most efficiently. Maintaining homeostasis in living things depends on the feedback mechanisms that provide for constant monitoring of the biochemical condition of the body's metabolic activity. In this case as the body detects the relatively high concentration of water in the blood tissue, the kidneys respond by withdrawing some of that water, mixing it with other materials to produce urine, and storing the urine temporarily in the urinary bladder for elimination.

WRONG CHOICES EXPLAINED

(1) The change in urine production during this 3-hour period was *not* most likely the result of *antibody production*. No information is presented in the question concerning the effect of antibodies on this process. This is a nonsense distracter.

(3) The change in urine production during this 3-hour period was *not* most likely the result of *enzymatic breakdown of the water consumed*. Water is not broken down enzymatically in the body. Rather, it is used in its full chemical form to facilitate the body's metabolic activities.

(4) The change in urine production during this 3-hour period was *not* most likely the result of *nerve cell malfunctions of the kidneys*. No information is presented in the question concerning the effect of nerve cells in the kidney on this process. This is a nonsense distracter.

51. One credit is allowed for correctly identifying a structure, in organisms that do not have kidneys, that is adapted to regulate water balance. Acceptable responses include but are not limited to: [1]

- *Cell membrane*
- *Guard cells/stomate*
- *Contractile vacuole*
- *Nephridium/nephridia*

52. One credit is allowed for correctly stating the length of time, *in minutes*, for the body to return to normal after the intake of water. Acceptable responses include but are not limited to: [1]

- *150 minutes*
- *160 minutes*
- *170 minutes*
- *180 minutes*
- *190 minutes*

53. One credit is allowed for correctly identifying the type of vegetation that appears to have the lowest soil nutrient requirements and for supporting the answer. Acceptable responses include but are not limited to: [1]

- *Grasses—They are the first plants/vegetation to grow on the abandoned field.*
- *Weeds—They grow in soil depleted of nutrients.*
- *Weeds and grasses—Just after the field was abandoned, the nutrient concentration was at its lowest. So it can be inferred that the plants that dominated for the first 18 years were those most tolerant of the depleted soil conditions.*

54. One credit is allowed for correctly identifying the type of vegetation that would be most common in this area 200 years after it was first abandoned and for supporting the answer. Acceptable responses include but are not limited to: [1]

- *Hardwood forest—This vegetation type makes up the climax community and will remain the most common plant until the environment changes.*
- *Hardwood forest—The hardwood forest stage will remain because it is the most stable plant community.*
- *Hardwood forest—Hardwood trees became the climax community because they replaced all the other species from year 70 to the present. They will likely continue to dominate through year 200.*

55. One credit is allowed for correctly describing how the types of vegetation present on this farmland would change if fire burned down all the trees 120 years after the land was abandoned. Acceptable responses include but are not limited to: [1]

- *Grasses and weeds would become the most common plants once again.*
- *Ecological succession would start again with grasses and weeds growing.*
- *The trees would be gone and would be replaced by grasses and weeds.*
- *Because the soil would no longer be depleted of nutrients, it is likely that ecological succession would progress rapidly from weeds and grasses to hardwood forest in a much shorter period of time.*
- *The seeds of all types of plants from nearby unburned areas would be transported to the burned area by wind and animals; these seeds would germinate in the fertile soil and begin the process of succession.*

PART C

56–58. Three credits are allowed for correctly explaining why these claims are valid. In your answer, be sure to:

- Explain why both plants and solar panels provide renewable energy rather than nonrenewable energy. [1]
- State how the widespread use of solar panels to generate electricity can help reduce global warming. [1]
- State how the energy-capturing process used by plants worldwide can help to reduce global warming. [1]

Acceptable responses include but are not limited to: [3]

The Sun constantly releases energy that is absorbed by both plants and solar panels. This energy is perpetually renewed by nuclear reactions in the Sun. [1] Using solar panels instead of fossil fuels to produce electricity will cut down on the release of greenhouse gases that cause global warming. [1] Carbon dioxide is a greenhouse gas that is removed from the atmosphere by green plants as they carry out photosynthesis. [1]

Solar panels and green plants are similar because they both absorb the Sun's energy and convert it to forms of energy that are useful to living things. The Sun's energy is constantly being produced, so is considered to be renewable. [1] Solar panels directly convert sunlight into electricity, reducing the need to burn fossil fuels for that purpose. This helps to reduce the release of carbon dioxide that contributes to global warming. [1] Green plants carry on photosynthesis, a process that removes the greenhouse gas carbon dioxide from the air. [1]

59. One credit is allowed for correctly describing the role of a drug like cyclosporine when transplanting organs from one person to another person. Acceptable responses include but are not limited to: [1]

- *The transplanted organ contains foreign antigens. Cyclosporine suppresses the immune response to these antigens.*
- *Cyclosporine stops the rejection of the organ.*
- *Drugs like cyclosporine suppress the response of the immune system.*

60. One credit is allowed for correctly explaining the difference between an infection caused by a fungus and an allergy caused by a fungus. Acceptable responses include but are not limited to: [1]

- *An infection involves an actual pathogen that attacks the human organism. An allergy is an immune response to a harmless substance in the environment.*

- *An infection is caused by a microbe, and an allergy is an immune response to an antigen or irritant.*
- *An infection is usually caused by a harmful organism, and an allergy is a response to a substance that is usually not harmful.*

61. One credit is allowed for correctly identifying the process most likely responsible for a mosquito initially becoming resistant to DEET. Acceptable responses include but are not limited to: [1]

- *Mutated gene that is passed on from one or both parents*
- *Recombination during fertilization involving the DEET-resistant gene*
- *Sexual reproduction involving a gamete containing the gene for DEET resistance*
- *Genetic inheritance from a parent with the gene for resistance*

62. One credit is allowed for correctly explaining how the continued use of repellant may cause the percentage of these resistant mosquitoes to increase in the future. Acceptable responses include but are not limited to: [1]

- *Mosquitoes resistant to DEET will land on people, extract blood, and successfully reproduce offspring with the DEET-resistant gene, increasing the percentage of the resistant gene in the mosquito gene pool.*
- *The presence of DEET on people will prevent mosquitoes not resistant to DEET from biting humans, thereby reducing the chances of successful reproduction and reducing the frequency of the non-resistant gene compared to the resistant gene.*
- *Mosquitoes that do not detect DEET will land on people, bite them, and then be able to reproduce more DEET-resistant mosquitoes.*

63. One credit is allowed for correctly identifying the process by which ancestral iguanas developed into the present-day marine iguanas and land iguanas of the Galapagos Islands. Acceptable responses include but are not limited to: [1]

- *Evolution*
- *Natural selection*
- *Adaptive radiation*
- *Speciation*
- *Geographic isolation*
- *Variation with selection pressure*

64. One credit is allowed for correctly identifying *one* organism in the Galapagos Islands that directly limits the population of both the marine iguanas and land iguanas. Acceptable responses include: [1]

- *Cats*
- *Rats*

[NOTE: No credit is allowed for goats because they do not compete with marine iguanas for food.]

65. One credit is allowed for correctly identifying which population of iguanas, marine or land, is expected to be larger and for supporting the answer. Acceptable responses include but are not limited to: [1]

- *Marine—The population is larger because its food supply is greater than that of land iguanas.*
- *Marine—Goats compete with the land iguanas for the limited supply of cacti.*

66. One credit is allowed for correctly stating whether the introduction of goats would have a greater effect on the population of the marine iguanas or the land iguanas and for supporting the answer. Acceptable responses include but are not limited to: [1]

- *Land—These iguanas would be affected more because the goats are land animals and would compete with land iguanas for cacti.*
- *Land—Goats will feed on the cacti upon which the land iguanas depend for food.*

67. One credit is allowed for correctly identifying *one* technique that can be used to support the conclusion that these two species of iguana developed from a common ancestor. Acceptable responses include but are not limited to: [1]

- *Electrophoresis*
- *Gel electrophoresis*
- *Chromatography*
- *DNA fingerprinting/analysis*
- *Comparative anatomy*
- *Comparative biochemistry*
- *Fossil studies of extinct iguana species*

68–72. Five credits are allowed for correctly designing an experiment to test the effect of aeration on the growth of roots in bean seedlings. In your answer, be sure to:

- State *one* hypothesis the experiment would test.* [1]
- Describe how the control group will be treated differently from the experimental group. [1]
- Identify the dependent variable in the experiment. [1]
- State *one* reason why many setups should be used in both the experimental and control groups. [1]
- State *one* reason why several different kinds of seedlings were *not* tested in this experiment. [1]

[*NOTE: The hypothesis must *not* be stated in the form of a question.]

Acceptable responses include but are not limited to: [5]

Hypothesis: Over a two-week period, bean seedling roots growing in aerated nutrient solution will gain twice the length of bean seedling roots growing in nonaerated nutrient solution. [1] Experimental vs. control groups: Both groups will be set up under identical conditions except that the control group will have no air bubbled through the rubber tube in the nutrient solution. [1] Dependent variable: The dependent variable in this experiment (the variable being measured) is the length of root growth in the two setups. [1] Because bean plants may vary in their genetically determined rates of growth, many setups will be used in both groups to ensure valid experimental results. [1] Because different species of plants can have very different growing habits, only bean plants will be used in both experimental groups. [1]

My hypothesis is that bean seedlings grown in an aerated nutrient solution will produce at least four times the root mass of bean seedlings grown in a nonaerated nutrient solution over a 30-day period. [1] It is important to ensure that both the experimental and control group setups are exactly the same in every respect, except for the presence or absence of aeration in the experimental group. [1] On days 0, 10, 20, and 30, I will measure and record the mass of roots (the dependent variable in the experiment) in every setup. [1] In order to minimize experimental statistical error, I will create 20 experimental (aerated) and 20 control (nonaerated) setups. [1] Only bean seedlings will be used in this experiment because other species of plants differ genetically from beans and their use would introduce additional variables that would confuse or invalidate the experimental results. [1]

PART D

73. **3** The buildup of waste products in muscle cells that are active might cause *increased fatigue*. Muscle fatigue is the body's reaction to elevated levels of lactic acid in muscle tissues. Lactic acid is a substance that results from the incomplete respiratory breakdown of glucose in the absence of molecular oxygen. Nerve endings in the muscle tissue detect this lactic acid. These nerve endings signal the brain that rest and an increased heart rate are needed to restore oxygen levels to and to remove excess lactic acid from the muscles via the bloodstream.

WRONG CHOICES EXPLAINED·

(1), (2) It is *not* true that the buildup of waste products in muscle cells that are active might cause *digestion* or *cellular respiration*. Digestion and cellular respiration are essential life functions that occur without stimulation from waste product buildup.

(4) It is *not* true that the buildup of waste products in muscle cells that are active might cause *decreased heart rate*. In fact, elevated levels of lactic acid in the muscles will trigger an increased, not a decreased, heart rate so as to enable the body to carry oxygen to the muscle cells more efficiently.

74. **2** A *whale* is the mammal that would be most closely related to a hippopotamus. A review of the chart indicates that the whale and hippopotamus both evolved from a common ancestor in the relatively recent past. This indicates that the whale and the hippopotamus are more closely related to each other than either is to any other mammal illustrated in the chart.

WRONG CHOICES EXPLAINED

(1), (4) A *deer* and a *cow* are *not* the mammals that would be most closely related to a hippopotamus. The chart illustrates the cow and the deer as having evolved from a common ancestor about the same time as did the whale and the hippopotamus. This indicates that the cow and the deer are more closely related to each other than either is to the hippopotamus.

(3) A *pig* is *not* the mammal that would be most closely related to a hippopotamus. The chart shows that the evolutionary line leading to the modern-day pig branched off from the line leading to the modern-day hippopotamus in the far distant past. This indicates that the pig is not closely related to the hippopotamus.

75. **4** *The pulse rate of student F increased by 30 beats per minute* is the statement that is an example of an observation the students could have made. This direct observation can be made by examining the data from the table that indicate that the pulse rate of person *F* before exercise was 68 beats per minute and after exercise was 98 beats per minute, a difference of 30 beats per minute.

WRONG CHOICES EXPLAINED

(1) *Pulse rates in beats per minute decrease for all people after exercise* is *not* the statement that is an example of an observation the students could have made. No data are provided in the table that indicate what happens to students' pulse rates after the completion of exercise. So this observation cannot be made based on the data provided.

(2) *Student A most likely exercises regularly* is *not* the statement that is an example of an observation the students could have made. No data are provided in the table that indicate what exercise habits each subject maintains. So this observation cannot be made based on the data provided.

(3) *The pulse rate of student C was dangerously low* is *not* the statement that is an example of an observation the students could have made. No data are provided in the table concerning the normal range of human pulse rate. So this observation cannot be made based on the data provided.

76..**3** *Respiratory and circulatory* are the two body systems that were most actively involved in this experiment. Data provided in the table give information about pulse rate, which directly involves the circulatory system. The student's general knowledge of biology is needed to understand that the respiratory system functions to provide oxygen needed for cellular respiration as well as to remove carbon dioxide and other metabolic wastes from muscles involved in the exercise regimen.

WRONG CHOICES EXPLAINED

(1), (4) *Respiratory and immune* and *immune and circulatory* are *not* the two body systems that were most actively involved in this experiment. The immune system provides the body with protection from foreign invaders and potentially dangerous antigens. This system was not directly involved with this experiment.

(2) *Digestive and endocrine* are *not* the two body systems that were most actively involved in this experiment. The digestive system functions to break down complex foods into soluble molecules. The endocrine system manufactures hormones that function to direct the activities of specific target tissues. These systems were not directly involved with this experiment.

77. One credit is allowed for correctly identifying a substance that could have been added to the slide to bring about the change shown. Acceptable responses include but are not limited to: [1]

- *Salt*
- *Salt solution*

78. One credit is allowed for correctly identifying which *two* species are more closely related and for supporting the answer. Acceptable responses include but are not limited to: [1]

- *A and B—They have more bases in common.*
- *A and B—Their DNA sequences are more similar to each other than either is to C.*
- *A and B—They differ in only one codon, whereas species C differs in three or more codons.*

79. One credit is allowed for correctly stating *one* action the student could take to improve the detail. Acceptable responses include but are not limited to: [1]

- *Add stain to the specimen.*
- *Adjust the diaphragm.*
- *Use a higher magnification.*
- *Adjust the fine/coarse adjustment knob.*
- *Add water to the slide.*
- *Clean the lens.*

80. One credit is allowed for correctly identifying *one* pigment molecule common to all five organisms. Acceptable responses include but are not limited to: [1]

- *Chlorophyll a*
- *Beta carotene*
- *The molecule in the blue-green band*
- *The pigment molecule that causes the yellow-orange band*

81. **4** *Red algae and cyanobacteria* are the two organisms that are most closely related. Of the choices given, these two species share the most pigments in common, indicating that they also share the most DNA in common. This fact makes it reasonable to infer that they are closely related compared to the other pairings.

WRONG CHOICES EXPLAINED
(1), (2), (3) *Cyanobacteria and green algae, red algae and spinach,* and *brown algae and red algae* are *not* the two organisms that are most closely related. These pairings of organisms differ significantly from each other in terms of the pigments they produce, indicating that they also differ significantly in their DNA. These organisms are therefore less closely related than red algae and cyanobacteria are to each other.

82. **1** In this diagram, the variety of beak sizes and shapes are adaptations directly related to successful *feeding*. An examination of the diagram reveals data concerning the types of foods consumed by these finch species, a clear indication that the variety of beak sizes and shapes are adaptations that assist the nutritional (feeding) processes in these species.

WRONG CHOICES EXPLAINED

(2), (3), (4) In this diagram, the variety of beak sizes and shapes are *not* adaptations directly related to successful *camouflage, defense*, or *singing*. No data are provided in the diagram that would indicate that the variety of beak sizes and shapes are adaptations that assist camouflage, defense, or singing in these species.

83. One credit is allowed for correctly stating *one* reason why the large ground finch and the woodpecker finch can live successfully on the same island. Acceptable responses include but are not limited to: [1]

- *The two finches eat different kinds of food.*
- *The woodpecker finch eats insects, and the large ground finch eats seeds.*
- *They do not compete for the same food resources.*

84. One credit is allowed for correctly identifying *one* finch in the diagram that is *least* likely to compete with any of the other finches and supporting the answer. Acceptable responses include but are not limited to: [1]

- *The cactus ground finch—It is the only one that eats cactus.*
- *The vegetarian tree finch is the only one that is a fruit eater.*

85. One credit is allowed for correctly stating *one* reason why a molecule may *not* be able to pass into or out of a cell. Acceptable responses include but are not limited to: [1]

- *The molecule might be too large to pass through the cell membrane.*
- *The cell membrane may have no receptor site for this type of molecule.*
- *The receptor site on the cell membrane may be damaged/faulty/inactive.*
- *The molecule may have an electrical charge that makes it incompatible with the cell membrane.*

STANDARDS/KEY IDEAS	JUNE 2015 QUESTION NUMBERS	NUMBER OF CORRECT RESPONSES
STANDARD 1		
Key Idea 1: The central purpose of scientific inquiry is to develop explanations of natural phenomena in a continuing and creative process.		
Key Idea 2: Beyond the use of reasoning and consensus, scientific inquiry involves the testing of proposed explanations involving the use of conventional techniques and procedures and usually requiring considerable ingenuity.	67, 68, 69, 70, 71, 72	
Key Idea 3: The observations made while testing proposed explanations, when analyzed using conventional and invented methods, provide new insights into natural phenomena.	44, 45, 46, 48	
Laboratory Checklist	67, 68, 69, 70, 71, 72	
STANDARD 4		
Key Idea 1: Living things are both similar to and different from each other and from nonliving things.	1, 2, 3, 6, 7, 21, 22, 41, 42, 43, 47	
Key Idea 2: Organisms inherit genetic information in a variety of ways that result in continuity of structure and function between parents and offspring.	11, 12, 18, 24, 25, 29, 34, 37	
Key Idea 3: Individual organisms and species change over time.	8, 13, 15, 26, 31, 36, 61, 62, 63	
Key Idea 4: The continuity of life is sustained through reproduction and development.	4, 5, 17, 20, 39	
Key Idea 5: Organisms maintain a dynamic equilibrium that sustains life.	16, 23, 27, 38, 49, 50, 51, 52, 58, 59, 60	
Key Idea 6: Plants and animals depend on each other and their physical environment.	30, 32, 33, 35, 53, 54, 55, 64, 65	
Key Idea 7: Human decisions and activities have a profound impact on the physical and living environment.	9, 10, 14, 19, 28, 40, 56, 57, 66	

STANDARDS/KEY IDEAS	JUNE 2015 QUESTION NUMBERS	NUMBER OF CORRECT RESPONSES
REQUIRED LABORATORIES		
Lab 1: "Relationships and Biodiversity"	74, 78, 80, 81	
Lab 2: "Making Connections"	73, 75, 76	
Lab 3: "The Beaks of Finches"	82, 83, 84	
Lab 5: "Diffusion Through a Membrane"	77, 79, 85	

Examination August 2015

Living Environment

PART A

Answer all questions in this part. [30]

Directions (1–30): For *each* statement or question, record in the space provided the *number* of the word or expression that, of those given, best completes the statement or answers the question.

1 Which statement describes a situation that would reduce the stability of a forest ecosystem?

   1 A fierce predator is removed from the ecosystem.
   2 The number of producers remains constant in the ecosystem.
   3 Organisms frequently interact within the ecosystem.
   4 The energy in the ecosystem flows from the Sun.   1

2 Although the digestive system is primarily responsible for the breakdown of food, this process can
be disrupted if the circulatory system malfunctions.
The best explanation for this disruption is that

    1 human body systems interact with each other to
       perform life functions
    2 the circulatory system is the control center of
       the body
    3 the digestive system and the circulatory system
       have many organs in common
    4 the circulatory system is responsible for the
       coordination of life functions, including the
       breakdown of food   2 __1__

3 When an organism reproduces asexually, it usually
has

    1 only one parent, and half as much DNA as the
       parent
    2 only one parent, and the same chromosome
       number as the parent
    3 two parents, and twice as much DNA as either
       parent
    4 two parents, and the same chromosome number
       as each parent   3 __2__

4 The diagram below represents a food pyramid in an ecosystem.

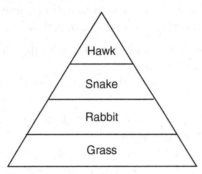

The best explanation for the decrease in the amount of energy transferred to each succeeding level is that much of the energy is

1 consumed by predators
2 released as heat
3 stored within inorganic materials
4 used in photosynthesis   4 _2_

5 The corn we eat today is larger and has more kernels than the corn people first grew thousands of years ago. Which process is most likely responsible for the changes that have occurred?

1 mitosis   3 direct harvesting
2 succession   4 selective breeding   5 _4_

6 Which statement is correct concerning hereditary information?

1 A chromosome is composed of many genes.
2 A gene is composed of many chromosomes.
3 Each chromosome carries the same information.
4 Each gene carries the same information.   6 _1_

7 Which process is most closely associated with the regulation of water loss from the leaves of trees?

  1 digestion of water within the cytoplasm in the leaf cells of the trees
  2 synthesis of protein by the chloroplasts in the leaf cells of the trees
  3 movement of water through leaf openings controlled by the guard cells
  4 absorption of nitrogen through leaf openings controlled by the guard cells   7 _3_

8 A mutation occurring in a human can be passed from parent to offspring when it occurs in a

  1 lung cell, due to exposure to a toxic gas
  2 gamete formed in the ovary
  3 body cell undergoing mitosis
  4 heart cell with chromosome damage   8 _2_

9 If the concentration of sodium is greater outside a cell than inside the cell, which process could move sodium out of the cell?

  1 diffusion
  2 carbohydrate synthesis
  3 active transport
  4 digestion   9 _3_

10 The basic building blocks of a protein are

  1 glucose molecules   3 hormones
  2 amino acids   4 fats   10 _2_

11 Over time, data that support the successful evolution of a species would include observations that describe

1 an increase in the genetic changes occurring in body cells

2 a decrease in the genetic variety carried in sex cells

3 an increase in the proportion of offspring that have favorable characteristics

4 a decrease in the proportion of the population that has beneficial traits   11 __1__

12 Caffeine is a compound found in the seeds of many different plants, such as coffee beans, cola nuts, and cacao beans (the source of chocolate). The presence of this chemical in all three types of plants suggests that these plants

1 inherited identical mutations

2 share a common ancestry

3 were exposed to the same type of radiation in the past

4 were cloned from a caffeine plant   12 __2__

13 Male turkeys are birds that naturally strut and display their large tail feathers, which attracts female turkeys. This display is an example of

1 a behavioral adaptation

2 selective breeding

3 asexual reproduction

4 a learned behavior   13 __1__

14 A scientist at a large natural history museum has a collection of fossils that were found throughout the world. Only a few of the fossils represent species that are still alive on Earth today. One reason for this is that

  1 most of the species that have ever lived on Earth are alive today

  2 most of the species that have ever lived on Earth are extinct

  3 fossils of only extinct species have been found

  4 species alive today will not form any fossils for future discovery by scientists   14 _2_

15 Which statement concerning sexual reproduction is correct?

  1 It is not necessary in order for the individual to survive.

  2 The offspring are identical to the parent.

  3 It is necessary in order for the individual to survive.

  4 The offspring are identical to each other.   15 _1_

16 When a paramecium, a single-celled organism, is living under stressful conditions, it sometimes switches from asexual to sexual reproduction. The main advantage when this switch occurs is that the paramecium is most likely to

  1 produce fewer offspring

  2 increase variation among its offspring

  3 avoid having to find a mate

  4 produce clones of itself   16 _2_

17 The diagram below represents some processes in the early development of a multicellular organism.

Which statement describing this diagram is correct?

1 The cell represented by structure 3 has the same genetic content as structure 2.

2 Process A represents the process of meiosis.

3 Each cell in structure 4 has the same genetic content as that in structure 3.

4 Processes A and B both occur in the placenta.

17 _3_

18 Which statement describes a function of the hormone estrogen?

1 It regulates the secretion of digestive enzymes.

2 It promotes sperm production in males.

3 It influences the development of adult sex characteristics.

4 It maintains blood sugar levels.

18 _3_

19 The primary function of the human male reproductive system is to

1 provide a site for fertilization

2 produce and transport gametes

3 protect and nourish the embryo

4 prevent urine from leaving the body

19 _2_

20 In an embryo, the formation of many types of tissues and organs occurs as a result of the process of

1 fertilization
2 genetic sorting
3 differentiation
4 gene recombination

20 _3_

21 Which activity would be an appropriate first step when designing an experiment?

1 reporting a conclusion based on multiple experimental trials
2 researching the problem, using information from a variety of sources
3 creating a data table to organize experimental observations
4 repeating the experiment with a different hypothesis

21 _2_

22 Every time a child visited a cousin who has two cats, the child's eyes turned red, itched, and began to water. Then, the child began to have trouble breathing. It is most likely that the child reacted this way because

1 normally harmless cat antigens stimulated the immune system
2 it is difficult for the respiratory system to filter cat antigens out of the inhaled air
3 cat antigens are a health hazard, since they always cause disease
4 cat antigens stop the immune system from making antibodies, so bacteria cause these responses

22 _1_

23 Shrimp that live in the cold waters off Alaska will die if introduced into warm water. One likely reason these shrimp do not survive is that enzymes in the shrimp

1 start to replicate
2 change shape
3 are composed of fat molecules that melt
4 break down into small starch molecules   23 _2_

24 A DNA segment removed from neurospora (a pink mold) contained the base sequence G-T-C-C-A-T-G-C-A. A similar segment of DNA removed from neurospora that had been exposed to radiation for several hours had the base sequence G-T-C-C-A-T. This change in the base sequence is an example of

1 a deletion   3 a substitution
2 an insertion   4 a replication   24 _1_

25 Farmers in India have increased the harvest yield of food crops like eggplant by growing them from seeds that have been modified to produce a bacterial toxin that is harmful to pest insects. This is an example of

1 selective breeding of the insects
2 spraying an insecticide on plants
3 selective breeding of the eggplant
4 an application of biotechnology   25 _4_

26 The graph below shows changes in human population numbers over time.

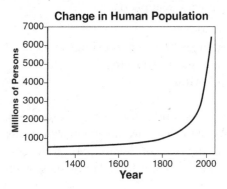

Change in Human Population

A consequence of these changes is

1 an increase in the numbers and kinds of organisms worldwide
2 a decrease in the availability of natural resources
3 a decrease in deforestation due to technological improvements
4 an increase in biosphere stability

26 __2__

27 In the fall, some farmers plow the remains of corn plants into the ground. This activity contributes most directly to the

1 increase in the biodiversity of their fields
2 depletion of nonrenewable resources
3 destruction of natural habitats
4 recycling of organic matter

27 __4__

28 A person usually experiences small variations in body temperature over a 24-hour period. These variations in temperature are an example of

1 an immune response
2 genetic differences between individuals
3 an adaptation to global warming
4 dynamic equilibrium   28 __4__

29 Fossil fuels have been used for years as a source of energy. Even though there are many negative issues associated with the use of fossil fuels, they continue to be used to a great extent. This is most likely because

1 they have been commercially available as an energy source
2 there are alternatives to these types of fuels
3 they have had a positive effect on global temperatures
4 fossil fuels can be burned to produce large quantities of carbon dioxide   29 __1__

30 Sometimes, a person is born with one or more extra chromosomes in each cell. This usually results in abnormalities because the affected person has

1 a reduced number of genes in cell nuclei
2 fewer cell mutations than a person with a normal chromosome number
3 more genes in each cell than a person with a normal chromosome number
4 less DNA in cell nuclei, but more proteins in cell mitochondria   30 __3__

PART B–1

Answer all questions in this part.   [13]

Directions (31–43): For *each* statement or question, record in the space provided the *number* of the word or expression that, of those given, best completes the statement or answers the question.

Base your answers to questions 31 through 33 on the information and graph below and on your knowledge of biology.

A population composed of tan snails and black snails inhabits the same sandy beach. A nearby volcano erupted, and black lava particles washed down to the beach. The once tan beach was now black. The graph below shows the population of tan snails and black snails before and after the volcanic eruption.

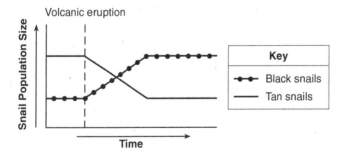

31 Which statement concerning the snails is correct?

  1 The lava particles turned the tan snails black.
  2 The tan snails will become extinct.
  3 The black snails had an adaptive advantage.
  4 The tan snails preyed on the black snails.   31 __3__

32 The increase in the number of black snails can best be explained by

  1 natural selection after an environmental change
  2 climatic change followed by ecological succession
  3 increased stability due to a decrease in variation
  4 an increase in mutation rate   32 __1__

33 Variation in snail color is an example of

  1 environmental stability
  2 a natural limitation
  3 equilibrium
  4 diversity   33 __4__

---

34 The diagram below represents relationships in a community. After a pathogen reduced the population of grasshoppers, the number of mice increased, while the numbers of toads and rabbits decreased.

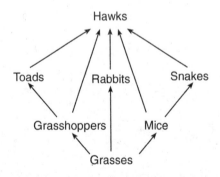

These changes in the community demonstrate that

  1 ecosystems are shaped by nonliving factors
  2 autotrophs convert solar energy into food
  3 grasshoppers are producers that are essential for ecosystem stability
  4 populations are linked with many others in the ecosystem   34 __4__

Base your answers to questions 35 and 36 on the diagram below, which represents a metabolic process, and on your knowledge of biology.

35 This process best represents

1 the bonding of amino acids to form a starch molecule
2 the digestion of amino acids to form a starch molecule
3 the bonding of simple sugars to form a starch molecule
4 the digestion of simple sugars to form a starch molecule   35 __3__

36 The letter X in the process represents

1 an antibody
2 a hormone
3 a receptor
4 an enzyme   36 __4__

37 The evolutionary pathways of ten different species are represented in the diagram below.

Which statement would most likely be correct, based on the information in the diagram?

1 Species C had many variations and lived in a stable, unchanging environment.
2 Species D, C, and J are extinct.
3 Species F evolved from species D.
4 Species J had little variation and lived in a changing environment.

37 __4__

38 Blockages caused by a condition known as Pelvic Inflammatory Disease (PID) are represented in the diagram of the female reproductive system below.

Blockages

If blockages of this type occur, the most likely result would be that

1 the egg would remain in the uterus and not travel upward

2 the female gamete would not be able to unite with the male gamete

3 hormones could not be produced by the ovaries

4 the process of asexual reproduction would be prevented or interrupted

38 _2_

Base your answers to questions 39 and 40 on the information below and on your knowledge of biology. The graph below shows the growth of *Paramecium aurelia* in the same culture dish for 14 days.

Growth of *Paramecium aurelia*

39 If no additional materials were added to the culture dish, after day 14, the paramecium population would most probably

1 remain the same, since it has reached carrying capacity and has an unlimited food supply
2 begin to increase as they continue to reproduce
3 begin to increase, since they have not yet reached carrying capacity
4 begin to decrease as finite resources are used up

39 ___

40 In another experiment, a second species of paramecium was introduced into a culture dish with *Paramecium aurelia*. Which statement describes a possible result as the populations interact over the next 14 days?

  1 The population numbers of *Paramecium aurelia* would be lower than 250, since the new species is competing with it for resources.

  2 The population of *Paramecium aurelia* would increase above 250, since they would mate with the new species.

  3 The population of *Paramecium aurelia* would increase above 250, since the two species occupy the same niche.

  4 The population of *Paramecium aurelia* would remain at 250, since the species compete with each other for the same resources.   40 __1__

---

41 The chart below lists substances involved in the process of photosynthesis.

Substance

A	glucose
B	oxygen
C	carbon dioxide
D	water

Which statement best describes how these substances interact in photosynthesis?

  1 *A* and *B* combine to produce *C* and *D*.
  2 *B* and *C* combine to produce *A* and *D*.
  3 *C* and *D* combine to produce *A* and *B*.
  4 *A* and *C* combine to produce *B* and *D*.   41 __3__

Base your answers to questions 42 and 43 on the diagram below and on your knowledge of biology. The diagram represents the current percentage of each population by age and gender (male/female) for two countries.

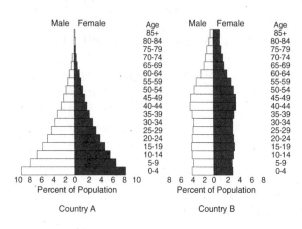

Adapted from: *Campbell Biology, 8th edition*

42 At the present time, both populations have the same number of individuals. In which of these countries will the population growth over the next 20 years place the greatest strain on the environment?

1 Country *A*, since the larger percentage of young could result in rapid population growth
2 Country *B*, since the smaller percentage of young could result in rapid population growth
3 Country *A*, since the smaller percentage of people over 60 uses the most resources
4 Country *B*, since the larger percentage of people over 60 uses the fewest resources

42 ___1___

43 Approximately what percent of the population of Country *A* is less than 10 years old?

1 8%   3 32%
2 16%   4 64%

43 ___3___

PART B–2

Answer all questions in this part. [12]

Directions (44–55): **For those questions that are multiple choice, record in the space provided the** *number* **of the choice that, of those given, best completes each statement or answers each question. For all other questions in this part, follow the directions given and record your answers in the spaces provided.**

Base your answers to questions 44 through 47 on the information below and on your knowledge of biology.

An experiment was carried out to determine the effect of exposure to UV light on the growth of bacteria. Equal quantities of bacteria were spread on 5 petri dishes containing nutrient agar. Half of each petri dish was exposed to UV light for various amounts of time, and the other half was protected from the UV light with a UV screen. After the UV treatment, the bacteria were grown in an incubator for 24 hours, and the number of colonies was counted. The diagram below represents the initial set up.

The table below contains the data collected by counting the number of bacterial colonies growing on both the screen-covered side and the unscreened side.

Growth of Bacterial Colonies

Petri Dish	Exposure Time to UV Light	Number of Bacterial Colonies on Screened Side	Number of Bacterial Colonies on Unscreened Side
1	No exposure (0.0 minutes)	17	18
2	1.0 minute	18	15
3	2.0 minutes	17	11
4	5.0 minutes	18	4
5	10.0 minutes	16	1

Directions (44–46): Using the information in the data table, construct a line graph on the grid on the next page, following the directions below.

44 Mark an appropriate scale, without any breaks in the data, on each labeled axis. [1]

45 Plot the data for the number of bacterial colonies on the screened side. Connect the points and surround each point with a small circle. [1]

Example:

46 Plot the data for the number of bacterial colonies on the unscreened side. Connect the points and surround each point with a small triangle. [1]

Example:

Growth of Bacterial Colonies

Key

⊙ Screened side

△ Unscreened side

Exposure Time to UV Light (minutes)

Number of Bacterial Colonies

Note: The answer to question 47 should be recorded in the space provided.

47 The diagram below represents cellular growth that can occur in human skin after prolonged exposure to ultraviolet light.

Skin cells

Which statement provides a possible explanation for this growth pattern?

1 Manipulation of genes caused the movement of embryonic skin cells.

2 Exposure to light stimulated the development of cells containing ozone.

3 Uncontrolled mitotic division occurred as a result of gene mutations.

4 An immune reaction triggered the formation of excess blood cells.

47 _3_

Base your answers to questions 48 and 49 on the information and graphs below and on your knowledge of biology. The graphs show the relative enzymatic activity of four different enzymes in acidic (below pH 7) and basic (above pH 7) environments.

pH and Enzyme Activity

Part of Digestive System	pH Range
mouth	6.5 – 7.5
stomach	1.5 – 4.0
small intestine	4.0 – 7.0
large intestine	4.0 – 7.0

48 Which enzyme would most likely function in the stomach? Support your answer. [1]

Enzyme: _Pepsin_

Pepsin works best in the pH of the Stomach (1.5 - 4).

Note: The answer to question 49 should be recorded in the space provided.

49 The activity of which enzyme decreases in both acidic and basic environments?

1 chymotrypsin   3 cholinesterase
2 pepsin   4 papain

49 __1__

Base your answers to questions 50 and 51 on the information below and on your knowledge of biology.

Head Start for Hellbenders

The hellbenders (a species of large salamander) at the Bronx Zoo are now approximately seven inches in length and will grow to full size in about five years. Once they are about two-and-a-half years old, they will be returned to the wild in western New York State. Hellbender populations are declining due to several factors including over-collection for the pet trade, disease, pollution, and habitat destruction. Juvenile hellbenders in the wild currently face great difficulties in reaching adulthood, so the "head start" provided by the reintroduction of the 41 animals will help boost local populations.

Source: Wildlife Conservation Society, *Members News*
May/June 2011

Note: The answer to question 50 should be recorded in the space provided.

50 Which statement best explains the hellbender population decline in western New York State?

  1 Human activities had the unexpected consequence of decreasing the hellbender population to a dangerous level.
  2 Humans have purposefully removed the hellbender from its habitat due to its aggressive behavior.
  3 The decline of the hellbender population is due mainly to natural causes that humans cannot control.
  4 The hellbender population decreased because salamanders are very resistant to climate change.

50 __1__

51 Describe *one* potential ecological effect, other than the loss of the hellbender from western New York State, of the hellbender population continuing to decline. [1]

The population that feeds on hellbender will decrease.

52–54 Animals eat and digest food to obtain the energy available for life activities. Discuss energy use in animals. In your discussion, be sure to:

- identify the type of protein molecules used to digest food [1]
- identify the organelle where energy from nutrients is released [1]
- state *one* inference that can be made concerning a cell that has many of these organelles [1]

Enzymes can help digest food. The mitochondria is where energy from nutrients is released by cellular respiration. The cell is active and requires a lot of energy.

Base your answer to question 55 on the diagram below and on your knowledge of biology. The diagram represents four types of bacteria.

Bacillus   Spirillum   Type C   Type D

55 A dichotomous key to these bacterial types is shown below. Complete the missing information for sections 3a. and 3b. so that the key is complete for all four types. [1]

1a. Is rod shaped .... bacillus
1b. Is not rod shaped .... go to 2

2a. Is spiral shaped.................................... spirillum
2b. Is not spiral shaped.............................. go to 3

3a. <u>Single celled</u>   type C
3b. <u>Multi-cellular</u>   type D

PART C

Answer all questions in this part.   [17]

Directions **(56–72): Record your answers in the spaces provided.**

Base your answers to questions 56 and 57 on the information below and on your knowledge of biology.

The fight-or-flight response in humans prepares the body to fight off or run away from a potential threat. This response results from a series of nerve and chemical signals that direct how cells function. This, in turn, determines the actions of organs in these situations.

Some of the changes experienced by the individuals as part of this response include:

increased pulse rate
increased blood glucose levels
increased breathing rate

56 Select *one* of the listed changes experienced by the individual and write it on the line below. Explain how the change you chose allows the individual to effectively respond to a threat. [1]

Change: _increased glucose_

When cells recieve more glucose they can perform cellular respiration faster, releasing more energy to the body, readying for more movement

57 Once the threat has passed, another series of changes returns the body to its original state. Why must this occur? [1]

This must take place in order to maintain homeostasis.

Base your answers to questions 58 and 59 on the information below and on your knowledge of biology.

Bats Devastated by Deadly Fungus

The most common bat species in North America, the little brown bat, could be facing extinction because of a fungus. The fungus, called white-nose syndrome, grows on the exposed skin of bats as they hibernate in cool caves or mines. Infected bats develop lesions (sores) on their wings, which play important roles in water balance, circulation and heat regulation. These lesions on a bat's wings or on its nose cause the bat to wake up during hibernation. Waking up early forces the bat to use up the energy it has stored as fat for its long sleep, exhausting the animal and eventually killing it.

In some infected caves, 90 percent to 100 percent of bats die. On average, the disease takes out 73 percent of the bat population at a given hibernation site. If infection continues at current rates, the researchers predict that the little brown bat population will drop below 0.01 percent of its current numbers by 2026.

The loss of the little brown bat would be harmful for humans because bats eat their body weight in insects each night. Many of these bugs are agricultural pests or carriers of human disease.

One way to decrease the spread of the disease would be for the researchers who visit infected caves to decontaminate their clothes and gear with antiseptics. It has also been suggested that a small number of these bats could be placed in an artificial hibernating area and medicated to protect them.

58 Describe *one* way that an infection with the white-nose fungus can cause death in little brown bats. [1]

It can cause the bat to wake up from hibernation, and use up energy, exhausting and killing the bat,

59 Describe *one* way that the little brown bats can be helped. [1]

They can be placed in an artificial, sterile cave to hibernate.

60 Describe how a student could use a microscope to compare the size of frog skin cells to the size of human skin cells. [1]

They would place each type of cell on a slide and compare structure, organelles, functions, etc. at a microscopic level.

Base your answers to questions 61 through 63 on the information below and on your knowledge of biology.

Chickenpox vaccine is the best way to prevent chickenpox. Vaccination not only protects vaccinated persons, it also reduces the risk for exposure in the community for persons unable to be vaccinated because of illness or other conditions, including those who may be at greater risk for severe disease. While no vaccine is 100% effective in preventing disease, the chickenpox vaccine is very effective: about 8 to 9 of every 10 people who are vaccinated are completely protected from chickenpox. In addition, the vaccine almost always prevents against severe disease. If a vaccinated person does get chickenpox, it is usually a very mild case lasting only a few days and involving fewer skin lesions (usually less than 50), mild or no fever, and few other symptoms.

Source: *www.cdc.gov*

61 Identify the component of a vaccine that makes it effective. [1]

Weak or dead pathogen

62 Describe how a vaccination prevents disease in the individual who receives a vaccination. [1]

When the vaccination is injected, the body creates antibodies against the virus, readying the body for a healthy pathogen.

63 Describe how vaccinations help to prevent the spread of disease, even in people who have not received the vaccination. [1]

The quicker one person can fight off the pathogen, the less people it will spread to.

Base your answer to question 64–66 on the diagram below and on your knowledge of biology. The diagram represents a food web typical of the Great Lakes area of New York State.

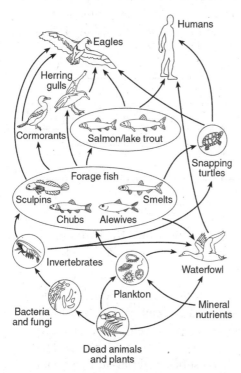

Humans

Eagles

Herring gulls

Cormorants

Salmon/lake trout

Snapping turtles

Forage fish

Sculpins   Smelts

Chubs   Alewives

Invertebrates

Waterfowl

Plankton

Mineral nutrients

Bacteria and fungi

Dead animals and plants

Adapted from: http://www.uwsp.edu/geo/ faculty/ritter/
geog101/textbook/title_page.html

64–66 Some people have argued for the removal of cormorants from the eastern shores of Lake Ontario
because of their negative effects on the fishing
industry. Describe the consequences of this action.
In your answer, be sure to:

- state *one* reason why removing the cormorants
  from the food web could have a positive impact
  on the fishing industry [1]
- state *one* possible effect of removing the cormorants on a species other than fish and support
  your answer [1]
- describe *one* action, other than removing a population of organisms from the environment, that
  humans could take to preserve the fishing industry in Lake Ontario [1]

_____

_____

_____

_____

_____

_____

Base your answers to questions 67 through 69
on the passage below and on your knowledge of
biology.

A field in New York State is mowed all summer
long for a number of years. The field is sold, and the
new owner decides to stop mowing. Over a number
of years, the ecosystem begins to undergo ecological succession. After a series of different plant communities are present, the area eventually becomes a
stable forest ecosystem.

67 Explain why *not* mowing the field allowed the ecosystem to undergo ecological succession. [1]

Trees grow from the
nutrients of weeds/grass.

68 Identify *one* specific human activity, other than mowing, that could alter this succession and explain how this activity affects biodiversity. [1]

Human activity: _Deforestation_
Effect on biodiversity: _Loss of habitat/_
Oxygen lessens

69 Describe how this forest ecosystem would respond to a natural disaster, such as a flood, that resulted in the destruction of the plant community. [1]

Ecological succession
would have to restart

70 In a laboratory, spinach leaves exposed to continuous fluorescent light increased in vitamin content by 50 to 100 percent. Spinach leaves kept in darkness for a similar period of time either lost vitamin content or produced no gain. Describe how these findings could influence the way in which spinach is displayed for sale in supermarkets. [1]

Could be labeled
according to light
absorption

Base your answers to questions 71 and 72 on the passage below and on your knowledge of biology.

On April 20, 2010, an explosion occurred at an oil well in the Gulf of Mexico, causing millions of gallons of oil to escape into the water over the next few months. Large areas of the Gulf were covered by oil. As the oil washed ashore, many areas along the coastline that were breeding grounds for various bird species were contaminated. By November 2010, researchers along the coast and in the Gulf had collected 6104 dead birds, 609 dead turtles, and 100 dead mammals. Although the oil well had provided oil for energy for a large number of people, the oil spill had a great effect on the ecosystems in and around the Gulf of Mexico.

71 Explain how the original decision to drill for oil in the Gulf of Mexico could be considered a trade-off. [1]

Dead animals + loss
of biodiversity / energy
for people

72 State *one* possible reason why it will most likely take the bird populations more time to recover from this oil spill than it will mammal populations. [1]

More birds were lost
than mammals.

PART D

Answer all questions in this part. [13]

Directions (73–85): For those questions that are multiple choice, record in the space provided the *number* of the choice that, of those given, best completes the statement or answers the question. For all other questions in this part, follow the directions given and record your answers in the spaces provided.

73 A plant was discovered that contained a compound that was found to have potential medicinal value. However, the plant is rare, so it is important to see if a related plant might also produce the same compound. The chart shows some characteristics of the plant and four possible relatives.

Plant	Flower	Leaves	Amino Acid Sequence
Medicinal Plant	Red, 6 petals	simple, parallel veins	Ile–Ile–Try–Gly–Glu–Asp–Pro
A	Red, 9 petals	simple, parallel veins	Ile–Arg–Try–Gly-Glu-Asp–Ser
B	Yellow, 8 petals	compound, branched veins	Ile–Arg–Ala–Gly–Glu–Asp–Pro
C	Pink, 6 petals	simple, parallel veins	Ile–Ile–Try–Gly–Glu–Asp–Ser
D	Yellow, 6 petals	compound, parallel veins	Ile–Arg–Try–Gly–Glu–Asp–Pro

Note: The answer to question 73 should be recorded in the space provided.

Which plant in the chart would be selected as most similar to the medicinal plant?

1 *A*   3 *C*

2 *B*   4 *D*   73 __3__

Note: The answer to question 74 should be recorded in the space provided.

74 A drug company has discovered an endangered plant that produces a chemical that might be used to cure cancer. A first step in developing this cancer cure could be to

1 preserve the habitat where the species is found
2 introduce a new plant species that will share the habitat of the plant
3 eliminate every species that eats this plant
4 harvest all of the plants of this species and use them to treat cancer patients

74 __1__

Note: The answer to question 75 should be recorded in the space provided.

75 Students in a science class took their pulse rates before and after they ran in place for one minute. The class data showed that pulse rates increased with exercise. A graph of the data would look most like

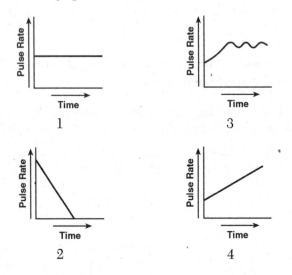

75 __4__

Base your answers to questions 76 and 77 on the information below and on your knowledge of biology.

Caretakers at a zoo are trying to determine which of two male tigers fathered the newest cub. They obtained DNA from the tiger cub, the mother tiger, and the two male tigers. The DNA was analyzed. The results of the analysis are shown below.

Male 1	Male 2	Cub	Female

Note: The answer to question 76 should be recorded in the space provided.

76 The technique used to separate the DNA for analysis is

1 genetic engineering
2 electrophoresis
3 chromatography
4 protein synthesis

76 __2__

77 Which male tiger is the father of the newborn cub? Support your answer. [1]

Male tiger: ____1____

The DNA bands of 1 are more like the cub than 2

78 Some roads are salted heavily in winter. Describe *one* way plants growing near these roads could be harmed by the salt. [1]

<u>Salt could drain the</u>
<u>water from the soil,</u>
<u>dehydrating the plant.</u>

Base your answer to question 79 on the diagram below and on your knowledge of biology.

Variations in Beaks of Galapagos Islands Finches

from: *Galapagos: A Natural History Guide*

79 Several populations of finches migrated to an island that had mostly large seeds with tough outer coverings. Identify a finch population that would most likely survive on the island. Support your answer. [1]

Large ground finch could break seeds easily w/ big beak

80 The diagram below shows how a coverslip should be placed on a drop of pond water during the preparation of a wet mount.

State *one* reason why this is the recommended procedure for placing a coverslip. [1]

Water will help the coverslip stay in place

Base your answers to questions 81 through 83 on the information below and on your knowledge of biology.

There are two different species of finch that live on the same small island, species *A* and species *B*. Both species successfully feed and reproduce on the island. Species *A* nests in pine trees and eats large seeds. Species *B* nests in hollowed-out dead logs and eats small insects.

Note: The answer to question 81 should be recorded in the space provided.

81 Both bird species *A* and species *B* can most likely survive on the same small island because they

1 use different resources and, therefore, they do not compete
2 mate with each other, keeping both populations constant
3 compete for food, but do not compete for shelter
4 eat the same food, but feed at different times of the day

81 __1__

Note: The answer to question 82 should be recorded in the space provided.

82 The factor most often acting as a selecting agent for the survival of a species in a particular location is the

1 strength of the organism
2 new mutations within the individual
3 speed of the organism
4 environment they inhabit

82 __4__

83 A third species of finch, species C, migrates to the island. It nests in pine trees and eats small insects. Predict what most likely will happen to the populations of both species A and species B if species C successfully survives on the island. Support your answer. [1]

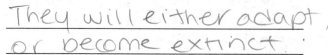

They will either adapt, or become extinct.

Base your answer to question 84 on the information below and on your knowledge of biology.

Biology students conducted a preliminary survey to study the relationship between body height and resting pulse rate. The students collected data from 10 classmates and the results are shown in the graph below.

Resting Pulse Rate

Height (in)

84 Is there a relationship between height and resting pulse rate? Support your answer. [1]

No, there is no data correlation

85 The table below shows the number of individual molecules obtained when a DNA molecule from a bacterial species is broken down. [1]

Molecules from Bacterial DNA

Molecule	Number
sugar	4.6 million
phosphate	4.6 million
adenine (A)	1.75 million
cytosine (C)	0.55 million
guanine (G)	0.55 million
thymine (T)	1.75 million

What data in the data table indicate that adenine pairs with thymine in a DNA molecule? [1]

There are the same number of adenine and thymine molecules.

Answers
August 2015
Living Environment

Answer Key

PART A

1. 1	6. 1	11. 3	16. 2	21. 2	26. 2
2. 1	7. 3	12. 2	17. 3	22. 1	27. 4
3. 2	8. 2	13. 1	18. 3	23. 2	28. 4
4. 2	9. 3	14. 2	19. 2	24. 1	29. 1
5. 4	10. 2	15. 1	20. 3	25. 4	30. 3

PART B–1

31. 3	34. 4	36. 4	38. 2	40. 1	42. 1
32. 1	35. 3	37. 4	39. 4	41. 3	43. 3
33. 4					

PART B–2

44. *See* Answers Explained.
45. *See* Answers Explained.
46. *See* Answers Explained.
47. 3
48. *See* Answers Explained.
49. 1

50. 1
51. *See* Answers Explained.
52. *See* Answers Explained.
53. *See* Answers Explained.
54. *See* Answers Explained.
55. *See* Answers Explained.

PART C. *See* **Answers Explained**.

PART D

73. 3
74. 1
75. 4
76. 2
77. *See* Answers Explained.
78. *See* Answers Explained.
79. *See* Answers Explained.

80. *See* Answers Explained.
81. 1
82. 4
83. *See* Answers Explained.
84. *See* Answers Explained.
85. *See* Answers Explained.

Answers Explained

PART A

1. **1** *A fierce predator is removed from the ecosystem* is the statement that describes a situation that would reduce the stability of the forest ecosystem. In any balanced ecosystem, the removal of any species tends to reduce the stability of that system. The removal of a predatory species would allow populations of prey organisms to increase in number. This increase could result in overpopulation of the prey species that could, in turn, lead to unintended effects on other species in the environment and thus destabilize the environment.

WRONG CHOICES EXPLAINED

(2) *The number of producers remains constant in the ecosystem* is *not* the statement that describes a situation that would reduce the stability of the forest ecosystem. Constant numbers of any species group such as producers in an environment tend to enhance the stability of that environment, not reduce it.

(3) *Organisms frequently interact within the ecosystem* is *not* the statement that describes a situation that would reduce the stability of the forest ecosystem. Interaction among the varied species in an environment tends to enhance the stability of that environment, not reduce it.

(4) *The energy in the ecosystem flows from the Sun* is *not* the statement that describes a situation that reduces the stability of the forest ecosystem. A constant source of energy, such as sunlight, that can be absorbed by producers in an environment tends to enhance the stability of that environment, not reduce it.

2. **1** The best explanation for this disruption is that *human body systems interact with each other to perform life functions*. In humans, these systems are cooperative in their operation and allow for coordination and homeostatic balance of the body.

WRONG CHOICES EXPLAINED

(2), (4) The best explanation for this disruption is *not* that *the circulatory system is the control center of the body* or that *the circulatory system is responsible for the coordination of life functions, including the breakdown of food*. In humans, control and coordination are functions provided by the nervous system, not the circulatory system.

(3) The best explanation for this disruption is *not* that *the digestive system and the circulatory system have many organs in common*. In humans, these

systems are separate and distinct. They are cooperative in their operation and allow for coordination and homeostatic balance of the body.

3. **2** When an organism reproduces asexually, it usually has *only one parent, and the same chromosome number as the parent.* The offspring produced as a result of asexual reproduction derives all of its genetic material from its single parent. For this reason, the offspring exactly resembles the parent in all genetically determined traits, including chromosome number.

WRONG CHOICES EXPLAINED
(1) It is *not* true that when an organism reproduces asexually, it usually has *only one parent, and half as much DNA as the parent.* The amount of DNA in the offspring resulting from asexual reproduction is exactly the same as that found in the parent organism, not half as much.

(3), (4) It is *not* true that when an organism reproduces asexually, it usually has *two parents, and twice as much DNA as either parent* or *two parents, and the same chromosome number as each parent.* The offspring of asexual reproduction arise by mitotic cell division of the cell(s) of a single parent, not from two parents. The amount of DNA and the number of chromosomes of the offspring are exactly like those of the parent.

4. **2** The best explanation for the decrease in the amount of energy transferred to each succeeding level is that much of the energy is *released as heat.* The net transfer of energy within an ecosystem is typically represented in the form of a food pyramid. Producer organisms (such as grasses) at the base of the pyramid absorb and store the greatest amount of solar energy through the process of photosynthesis. Producers are followed by herbivores (such as rabbits) in the second tier of the pyramid. Herbivores collectively contain only about 90% of the energy held by producers. Carnivores (such as snakes and hawks) at the top of the pyramid collectively contain only about 90% of the energy held by the herbivores. The energy lost between each level of the pyramid is dissipated into the environment as heat.

WRONG CHOICES EXPLAINED
(1) The best explanation for the decrease in the amount of energy transferred to each succeeding level is *not* that much of the energy is *consumed by predators.* Predators do consume energy in the form of food provided by prey organisms. This energy is held by the predators only temporarily and is eventually released to the environment as heat.

(3) The best explanation for the decrease in the amount of energy transferred to each succeeding level is *not* that much of the energy is *stored within inorganic materials.* Inorganic molecules (such as water and minerals) do not store energy represented in a food pyramid. Only organic molecules (such as carbohydrates, proteins, and fats) contain this energy in their chemical bonds.

(4) The best explanation for the decrease in the amount of energy trans-ferred to each succeeding level is *not* that much of the energy is *used in pho-tosynthesis*. A small amount of energy is used in the process of photosynthesis to activate its chemical reactions. The net result of photosynthesis is the syn-thesis of energy-rich organic compounds such as glucose.

5. **4** *Selective breeding* is the process that is most likely responsible for the changes that have occurred. Selective breeding is a technique used by animal and plant breeders in which organisms with desirable traits are cross-bred with the hope of producing offspring that also display those traits. Fossilized ears of corn (maize) discovered in Mexico were only about an inch (2.5 cm) long and contained only a small number of seeds. Mesoamerican peoples that cultivated this prehistoric corn selected the seeds of the most productive plants to begin the next generation. Over many centuries, this selective-breeding process resulted in larger ears containing more seeds. Modern agricultural scientists have continued this process that has resulted in the corn plants in common use by farmers today.

WRONG CHOICES EXPLAINED
(1) *Mitosis* is *not* the process that is most likely responsible for the changes that have occurred. Mitosis is a process by which an individual cell replicates the DNA in its nucleus and divides into two genetically identical daughter cells. Mitosis leads to the maintenance of genetic characteristics such as corn ear size, not to changes in these characteristics.

(2) *Succession* is *not* the process that is most likely responsible for the changes that have occurred. Succession is a process by which one plant com-munity replaces another in an ecosystem until a stabile climax community is established. The cultivation and modification of a food plant such as corn does not represent succession.

(3) *Direct harvesting* is *not* the process that is most likely responsible for the changes that have occurred. Direct harvesting is a human practice in which plants or animals are removed (harvested) from a natural environment for their economic value. The cultivation and modification of a food plant such as corn does not represent direct harvesting.

6. **1** *A chromosome is composed of many genes* is the statement that is correct concerning hereditary information. Chromosomes are microscopic structures located inside the cell nucleus. Chromosomes are composed of many hundreds or thousands of submicroscopic molecules of DNA. DNA is the chemically active component of genes. Each chromosome therefore contains hundreds or thousands of genes. Each gene carries a specific set of instructions for the production of cell proteins.

WRONG CHOICES EXPLAINED

(2) *A gene is composed of many chromosomes* is *not* the statement that is correct concerning hereditary information. This is the exact opposite relationship that scientists know to be true about chromosomes and genes.

(3) *Each chromosome carries the same information* is *not* the statement that is correct concerning hereditary information. Each of the several chromosomes in a cell carries a set of genes that is distinct in its genetic effects on the operation of the cell. Therefore, no two chromosomes in a cell carry the same genetic information.

(4) *Each gene carries the same information* is *not* the statement that is correct concerning hereditary information. Each of the thousands of genes in a cell nucleus codes for the production of a unique protein. Therefore, no two genes in a cell carry the same genetic information.

7. **3** *Movement of water through leaf openings controlled by the guard cells* is the process that is most closely associated with the regulation of water loss from the leaves of trees. The guard cells surrounding these leaf openings (known as stomates) change shape to open or close the openings. This shape change is made possible by the photosynthetic process that occurs in chloroplasts located in the guard cells. When conditions are dry and hot, guard cells close the stomates to conserve water in the leaf. When conditions are moist and cool, guard cells open the stomates to allow excess water to escape the leaf interior.

WRONG CHOICES EXPLAINED

(1) *Digestion of water within the cytoplasm in the leaf cells of the tree* is *not* the process that is most closely associated with the regulation of water loss from the leaves of trees. Water is an inorganic compound that cannot be digested into simple food molecules by any living thing. This is a nonsense distracter.

(2) *Synthesis of protein by the chloroplasts in the leaf cells of the tree* is *not* the process that is most closely associated with the regulation of water loss from the leaves of trees. Proteins synthesized by the chloroplasts in the leaf cells of the tree may be most closely associated with the process of photosynthesis, not directly with the regulation of water loss.

(4) *Absorption of nitrogen through leaf openings controlled by the guard cells* is *not* the process that is most closely associated with the regulation of water loss from the leaves of trees. Nitrogen gas (N_2) is abundant in Earth's atmosphere and passes readily into the leaf interior through the stomates. However, plants cannot use nitrogen in its gaseous form for any process, including the regulation of water loss.

8. **2** A mutation occurring in a human can be passed from parent to offspring when it occurs in a *gamete formed in the ovary*. Mutations are random events that alter the base sequence of DNA and therefore the cell's genetic

information. Mutations that occur in gametes (sex cells) may be passed on to offspring through the process of sexual reproduction.

WRONG CHOICES EXPLAINED

(1), (3), (4) A mutation occurring in a human *cannot* be passed from parent to offspring when it occurs in a *lung cell, due to exposure to a toxic gas*; a *body cell undergoing mitosis*; or a *heart cell with chromosome damage*. Mutations that occur in body cells (such as lung cells and heart cells) cannot be passed on to future generations. Only those mutations that occur in sex cells can be passed along to offspring.

9. **3** If the concentration of sodium is greater outside a cell than inside the cell, *active transport* is the process that could move sodium out of the cell. In this situation, the concentration gradient would tend to push sodium into the cell from an area of relatively high concentration to an area of relatively low concentration by simple diffusion. In order to move sodium out of the cell in this situation, the cell must expend energy. The expenditure of energy by a living cell to move a substance across a membrane from an area of relatively low concentration to an area of relatively high concentration is known as active transport.

WRONG CHOICES EXPLAINED

(1) It is *not* true that if the concentration of sodium is greater outside a cell than inside the cell, *diffusion* is the process that could move sodium out of the cell. In this situation, the concentration gradient tends to push sodium into the cell from an area of relatively high concentration outside the cell to an area of relatively low concentration inside the cell by simple diffusion. This is the exact opposite effect from the one described in the question.

(2) It is *not* true that if the concentration of sodium is greater outside a cell than inside the cell, *carbohydrate synthesis* is the process that could move sodium out of the cell. Carbohydrate synthesis is a biochemical process by which complex carbohydrates (e.g., starch, cellulose) are manufactured from simple sugars. Carbohydrate synthesis has no direct role in the movement of sodium into or out of the cell.

(4) It is *not* true that if the concentration of sodium is greater outside a cell than inside the cell, *digestion* is the process that could move sodium out of the cell. Digestion is a process in which complex food molecules are broken down into simple food molecules by enzymatic hydrolysis. Digestion has no direct role in the movement of sodium into or out of the cell.

10. **2** The basic building blocks of a protein are *amino acids*. Approximately 20 different amino acids can be linked together via peptide bonds in different configurations to produce an almost infinite number of unique polypeptides (proteins).

WRONG CHOICES EXPLAINED

(1) It is *not* true that the basic building blocks of a protein are *glucose molecules*. Glucose molecules form the basic structure of complex carbohydrates such as starch, not of proteins.

(3) It is *not* true that the basic building blocks of a protein are *hormones*. Hormones are specialized proteins that perform a regulatory function in complex plants and animals. Hormones do not act as building blocks of other compounds.

(4) It is *not* true that the basic building blocks of a protein are *fats*. Fats represent a class of organic compounds found in complex plants and animals. Fats do not act as building blocks of other compounds.

11. **3** Over time, data that support the successful evolution of a species would include observations that describe *an increase in the proportion of offspring that have favorable characteristics*. Favorable characteristics are those phenotypes that promote survival of the individuals that display them. As more individuals displaying such phenotypes survive and reproduce in a population, the proportion of such favorable characteristics in the population will increase over time.

WRONG CHOICES EXPLAINED

(1) Over time, data that support the successful evolution of a species would *not* include observations that describe *an increase in the genetic changes occurring in body cells*. Genetic changes that occur in body cells cannot be passed on to future generations. So they do not affect the evolution of the species and therefore are not observable.

(2) Over time, data that support the successful evolution of a species would *not* include observations that describe *a decrease in the genetic variety carried in sex cells*. Genetic variety is a key factor in biological evolution. Decreasing genetic variety tends to decrease the rate of successful evolution in a species by reducing the opportunity for the creation of favorable phenotypes in the population.

(4) Over time, data that support the successful evolution of a species would *not* include observations that describe *a decrease in the proportion of the population that has beneficial traits*. Successful evolution of a species is characterized by an increase in the proportion of individuals that display beneficial traits, not a decrease in that proportion.

12. **2** The presence of this chemical in all three types of plants suggests that these plants *share a common ancestry*. The synthesis of caffeine by coffee, cola, and cacao plants is a genetic characteristic made possible by biochemical reactions catalyzed by specific enzymes. The fact that the genes that code for these specific enzymes are found in common in modern-day coffee, cola, and cacao plants is strong evidence that all three plants evolved from a

common ancestral plant that had developed this genetic ability in the remote past through the processes of mutation and natural selection.

WRONG CHOICES EXPLAINED

(1) It is *not* true that the presence of this chemical in all three types of plants suggests that these plants *inherited identical mutations*. It is very unlikely that all three plants received these identical mutations independent of the reproductive process. However, it is very likely that these three plants did inherit identical mutations from a long line of parental plants leading back to a common ancestor.

[NOTE: This is a possible correct answer to the question as written, although it is not the best answer.]

(3) It is *not* true that the presence of this chemical in all three types of plants suggests that these plants *were exposed to the same type of radiation in the past*. Radiation is a general term for various types of radiant energy that have always been present in the ecosystems inhabited by these plants. The presence of caffeine in these plants is not directly affected by their exposure to radiation in the past.

(4) It is *not* true that the presence of this chemical in all three types of plants suggests that these plants *were cloned from a caffeine plant*. Cloning is a laboratory technique developed in the mid–late twentieth century. So it cannot be responsible for the creation of plants whose ancestry goes back hundreds of thousands or millions of years.

13. 1 This display is an example of *a behavioral adaptation*. Such adaptations are instinctive (inborn) actions that are genetically programmed. Assume that the behavior provided an ancestral individual with some adaptive advantage. Therefore, it is probable that the behavior was passed on to subsequent generations and became increasingly common in the turkey species population until it became universally displayed within the species.

WRONG CHOICES EXPLAINED

(2) This display is *not* an example of *selective breeding*. Selective breeding is a technique used by animal and plant breeders in which organisms with desirable traits are cross-bred with the hope of producing offspring that also display those traits. Selective breeding techniques have been developed over a few thousand years of human history. So these techniques cannot be responsible for the behavioral traits of turkeys whose ancestry goes back hundreds of thousands or millions of years.

(3) This display is *not* an example of *asexual reproduction*. Turkeys are birds. As such, they carry on sexual reproduction, not asexual reproduction.

(4) This display is *not* an example of *a learned behavior*. Learned behaviors are typically displayed by members of a species that live within a limited geographical region, not by all members of the species. Behaviors that are displayed by all members of a species regardless of region are genetically determined behaviors, not learned behaviors.

14. **2** One reason for this is that *most of the species that have ever lived on Earth are extinct.* Over the more than 4 billion years of Earth's history, many species have come into existence and reproduced successfully for many generations. However, drastic changes in the environmental conditions during that same period have periodically eliminated species unable to adapt to these changes. Scientists estimate that most of the species that have ever lived on Earth have been eliminated (become extinct) by the process of natural selection, making their fossils fairly common. Species alive on Earth today have either survived these changes or have evolved relatively recently in Earth's history. So their fossils are relatively uncommon compared with those of ancient species.

WRONG CHOICES EXPLAINED

(1) One reason for this is *not* that *most of the species that have ever lived on Earth are alive today.* This statement is contrary to the findings of competent scientists as stated in the correct answer above.

(3) One reason for this is *not* that *fossils of only extinct species have been found.* Fossils are the preserved remains of individual organisms that lived in the past, not necessarily only those that have become extinct. Fossils of extant (living) species are routinely discovered and studied by scientists.

(4) One reason for this is *not* that *species alive today will not form any fossils for future discovery by scientists.* There is no scientific reason to assume that the process of fossilization that has occurred for billions of years of Earth's history will not continue to operate in the future.

15. **1** *It is not necessary in order for the individual to survive* is the statement concerning sexual reproduction that is correct. An individual organism is fully capable of surviving its entire life without carrying out the life function of reproduction. This is not true of other major life functions, such as respiration or excretion.

WRONG CHOICES EXPLAINED

(2), (4) *The offspring are identical to the parent* and *the offspring are identical to each other* are *not* the statements concerning sexual reproduction that are correct. By its nature, sexual reproduction introduces variation into the genetic characteristics of offspring resulting from this process. Only in the event of identical twins will the offspring of sexual reproduction be genetically identical to each other.

(3) *It is necessary for the individual to survive* is *not* the statement concerning sexual reproduction that is correct. This statement is contrary to the findings of competent scientists as stated in the correct answer above.

16. **2** The main advantage when this switch occurs is that the paramecium is most likely to *increase variation among its offspring.* Asexual reproduction limits genetic variation, whereas sexual reproduction promotes it.

The paramecium's genetic structure is normally copied exactly through the process of mitotic cell division (asexual reproduction). When the paramecium engages in conjugation (sexual reproduction), genes are exchanged between participating cells. This process allows new genetic traits to be taken on by both cells and may give one or both of them an enhanced ability to survive difficult environmental conditions.

WRONG CHOICES EXPLAINED

(1) It is *not* true that the main advantage when this switch occurs is that the paramecium is most likely to *produce fewer offspring*. It is unlikely that the paramecium's occasional tendency to engage in sexual reproduction would have a significant effect on the numbers of offspring that result.

(3) It is *not* true that the main advantage when this switch occurs is that the paramecium is most likely to *avoid having to find a mate*. In fact, a paramecium engaging in sexual reproduction may do so only if it is able to find a mate with which to share genetic information.

(4) It is *not* true that the main advantage when this switch occurs is that the paramecium is most likely to *produce clones of itself*. Cloning is a complex laboratory process that involves producing exact genetic duplicates of sexually reproducing animals or plants. Cloning is not carried on by naturally occurring species.

17. **3** *Each cell in structure 1 has the same genetic content as that in structure 3* is the statement that is correct describing this diagram. Structure 4 illustrates a two-cell mass that has been derived from structure 3 (a zygote) by means of process *B* (mitotic cell division). Structure 3 is a diploid ($2n$) cell. So the cells in structure 4 (derived from structure 3 by mitotic cell division) must also be diploid and must contain the exact genetic content as that found in structure 3.

WRONG CHOICES EXPLAINED

(1) *The cell represented by structure 3 has the same genetic content as structure 2* is *not* the statement that is correct describing this diagram. Structure 3 is a diploid ($2n$) zygote and structure 2 is a haploid (n) egg cell. So the genetic content of these two cells cannot be the same.

(2) *Process A represents the process of meiosis* is *not* the statement that is correct describing this diagram. Process *A* illustrates the fusion of structure 1 (sperm cell) and structure 2 (egg cell). This process is known as fertilization, not meiosis.

(4) *Processes A and B both occur in the placenta* is *not* the statement that is correct describing this diagram. Processes *A* and *B* both occur in the oviduct (Fallopian tube), not the placenta.

18. **3** *It influences the development of adult sex characteristics* is the statement that describes a function of the hormone estrogen. Estrogen is a hormone that is produced by the ovaries of an adult female human. In addi-

tion to regulating certain aspects of the female reproductive cycle, it is also known to be responsible for the production of secondary sex characteristics in young women, such as enlargement of breasts and growth of body hair.

WRONG CHOICES EXPLAINED

(1) *It regulates the secretion of digestive enzymes* is *not* the statement that describes a function of the hormone estrogen. This function is the responsibility of peptic hormones such as gastrin and secretin, not estrogen.

(2) *It promotes sperm production in males* is *not* the statement that describes a function of the hormone estrogen. This function is the responsibility of the male testicular hormone testosterone, not estrogen.

(4) *It maintains blood sugar levels* is *not* the statement that describes a function of the hormone estrogen. This function is the responsibility of pancreatic hormones such as insulin and glucagon, not estrogen.

19. **2** The primary function of the human male reproductive system is to *produce and transport gametes*. Male gametes (sperm cells) are produced in the testes and are transported through the vasa deferentia (plural for vas deferens) and urethra to the female reproductive tract.

WRONG CHOICES EXPLAINED

(1), (3) The primary function of the human male reproductive system is *not* to *provide a site for fertilization* or to *protect and nourish the embryo*. These functions are carried on by the female reproductive system, not the male reproductive system.

(4) The primary function of the human male reproductive system is *not* to *prevent urine from leaving the body*. No bodily system carries on this function. This is a nonsense distracter.

20. **3** In an embryo, the formation of many types of tissues and organs occurs as a result of the process of *differentiation*. Differentiation is a process by which unspecialized embryonic cells take on the characteristics of specialized tissues. Specialized tissues carry on a limited number of metabolic activities, allowing those tissues to conserve energy and maximize efficiency.

WRONG CHOICES EXPLAINED

(1) In an embryo, the formation of many types of tissues and organs does *not* occur as a result of the process of *fertilization*. Fertilization is a process by which male and female gametes fuse to form a zygote. The zygote represents a completely undifferentiated and unspecialized cell type.

(2), (4) In an embryo, the formation of many types of tissues and organs does *not* occur as a result of the processes of *genetic sorting* or *gene recombination*. These processes occur during meiotic cell division when chromosomes are separated (assorted) into haploid (n) gametes then later reunited (recombined) with corresponding homologous chromosomes during fertilization in the diploid $(2n)$ zygote. Although gametes are specialized reproductive cells, the zygote that results from these processes is completely unspecialized. These processes occur in a mature adult, not in an embryo.

21. **2** *Researching the problem, using information from a variety of sources* is the activity that would be an appropriate first step when designing an experiment. Good science always begins with a review of available scientific literature that describes the results of prior experimental research in the area of interest. Such reviews provide the experimenter with insight that will improve the efficiency and effectiveness of his/her own research efforts.

WRONG CHOICES EXPLAINED
(1), (3), (4) *Reporting a conclusion based on multiple experimental trials, creating a data table to organize experimental observations,* and *repeating the experiment with a different hypothesis* are *not* the activities that would be an appropriate first step when designing an experiment. Each of these activities would logically be carried out after other activities such as designing and conducting the experiment, not as the first step.

22. **1** It is most likely that the child reacted this way because *normally harmless cat antigens stimulated the immune system.* An allergy is an immune response to an antigen or irritant that may be harmless. In this case, antigens associated with harmless cat dander irritating the child's immune system likely caused the child's allergic reaction.

WRONG CHOICES EXPLAINED
(2) It is *not* most likely that the child reacted this way because *it is difficult for the respiratory system to filter cat antigens out of the inhaled air.* The human respiratory system is designed to filter out large particles to prevent them from entering the lungs. These filtering mechanisms (such as nasal hair) are ineffective against antigens that stimulate allergic reactions. So they may allow entry of these antigens into the body interior, where the antigens may stimulate an allergic reaction.
[NOTE: This is a possible correct answer to the question as written, although it is not the best answer.]
This question contains ambiguities that make this distractor, as written, a possible correct answer. It is absolutely true that "it is difficult for the respiratory system to filter cat antigens out of the inhaled air." This fact makes it entirely likely that cat antigens infiltrated the lung tissues of this "child" and triggered an immune (allergic) response in his/her blood. In this respect, the distracter can be considered a correct response to the question.
(3), (4) It is *not* most likely that the child reacted this way because *cat antigens are a health hazard, since they always cause disease* or because *cat antigens stop the immune system from making antibodies, so bacteria cause these responses.* No competent science would support either of these statements. These are nonsense distracters.

23. **2** One likely reason these shrimp do not survive is that enzymes in the shrimp *change shape*. Most enzymes operate optimally in a fairly narrow range of temperatures specific to each type of enzyme. The enzymes present in cold-water shrimp likely operate best at relatively low environmental temperatures. When the environmental temperature rises above a certain level, the reaction rates of these enzymes may fall far below their optimum range. Extremely high temperatures may even cause the enzymes to change their shapes permanently (denature). This phenomenon would likely result in reduced metabolic efficiency and eventual death for these shrimp.

WRONG CHOICES EXPLAINED

(1) One likely reason these shrimp do not survive is *not* that enzymes in the shrimp *start to replicate*. The only biochemical known to replicate is DNA (deoxyribonucleic acid). Enzymes are a type of protein, not a type of nucleic acid.

(3), (4) One likely reason these shrimp do not survive is *not* that enzymes in the shrimp *are composed of fat molecules that melt* or that they *break down into small starch molecules*. Enzymes are a type of protein, not a type of fat or carbohydrate (starches).

24. **1** This change in the base sequence is an example of *a deletion*. The irradiated neurospora DNA sequence contains three fewer bases in its sequence than the original sequence. The bases missing from the original sequence are G-C-A. The loss of these three bases represents the removal of an entire triplet codon and is known as a deletion.

WRONG CHOICES EXPLAINED

(2) This change in the base sequence is *not* an example of *an insertion*. An insertion in the context of DNA mutation could be illustrated by the addition of bases to the irradiated DNA, not the removal of bases.

(3) This change in the base sequence is *not* an example of *a substitution*. A substitution in the context of DNA mutation could be illustrated by the exchange of bases within the irradiated DNA, not the removal of bases.

(4) This change in the base sequence is *not* an example of *a replication*. Replication is the process by which DNA produces an exact copy of itself during cell division. Replication could be illustrated by no change of bases compared to the original DNA.

25. **4** This is an example of *an application of biotechnology*. The specific type of biotechnology illustrated is known as genetic engineering. Genetic engineering is a laboratory technique in which a gene for a desired trait (such as production of a biochemical toxic to insects) is snipped from the DNA of a donor (bacterial) cell and inserted into the genome of a recipient (eggplant) egg cell. The recipient egg cell is then cultured to produce a mature plant capable of producing the desired biochemical. In this case, the genetically modified

eggplants are able to ward off potentially crop-destroying insects without the need to apply chemical pesticides. An unintended consequence of this process may be that the biochemical may also prove to be toxic to beneficial insects, such as honeybees, that are needed to pollinate the flowers of the eggplant.

WRONG CHOICES EXPLAINED

(1), (3) This is *not* an example of *selective breeding of the insects* or *selective breeding of the eggplant*. Selective breeding is a technique used by animal and plant breeders in which organisms with desirable traits are cross-bred with the hope of producing offspring that also display those traits. Selective breeding is not the process that is described in this question.

(2) This is *not* an example of *spraying an insecticide on plants*. This process is actually designed to avoid spraying chemical pesticides onto the genetically modified plants.

26. **2** A consequence of these changes is *a decrease in the availability of natural resources.* As the human population has grown, human consumption of finite resources has depleted materials used for industrial and economic purposes, such as minerals, petrochemicals, and other nonrenewable resources. Naturally occurring materials that used to be in abundant supply have become increasingly difficult and expensive to extract.

WRONG CHOICES EXPLAINED

(1) It is *not* true that a consequence of these changes is *an increase in the numbers and kinds of organisms worldwide*. As the human population has grown, the diversity of natural ecosystems has been greatly diminished. This has led to a decrease in the numbers and kinds of organisms worldwide, not an increase.

(3) It is *not* true that a consequence of these changes is *a decrease in deforestation due to technological improvements*. As the human population has grown, natural habitats—including forests—have been destroyed at an alarming rate. This has led to an increase in deforestation, not a decrease.

(4) It is *not* true that a consequence of these changes is *an increase in biosphere stability*. As the human population has grown, the dynamic equilibrium of healthy ecosystems has been disrupted. This has led to a decrease in biosphere stability, not an increase.

27. **4** This activity contributes most directly to the *recycling of organic matter*. When the spent corn stalks are plowed back into a corn field, their complex biochemical structures are broken down by bacteria and fungi into inorganic substances that can be taken up by new plants grown in the field the following year. This practice helps to ensure the sustained fertility of the farm field and its continued ability to support crops for human and animal consumption. It also reduces the need to apply artificial fertilizers that may contribute to environmental pollution.

WRONG CHOICES EXPLAINED

(1) This activity does *not* contribute most directly to the *increase in the biodiversity of their fields*. When fields are cleared of native plants and used for the cultivation of a single crop, biodiversity is decreased, not increased. As long as the field is maintained for this purpose, biodiversity will not increase.

(2) This activity does *not* contribute most directly to the *depletion of non-renewable resources*. Nonrenewable resources include substances such as minerals and fossil fuels. These substances are not normally present in significant quantities in a farm field or in the operations needed to plow under spent corn stalks.

(3) This activity does *not* contribute most directly to the *destruction of natural habitats*. The activities necessary to clear and prepare a field for the cultivation of a single crop are inherently destructive of the natural habitats that existed in the area prior to these activities. However, the action of plowing spent corn stalks back into the field will likely extend the field's productivity. This will reduce the need to destroy additional natural habitats to create a new corn field to replace it in the future.

28. **4** These variations in temperature are an example of *dynamic equilibrium*. Human body temperature normally varies within a range just above and below the typical human body temperature of 98.6°F (37°C). These variations are controlled by feedback mechanisms monitored by the central nervous system in a constant effort to maintain dynamic equilibrium (also known as homeostasis or steady state) in the body. This steady state is necessary for the efficient operation of the body's metabolic activities.

WRONG CHOICES EXPLAINED

(1) These variations in temperature are *not* an example of *an immune response*. The immune response involves the body's reaction to foreign antigens, including pathogens. The immune response has no direct effect on normal body temperature variations.

(2) These variations in temperature are *not* an example of *genetic differences between individuals*. Genetic differences among individual humans do result in different average body temperatures for those individuals. However, the feedback mechanisms that regulate variations in body temperature represent a genetic trait that all humans share in common.

(3) These variations in temperature are *not* an example of *an adaptation to global warming*. Global warming is a term relating to the effects of relatively recent human activities on weather systems that are manifested as increasing worldwide temperatures. The feedback mechanisms that regulate variations in body temperature have been part of the human genome for hundreds of thousands of years.

29. **1** This is most likely because *they have been commercially available as an energy source*. Companies that make a profit from promoting the use and sale of fossil fuels have been very successful—until recently—at lobbying

to block the development of alternate energy sources (e.g., solar, wind, tidal, geothermal, fuel cell, and fusion). Energy consumers also have been slow to demand the implementation of such alternatives. The longer fossil fuel companies are allowed to dictate energy policy here and abroad, the longer it will take for the use of fossil fuel for energy production to become a thing of the past.

WRONG CHOICES EXPLAINED

(2) This is *not* most likely because *there are alternatives to these types of fuels*. The fact that alternative energy sources exist and are commercially available should be an incentive to reduce our use of fossil fuel for energy production.

(3), (4) This is *not* most likely because *they have had a positive effect on global temperatures* or because *fossil fuels can be burned to produce large quantities of carbon dioxide*. When fossil fuels are burned, carbon dioxide is released. Competent scientists know that carbon dioxide is a greenhouse gas that contributes to the phenomenon known as global warming (climate change). The burning of fossil fuels has a negative effect on global temperatures, not a positive effect. The fact that the burning of fossil fuels releases carbon dioxide that contributes to global warming should be ample reason to reduce this practice for energy production.

30. **3** This usually results in abnormalities because the affected person has *more genes in each cell than a person with a normal chromosome number*. Because each gene codes for a specific polypeptide (protein), the presence of additional genes in the cell results in the production of more than the amount of each protein actually required by the cell. This overabundance of protein in the cell can result in an imbalance of the cell's metabolic processes and consequent abnormalities.

WRONG CHOICES EXPLAINED

(1) It is *not* true that this usually results in abnormalities because the affected person has *a reduced number of genes in cell nuclei*. Extra chromosomes in a cell introduce a greater number of genes into the cell, not a reduced number.

(2) It is *not* true that this usually results in abnormalities because the affected person has *fewer cell mutations than a person with a normal chromosome number*. Chromosomal abnormalities such as trisomy (three instead of the normal two homologous chromosomes) and polyploidy (multiple chromosomes) are chromosomal mutations. For this reason, a person with additional chromosomes has more mutations than a normal person, not fewer.

(4) It is *not* true that this usually results in abnormalities because the affected person has *less DNA in cell nuclei, but more proteins in cell mitochondria*. A person who has additional chromosomes has more DNA in his/her cells, not less. Mitochondrial DNA synthesizes proteins independently of nuclear DNA and is therefore not affected by this condition.

PART B–1

31. **3** *The black snails had an adaptive advantage* is the statement concerning the snails that is correct. The beach environment was blackened due to the deposition of black volcanic material. Members of the snail species that displayed black coloration blended in better with the environment and became less visible to predators. Because the black snails were taken less often by predators, they survived to produce offspring (that are also black) at a higher rate than the tan snails. The black snails gained an advantage over the tan snails. The species survived as a result of the black snails' ability to adapt to a changing environmental condition. This phenomenon is recognized by competent scientists as representing the process of natural selection.

WRONG CHOICES EXPLAINED

(1) *The lava particles turned the tan snails black* is *not* the statement concerning the snails that is correct. Characteristics obtained during the lifetime of an organism cannot be passed on genetically to the next generation.

(2) *The tan snails will become extinct* is *not* the statement concerning the snails that is correct. It is very likely that the gene controlling the tan color will continue to be represented in the gene pool of this species, though at a lower frequency.

(4) *The tan snails preyed on the black snails* is *not* the statement concerning the snails that is correct. No information is presented in the question that indicates that some members of this snail species prey upon other members.

32. **1** The increase in the number of black snails can best be explained by *natural selection after an environmental change*. The beach environment blackened due to the deposition of black volcanic material. Members of the snail species that displayed black coloration blended in better with the environment and became less visible to predators. Because the black snails were taken less often by predators, the black snails survived to produce offspring (that are also black) at a higher rate than did the tan snails. The black snails gained an advantage over the tan snails. The species survived as a result of the black snails' ability to adapt to a changing environmental condition. This phenomenon is recognized by competent scientists as representing the process of natural selection.

WRONG CHOICES EXPLAINED

(2) The increase in the number of black snails *cannot* best be explained by *climatic change followed by ecological succession*. Ecological succession refers to the series of changes that occur to the dominant plant community of an area that has been significantly affected by an environmental change. Snails are animals, not plants, so ecological succession cannot be at work in this example.

(4) The increase in the number of black snails *cannot* best be explained by *increased stability due to a decrease in variation*. No mention is made in the question concerning the relative stability of this situation. Variations in this snail species are neither increasing nor decreasing. This is a nonsense distracter.

(3) The increase in the number of black snails *cannot* best be explained by *an increase in mutation rate*. The question refers to a snail population with both tan and black members. The information indicates that any mutation in this snail species (for black or tan coloration) occurred in the past and not as a result of an increased mutation rate in the present.

33. **4** Variation in snail color is an example of *diversity*. "Diversity" is a term that by biological convention refers to the number of different plant and animal species that inhabit a particular ecosystem. By this convention, the term "diversity" should not be used interchangably with the terms *genetic variation* or *adaptation*, which occur among the members of an individual species.

WRONG CHOICES EXPLAINED

(1), (3) Variation in snail color is *not* an example of *environmental stability* or *equilibrium*. These terms relate to situations in which the conditions surrounding life are relatively unchanging. Whether these conditions are environmental or physiological in nature, they do not normally relate to genetic variation within a single species.

(2) Variation in snail color is *not* an example of *a natural limitation*. Limiting factors are those biotic and abiotic conditions that impose upper limits on the growth of populations. Variation within a species is not a limiting factor.

34. **4** These changes in the community demonstrate that *populations are linked with many others in the ecosystem*. As the population of grasshoppers decreases, less food is available for toads and hawks. The population of toads may decline as a result of less available food. As the populations of grasshoppers and toads decrease, fewer food choices are available for hawks. The hawks may then be forced to hunt more aggressively for rabbits. As grasshoppers decrease in number and more rabbits are taken by hawks, mice have more food available. As a result of increased food supplies, the population of mice then increases. These effects collectively demonstrate the interrelatedness of populations inhabiting an ecosystem.

WRONG CHOICES EXPLAINED

(1) These changes in the community do *not* demonstrate that *ecosystems are shaped by nonliving factors*. It is true that ecosystems are shaped by nonliving factors. However, this fact does not relate to the specific population changes noted in the question.

(2) These changes in the community do *not* demonstrate that *autotrophs convert solar energy into food*. It is true that autotrophs convert solar energy into food. However, this fact does' not relate to the specific population changes noted in the question.

(3) These changes in the community do *not* demonstrate that *grasshoppers are producers that are essential for ecosystem stability*. Grasshoppers are animals that consume plant matter for food. Grasshoppers are consumers, not producers.

35. **3** This process best represents *the bonding of simple sugars to form a starch molecule*. Starch is a complex carbohydrate composed of multiple simple sugar molecules such as glucose (□) bonded together chemically with the catalytic assistance of an enzyme (*X*).

WRONG CHOICES EXPLAINED

(1) This process does *not* best represent *the bonding of amino acids to form a starch molecule*. The bonding of amino acids forms a protein molecule, not a starch molecule.

(2), (4) This process does *not* best represent *the digestion of amino acids to form a starch molecule* or *the digestion of simple sugars to form a starch molecule*. Digestion is a metabolic process by which complex molecules are broken down by enzymatic hydrolysis, not built up. Amino acids are subcomponents of proteins, not starches.

36. **4** The letter *X* in the process represents *an enzyme*. By convention, a symbol placed above an arrow in a chemical pathway represents a catalyst. Enzymes are biological catalysts that enable chemical reactions between substrates in a biochemical reaction pathway.

WRONG CHOICES EXPLAINED

(1) The letter *X* in the process does *not* represent *an antibody*. Antibodies are specialized proteins that are synthesized in response to an invasion of the body by foreign antigens. Antigens do not catalyze biochemical reactions.

(2) The letter *X* in the process does *not* represent *a hormone*. Hormones are specialized proteins that perform specific regulatory functions in complex plants and animals. Hormones do not catalyze biochemical reactions.

(3) The letter *X* in the process does *not* represent *a receptor*. Receptors are specialized proteins that are embedded in cell membranes and regulate the passage of hormones and other substances into and out of cells. Receptors do not catalyze biochemical reactions.

37. **4** *Species J had little variation and lived in a changing environment* is the statement that would most likely be correct based on the information in the diagram. The diagram shows that species *J* branched off the evolutionary line as a descendant of species *B*. Species *J* existed for a relatively short

period of time before becoming extinct. Species *J* did not serve as an ancestor of other related species. This indicates that species *J* most likely became extinct in the face of a changing environment to which it was unable to adapt.

WRONG CHOICES EXPLAINED

(1) *Species C had many variations and lived in a stable, unchanging environment* is *not* the statement that would most likely be correct based on the information in the diagram. The diagram indicates that species *C* descended from species *H* and existed for a relatively long period of time before becoming extinct. Species *C* did not serve as an ancestor of other related species. The relatively long existence of species *C* is an indication that it thrived for a long time in a stable environment before becoming extinct due to the lack of variation when the environment finally changed. Species *C* had few variations, not many variations.

(2) *Species D, C, and J are extinct* is *not* the statement that would most likely be correct based on the information in the diagram. The diagram indicates that species *D* is extant (existing in the present day), not extinct.

(3) *Species F evolved from species D* is *not* the statement that would most likely be correct based on the information in the diagram. According to the diagram, species *F* and *D* share a common ancestor in species *G*. This is an indication that species *F* evolved from species *G*, not from species *D*.

38. **2** If blockages of this type occur, the most likely result would be that *the female gamete would not be able to unite with the male gamete*. Female gametes (eggs) are formed in the ovaries and are transported through the oviducts (Fallopian tubes) toward the uterus. Male gametes (sperm) implanted in the female reproductive tract by the male move through the cervix and uterus into the oviducts. In normal human reproduction, the egg and sperm unite (fuse) in the oviducts. If the oviducts are blocked, this fusion cannot occur.

WRONG CHOICES EXPLAINED

(1) It is *not* true that if blockages of this type occur, the most likely result would be that *the egg would remain in the uterus and not travel upward*. Eggs travel downward from the ovaries to the uterus, not upward from the uterus to the ovaries.

(3) It is *not* true that if blockages of this type occur, the most likely result would be that *hormones could not be produced by the ovaries*. Ovaries are endocrine glands that produce female reproductive hormones, including estrogen and progesterone. This endocrine function would not be interrupted by blockages in the oviducts.

(4) It is *not* true that if blockages of this type occur, the most likely result would be that *the process of asexual reproduction would be prevented or interrupted*. As mammals, humans reproduce by sexual means, not asexual means.

39. **4** If no additional materials were added to the culture dish after day 14, the paramecium population would most probably *begin to decrease as finite resources are used up*. We must assume that the culture dish contained enough nutrient material to sustain this population for the 14-day period indicated in the graph. We must also assume that at some time after day 14, the nutrient material will become depleted. In this case, it is likely that the paramecium population will experience a significant decline when food supplies are exhausted. (In reality, the paramecium population in the closed environment of a culture dish will also experience decline due to exposure to toxic waste that is not removed from the dish in the way it would be in the species' natural aquatic habitat.)

WRONG CHOICES EXPLAINED
(1), (3) If no additional materials were added to the culture dish after day 14, the paramecium population would *not* most probably *remain the same, since it has reached carrying capacity and had an unlimited food supply* or *begin to increase, since they have not yet reached carrying capacity*. The term "carrying capacity" relates to the number of individuals in a natural population that can be sustained by the continually renewable but limited resources present in the natural habitat. It is unlikely that the paramecium population would either remain the same or increase under the conditions described in the question.

(2) If no additional materials were added to the culture dish after day 14, the paramecium population would *not* most probably *begin to increase as they continue to reproduce*. Reproduction is an energy-dependent and a material-dependent process. Without a sustained food (energy and material) supply, it is unlikely that the paramecium population would increase by means of reproduction.

40. **1** *The population numbers of* Paramecium aurelia *would be lower than 250, since the new species is competing with it for resources* is the statement that describes a possible result as the populations interact over the next 14 days. In this event, *Paramecium aurelia* and the second paramecium species would compete for limited resources in the culture dish. Interspecies competitions of this sort normally result in the elimination of one of the competing species from the environment. Under these assumptions, it is likely that the numbers of *Paramecium aurelia* in the culture dish would fall below 250 over the 14 days of the experiment as a direct result of this competition.

(The student must assume that the second species of paramecium has requirements for resources identical to those of *Paramecium aurelia*. In other words, the second species occupies the same environmental niche as *P. aurelia*.)

WRONG CHOICES EXPLAINED

(2) *The population of* Paramecium aurelia *would increase above 250, since they would mate with the new species* is *not* the statement that describes a possible result as the populations interact over the next 14 days. Paramecia are unicellular organisms that reproduce primarily by asexual means, not sexual means. For this reason, the two species cannot mate.

(3), (4) *The population of* Paramecium aurelia *would increase above 250, since the two species occupy the same niche* and *the population of* Paramecium aurelia *would remain at 250, since the species compete with each other for the same resources* are *not* the statements that describe a possible result as the populations interact over the next 14 days. Because the two populations occupy the same niche and compete virtually equally for limited resources, the populations of both species are expected to be negatively affected.

41. **3** C *and* D *combine to produce* A *and* B is the statement that best describes how these substances interact in photosynthesis. In the photosynthetic reactions, the bonds between the atoms that make up molecules of carbon dioxide (*C*) and water (*D*) are broken with the assistance of light energy. Atoms of carbon, hydrogen, and oxygen are released. The carbon, hydrogen, and some of the oxygen atoms are rearranged to form molecules of glucose (*A*). The remaining oxygen atoms are rearranged to form molecules of oxygen gas (*B*).

WRONG CHOICES EXPLAINED

(1) A *and* B *combine to produce* C *and* D is *not* the statement that best describes how these substances interact in photosynthesis. This statement describes the process of aerobic respiration, not photosynthesis.

(2), (4) B *and* C *combine to produce* A *and* D and A *and* C *combine to produce* B *and* D are *not* the statements that best describe how these substances interact in photosynthesis. These combinations of reactants and products do not represent any known biochemical reaction and, in fact, are chemically impossible. These are nonsense distracters.

42. **1** *Country* A, *since the larger percentage of young could result in rapid population growth* is the country/situation in which the population growth over the next 20 years will place the greatest strain on the environment. A review of the graphic reveals that the population of young people (0–14 years of age) is much greater in country *A* (about 45%) than the same age group in country *B* (about 20%). However, the death rate for country *A* is also greater than that of country *B* for all age groups. The result is that the percentages of the populations of childbearing age (15–49 years of age) are nearly equal in both countries (about 48% for country *A* and about 45% for country *B*). Although country *A* will probably continue to experience a high birth rate in the future, it is unlikely that its overall population will increase significantly faster than that of country *B* unless something is done to reduce its high death rate.

[NOTE: There is no correct answer to the question as written. However, this answer is the best of those presented.]

This question contains ambiguities that make this answer, as written, an incorrect response. The data in the graphic for country A illustrate a pattern typical of a developing ("Third World") country with very high birth rates (hence, the wide bars for ages 0–4) followed by very high death rates (hence, the rapidly narrowing bars for ages 5–85). The sum of the data for the reproductively active segment of the population (ages 15-49) are nearly the same for country A and country B, indicating that their actual capacities to produce offspring are nearly identical. For this reason, the data presented do not conclusively demonstrate that the total population of country A will grow significantly over the next 20 years unless its birth rate remains high AND its death rate is greatly reduced. It is more likely that the population and demographic distribution by age grouping of country A will remain relatively unchanged during this period. Because these factors are not included in the information provided, the answer given cannot be considered to be a correct response to the question. However, it IS the best answer of those given.

WRONG CHOICES EXPLAINED

(2) *Country* B, *since the smaller percentage of young could result in rapid population growth* is *not* the country/situation in which the population growth over the next 20 years will place the greatest strain on the environment. Although the percentage of young people (0–14 years of age) in country *B* is significantly smaller than the same age group in country *A*, the death rate for all age groups in country *B* is significantly lower than in country *A*. Country *B* displays the characteristics of a stable human population with adequate resources and a superior health care system. Country *B* is unlikely to experience rapid population growth in the future as long as environmental and economic conditions remain unchanged.

(3), (4) *Country* A, *since the smaller percentage of people over 60 uses most of the resources* or *country* B, *since the larger percentage of people over 60 uses the fewest resources* are *not* the countries/situations in which the population growth over the next 20 years will place the greatest strain on the environment. No data are provided in the question concerning the relative use of resources by the age groups over 60 years of age.

43. **3** Approximately *32%* of the population of country *A* is less than 10 years old. This value can be determined by reading the percentages of females and males aged 0–4 and 5–9 from the graph titled "Country A" and adding them together (8% + 10% + 6% + 8% = 32%) to arrive at the correct result.

WRONG CHOICES EXPLAINED

(1), (2), (4) It is *not* true that approximately *8%, 16%,* or *64%* of the population of country *A* is less than 10 years old. These values are not consistent with the method described above.

PART B–2

44. One credit is allowed for correctly marking an appropriate scale, without any breaks in the data, on each labeled axis. [1]

45. One credit is allowed for correctly plotting the data for the number of bacterial colonies on the screened side, connecting the points, and surrounding each point with a small circle. [1]

46. One credit is allowed for correctly plotting the data for the number of bacterial colonies on the unscreened side, connecting the points, and surrounding each point with a small triangle. [1]

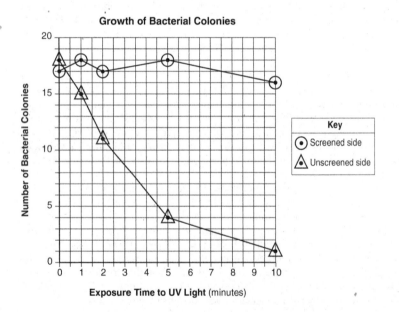

Growth of Bacterial Colonies

Exposure Time to UV Light (minutes)

47. **3** *Uncontrolled mitotic division occurred as a result of gene mutations* is the statement that provides a possible explanation for this growth pattern. The diagram illustrates the growth of cancerous cells within healthy skin tissues. As the mutated cancer cells multiply rapidly by means of uncontrolled and abnormal cell division, they push aside normal tissues and interfere with the homeostatic balance of the body.

WRONG CHOICES EXPLAINED

(1) *Manipulation of genes caused the movement of embryonic skin cells* is *not* the statement that provides a possible explanation for this growth pattern. The diagram illustrates the addition of cells to the original normal skin cells, not the movement of existing cells due to an outside force.

(2) *Exposure to light stimulated the development of cells containing ozone* is *not* the statement that provides a possible explanation for this growth pattern. Normal human skin cells do not produce or contain ozone under any circumstances. This is a nonsense distracter.

(4) *An immune reaction triggered the formation of excess blood cells* is *not* the statement that provides a possible explanation for this growth pattern. The diagram does not indicate the presence of blood cells within the mass of skin cells. Forming blood cells is not a function of the immune reaction.

48. One credit is allowed for correctly identifying the enzyme that would most likely function in the stomach and for supporting the answer. Acceptable responses include but are not limited to: [1]

- *Pepsin—The pH of the stomach is acidic, and pepsin operates best at an acidic pH.*
- *Pepsin—At a pH value of 1.5–4.0, the acidity of the stomach most closely matches the optimum pH for the operation of this enzyme as shown in the graph.*
- *Pepsin—This enzyme performs its hydrolytic function most efficiently in the very acidic environment of the stomach.*

49. **1** The activity of *chymotrypsin* decreases in both acidic and basic environments. The graph for chymotrypsin indicates that the optimum pH value for this enzyme is about 7.8 (very nearly neutral). The graph also indicates that the rate of the reaction controlled by chymotrypsin drops off steadily as the pH of its environment approaches either the acidic or basic ends of the pH spectrum.

WRONG CHOICES EXPLAINED

(2) It is *not* true that the activity of *pepsin* decreases in both acidic and basic environments. The graph for pepsin indicates that this enzyme functions most efficiently at a very low (acidic) pH and at much reduced efficiency at pH values above 2.0.

(3) It is *not* true that the activity of *cholinesterase* decreases in both acidic and basic environments. The graph for cholinesterase indicates that this enzyme functions most efficiently at neutral and basic pH values and at much reduced efficiency at acidic pH values.

(4) It is *not* true that the activity of *papain* decreases in both acidic and basic environments. The graph for papain indicates that this enzyme functions most efficiently at all pH values.

50. **1** *Human activities had the unexpected consequence of decreasing the hellbender population to a dangerous level* is the statement that best explains the hellbender population decline in western New York State. The passage cites overcollection, pollution, and habitat destruction as being partially responsible for this decline. These conditions are all the consequence of human activities.

WRONG CHOICES EXPLAINED

(2) *Humans have purposefully removed the hellbender from its habitat due to its aggressive behavior* is *not* the statement that best explains the hellbender population decline in western New York State. The passage cites the fact that humans have purposefully removed the animal from its habitat for the pet trade, not due to its aggressiveness.

(3) *The decline of the hellbender population is due mainly to natural causes that humans cannot control* is *not* the statement that best explains the hellbender population decline in western New York State. The passage cites four reasons for the animal's decline in the wild, three of which are human in origin.

(4) *The hellbender population decreased because salamanders are very resistant to climate change* is *not* the statement that best explains the hellbender population decline in western New York State. No information is provided in the passage concerning the resistance of salamanders to climate change. Climate change is not cited in the passage as a factor in the animal's decline.

51. One credit is allowed for correctly describing *one* potential ecological effect, other than the loss of the hellbender from western New York State, of the hellbender population continuing to decline. Acceptable responses include but are not limited to: [1]

- *The population size of organisms that feed on the hellbender will decline.*
- *The population size of organisms that the hellbender feeds upon will increase.*
- *The biodiversity of the western New York ecosystem will decrease, causing instability.*
- *Food webs in the ecosystem will be disrupted.*
- *The ecological niche of the hellbender will go unfilled until a new organism moves into the area to fill it.*

52–54. Three credits are allowed for correctly discussing energy use in animals. In your answer, be sure to:

- Identify the type of protein molecules used to digest food. [1]
- Identify the organelle where energy from nutrients is released. [1]

- State *one* inference that can be made concerning a cell that has many of these organelles. [1]

Acceptable responses include but are not limited to: [3]

Enzymes are specialized proteins that catalyze reactions in the cell, including the digestion of food molecules. [1] Within the cell, the organelle where energy is released is the mitochondrion. [1] It can be inferred that a cell that contains large numbers of mitochondria performs functions that require a lot of energy. [1]

The specialized proteins that help biochemical reactions occur in the cell are known as organic catalysts. [1] The cell organelles in which the energy-releasing biochemical reactions (aerobic respiration) occur are the mitochondria. [1] When cells (such as muscle cells) contain large quantities of mitochondria, it can be inferred that those cells perform functions that require high energy outputs. [1]

55. One credit is allowed for correctly completing the missing information for sections 3a. and 3b. so that the key is complete for all four types. Acceptable responses include but are not limited to: [1]

1a. Is rod shaped	bacillus
1b. Is not rod shaped	go to 2
2a. Is spiral shaped	spirillum
2b. Is not spiral shaped	go to 3
3a. **Exists singly or Single cells**	type C
3b. **Exists in a group *or* colony *or* Clumps of cells *or* Chains of cells**	type D

PART C

56. One credit is allowed for selecting *one* of the listed changes experienced by the individual and correctly explaining how the change allows the individual to effectively respond to a threat. Acceptable responses include but are not limited to: [1]

- *Increased pulse rate—moves blood rich in oxygen to the muscle cells faster*
- *Increased pulse rate—allows the body to get rid of carbon dioxide faster*
- *Increased blood glucose levels—provide cells with a ready source of energy more rapidly*
- *Increased blood glucose levels—gets energy-rich glucose to the muscles, where it can be metabolized to release energy for the formation of ATP*
- *Increased breathing rate—eliminates carbon dioxide from the body faster*
- *Increased breathing rate—allows the body to get oxygen into the blood faster*

57. One credit is allowed for correctly explaining why, once the threat has passed, another series of changes returns the body to its original state. Acceptable responses include but are not limited to: [1]

- *The fight-or-flight response includes many changes in body activity that could result in damage to the body if not returned to normal levels.*
- *The fight-or-flight response sped up activities that now must be slowed down to normal rates.*
- *Once the danger is passed, the high activity levels of cells and organs are reduced to normal levels in order to prevent damage to the body.*
- *The metabolic changes of the defense responses upset the body's homeostatic balance. This balance must be restored so that all systems and processes operate normally.*

58. One credit is allowed for correctly describing *one* way that an infection with the white-nose fungus can cause death in little brown bats. Acceptable responses include but are not limited to: [1]

- *Lesions interfere with water balance.*
- *The fungus can make the bats wake up during hibernation, which makes them use up too much energy.*
- *Lesions can interfere with heat regulation.*
- *Sores interfere with the bats' blood circulation.*
- *Bats can become exhausted and die during hibernation due to the effects of the fungus.*

59. One credit is allowed for correctly describing *one way* that the little brown bats can be helped. Acceptable responses include but are not limited to: [1]

- *Research to find a way to stop the spread of the disease.*
- *Decontaminate the clothing of researchers who are studying the caves.*
- *Make artificial hibernation areas for the bats.*
- *Decontaminate the caves before the winter hibernation season.*
- *Prevent people from entering the caves used by the bats as hibernation sites.*
- *Medicate the bats.*

60. One credit is allowed for correctly describing how a student could use a microscope to compare the size of frog skin cells to the size of human skin cells. Acceptable responses include but are not limited to: [1]

- *The student could use a microscope with a known field diameter to measure the size of the two kinds of cells in an appropriate unit of measure (microns).*
- *The student could place both types of cells side by side on a slide and compare their relative sizes.*

61. One credit is allowed for correctly identifying the component of a vaccine that makes it effective. Acceptable responses include but are not limited to: [1]

- *The effective components of a vaccine are antigens of the infectious agent.*
- *Weakened/dead/fragmented pathogens that will stimulate the immune response are the effective components of a vaccine.*
- *In a chickenpox vaccine, the effective component is inactive chickenpox virus.*

62. One credit is allowed for correctly describing how a vaccination prevents disease in the individual who receives a vaccination. Acceptable responses include but are not limited to: [1]

- *The vaccination activates the immune system.*
- *A vaccination stimulates the production of antibodies against the pathogen.*
- *Inactive chickenpox virus in the vaccine can cause the immune system to form antibodies to fight off the chickenpox disease.*
- *Once the body has produced antibodies as a result of receiving the vaccination, the body can respond quickly to produce antibodies in response to an actual infection.*

63. One credit is allowed for correctly describing how vaccinations help to prevent the spread of disease, even in people who have not received the vaccination. Acceptable responses include but are not limited to: [1]

- *Not as many people get sick, so the disease does not spread to as many other people.*
- *It is less likely that the disease will spread since fewer people are going to have the disease.*
- *Vaccination reduces the risk for exposure to the disease in the general community.*

64–66. Three credits are allowed for correctly describing the consequences of removing cormorants from the eastern shores of Lake Ontario. In your answer, be sure to:

- State *one* reason why removing the cormorants from the food web could have a positive impact on the fishing industry. [1]
- State *one* possible effect of removing the cormorants on a species other than fish and support the answer. [1]
- Describe *one* action, other than removing a population of organisms from the environment, that humans could take to preserve the fishing industry in Lake Ontario. [1]

Acceptable responses include but are not limited to: [3]

A positive effect on fishing from removing cormorants could be that temporarily there would be more salmon and lake trout to be caught by humans. [1] The removal of the cormorants could reduce the eagle population because an important food supply for the eagles would be gone. [1] Another way to protect the fishing industry would be to institute catch-and-release programs so game fish populations can be preserved long term. [1]

A short-term positive effect on the fishing industry of removing cormorants is that forage fish populations on which the cormorants feed would temporarily increase. This would provide more food for game fish and help increase their populations. [1] With more forage fish in the lake, invertebrates would be taken more frequently, which would decrease their populations and their availability as food for other lake species. [1] Instead of removing cormorants because they are believed responsible for the Lake Ontario fishery problem, scientifically study the causes of declining game fish populations, such as overfishing by humans or heat pollution from lakeside nuclear power plants. Then take steps to reduce or eliminate those causes. [1]

67. One credit is allowed for correctly explaining why *not* mowing the field allowed the ecosystem to undergo ecological succession. Acceptable responses include but are not limited to: [1]

- *Not mowing allowed taller plants, such as shrubs and trees, to grow.*
- *Mowing favored the growth of short plants such as herbs and grasses. Not mowing allowed all plants to grow, to compete, and to follow the natural progression from field to forest.*
- *Trees and bushes that were cut short or killed by the mower could reach their full height and maturity when mowing was stopped.*
- *Once mowing activities ended, the plant community could alter the conditions in the field that favored field plants and make the area less favorable to them and more favorable for the growth of the next stages of succession.*

68. One credit is allowed for correctly identifying *one* specific human activity, other than mowing, that could alter this succession and for correctly explaining how this activity affects biodiversity. Acceptable responses include but are not limited to: [1]

- *Cut down the forest—Many species that live in the forest will be forced out or killed, reducing biodiversity. Different plant and animal species would then move in and establish a new biodiverse community.*
- *Develop the area for a mall/commercial purpose—This would remove nearly all natural species from the environment by clearing and paving large areas, eliminating biodiversity altogether.*
- *Use the area as a dump for toxic materials—Doing so would interrupt the natural balance of the ecosystem, would kill or sicken many natural species, and would reduce the diversity of the species living in the environment.*
- *Introduce invasive species—Many invasive species compete with native species by crowding them out or by competing for limited resources and eliminating their ecological niche. This could negatively impact other species that depend on the eliminated native species and reduce the biodiversity of the area.*
- *Plant native wildflowers—Doing so would help to increase the number of different species and would temporarily increase biodiversity until interspecies competition eliminated less successful species.*

69. One credit is allowed for correctly describing how this forest ecosystem would respond to a natural disaster, such as a flood, that resulted in the destruction of the plant community. Acceptable responses include but are not limited to: [1]

- *The ecosystem would undergo succession again and eventually return to a stable state.*
- *The ecosystem would gradually change back to the climax community that was there before the flood.*

- *The ecosystem would slowly revert to what it was like before the flood.*
- *Succession would begin with pioneer plants invading the bare soil or rock.*

70. One credit is allowed for correctly describing how these findings could influence the way in which spinach is displayed for sale in supermarkets. Acceptable responses include but are not limited to: [1]

- *Spinach could be displayed under fluorescent lights.*
- *The grocer could display spinach in transparent packages that allow exposure to fluorescent lighting.*
- *Spinach should not be sold in containers that block light.*
- *The store manager should be sure that the lighting over the spinach display is fluorescent, not incandescent.*

71. One credit is allowed for correctly explaining how the original decision to drill for oil in the Gulf of Mexico could be considered a trade-off. Acceptable responses include but are not limited to: [1]

- *People could get oil to be used for energy, but they might damage the environment while doing it.*
- *Oil companies provide many jobs for people, but there could be negative effects on the environment that cost jobs.*
- *The oil industry provides energy to run the nation's economy, but other industries such as commercial fishing, tourism, and real estate that depend on a clean Gulf environment can be negatively affected by oil spills.*
- *Owners and stockholders of oil companies make a lot of money in a short time, but not enough money is spent on safeguarding the animals and plant communities that have lived in the Gulf for many thousands of years.*

72. One credit is allowed for correctly stating *one* possible reason why it will most likely take bird populations more time to recover from this oil spill than it will mammal populations. Acceptable responses include but are not limited to: [1]

- *More birds were killed than mammals.*
- *The birds' breeding grounds were contaminated, but the mammals' were not.*
- *Many marine bird species build their nests along the beach where oil can reach them and kill young birds before they can hatch. Mammals don't lay eggs.*
- *Mammals carry their young inside their bodies as they develop. So they are better protected from oil contamination than the birds' offspring that hatch from eggs on the beach.*

PART D

73. **3** Plant *C* is the plant in the chart that would be selected as most similar to the medicinal plant. Plant *C* is phenotypically similar to the medicinal plant (produces flowers with the same number of petals and leaves with the same shape and vein pattern). This is a strong indication that they share a common ancestor. In addition, plant *C*'s amino acid sequence differs from that of the medicinal plant by only one amino acid. This is a strong indication that these two plants carry on very similar biochemical reactions, possibly including the synthesis of the desired medicinal compound.

WRONG CHOICES EXPLAINED
(1), (2), (4) Plants *A, B,* and *D* are *not* the plants in the chart that would be selected as most similar to the medicinal plant. Each of these plants differs significantly from the medicinal plant either (or both) in phenotypic characteristics or in amino acid sequence.

74. **1** A first step in developing the cancer cure could be to *preserve the habitat where the species is found.* Endangered plants are particularly susceptible to damage from environmental changes, including human activities. By preserving the habitat of the plant, the researchers help to ensure that the plant species will survive and reproduce at least until a means can be found to cultivate it commercially or to synthesize artificially the desired medicinal compound it produces.

WRONG CHOICES EXPLAINED
(2) It is *not* true that a first step in developing the cancer cure could be to *introduce a new plant species that will share the habitat of the plant.* Doing so could introduce an invasive species that may compete with the medicinal plant for needed resources such as space, light, or soil nutrients. This represents an environmental change that could negatively affect the endangered plant species.
(3) It is *not* true that a first step in developing the cancer cure could be to *eliminate every species that eats this plant.* Doing so would upset the ecological balance in the habitat of the plant by reducing biodiversity. This represents an environmental change that could negatively affect the endangered plant species.
(4) It is *not* true that a first step in developing the cancer cure could be to *harvest all of the plants of this species and use them to treat cancer patients.* Doing so would remove breeding members of the plant species from its habitat and prevent the species from producing offspring for further research or commercial cultivation. Following this activity, the species would be effectively extinct and no further health benefits could be derived from it.

75. **4** A graph of this data would look most like graph *4*. If measurements of pulse rate are taken at only two points (before exercise and after exercise), the first value will be lower than the second value. The line connecting these two data points will be an ascending straight line as shown in graph 4.

WRONG CHOICES EXPLAINED

(1), (2), (3) A graph of this data would *not* look most like graphs *1, 2,* or *3*. The shapes of these graphs do not match the predicted shape based on the data description provided in the question.

76. **2** The technique used to separate the DNA for analysis is *electrophresis*. In this laboratory technique, fragments of DNA from different donors are placed into a gel and then drawn through the gel by the electric current. The banding patterns that result are then analyzed to determine the relative number of bases in the DNA samples and, by extension, the degree of relatedness of the DNA sample donors.

WRONG CHOICES EXPLAINED

(1) It is *not* true that the technique used to separate the DNA for analysis is *genetic engineering*. Genetic engineering is a laboratory technique in which a gene for a desired trait is snipped from the DNA of a donor cell and inserted into the genome of a recipient cell. Genetic engineering is not the technique illustrated in the diagram.

(3) It is *not* true that the technique used to separate the DNA for analysis is *chromatography*. Paper chromatography is a laboratory technique used to help researchers study mixtures of organic chemicals such as plant pigments. Chromatography is not the technique illustrated in the diagram.

(4) It is *not* true that the technique used to separate the DNA for analysis is *protein synthesis*. In the process of protein synthesis, amino acids are linked by means of peptide bonds in specific sequences as determined by DNA codes. Protein synthesis is a cellular process, not a laboratory technique, and is not illustrated in the diagram.

77. One credit is allowed for correctly identifying the tiger that is the father of the newborn cub and for supporting the answer. Acceptable responses include but are not limited to: [1]

- *Male tiger 1—All of the cub's DNA fragments match fragments found in either the female or male 1.*
- *Male tiger 1—Half of the cub's DNA fragments matched DNA fragments from male 1.*
- *Male tiger 1—The cub's DNA has two matches with male 1 and only one match with male 2.*

78. One credit is allowed for correctly describing *one* way plants growing near these roads could be harmed by the salt. Acceptable responses include but are not limited to: [1]

- *Salt could cause water to diffuse out of the cells of the plants.*
- *The solution contains less water than was in the leaves, so water could diffuse out of the leaves.*
- *Water could leave the plant by osmosis.*
- *The plants could dehydrate.*

79. One credit is allowed for correctly identifying a finch population that would most likely survive on the island and for supporting the answer. Acceptable responses include but are not limited to: [1]

- *Large ground finch because it has a large crushing bill*
- *Medium ground finch because it has a crushing bill and eats mainly plant food*
- *Small ground finch (or sharp-billed ground finch) because it eats seeds, has a crushing bill, and may be very successful at consuming the available seeds*
- *Cactus finch because it would not have to compete with other finches for any cactus that are present on the island*
- *Warbler finch (or woodpecker finch) because it can use a probing bill to catch and consume wood-burrowing insects in the plants that produce the large seeds*

[NOTE: Insufficient information is given in the question concerning the variety of food choices available to the "several populations of finches" that migrate to the island. Although the large and medium ground finches would be the most likely to thrive on the "large seeds with tough outer coverings," not enough information is given to exclude the small and sharp-billed ground finches as successful competitors for these seeds as well. In addition, the question does not state that the *only* foods available are "large seeds with tough outer coverings." It is quite likely that other plant species such as cactus and fruiting plants are also present, as well as edible animal species such as insects. In fact, any of the finch populations would "most likely survive on the island" as long as ample food types and other resources were present to meet their specific requirements. Because of the ambiguous nature of this question as written, responses identifying any of these populations would be correct and credit due as long as the response is accompanied by a biologically correct supporting statement.]

80. One credit is allowed for correctly stating *one* reason why this is the recommended procedure for placing a coverslip. Acceptable responses include but are not limited to: [1]

- *This procedure reduces the chance of trapping air bubbles.*
- *By using forceps to hold the coverslip, any contamination present on the student's fingers will not get into the pond water specimen.*
- *This technique will prevent the fragile coverslip from cracking.*
- *Any relatively large specimens in the pond water will not be crushed if the coverslip is added in this fashion.*
- *It keeps fingerprints from getting on the coverslip.*
- *It ensures even distribution of the pond water sample under the field of view.*

81. **1** Both bird species *A* and species *B* can most likely survive on the same small island because they *use different resources and, therefore, they do not compete*. The question clearly states that these two species have different food (large seeds vs. small insects) and nesting (pine trees vs. hollowed-out dead logs) requirements. The lack of competition helps to ensure that these two species can coexist on the island and can inhabit different niches.

WRONG CHOICES EXPLAINED

(2) It is *not* true that both bird species *A* and species *B* can most likely survive on the same small island because they *mate with each other, keeping both populations constant*. By definition, organisms that are considered to be members of separate species cannot mate and produce fertile offspring.

(3), (4) It is *not* true that both bird species *A* and species *B* can most likely survive on the same small island because they *compete for food, but do not compete for shelter* or because they *eat the same food, but feed at different times of the day*. These species do not compete for food. Species *A* eats large seeds, while species *B* eats small insects. No information is given in the question concerning the times of day that these species eat.

82. **4** The factor most often acting as a selecting agent for the survival of a species in a particular location is the *environment they inhabit*. The biotic and abiotic factors present in any natural environment establish conditions that either promote or limit the survival of the various species that inhabit it. Stable environmental conditions result in stable plant and animal communities whose populations interact successfully. When environmental conditions change, a new set of factors may select some species varieties to survive and others to die out.

[NOTE: Although test directions state that this question is based on the information presented, it is *not*. Instead, this is a stand-alone question that must be answered based on the student's knowledge of biology only.]

WRONG CHOICES EXPLAINED

(1), (3) The factor most often acting as a selecting agent for the survival of a species in a particular location is *not* the *strength of the organism* or the *speed of the organism*. Strength and speed are characteristics of certain species that may provide them with an adaptive advantage in some circumstances and environments but not in others.

(2) The factor most often acting as a selecting agent for the survival of a species in a particular location is *not* the *new mutations within the individual.* Mutations received by an individual member of a species have no impact on the species population unless they are received in gametic tissues, are passed on to future generations, and provide adaptive advantages to the offspring.

83. One credit is allowed for correctly predicting what will most likely happen to the populations of both species *A* and species *B* if species *C* successfully survives on the island. Acceptable responses include but are not limited to: [1]

- *Both species A and species B will decrease in number since species C competes with both of them in different ways.*
- *Species A will compete with species C for nesting sites, and species B will compete with species C for food. Therefore, the populations of both species A and species B will decrease.*
- *As long as there are adequate sources of food and shelter on the island, both species A and species B will survive because their competition with species C will be minimized.*

84. One credit is allowed for correctly stating if there is a relationship between height and resting pulse rate and for supporting the answer. Acceptable responses include but are not limited to: [1]

- *No, the data points are scattered.*
- *No, the data do not show a trend.*
- *No relationship can be identified from the data collected because too few experimental subjects were tested.*

85. One credit is allowed for correctly stating what data in the table indicate that adenine pairs with thymine in a DNA molecule. Acceptable responses include but are not limited to: [1]

- *Adenine and thymine are present in equal numbers as indicated in the chart.*
- *The number of molecules of adenine and thymine are the same in the chart (1.75 million).*
- *When the DNA was broken down chemically, there was one thymine molecule for every adenine molecule.*
- *The data show a 1:1 correspondence between adenine and thymine molecules.*

STANDARDS/KEY IDEAS	AUGUST 2015 QUESTION NUMBERS	NUMBER OF CORRECT RESPONSES
STANDARD 1		
Key Idea 1: The central purpose of scientific inquiry is to develop explanations of natural phenomena in a continuing and creative process.	21	
Key Idea 2: Beyond the use of reasoning and consensus, scientific inquiry involves the testing of proposed explanations involving the use of conventional techniques and procedures and usually requiring considerable ingenuity.	60, 70	
Key Idea 3: The observations made while testing proposed explanations, when analyzed using conventional and invented methods, provide new insights into natural phenomena.	44, 45, 46	
Laboratory Checklist	48, 49, 55, 60	
STANDARD 4		
Key Idea 1: Living things are both similar to and different from each other and from nonliving things.	1, 2, 9, 10, 30, 34, 35, 36, 52, 53, 54, 55, 56, 57	
Key Idea 2: Organisms inherit genetic information in a variety of ways that result in continuity of structure and function between parents and offspring.	3, 5, 6, 8, 24, 25	
Key Idea 3: Individual organisms and species change over time.	11, 12, 13, 14, 16, 31, 32, 33, 37	
Key Idea 4: The continuity of life is sustained through reproduction and development.	15, 17, 18, 19, 20, 38	
Key Idea 5: Organisms maintain a dynamic equilibrium that sustains life.	7, 22, 23, 28, 41, 47, 48, 49, 58, 59, 61, 62, 63	
Key Idea 6: Plants and animals depend on each other and their physical environment.	4, 39, 40, 64, 65, 66, 67, 68, 69	
Key Idea 7: Human decisions and activities have a profound impact on the physical and living environment.	26, 27, 29, 42, 43, 50, 51, 71, 72	

STANDARDS/KEY IDEAS	AUGUST 2015 QUESTION NUMBERS	NUMBER OF CORRECT RESPONSES
REQUIRED LABORATORIES		
Lab 1: "Relationships and Biodiversity"	73, 74, 76, 77, 85	
Lab 2: "Making Connections"	75, 84	
Lab 3: "The Beaks of Finches"	79, 81, 82, 83	
Lab 5: "Diffusion Through a Membrane"	78, 80	

Examination
June 2016
Living Environment

PART A

Answer all questions in this part.   [30]

Directions **(1–30): For *each* statement or question, record in the space provided the *number* of the word or expression that, of those given, best completes the statement or answers the question.**

1 Which statement is an example of the interdependence of organisms?

  1 Owls hunt at night.

  2 Ants get food from insects and protect insects from predators.

  3 Ticks feed on the blood of animals and the ticks grow larger.

  4 Crows feed on dead mice.

1 __2__

2 Residents of a town are concerned that a recently built factory could pose health risks. Scientists were asked to investigate the effects of the factory on the health of local residents. The most relevant information they reported was that

  1 in a survey, residents felt that the air in town looks dirtier now

  2 there have been reports that other types of factories have been linked with health issues

  3 residents have occasionally seen smoke coming from the factory

  4 local medical facilities have recently reported a 15% increase in the number of patients treated for asthma   2 _4_

3 Farmers may someday clone their best milk-producing cow into a whole herd. What potential disadvantage might be important to consider in having such a large group of clones on one farm?

  1 It may be difficult to tell the animals apart.

  2 Lack of variation may limit survival in the herd.

  3 The cows could be fertilized by only one type of bull.

  4 The cows could be mated only with each other.   3 _2_

4 DNA replication occurs in preparation for

  1 mitosis, only

  2 meiosis, only

  3 both mitosis and meiosis

  4 neither mitosis nor meiosis   4 _1_

5 An individual eats a hamburger. Which two systems must interact to transfer the nutrients in the hamburger to human muscle tissue?

1 respiratory and excretory
2 digestive and immune
3 digestive and circulatory
4 circulatory and respiratory

5

6 The diagram below shows cell *A* completing a life process.

Cell A

Cell *A* performs functions similar to the tissues and systems in complex, multicellular organisms. This process results in

1 increased genetic variation
2 the maintenance of homeostasis
3 a reduction in competition
4 increased autotrophic nutrition

6 _2_

7 The shape of a protein molecule directly determines its

1 movements through the cytoplasm
2 functions inside and outside of cells
3 roles in building water molecules
4 circulation throughout the body

7 _2_

8 The diagram below represents a cell and some molecules in its environment.

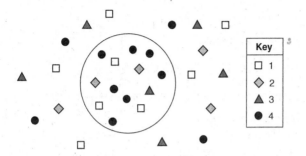

Which molecule would require the use of energy in order to be brought into the cell?

1 1   3 3

2 2   4 4   8 __4__

9 Many domestic plants that are currently used for food by humans share a wild plant ancestor. The changes that have occurred in four common plants and the results are shown in the chart below.

Wild Plant Ancestor	Change That Occurred	Resulting Modern Plant
wild mustard	reduced flower development	broccoli
wild mustard	sterile flowers	cauliflower
wild mustard	enlargement of leaves	kale
wild mustard	shortened stem length	cabbage

What event most likely produced the changes that occurred in the wild plant ancestor?

1 Mutations in wild mustard sex cells were passed on to offspring.

2 Humans did not like to eat wild mustard.

3 Competition for survival occurred in all ecosystems of the world.

4 Ancient herbivores overgrazed wild mustard.   9 __1__

10 A strand of DNA in a skin cell contains the bases:

A-T-G-C-C-A-T-C-G-G-T-A

After the cell is exposed to ultraviolet light, the strand contains the bases:

A-T-G-G-C-C-A-T-C-G-G-T-A

Which statement describes the result of this exposure?
1 A new base has been inserted.
2 A base has been deleted.
3 One base has been substituted for another.
4 There have been no changes in the bases. 10 __1__

11 An ameba, a one-celled organism, can move, ingest, and transport materials within the cell, because it has

1 organs   3 tissues
2 organelles   4 systems   11 __2__

12 In humans, the placenta is essential to the embryo for

1 nutrition, excretion, and reproduction
2 respiration, nutrition, and excretion
3 movement, reproduction, and nutrition
4 coordination, movement, and growth   12 __2__

13 A student infected by a common cold virus ran a low-grade fever. After a few days, the student's temperature returned to normal and the student was free of cold symptoms. The fever served as

1 an antigen in the circulatory system
2 an immune response to a pathogen
3 a biological catalyst
4 a weakened pathogen   13 __3__

14 Many animals have developed courtship behaviors. Males will often dance, swim, or sing in a particular way to attract a female. Males who are more successful at the courtship behavior will have a greater chance of having more offspring. This behavior is a result of

  1 natural selection
  2 genetic engineering
  3 asexual reproduction
  4 gene manipulation   14 ___1___

15 After a zygote is formed, specialization of cells occurs. Through which process do the cells of a zygote become specialized?

  1 sexual reproduction   3 fertilization
  2 meiosis   4 differentiation   15 ___4___

16 A farmer wanted to rid his apple trees of a particular leaf-eating insect. He sprayed his trees with an insecticide that killed 98% of the insects. The survival of 2% of this population of insects is most likely due to

  1 genes obtained from another species
  2 certain chemicals that stimulated overproduction
  3 variations that resulted from sexual reproduction
  4 their ability to produce food from the pesticide   16 ___3___

17 Which occurrence represents an example of evolution?

  1 Exposure to radiation reduces the rate of mutation in leaf cells.

  2 A mutation in a liver cell causes a person to produce an enzyme that is less efficient.

  3 Cells in a zygote eventually change into bone cells or skin cells.

  4 Some antibiotics are almost useless, because pathogens have developed a resistance to these antibiotics.   17 __4__

18 Populations of aspen trees in the western United States are being destroyed by an unexplained illness. The altered landscape is affecting the animals that live there. Populations of deer mice are increasing greatly in these areas. Unfortunately, these mice often carry a virus that is deadly to humans. This scenario best illustrates that

  1 a change in the environment always results in disease

  2 humans are the cause of the breakdown of this ecosystem

  3 the stability of this ecosystem is limited by the amount of water available

  4 every population in an ecosystem is linked with other populations   18 __4__

19 The paramecium is a single-celled organism that reproduces asexually. The offspring of a paramecium usually contain

  1 only half of the genes of the parent cells

  2 more DNA than the parent cell

  3 genetic material identical to that of the parent cell

  4 fewer mutations than the parent cell   19 __3__

20 A dead or weakened pathogen used to establish immunity would most likely be found in

  1 a pesticide   3 a vaccine

  2 an antibiotic   4 a toxin   20 _3_

21 Which statement is true for all of the organisms in the ecosystem represented in the diagram below?

  1 They use energy to combine the inorganic molecules carbon dioxide and water into energy-rich organic compounds.

  2 Stored energy cannot be used by these organisms as a source of energy for life processes.

  3 Energy stored in inorganic molecules is released during cellular respiration in these organisms.

  4 Energy is used by the organisms to obtain and transport materials, and to eliminate wastes.   21 ____

22 When a natural disaster destroys a stable ecosystem, the area is temporarily less stable than before. This is most likely due to

  1 a decrease in biodiversity

  2 an increase in the number of food chains

  3 an increase in the number of species

  4 a decrease in the rate of mutation   22 ____

23 An individual walks out of his air-conditioned (75°F) home into the hot outside environment (85°F). His ability to adjust to this changing environment involves a mechanism similar to

1 the regulation of water loss by guard cells in plant leaves
2 the digestion of carbohydrates by enzymes
3 using ATP for the diffusion of water
4 glucose production in the pancreas   23 _____

24 Nonrenewable resources are

1 not finite and are not depleted over time
2 not finite and are depleted over time
3 finite and are not depleted over time
4 finite and are depleted over time   24 _____

25 Dodder, a plant with no chlorophyll, grows on a living plant of a different species from which it obtains nutrients. Which pair of terms describes this relationship?

1 parasite and host
2 predator and prey
3 producer and decomposer
4 consumer and scavenger   25 _____

26 Three human hormones most directly involved in sexual reproduction are

1 estrogen, insulin, and progesterone
2 testosterone, estrogen, and insulin
3 progesterone, ATP, and testosterone
4 estrogen, progesterone, and testosterone   26 _____

27 At one point, scientists observed that the ozone
shield was getting thinner. They warned that the
loss of the effectiveness of this shield may lead to an
increase in

1 allergies to ozone
2 mutations that lead to cancer
3 viral diseases, such as AIDS
4 ice formation at the poles   27 _____

28 As it grows from a seed to a mature plant, a plant
will grow taller and thicker. Which are abiotic factors most responsible for the increase in the mass of
the plant?

1 water, minerals, bacteria
2 sunlight, oxygen, plant receptors
3 minerals, water, plant enzymes
4 water, sunlight, carbon dioxide   28 _____

29 Structures in an animal cell are represented in the
diagram below.

Which row in the chart correctly identifies the functions of structures *A*, *B*, and *C*?   29 _____

Row	Structure A	Structure B	Structure C
(1)	waste removal	extract energy from nutrients	protein synthesis
(2)	information storage	transport of materials	storage of liquids
(3)	protein synthesis	storage of wastes	reproduction
(4)	cell communication	transport of materials	waste removal

30 The diagram below represents how air pollution may move across the eastern United States.

Movement of Air Pollution

In order to reduce the amount of air pollution in Pennsylvania, which change is necessary?

1 Laws must be passed to protect endangered species.
2 The use of natural resources must be increased.
3 More coal-burning power plants must be built.
4 The cooperation between the different states must be improved. 30 _____

PART B–1

Answer all questions in this part. [13]

Directions (31–43): For *each* statement or question, record in the space provided the *number* of the word or expression that, of those given, best completes the statement or answers the question.

Base your answers to questions 31 and 32 on the diagram below and on your knowledge of biology. The diagram represents part of a food web.

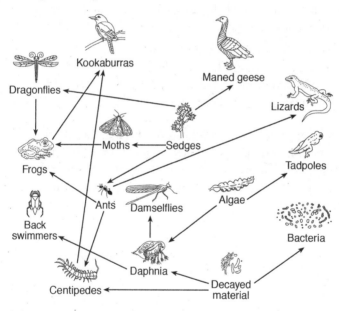

31 Which sequence of organisms represents a food chain within this food web?

1 tadpoles → algae → daphnia → back swimmers
2 sedges → ants → frogs → kookaburras
3 algae → daphnia → decayed material → bacteria
4 dragonflies → sedges → ants → centipedes

31 __2__

32 Which population would be most immediately affected by the removal of the lizard population

  1 sedges   3 ants

  2 algae   4 centipedes   32 _3_

---

33 During an investigation, a student measures out 15 grams of salt. Then, he measures 15 milliliters of water and adds the salt to it. Next, he measures a 1 centimeter wide by 4 centimeters long section of plant leaf. Which list of tools is arranged in the order that the student used them?

  1   graduated cylinder, ruler, balance

  2   balance, ruler, graduated cylinder

  3   graduated cylinder, balance, ruler

  4   balance, graduated cylinder, ruler   33 _4_

Base your answers to questions 34 through 36 on the graph below and on your knowledge of biology. The graph shows the masses of different types of plants found in an area of the Adirondack Mountains after a forest fire occurred.

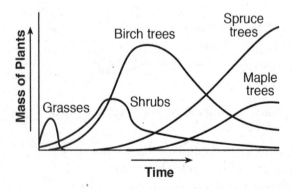

34 Based on the information provided in the graph, the process that is occurring is

1 ecological succession   3 selective breeding
2 biological evolution   4 genetic engineering   34 __1__

35 The time shown in the graph is most likely measured in

1 days   3 months
2 weeks   4 years   35 __4__

36 The mass of plants shown in the graph refers to the mass of a number of

1 populations   3 ecosystems
2 decomposers   4 communities   36 __1__

Base your answers to questions 37 and 38 on the diagram below and on your knowledge of biology. The diagram represents a portion of a starch molecule.

37 The building blocks for this molecule are

1 amino acids   3 fats
2 simple sugars   4 molecular bases   37 __2__

38 The energy in this molecule is stored

1 in the bonds between atoms
2 in the oxygen found in the molecule
3 when the carbon atoms break off
4 when water breaks this molecule apart   38 __1__

Base your answer to question 39 on the information below and on your knowledge of biology.

A genetically modified pig, nicknamed the "enviropig," has the ability to produce a bacterial enzyme in its saliva that helps reduce the amount of phosphorus in its wastes. Phosphorus pollution is a serious environmental concern. Enviropigs are expensive, but the cost is balanced against the benefit to the environment. There is also a concern that the US Department of Agriculture still has not cleared enviropig meat for human consumption.

39 Government agencies and citizens should propose the use of enviropig in the future only after

1 developing ways to remove the bacterial enzyme
2 assessing risks, costs, and benefits
3 people have eaten lots of enviropig meat and determined the effects
4 a different, cheaper pig can be produced regardless of the output of phosphorus

39 __2__

Base your answers to questions 40 and 41 on the information below and on your knowledge of biology.

Researchers have produced rice plants that can withstand being completely submerged for up to two weeks. This is good news for farmers in the flood regions of Southeast Asia. The farmers in this region rely heavily on this crop. The diagram below illustrates the process used to genetically modify plants, such as rice.

40 The molecules used to cut, copy, and connect the DNA segments used in this process are

1 sugars   3 indicators

2 enzymes   4 antigens   40 _2_

41 The best explanation for these modified rice plants being flood resistant is that

1 the gene for flood resistance was inserted into plant cells, which grew into plants whose cells are expressing this gene

2 they were produced by fertilization, using gametes from two flood-resistant bacterial cells

3 there was a mutation in the bacterial DNA after it was inserted into the plant that caused it to be flood resistant

4 the researchers used selective breeding for the flood-resistance trait

41 ____

42 The diagram below represents two cells viewed using the same magnification with the same microscope.

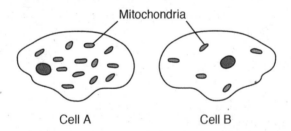

Mitochondria

Cell A   Cell B

One possible conclusion that can be drawn about the activity of these two cells is that

1 more active transport occurs in cell *B* than in cell *A*

2 more active transport occurs in cell *A* than in cell *B*

3 cell *B* uses some of the extra mitochondria to make food

4 cell *A* is a plant cell since it has a cell wall

42 ____

43 The diagram below represents the processes leading to the formation of a human embryo.

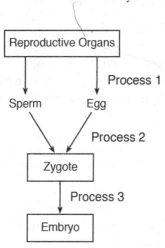

The correct sequence for processes 1, 2, and 3 represented in the diagram is   43 _3_

Row	Process 1	Process 2	Process 3
(1)	gamete formation	cell division	fertilization
(2)	cell division	gamete formation	fertilization
(3)	gamete formation	fertilization	cell division
(4)	fertilization	gamete formation	cell division

PART B–2

Answer all questions in this part. [12]

Directions (44–55): **For those questions that are multiple choice, record in the space provided the *number* of the choice that, of those given, best complete each statement or answers each question. For all other questions in this part, follow the directions given and record your answers in the spaces provided.**

Base your answers to questions 44 through 47 on the information and data table below and on your knowledge of biology.

For most animals, the sex of the offspring is determined by sex chromosomes. In some species of reptiles, such as the painted turtle, there are no sex chromosomes. It has been discovered that the sex of the offspring is determined by the temperature of the nest in which the egg develops.

**Sex of Painted Turtle Offspring
at Various Nest Temperatures**

Temperature (°C)	Sex of Offspring	
	Males (%)	Females (%)
19	0	100
20	5	95
21	20	80
22	25	75
23	0	100
24	0	100
25	0	100

Directions **(44–46): Using the information in the data table, construct a line graph on the grid below, following the directions below.**

44   Mark an appropriate scale, without any breaks in the data, on each axis.   [1]

45   Plot the data for percent males on the grid. Connect the points and surround each point with a small circle.   [1]

     Example:

46   Plot the data for percent females on the grid. Connect the points and surround each point with a small triangle.   [1]

     Example:

Sex of Painted Turtles at Various Nest Temperatures

Offspring (%)

Key
⊙ Males
△ Females

Temperature (°C)

Note: The answer to question 47 should be recorded in the space provided.

47 The fact that the sex of the painted turtle offspring is controlled by the temperature of the nest is an example of

1 Manipulation of genes caused the movement of embryonic skin cells.
2 Exposure to light stimulated the development of cells containing ozone.
3 Uncontrolled mitotic division occurred as a result of gene mutations.
4 An immune reaction triggered the formation of excess blood cells.

47 __4__

Base your answers to questions 48 through 50 on the information and graphs below and on your knowledge of biology. The graphs represent the results of two investigations using leaf disks from spinach plants.

Small disks were cut from spinach leaves that had been treated to remove any air from inside the leaf. The disks were placed in a solution that allowed them to carry out photosynthesis. At first, all the disks sank to the bottom of the container. These disks were then used for two different investigations.

Investigation 1

The disks were divided into five groups. Each group was exposed to light of a different intensity, measured in watts per meter squared (W/m^2). Some of the disks began to float. The results of the first investigation are shown in the graph below.

48 State the relationship between increasing light intensity and the percentage of disks floating at the conclusion of Investigation 1.   [1]

As light intensifies,
Disks Floating increases.

Note: The answer to question 49 should be recorded in the space provided.

49 The substance produced inside the leaf disks that caused them to float to the surface of the solution is

1 ozone　　　　　3 water
2 oxygen　　　　4 nitrogen　　　　　49 _2_

Investigation 2

A number of freshly prepared disks were placed in five containers. These containers were then each exposed to light of a different color. The results of the second investigation are shown in the graph below.

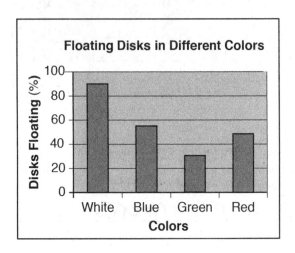

Note: The answer to question 50 should be recorded in the space provided.

50 Which color of light appears to be *least* effective for photosynthesis in spinach leaves?

1 white   3 green

2 blue   4 red   50 _3_

Base your answer to question 51 on the information and photograph below and on your knowledge of biology. The photograph shows an oriental hornet.

Oriental hornets are unique insects. A yellow pigment in the body of the insect converts solar energy to electrical energy. Plants also convert energy from the Sun.

Yellow pigment

51 Identify the organelle present in plants where this conversion takes place. [1]

<u>Chloroplast</u>

Base your answer to question 52 on the information below and on your knowledge of biology.

The SUNY Solar Car Model Racing Team's Sunhawk: Car of the Future?

The Sunhawk, a car built by students at SUNY New Paltz, prompted Forbes Magazine to ask "Is The $250,000 Sunhawk the Solar Car of the Future?" These cars show the most advanced solar technology and vehicle construction.

52 There are trade-offs involved in the use of solar-powered cars. Provide *one* advantage and *one* disadvantage of owning a solar car.   [1]

Advantage: _Renewable Energy_

Disadvantage: _Expense_

Base your answer to question 53 on the information below and on your knowledge of biology.

> Ten years ago, scientists discovered a well-preserved set of dinosaur remains in China. This dinosaur, which walked on Earth about 125 million years ago, had feathers and was about the same size as a turkey—but don't be fooled. This dino's bite was a lot worse than a turkey's gobble. After a close (and careful!) examination of the dino's teeth, scientists recently concluded that this dinosaur was probably poisonous. The study was led by David Burnham, who works and teaches at the University of Kansas in Lawrence.

Source: *www.sciencenewsforkids.org*

53 State *one* inference that could be made based on the fact that this dinosaur had feathers.   [1]

_____

_____

Base your answer to questions 54 and 55 on the information below and on your knowledge of biology.

A captive New Caledonian crow forages for food
using a stick tool. (Credit: Dr. Simon Walker)

A Great Larvae Meal

New Caledonian crows consume a wide range of foods. These crows require tools to extract the larvae of wood boring beetles from their burrows. A bird pokes a larva with a stick until the larva is disturbed enough to bite the stick and hang on to it. The bird is then able to pull the larva out of its burrow. These larvae, with their unusual diet, have a distinct chemical that can be found in the feathers and blood of crows—allowing scientists to determine the percentage of the crows' diet that is made up of beetle larvae. Scientists found that the beetle larvae are so energy-rich that just a few could satisfy the daily energy requirement for a crow. The crows with the greatest skill in using a twig as a tool benefit most in terms of nutrition.

54 State *one* reason why the offspring of crows skilled at using twigs as tools would have the greatest chance of survival. [1]

_____

_____

55 State *one* reason why some members of a population of crows equally skilled in the use of twigs have different rates of survival. [1]

_____

_____

PART C

Answer all questions in this part.   [17]

Directions (56–72): **Record your answers in the spaces provided.**

Base your answers to question 56–57 on the information below and on your knowledge of biology.

African violet plants are grown for their delicate, colorful flowers and furry, soft leaves. People often want to touch the leaves and brush the hairy leaves with their fingers. Growers and plant owners were concerned that this could negatively affect the plant. Of particular concern was the presence of body lotion or other skin products on the hands of persons touching the leaves.

A student thought this might be the basis of a science project. He selected two African violet plants. Ten leaves on each of the two plants were brushed with a gloved hand for 30 seconds, once a day, for a period of five days. The difference was that leaves of the second plant were brushed with a gloved hand that had hand lotion applied to the glove.

56–57 As part of the peer review process, evaluate the student's experiment. As part of your evaluation, be sure to:

- state *one* possible hypothesis for the experiment proposed by the student [1]
- describe the type of data that should be collected to determine if the brushing with lotion was having a *negative* effect on the African violet plant [1]

The African Violet will wilt when brushed with lotion. Any color changes, texture and health changes of the plant should be recorded.

Base your answers to questions 58 and 59 on the information and chart below and on your knowledge of biology.

Scientists studied the distribution of a species of pocket mouse that lived in the sandy desert regions of the southwestern United States. They are eaten by a variety of predators. Pocket mice are active at night, and feed on seeds and grasses. A single female mouse can reproduce several times each year, producing a litter of 3 to 13 offspring each time. Each new litter is considered a generation.

A volcanic eruption that resulted in lava flows changed the color of the area that the mice inhabit from light brown to black. Data from the scientists' research of the population are shown in the chart below.

Changes in Pocket Mouse Fur Color after a Volcanic Eruption

Number of Generations	Percentage of Pocket Mice with Light Brown Fur	Percentage of Pocket Mice with Black Fur
10	95%	5%
25	90%	10%
50	75%	25%
100	5%	95%

58 State the role of mutation *or* recombination in the appearance of the trait for black fur color in the pocket mouse population. [1]

One mutation that turned mice black, ended up being helpful, and now almost the whole population has it.

59 Explain why the percentage of black pocket mice changed so much after the volcanic eruption. [1]

When the lava dries, it leaves a hard, black crust, which is easy for black mice to hide on.

Base your answers to questions 60 and 61 on the information below and on your knowledge of biology.

In 2003, as a result of the Human Genome Project, the complete sequence of all the bases in human DNA was released to the public. Although knowing the entire sequence of bases has proven valuable, scientists are currently working to map genes. Mapping genes involves determining the exact location of each gene. Since much of human DNA does not code for a protein, it is challenging to figure out which segments are actual genes. Often, scientists look at the percent composition of bases in a segment of DNA. If the segment of DNA has a large percentage of C and G bases (together over 50%), it is likely that it is a gene and codes for a protein.

60 A scientist analyzes the bases in a segment of DNA from a human skin cell to determine if it codes for a protein. The base *A* is 12% of the bases in this segment of DNA. Calculate the percentage of bases that would be *C*. [1]

_____38_____ %

61 Is it likely this segment of DNA codes for a protein? Circle yes *or* no and support your answer. [1]

Circle one: Yes   *or*   No

C and G make up 76%
which is greater than 50%.

Base your answers to questions 62 through 64 on the illustration and information below and on your knowledge of biology. The illustration is of a Tasmanian devil.

Source: http://www.statelibrary.tas.gov.au

The Tasmanian devil is the largest surviving carnivorous marsupial in Australia. It is in danger of extinction due to an unusual type of cancer called Devil Facial Tumor Disease (DFTD). It can be passed from one individual to another through wounds that occur when they fight over food. Tumor cells in the mouth of an infected animal break off and enter the wound on an uninfected animal. The tumor cells multiply in the body of the newly infected devil, forming new tumors that eventually kill the animal.

Recent research has shown that the immune system of a Tasmanian devil accepts tumor cells from another devil as if they were cells from its own body. The tumor cells are ignored by the immune system. No immune response develops against them, and the cancerous cells multiply. Scientists predict that DFTD could wipe out all the remaining Tasmanian devils in 25 years, unless a treatment is developed.

62 Using the terms antigens and antibodies, explain why the tumor cells are ignored by the immune system in Tasmanian devils. [1]

The antibodies of the host do not recognize the foriegn antigens, as their own

63 Explain how cancer cells differ from normal cells. [1]

Cancer cells uncontrolabuy multiply because of mutation in the DNA.

64 Describe *one* possible way to maintain a population of healthy, uninfected Tasmanian devils until a treatment or cure can be found. [1]

Prevent them from fighting by leaving fresh, dead carcusses out for claim.

Base your answers to questions 65 through 67 on the information below and on your knowledge of biology. The photograph below shows a Canada lynx, a mammal native to North America.

Source: http://www.allposters.com

Lynx are found in areas where there is deep, soft snow cover during the winter months. The body design of the Canada lynx helps keep the animal on top of the soft snow. Several unique characteristics, such as the design of its feet and its weight, enable the cat to successfully chase and catch snowshoe hares, its primary source of food. Snowshoe hares are also able to remain on top of the snow.

Increased winter recreation has created packed snow trails in lynx habitat. This allows coyotes and cougars to compete with lynx.

65 Explain why coyotes were *not* in competition with the lynx prior to the presence of packed snow trails. [1]

_____

_____

66 Describe the niche that the lynx, coyote, and cougar are competing to fill. [1]

_____

_____

67 Explain how the carrying capacity affects the number of predators in an area. [1]

_____

_____

Base your answers to questions 68 and 69 on the information below and on your knowledge of biology.

Cowherds Discovering Ticks Are for the Birds

South African cowherds [cowboys] are discovering that when it comes to debugging their cattle, nature knows best. Generations of cattle owners who dipped their livestock in pesticides ended up killing not only the ticks that feast on them, but also the red-billed oxpeckers [birds] that eat the ticks. Now environmentalists want to cut out the pesticides, hand the job back to the birds and in the process save them from extinction....

The bird is famous for its bright red bill, yellow ringed eyes and voracious appetite for ticks. An oxpecker can eat 13,000 of them [ticks] in a day, and meals are everywhere— on antelope, horses, cattle, buffalo, rhino, lion, elephant and leopard. The ticks carry a host of illnesses, including red-water disease, a common killer of cattle, but [ticks] are harmless to oxpeckers....

Source: Eric Naki, Associated Press writer

68 State *one* way that the use of pesticides to kill ticks could lead to the decline of the oxpecker population.   [1]

_____

_____

69 State *one* ecological advantage of using oxpeckers to solve the problem with these ticks.   [1]

_____

_____

Base your answers to questions 70 through 72 on the information below and on your knowledge of biology.

Over the past few decades, researchers have observed declining numbers in two species of penguins native to the West Antarctic peninsula. New evidence is pointing to a decline in their food supply as the primary cause for the recent drops in their numbers. These penguins feed on krill, small animals that grow and develop under ice masses. The graphs below show data related to two factors: atmospheric carbon dioxide (CO_2) levels and Antarctic ice mass. The diagram of a generalized Antarctic food web illustrates the role of the penguins.

Direct CO_2 Measurements: 2005-2011

Antarctica Mass Variation Since 2002

Source: http://www.nasa.gov

Antarctic Food Web

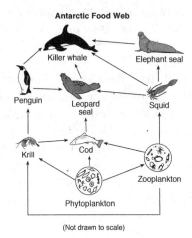

(Not drawn to scale)

Source: School Improvement in Maryland, www.mdk12.org

70 State *one* possible relationship between CO_2 levels and the change in Antarctic ice mass. [1]

_____

_____

71 Explain why the change in ice mass is resulting in a decline in the penguin populations. [1]

_____

_____

72 State *one* specific way in which humans might have caused the changes in atmospheric CO_2 levels. [1]

_____

_____

PART D

Answer all questions in this part.   [13]

Directions (73–85): **For those questions that are multiple choice, record in the space provided the *number* of the choice that, of those given, best completes the statement or answers the question. For all other questions in this part, follow the directions given and record your answers in the spaces provided.**

Base your answers to questions 73 through 75 on the diagram below and on your knowledge of biology. The diagram represents an experimental setup.

Starch indicator and water
Dialysis tube
Starch solution

Note: The answer to question 73 should be recorded in the space provided.

73 Which statement best describes what would most likely be observed after 20 minutes?

1 The contents of the dialysis tube would turn blue-black.

2 The liquid in the beaker would turn blue-black.

3 The dialysis tube would burst.

4 There would be no change visible.

73 __1__

Note: The answer to question 74 should be recorded in the space provided.

74 Which term correctly identifies the process by which molecules move through the dialysis tube membrane?

1 paper chromatography   3 diffusion
2 active transport   4 digestion

74 _3_

Note: The answer to question 75 should be recorded in the space provided.

75 A student filled a dialysis tube with 97% water solution and sealed the ends. The tube and its contents had a mass of 55 grams. The student placed the tube in a solution, and the mass of the tube and its contents increased to 60 grams. Into which solution was the dialysis tube placed?

1   0% water   3   97% water
2   95% water   4   99% water

75 _4_

Base your answers to questions 76 and 77 on the information and diagram below and on your knowledge of biology.

Scientists attempted to determine the evolutionary relationships between three different finch species, *A*, *B*, and *C*. In order to do this, they examined the physical characteristics and DNA of these species. DNA was extracted from all three species and analyzed using gel electrophoresis. The results are shown in the diagram.

Note: The answer to question 76 should be recorded in the space provided.

76 Which statement best describes the method used above to determine the evolutionary relationships between three species of finches?

  1 Examine the structure of the beaks and compare them.

  2 Observe behavioral and physical characteristics of all the finches and group them by similarities.

  3 Obtain molecular evidence from all three species and identify similarities.

  4 Compare common ancestors of all three of the species to see if they are the same.

76 _3_

77 Based on the data they collected using gel electrophoresis, label the branching tree diagram below. Write the letters *A*, *B*, and *C*, to represent the possible evolutionary relationships between species *A*, *B*, and *C*.   [1]

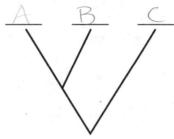

Base your answers to questions 78 through 80 on the information and diagram below and on your knowledge of biology.

Two species of finches found on a particular Galapagos island eat the seeds of a certain variety of plant. The relative strength of their beaks is shown in the graph below.

78 One of the finch species has a slightly smaller, weaker beak. Is this species 1 or species 2? Support your answer with information from the graph. [1]

Species: _____1_____

_____

_____

79 Select the point on the graph where beak strength of the two bird species is equal. Support your answer. [1]

Point: _____C_____

_____

_____

80 If the environment on the island changed and the seeds of more of the plants became harder to crack open, describe what the graph might look like after many years have passed. [1]

_____

_____

Base your answers to questions 81 and 82 on the information and diagram below and on your knowledge of biology.

Using a microscope and a wet-mount slide, a student observed a pond water sample containing paramecia, which are single-celled freshwater organisms. He noticed that there was a structure within each living paramecium that contracted regularly—about four times each minute. He researched the organism in his science textbook and found that the structure was a contractile vacuole and its function was to remove excess water from the paramecium.

In the diagram below, a paramecium is represented as seen through a microscope. The function of the contractile vacuole is described.

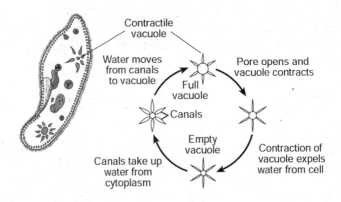

He decided to determine if the concentration of salt in the environment of the paramecium would affect the rate at which the contractile vacuole would contract.

Note: The answer to question 81 should be recorded in the space provided.

81 The process used to remove excess water from the paramecium by the contractile vacuole is

| 1 synthesis | 3 active transport |
| 2 digestion | 4 passive transport |

81 __3__

Note: The answer to question 82 should be recorded in the space provided.

82 The student predicted that the contractile vacuole would contract fewer times in one minute in a solution that had a higher concentration of salt than that found in typical pond water. This prediction is most likely

1 correct, because a high concentration of salt in the environment will force water into the cell, causing the contractile vacuole to pump

2 incorrect, because salt would be entering the cell, and the contractile vacuole would have to pump it out

3 incorrect, because the concentration of salt in the environment should not affect a cell

4 correct, because water would be moving out of the cell into the salt solution

82 __4__

Base your answers to questions 83 through 85 on the information below and on your knowledge of biology.

Five individuals had their pulses taken in beats per minute (bpm) before and after exercise. The data are shown in the chart below.

Pulse Rates

Individual	Pulse before Exercise (bpm)	Pulse after Exercise (bpm)
A	68	100
B	70	120
C	54	130
D	64	122
E	75	115

83 State *one* reason why an individual's pulse rate increased during exercise.   [1]

Pumping blood to supply oxygen

84 Calculate the average pulse rate *before* exercise for this group, to the *nearest tenth*.   [1]

_____ **bpm**

85 State why the individuals in this group have different pulse rates before exercise.   [1]

Age, weight

Answers
June 2016

Living Environment

Answer Key

PART A

1. 2	**6.** 2	**11.** 2	**16.** 3	**21.** 4	**26.** 4
2. 4	**7.** 2	**12.** 2	**17.** 4	**22.** 1	**27.** 2
3. 2	**8.** 4	**13.** 2	**18.** 4	**23.** 1	**28.** 4
4. 3	**9.** 1	**14.** 1	**19.** 3	**24.** 4	**29.** 1
5. 3	**10.** 1	**15.** 4	**20.** 3	**25.** 1	**30.** 4

PART B–1

31. 2	**34.** 1	**36.** 1	**38.** 1	**40.** 2	**42.** 2
32. 3	**35.** 4	**37.** 2	**39.** 2	**41.** 1	**43.** 3
33. 4					

PART B–2

44. *See* Answers Explained.
45. *See* Answers Explained.
46. *See* Answers Explained.
47. 4
48. *See* Answers Explained.
49. 2

50. 3
51. *See* Answers Explained.
52. *See* Answers Explained.
53. *See* Answers Explained.
54. *See* Answers Explained.
55. *See* Answers Explained.

PART C. *See* **Answers Explained**.

PART D

73. 1
74. 3
75. 4
76. 3
77. *See* Answers Explained.
78. *See* Answers Explained.
79. *See* Answers Explained.

80. *See* Answers Explained.
81. 3
82. 4
83. *See* Answers Explained.
84. *See* Answers Explained.
85. *See* Answers Explained.

Answers Explained

PART A

1. **2** *Ants get food from insects and protect insects from predators* is the statement that is an example of the interdependence of organisms. The symbiotic relationship described is mutualism, in which two species, the ant and an insect, both derive benefit from the interdependent relationship.

WRONG CHOICES EXPLAINED

(1) *Owls hunt at night* is *not* the statement that is an example of the interdependence of organisms. The nutritional behavior described is predation, in which a predator species, the owl, hunts a prey species, such as mice (not mentioned), during the night.

(3) *Ticks feed on the blood of animals and the ticks grow larger* is *not* the statement that is an example of the interdependence of organisms. The symbiotic relationship described is parasitism, in which a parasitic species, the tick, lives in or on the body of a host species, such as a deer (not mentioned), and derives a benefit from the host while harming it.

(4) *Crows feed on dead mice* is *not* the statement that is an example of the interdependence of organisms. The nutritional behavior described is scavenging, in which a scavenger species, the crow, consumes the bodies of dead organisms, such as mice.

[NOTE: Predation, parasitism, and scavenging are all behaviors that involve interactions between species. So they can be considered to be examples of interdependence of organisms. However, because the effects on the prey/host organisms are not mentioned, they are not the best answers to the question.]

2. **4** The most relevant information they reported was that *local medical facilities have recently reported a 15% increase in the number of patients treated for asthma*. This information was drawn from the records of an independent scientific organization in the affected community, so it should be considered objective and scientifically accurate. Because the data describe an increase in a health-related condition in people living in this community, they provide researchers with valuable information directly relevant to the question of the possible effects of industrial pollution on human health in this community.

WRONG CHOICES EXPLAINED

(1), (3) It is *not* true that the most relevant information they reported was that *in a survey, residents felt that the air in the town looks dirtier now* or that *residents have occasionally seen smoke coming from the factory*. As opinion and observation offered by local residents, this information is derived from a source of questionable scientific objectivity. As such, it is not the most relevant information on this question available to researchers.

(2) It is *not* true that the most relevant information they reported was that *there have been reports that other types of factories have been linked with health issues*. These reports may be scientifically accurate for the factories and communities studied. However, they may not accurately describe the health effects on people for this factory in this community. As such, it is not the most relevant information on this question available to researchers.

3. **2** *Lack of variation may limit survival in the herd* is the potential disadvantage that might be important to consider in having such a large group of clones on one farm. Because all members of the cloned herd share identical genetic characteristics, a disease or severe environmental condition will likely affect all members of the herd equally. This situation may result in the disability or death of the entire herd.

WRONG CHOICES EXPLAINED

(1) *It may be difficult to tell the animals apart* is *not* the potential disadvantage that might be important to consider in having such a large group of clones on one farm. Although this is a true statement due to the cows' identical genetic information, not being able to tell individual animals apart on a farm does not usually represent a disadvantage to the farm's operation.

(3) *The cows could be fertilized by only one type of bull* is *not* the potential disadvantage that might be important to consider in having such a large group of clones on one farm. As sexually reproducing animals, cloned cows can be fertilized by any healthy bull, not by just one type of bull.

(4) *The cows could be mated only with each other* is *not* the potential disadvantage that might be important to consider in having such a large group of clones on one farm. As sexually reproducing animals, cloned cows can be fertilized only by a bull, not by other cows, which are females.

4. **3** DNA replication occurs in preparation for *both mitosis and meiosis*. Before any normal living cell can divide, it must first replicate (exactly self-duplicate) its DNA. In mitosis, the replicated DNA separates into two sets of double-stranded chromosomes in two diploid (2n) daughter nuclei that are housed in two somatic cells that are genetically identical to each other and to the parent cell. In meiosis, the replicated DNA separates into four sets of single-stranded chromosomes in four haploid (n) daughter nuclei that are housed in four gametic cells that are not genetically identical to each other or to the parent cell.

WRONG CHOICES EXPLAINED

(1), (2), (4) It is *not* true that DNA replication occurs in preparation for *mitosis, only; meiosis, only;* or *neither mitosis nor meiosis.* DNA replication occurs in both mitosis and meiosis as described above.

5. **3** *Digestive and circulatory* are the two systems that must interact to transfer the nutrients in the hamburger to human muscle tissue. Before they can reach the muscles, the proteins and fats found in the hamburger are digested into their component biochemicals. For protein, these components are amino acids; for fats, the components are fatty acids and glycerol. Amino acids, fatty acids, and glycerol are absorbed into the blood through capillaries within the small intestine and are transported to muscle tissues via the circulatory system.

WRONG CHOICES EXPLAINED

(1) *Respiratory and excretory* are *not* the two systems that must interact to transfer the nutrients in the hamburger to human muscle tissue. The respiratory system is specialized for the transfer of gases (oxygen and carbon dioxide) between the atmosphere and the body. The excretory system is specialized to remove toxic materials (including urea and salts) from the body and expel them to the environment.

(2) *Digestive and immune* are *not* the two systems that must interact to transfer the nutrients in the hamburger to human muscle tissue. The immune system is specialized to protect the body from infectious disease carried by foreign invaders (including bacteria and viruses).

(4) *Circulatory and respiratory* are *not* the two systems that must interact to transfer the nutrients in the hamburger to human muscle tissue. The respiratory system is specialized for the transfer of gases (oxygen and carbon dioxide) between the atmosphere and the body.

6. **2** This process results in *the maintenance of homeostasis.* The process illustrated is heterotrophic nutrition, in which cell *A* ingests food materials in preparation for digestion and incorporation of nutrients into the cell's biochemical structure. By so doing, cell *A* is able to maintain homeostatic balance, or steady state, within its environment.

WRONG CHOICES EXPLAINED

(1) This process does *not* result in *increased genetic variation.* Increased genetic variation is accomplished by the processes of mutation and genetic recombination. Genetic variation is not illustrated in the diagram.

(3) This process does *not* result in *a reduction in competition.* Competition between organisms involves situations in which these organisms both require the same limited resource. Competition is not illustrated in the diagram.

(4) This process does *not* result in *increased autotrophic nutrition.* Autotrophic nutrition is the process of self-feeding commonly carried on by

green plants and certain bacteria. Autotrophic nutrition is not illustrated in the diagram.

7. **2** The shape of a protein molecule directly determines its *functions inside and outside cells*. A protein molecule's shape, especially at its active site, governs the protein's ability to link with and catalyze reactions between specific substrate molecules. This catalytic function of the protein enables the cell to perform functions consistent with the cell's type and degree of specialization.

WRONG CHOICES EXPLAINED
(1) The shape of a protein molecule does *not* directly determine its *movements through the cytoplasm*. The cytoplasm is a watery medium that allows relatively free movement of all types of molecules through the cell interior. The shape of a protein does not affect its movement in cytoplasm.

(3) The shape of a protein molecule does *not* directly determine its *roles in building water molecules*. Water is an inorganic substance that is formed by chemically bonding two hydrogen atoms to a single oxygen atom. Proteins are not directly involved in the formation of water molecules.

(4) The shape of a protein molecule does *not* directly determine its *circulation throughout the body*. The blood fluid, which is responsible for circulation within the human body, is a watery medium that allows relatively free movement of dissolved and small suspended molecules through the body interior. The shape of a protein does not affect its movement in blood.

8. **4** Molecule *4* is the molecule that would require the use of energy in order to be brought into the cell. The diagram shows that molecule *4* exists in a relatively high concentration (seven molecules) inside the cell, compared to a relatively low concentration (four molecules) outside the cell. In this case, moving additional molecules *4* into the cell could be accomplished only via active transport, a process that requires ATP energy.

WRONG CHOICES EXPLAINED
(1), (2), (3) Molecules *1*, *2*, and *3* are *not* the molecules that would require the use of energy in order to be brought into the cell. The diagram shows that molecules *1*, *2*, and *3* exist in relatively low concentrations (three, two, and one molecules, respectively) inside the cell, compared to relatively high concentrations (five, three, and five molecules, respectively) outside the cell. In these cases, moving additional molecules *1*, *2*, and *3* into the cell could be accomplished via simple diffusion, a process that does not require ATP energy.

9. **1** *Mutations in wild mustard sex cells were passed on to offspring* is the event that most likely produced the changes that occurred in the wild plant ancestor. Mutations in ancestral plants can be passed on to future generations

only if they occur in sex cells (pollen and ova). When such mutations result in desirable traits, the resulting offspring may be cross-bred with the hope of producing additional generations that also display those traits.

WRONG CHOICES EXPLAINED

(2), (3), (4) *Humans did not like to eat wild mustard, competition for survival occurred in all ecosystems of the world,* and *ancient herbivores overgrazed wild mustard* are *not* the events that most likely produced the changes that occurred in the wild plant ancestor. These statements about wild mustard may or may not be true. They are certainly not scientifically verifiable. However, these events alone could not have been responsible for the genetic changes that have resulted in the desirable traits that make wild mustard's domestic descendants successful nutritional choices for humans.

10. **1** *A new base has been inserted* is the statement that describes the result of this exposure. Examination of the original DNA strand shows that it contains 12 bases, while the postexposure strand contains 13 bases. Further examination reveals that an extra base G exists in the postexposure strand that was not present in the original strand. This extra base was added to the original DNA sequence by means of a gene mutation known as insertion.

WRONG CHOICES EXPLAINED

(2) *A base has been deleted* is *not* the statement that describes the result of this exposure. Had a base been deleted from the original DNA strand, an examination of the postexposure strand would have shown 11 bases, not 13, as well as one missing base compared to the original sequence.

(3) *One base has been substituted for another* is *not* the statement that describes the result of this exposure. Had a base been substituted in the original DNA strand, an examination of the postexposure strand would have shown 12 bases, not 13, as well as a change in one base compared to the original sequence.

(4) *There have been no changes to the bases* is *not* the statement that describes the result of this exposure. A careful examination of the DNA strands reveals that a change has occurred as described in the correct answer above.

11. **2** An ameba, a one-celled organism, can move, ingest, and transport materials within the cell because it has *organelles*. Organelles are submicroscopic structures within cells that perform specialized life functions for one-celled organisms that enable these organisms to survive in their environment. Examples of organelles found in the ameba include pseudopods (locomotion), vacuoles (ingestion), and endoplasmic reticula (material transport).

WRONG CHOICES EXPLAINED

(1), (3), (4) It is *not* true that an ameba, a one-celled organism, can move, ingest, and transport materials within the cell because it has *organs, tissues,* or *systems*. Organs, tissues, and systems are complex levels of organization formed from many cells and are found in multicellular organisms such as the human, not in one-celled organisms such as the ameba.

12. **2** In humans, the placenta is essential to the embryo for *respiration, nutrition, and excretion*. Placental tissues allow dissolved oxygen and nutrients to pass through membranes from the mother's blood to the embryo's blood, providing materials essential for respiration and nutrition. Placental tissues also allow dissolved wastes such as carbon dioxide and urea to pass from the embryo's blood to the mother's blood, removing these toxic materials from the embryo by the process of excretion.

WRONG CHOICES EXPLAINED

(1) In humans, the placenta is *not* essential to the embryo for *nutrition, excretion, and reproduction*. Reproduction is a life function carried out by adult humans, not by human embryos.

(3) In humans, the placenta is *not* essential to the embryo for *movement, reproduction, and nutrition*. Movement in a human embryo is accomplished by muscles and bones developing within the embryo's body. Reproduction is a life function carried out by adult humans, not by human embryos.

(4) In humans, the placenta is *not* essential to the embryo for *coordination, movement, and growth*. Coordination in a human embryo is accomplished by the nervous system developing within its body. Movement in a human embryo is accomplished by muscles and bones developing within the embryo's body. Growth of the human embryo is accomplished as its tissues, organs, and systems develop within the embryo's body.

13. **2** The fever served as *an immune response to a pathogen*. An elevated body temperature (fever) is a symptomatic response to the presence of disease-causing foreign invaders in the body known as pathogens. The elevated body temperature creates an unfavorable condition that slows the growth of these pathogens in the body's tissues and, when coupled with the production of antibodies, helps to speed recovery from the cold.

WRONG CHOICES EXPLAINED

(1) The fever did *not* serve as *an antigen in the circulatory system*. An antigen is a foreign protein located on the cell membrane of a pathogen that is recognized by the body's immune system, which produces specific antibodies to neutralize the antigen. A fever is not an antigen.

(3) The fever did *not* serve as *a biological catalyst*. Biological catalysts, also known as enzymes, are produced in all living cells to regulate the cell's metabolism. A fever is not a biological catalyst.

(4) The fever did *not* serve as *a weakened pathogen*. A weakened pathogen is a disease-causing organism that has undergone laboratory techniques that have made it harmless to humans. Weakened pathogens are often incorporated into vaccines that are used to treat specific diseases caused by these pathogens. A fever is not a weakened pathogen.

14. **1** This behavior is a result of *natural selection*. Such behaviors are instinctive (inborn) actions that are genetically programmed. Assuming that the courtship behaviors provided ancestral individuals with some adaptive advantage, it is probable that the behaviors were passed on to subsequent generations and became increasingly common in these species populations until they were universally displayed within each species. This series of events is typical of natural selection.

WRONG CHOICES EXPLAINED
(2), (4) This behavior is *not* a result of *genetic engineering* or *gene manipulation*. These activities are examples of laboratory techniques that are used by scientists to alter genetic information in the genomes of selected species. They are not used to modify the instinctive courtship behaviors of animals.
(3) This behavior is *not* a result of *asexual reproduction*. Courtship and mating behaviors support sexual, not asexual, reproduction in animals.

15. **4** *Differentiation* is the process through which the cells of a zygote become specialized. Differentiation involves the switching on of genes that control the characteristics of specialized tissues while turning off other genes not needed in those specialized tissues.

WRONG CHOICES EXPLAINED
(1), (3) *Sexual reproduction* and *fertilization* are *not* the processes through which the cells of a zygote become specialized. Sexual reproduction and fertilization are processes by which haploid gametes fuse to form a diploid zygote containing all the genes for all the traits of the species. A zygote is an unspecialized cell, not a specialized cell.
(2) *Meiosis* is *not* the process through which the cells of a zygote become specialized. Meiosis is a process by which a diploid primary sex cell divides to form four haploid gametes, not specialized tissues.

16. **3** The survival of 2% of this population of insects is most likely due to *variations that resulted from sexual reproduction*. The gene controlling this variation resulted from a random mutation, not from sexual reproduction as such. Sexual reproduction involves the recombination of genes in offspring through the processes of meiosis and fertilization. A mutated gene that provides an adaptive advantage (such as resistance to a pesticide) to a sexually reproducing species (such as a leaf-eating insect) may allow the insect to survive in the presence of the insecticide. Surviving insects mate with other sur-

viving insects and produce offspring that inherit the gene and are themselves resistant to the insecticide.

WRONG CHOICES EXPLAINED

(1) The survival of 2% of this population of insects is *not* most likely due to *genes obtained from other species*. Successful sexual reproduction can be accomplished only between males and females of the same species. Genes may be obtained from other species only in the case of viral infection or gene manipulation.

(2) The survival of 2% of this population of insects is *not* most likely due to *certain chemicals that stimulated overproduction*. Overproduction is a phenomenon noted among all species that serves as a means of ensuring species survival. Overproduction is not known to be stimulated by any chemical substance.

(4) The survival of 2% of this population of insects is *not* most likely due to *their ability to produce food from the pesticide*. Pesticides are chemicals that are known to be poisonous to insects and other species. Pesticides are not used to produce foods.

17. **4** *Some antibiotics are almost useless, because pathogens have developed a resistance to these antibiotics* is the occurrence that represents an example of evolution. A mutated gene that provides an adaptive advantage (such as resistance to an antibiotic) to a species (such as a pathogen) may allow the pathogen to survive in the presence of the antibiotic. Surviving pathogens produce offspring that inherit the gene and are themselves resistant to the antibiotic. This series of events illustrates natural selection and, by extension, evolution.

WRONG CHOICES EXPLAINED

(1) *Exposure to radiation reduces the rate of mutation in leaf cells* is *not* the occurrence that represents an example of evolution. Radiation is a mutagenic agent that increases, not reduces, the rate of mutation in living tissues. Exposure to radiation does not represent evolution.

(2) *A mutation in a liver cell causes a person to produce an enzyme that is less efficient* is *not* the occurrence that represents an example of evolution. The process of evolution depends on the production of mutations that result in favorable, not unfavorable, variations.

(3) *Cells in a zygote eventually change into bone cells or skin cells* is *not* the occurrence that represents an example of evolution. The process described, in which unspecialized cells change into specialized cells, is known as differentiation, not evolution.

18. **4** This scenario best illustrates that *every population in an ecosystem is linked with other populations*. The scenario describes one way that the roles of aspen trees, deer mice, viruses, and humans are interwoven in the

nâtural environment. Although many other species could have been included in this scenario, those that are included illustrate the interdependence of all species in the ecosystem.

WRONG CHOICES EXPLAINED

(1) It is *not* true that this scenario best illustrates that *a change in the environment always results in disease*. Although an environmental change is linked to the spread of a specific disease-causing virus in this scenario, it is not reasonable to infer that all environmental changes result in disease.

(2) It is *not* true that this scenario best illustrates that *humans are the cause of the breakdown of this ecosystem*. The role of human technology is not specifically mentioned in the scenario, so it is not reasonable to infer that it is a factor in this case.

(3) It is *not* true that this scenario best illustrates that *the stability of this ecosystem is limited by the amount of water available*. The availability of water as a natural resource is not specifically mentioned in the scenario, so it is not reasonable to infer that it is a factor in this case.

19. **3** The offspring of a paramecium usually contain *genetic material identical to the parent cell*. The form of asexual reproduction carried out by a paramecium is mitotic cell division. Mitotic cell division is characterized by replication (exact self-duplication) of genetic material (DNA) followed by separation of the replicated material into two daughter nuclei/cells. The resulting daughter cells are exact genetic duplicates of each other and of the parent cell. This process helps ensure genetic continuity from one generation of paramecia to the next.

WRONG CHOICES EXPLAINED

(1), (2) The offspring of a paramecium do *not* usually contain *only half the genes of the parent cells* or *more DNA than the parent cell*. The process of replication ensures that the same number of genes and amount of DNA are present in the daughter cells as was present in the parent cell.

(4) The offspring of a paramecium do *not* usually contain *fewer mutations than the parent cell*. Mutations that were present in the parent cell would be duplicated during replication and would be present in the daughter cells.

20. **3** A dead or weakened pathogen used to establish immunity would most likely be found in *a vaccine*. Vaccines are produced in laboratories and are specialized to provide immunity against specific diseases. It is common that dead or weakened pathogens are used in preparing vaccines. These dead or weakened pathogens stimulate the body's immune system to produce antibodies against the antigens located on the pathogens.

WRONG CHOICES EXPLAINED

(1), (4) A dead or weakened pathogen used to establish immunity would *not* most likely be found in *a pesticide* or *a toxin*. Pesticides and·toxins are prepared using either inorganic or organic poisons that kill target organisms on contact. Pesticides and toxins are not normally prepared using dead or weakened pathogens.

(2) A dead or weakened pathogen used to establish immunity would *not* most likely be found in *an antibiotic*. Antibiotics are prepared from either naturally occurring or artificially synthesized organic substances that are toxic to bacteria and other pathogenic organisms (except viruses) but usually not toxic to humans. Antibiotics are not normally prepared using dead or weakened pathogens.

21. **4** *Energy is used by the organisms to obtain and transport materials, and to eliminate wastes* is the statement that is true for all of the organisms in the ecosystem represented in the diagram. The diagram contains illustrations of plants and animals typical of Northeast deciduous forest ecosystems similar to those found throughout New York State. All living things, both plant and animal, expend cellular energy to carry out basic life functions such as those included in this answer.

WRONG CHOICES EXPLAINED

(1) *They use energy to combine the inorganic molecules carbon dioxide and water into energy-rich organic compounds* is *not* the statement that is true for all of the organisms in the ecosystem represented in the diagram. The process described in this statement is photosynthesis, a process carried on by the green plants, but not by the animals, illustrated in this ecosystem.

(2) *Stored energy cannot be used by these organisms as a source of energy for life processes* is *not* the statement that is true for all of the organisms in the ecosystem represented in the diagram. This statement is not true for any of the plants and animals illustrated in this ecosystem. All plants and animals use energy stored in organic chemicals such as glucose and starch to aid them in the operation of their life processes.

(3) *Energy stored in inorganic molecules is released during cellular respiration in these organisms* is *not* the statement that is true for all of the organisms in the ecosystem represented in the diagram. This statement is not true for any of the plants and animals illustrated in this ecosystem. No plants or animals release energy from inorganic molecules during the process of respiration. Instead, energy is released from organic molecules, such as glucose, during respiration.

22. **1** The area is temporarily less stable than before is most likely due to *a decrease in biodiversity*. A natural disaster such as a flood, fire, or drought normally eliminates most life-forms that are not able to escape or withstand the disaster. The elimination of species from an ecosystem represents a

decrease in the biodiversity of that ecosystem. A decrease in the biodiversity of an ecosystem leads to a reduction in the stability of that ecosystem.

WRONG CHOICES EXPLAINED
(2), (3) The area is temporarily less stable than before is *not* most likely due to *an increase in the number of food chains* or *an increase in the number of species*. A natural disaster would tend to decrease, not increase, the number of species and the number of food chains in an ecosystem. Therefore, instability of the ecosystem cannot have been caused by these factors because they would not have occurred in this case.

(4) The area is temporarily less stable than before is *not* most likely due to *a decrease in the rate of mutation*. Mutation is a random process by which genetic material is altered by the effects of mutagenic agents. Natural disasters are not mutagenic agents and have little to no effect on mutation rates in the affected ecosystem.

23. **1** An individual's ability to adjust to this changing environment involves a mechanism similar to *the regulation of water loss by guard cells in plant leaves*. When an individual moves from a cooler to a hotter environment, sweat glands located in the skin open and sweat is released to the skin surface. As the sweat evaporates from the skin surface, body temperature is reduced, which in turn stimulates the body to close the sweat glands to reduce the amount of sweat released by sweat glands. This helps the individual to regulate the temperature of the body. In a similar manner, stomates in the leaf open when water content of the leaf is high and close when water content is low. This helps the plant to regulate (slow) the rate of water loss from the plant when the environment changes from cool, wet conditions to hot, dry conditions.

WRONG CHOICES EXPLAINED
(2) It is *not* true that an individual's ability to adjust to this changing environment involves a mechanism similar to *the digestion of carbohydrates by enzymes*. Carbohydrates are organic food molecules that are digested through the action of enzymes secreted from digestive glands in the human body. This biochemical mechanism is in no way similar to the manner by which the human body adjusts to changing environmental conditions.

(3) It is *not* true that an individual's ability to adjust to this changing environment involves a mechanism similar to *using ATP for the diffusion of water*. ATP is a substance produced during respiration that transports energy to mechanisms requiring the expenditure of cellular energy. Diffusion is a process by which materials move from a region of higher relative concentration to a region of lower relative concentration without the expenditure of cellular energy. These mechanisms are in no way similar to the manner by which the human body adjusts to changing environmental conditions.

(4) It is *not* true that an individual's ability to adjust to this changing environment involves a mechanism similar to *glucose production in the pancreas.* Glucose is produced in green plant cells through the process of photosynthesis. The pancreas is an organ found in humans and other animals that synthesizes digestive enzymes and hormones used in body processes. These mechanisms are in no way similar to the manner by which the human body adjusts to changing environmental conditions.

24. **4** Nonrenewable resources are *finite and are depleted over time.* The term "nonrenewable resources" is used to refer to materials that are useful to humans and other organisms and that are not being replaced by natural processes in the immediate period of time. The term "finite" means limited in quantity. As human technologies use more and more nonrenewable, finite resources (such as coal, oil, metals, and minerals), the amount of these resources available is reduced. When virtually all of a specific material is used, it is said to be depleted and is no longer available to support those technologies. Many such nonrenewable resources are both finite and are being depleted at an accelerating pace by human technologies.

WRONG CHOICES EXPLAINED

(1) It is *not* true that nonrenewable resources are *not finite and are not depleted over time.* This may be said for many renewable resources (such as solar and wind energy, timber, and sand) that are being constantly replaced by natural processes. However, it is not true for nonrenewable resources.

(2), (3) It is *not* true that nonrenewable resources are *not finite and are depleted over time* or are *finite and are not depleted over time.* These nonsense distracters simply manipulate words to create alternative answers to distract the student. They do not describe the characteristics of any known nonrenewable resources.

25. **1** *Parasite and host* is the pair of terms that describes this relationship. Parasitism is a type of symbiotic relationship in which one organism (the parasite) lives in or on the body of another organism (the host) and derives some benefit from the host while harming it in the process. Dodder's parasitic action produces small tubules that invade the vascular system of the host plant, robbing it of nutrients and reducing its growth potential. In this example, dodder is the parasite and the affected plants are the hosts.

WRONG CHOICES EXPLAINED

(2) *Predator and prey* is *not* the pair of terms that describes this relationship. A predator is an animal that hunts and kills other animals (prey) for food. This description does not match the example given in the question because dodder is not an animal predator and the host plants are not animal prey.

(3) *Producer and decomposer* is *not* the pair of terms that describes this relationship. A producer is an organism that manufactures its own food from inorganic materials and energy. A decomposer is an organism that derives its nutrition by digesting the remains of dead organisms and absorbing them into its own body. This description does not match the example given in the question because dodder is not a producer and the host plants are not decomposers.

(4) *Consumer and scavenger* is *not* the pair of terms that describes this relationship. A consumer is an organism that derives its nutrition from the bodies of other organisms. A scavenger is a type of consumer organism that consumes the remains of dead animals as its source of nutrition. Although dodder is a consumer organism, this description does not match the example given in the question because the host plants are not scavengers.

26. **4** Three human hormones most directly involved in sexual reproduction are *estrogen, progesterone, and testosterone*. Estrogen is a reproductive hormone that regulates the maturation and release of egg cells in human females. The female reproductive hormone progesterone regulates the condition of the uterus, maintaining it for the development of the embryo. Testosterone is a male reproductive hormone that regulates the maturation of sperm cells in the body.

WRONG CHOICES EXPLAINED

(1), (2) It is *not* true that three human hormones most directly involved in sexual reproduction are *estrogen, insulin, and progesterone* or *testosterone, estrogen, and insulin*. Insulin is a human hormone that regulates the concentration of glucose in the blood. Insulin is not considered to be a reproductive hormone.

(3) It is *not* true that three human hormones most directly involved in sexual reproduction are *progesterone, ATP, and testosterone*. ATP (adenosine triphosphate) is a substance produced during respiration that transports energy to mechanisms requiring the expenditure of cellular energy. ATP is not a human hormone.

27. **2** The scientists warned that the loss of the effectiveness of this shield may lead to an increase in *mutations that lead to cancer*. The ozone shield functions to screen out harmful elements of solar radiation that are known to be mutagenic agents. When the ozone shield is depleted, these harmful elements of solar radiation enter Earth's atmosphere and can induce mutations in the living tissues of any organism that are exposed to them, including humans.

WRONG CHOICES EXPLAINED

(1), (3), (4) The scientists did *not* warn that the loss of the effectiveness of this shield may lead to an increase in *allergies to ozone, viral diseases such as*

AIDS, or *ice formation at the poles*. Competent scientists have not warned of these as being effects of the depletion of the ozone shield. These are non-sense distracters.

28. **4** *Water, sunlight, carbon dioxide* are abiotic factors that are most responsible for the increase in plant mass. Abiotic factors are nonliving conditions encountered in the natural environment. These particular abiotic factors are important elements of the photosynthetic process, which uses the energy of sunlight to manufacture molecules of glucose from the atoms that make up water and carbon dioxide. Glucose is the building block of most plant tissues, so it is essential to the growth of the plant.

WRONG CHOICES EXPLAINED
(1) *Water, minerals, bacteria* are *not* abiotic factors that are most responsible for the increase in plant mass. Bacteria are living organisms and as such are biotic, not abiotic, factors in the environment.
(2), (3) *Sunlight, oxygen, plant receptors* and *minerals, water, plant enzymes* are *not* abiotic factors that are most responsible for the increase in plant mass. Plant receptors and plant enzymes are component parts of living organisms and as such are biotic, not abiotic, factors in the environment.

29. **1** Row *1* in the chart correctly identifies the functions of structures *A*, *B*, and *C*. Structure *A* in the diagram is the cell membrane, which bounds the cytoplasm and regulates the passage of materials, including wastes, out of the cell. Structure *B* in the diagram is a mitochondrion, which serves as the site for aerobic respiration that releases energy from molecules of glucose. Structure *C* in the diagram is a ribosome, which serves as the site for the manufacture of protein molecules based on a template provided by DNA.

WRONG CHOICES EXPLAINED
(2) Row *2* in the chart does *not* correctly identify the functions of structures *A*, *B*, and *C*. Information storage in the cell is provided by the nucleus, not the cell membrane. Transport of materials in the cell is provided by the endoplasmic reticulum and cell membrane, not mitochondria. Storage of liquids in the cell is provided by vacuoles, not ribosomes.
(3) Row *3* in the chart does *not* correctly identify the functions of structures *A*, *B*, and *C*. Protein synthesis in the cell is provided by the ribosomes, not the cell membrane. Storage of wastes in the cell is provided by vacuoles, not mitochondria. Reproduction in the cell is regulated by the nucleus, not ribosomes.
(4) Row *4* in the chart does *not* correctly identify the functions of structures *A*, *B*, and *C*. It is true that cell communication in the cell is provided by the cell membrane. Transport of materials in the cell is provided by the endoplasmic reticulum, not mitochondria. Waste removal in the cell is provided by vacuoles, not ribosomes.

30. **4** *The cooperation between the different states must be improved* is the change that is necessary in order to reduce the amount of air pollution in Pennsylvania. The diagram implies that air pollution produced in states such as Illinois, Indiana, and Ohio moves with the prevailing westerly winds into Pennsylvania. Cooperation among these states may help to reduce the quantity of air pollutants released into the air in states west of Pennsylvania and thereby reduce the amount of air pollution that reaches Pennsylvania and pollutes its air quality.

WRONG CHOICES EXPLAINED

(1) *Laws must be passed to protect endangered species* is *not* the change that is necessary in order to reduce the amount of air pollution in Pennsylvania. Laws of this type would neither increase nor decrease the amount of air pollution in Pennsylvania.

(2) *The use of natural resources must be increased* is *not* the change that is necessary in order to reduce the amount of air pollution in Pennsylvania. Increasing the use of many renewable and nonrenewable natural resources would neither increase nor decrease the amount of air pollution. However, increasing the use of certain natural resources (such as coal) would increase, not reduce, air pollution in Pennsylvania.

(4) *More coal-burning power plants must be built* is *not* the change that is necessary in order to reduce the amount of air pollution in Pennsylvania. The burning of coal to produce electricity is identified as one of the chief sources of air pollution in the industrialized world. Building additional coal-burning power plants would increase, not reduce, air pollution in Pennsylvania.

PART B–1

31. **2** *Sedges → ants → frogs → kookaburras* is the sequence of organisms that represents a food chain within this food web. An examination of the food web illustrated in the diagram reveals that these four organisms are connected by arrows leading from a producer (sedges), to an herbivore (ants), to a first-level carnivore (frogs), to a second-level carnivore (kookaburras). The arrows in the diagram indicate the direction of energy flow in this food web and must be followed closely to determine the food chains illustrated within the food web.

WRONG CHOICES EXPLAINED

(1), (3), (4) *Tadpoles → algae → daphnia → back swimmers, algae → daphnia → decayed material → bacteria,* and *dragonflies → sedges → ants → centipedes* are *not* the sequences of organisms that represent food chains within this food web. Each of these sequences contains at least one pairing in which no arrow connects them or in which the arrow is pointed in a direction inconsistent with the direction of energy flow in that sequence.

32. **3** The population of *ants* would be most immediately affected by the removal of the lizard population. The arrow that points from ants to lizards in this diagram indicates that ants serve as the principal source of food for lizards in this food web. If lizards were removed from the environment, pressure would be immediately removed from the ant population, allowing it to grow rapidly in numbers.

WRONG CHOICES EXPLAINED

(1), (2), (4) It is *not* true that the population of *sedges, algae,* or *centipedes* would be most immediately affected by the removal of the lizard population. No arrow connects lizards to sedges, algae, or centipedes in this diagram, so the removal of lizards would be unlikely to affect any of these populations either directly or immediately.

33. **4** *Balance, graduated cylinder, ruler* is the list of tools that is arranged in the order that the student used them. A balance is used to measure the mass of a material. So the student used the balance first to measure out 15 grams of salt. A graduated cylinder is used to measure the volume of a liquid. So the student next used the graduated cylinder to measure out 15 milliliters of water. A ruler is used to measure linear distance. So the student lastly used a ruler to measure a 1 centimeter by 4 centimeter section of a plant leaf.

WRONG CHOICES EXPLAINED

(1), (2), (3) *Graduated cylinder, ruler, balance; balance, ruler, graduated cylinder;* or *graduated cylinder, balance, ruler* are *not* the lists of tools that are arranged in the order that the student used them. Each of these choices contains one or more tools out of the correct order as explained above.

34. **1** Based on the information provided in the graph, the process that is occurring is *ecological succession.* Ecological succession refers to the series of changes that occur to the dominant plant community of an area that has been significantly affected by an environmental change. In this case, the environmental change that occurred was a forest fire. Over time, several different plant communities inhabited the area, each replacing the last until a stable climax forest community was established.

WRONG CHOICES EXPLAINED

(2) It is *not* true that based on the information provided in the graph, the process that is occurring is *biological evolution.* Biological evolution is a process by which members of a species displaying favorable characteristics survive harsh environmental conditions to reproduce and pass on their favorable characteristics to offspring. Biological evolution is not the process described in the graph.

(3) It is *not* true that based on the information provided in the graph, the process that is occurring is *selective breeding.* Selective breeding is a tech-

nique used by animal and plant breeders in which organisms with desirable traits are cross-bred with the hope of producing offspring that also display those traits. Selective breeding is not the process described in the graph.

(4) It is *not* true that based on the information provided in the graph, the process that is occurring is *genetic engineering*. Genetic engineering is a laboratory technique in which a gene for a desired trait is snipped from the DNA of a donor cell and inserted into the genome of a recipient cell. Genetic engineering is not the process described in the graph.

35. **4** The time shown in the graph is most likely measured in *years*. Ecological succession is a process that commonly occurs over decades or even centuries. For this reason, time on the horizontal axis of this graph would most likely be expressed in years.

WRONG CHOICES EXPLAINED

(1), (2), (3) The time shown in the graph is *not* most likely measured in *days*, *weeks*, or *months*. These units of time are much too small to represent accurately the time required for ecological succession to occur.

36. **1** The mass of plants shown in the graph refers to the mass of a number of *populations*. The plants identified in the graph are either single species or groups of related species. For example, the term "grasses" may refer to several different species of grasslike plants that grow in open fields or developing forest lands. The term "shrubs" indicates not one but several different species of low-growing, woody-stemmed, flowering plants often found in open fields or in the understory of a forest. Each of these plant species represents a population in the changing plant community illustrated in this graph.

WRONG CHOICES EXPLAINED

(2) The mass of plants shown in the graph does *not* refer to the mass of a number of *decomposers*. Decomposers are organisms such as bacteria and fungi that are known to break down the complex bodies of dead plants and animals and then convert them to simple nutrients. The living organisms illustrated in the graph are plants only, not decomposers.

(3), (4) The mass of plants shown in the graph does *not* refer to the mass of a number of *ecosystems* or *communities*. An ecosystem is the sum total of all the living communities that live in a particular environment under a particular set of abiotic environmental conditions. An ecological community is the sum total of all the living species populations that live within a given ecosystem. The living organisms illustrated in the graph are plants only, not animals, decomposers, other species, or the abiotic conditions surrounding them.

37. **2** The building blocks for this molecule are *simple sugars*. Starch is a complex carbohydrate produced in plant cells by linking many repeating units

of glucose (a simple sugar) by means of dehydration synthesis. This diagram illustrates a three-unit portion of a starch molecule.

WRONG CHOICES EXPLAINED

(1) It is *not* true that the building blocks for this molecule are *amino acids*. Amino acids are the chemical building blocks of proteins, not starches.

(3) It is *not* true that the building blocks for this molecule are *fats*. Fats are complex organic chemicals whose building blocks are glycerol and fatty acids.

(4) It is *not* true that the building blocks for this molecule are *molecular bases*. Molecular bases are the chemical building blocks of nucleic acids such as DNA, not starches.

38. **1** The energy in this molecule is stored *in the bonds between atoms*. When starch is digested into glucose, the glucose is absorbed into mitochondria and subjected to chemical breakdown by respiratory enzymes. During this process, small amounts of energy are released as the bonds linking carbon atoms to each other, and linking carbon atoms to hydrogen and oxygen atoms, are broken. This energy is then transferred to molecules of ATP for use in cellular processes.

WRONG CHOICES EXPLAINED

(2) The energy in this molecule is *not* stored *in the oxygen found in the molecule*. The atomic energy that holds atoms of oxygen together in a diatomic molecule cannot be released by normal biological processes such as respiration or hydrolysis.

(3) The energy in this molecule is *not* stored *when the carbon atoms break off*. The respiratory reaction that breaks carbon-carbon bonds results in the transfer and temporary storage of chemical bond energy in ATP molecules. [**Note:** This is a possible correct answer to the question as written, but it is not the best answer.]

(4) The energy in this molecule is *not* stored *when water breaks this molecule apart*. Enzymatic hydrolysis uses hydrogen and oxygen atoms that make up water to help break starch into glucose. The hydrolytic reaction is an energy-using reaction, not an energy-storing reaction.

39. **2** Government agencies and citizens should propose the use of enviropig in the future only after *assessing risks, costs, and benefits*. This is an example of a trade-off between an economic/environmental development and its potential effect on human health. It would be wise to delay widespread use of this development until independent scientific research has determined that the enviropig is safe for human consumption.

WRONG CHOICES EXPLAINED

(1) It is *not* true that government agencies and citizens should propose the use of enviropig in the future only after *developing ways to remove the bacterial enzyme*. Removing the phosphorus-reducing enzyme would defeat the primary reason for developing the enviropig.

(3) It is *not* true that government agencies and citizens should propose the use of enviropig in the future only after *people have eaten lots of enviropig meat and determined the effects*. Scientific research designed to determine the safety of the enviropig for human consumption should be conducted prior to the product being made available to the general public.

(4) It is *not* true that government agencies and citizens should propose the use of enviropig in the future only after *a different, cheaper pig can be produced regardless of the output of phosphorus*. This effectively describes pigs that are currently raised on farms and commonly used for food by the general public, so it would negate the need for enviropig technology.

40. **2** The molecules used to cut, copy, and connect the DNA segments in this process are *enzymes*. Molecules known as restriction enzymes are specialized to recognize a specific chemical sequence in the DNA and to attach at that point so that a cutting reaction can be catalyzed. Other enzymes are integral to the synthesis of segments of DNA that are identical to the cut segments. Still other enzymes catalyze the reactions that are needed to connect these segments to plant chromosomes.

WRONG CHOICES EXPLAINED

(1) It is *not* true that the molecules used to cut, copy, and connect the DNA segments in this process are *sugars*. Sugars are a class of organic chemicals composed of carbon, hydrogen, and oxygen in the ratio of 1:2:1. Sugars are not used to cut, copy, or connect DNA segments.

(3) It is *not* true that the molecules used to cut, copy, and connect the DNA segments in this process are *indicators*. Indicators are chemical agents used by scientists to detect the presence of chemicals in a mixture. Indicators are not used to cut, copy, or connect DNA segments.

(4) It is *not* true that the molecules used to cut, copy, and connect the DNA segments in this process are *antigens*. Antigens are proteins located on the cell membranes of pathogens and other cells that are foreign to the body of an individual and that trigger an immune response in that individual. Antigens are not used to cut, copy, or connect DNA segments.

41. **1** The best explanation for these modified rice plants being flood resistant is that *the gene for flood resistance was inserted into plant cells, which grew into plants whose cells are expressing this gene*. As long as the bacterial gene is inserted into a plant germ cell (fertilized ovum), it will replicate and be located in every cell of the adult rice plant.

[**Note:** The correct answer is expressed nearly word-for-word in the diagram.]

WRONG CHOICES EXPLAINED

(2) The best explanation for these modified rice plants being flood resistant is *not* that *they were produced by fertilization, using gametes from two flood-resistant bacterial cells.* Bacteria reproduce asexually, so they do not produce gametes.

(3) The best explanation for these modified rice plants being flood resistant is *not* that *there was a mutation in the bacterial DNA after it was inserted into the plant that caused it to be flood resistant.* The bacterial DNA that was transferred to the plant cell was selected because scientists already knew that it produced the desired characteristic. Mutations are random events that would be highly unlikely to produce the desired result after bacterial insertion.

(4) The best explanation for these modified rice plants being flood resistant is *not* that *the researchers used selective breeding for the flood-resistant trait.* Rice plants that have developed from germ cells (fertilized ova) that were engineered to receive this genetic modification can pass this trait on to future generations of rice plants. In the future, scientists can selectively breed these modified rice plants to create a pure-breeding strain of flood-resistant rice.

42. **2** One possible conclusion that can be drawn about the activity of these two cells is that *more active transport occurs in cell A than in cell* B. Mitochondria are cell organelles responsible for the operation of the aerobic respiration reactions that release chemical bond energy from glucose and transfer it to ATP (cellular energy). Active transport is one of many metabolic processes that require the expenditure of cellular energy in order to operate. Because of the relative abundance of mitochondria in cell *A* compared with cell *B*, it is reasonable to conclude that cell *A* is carrying on more metabolic activities (including active transport) that require the expenditure of cellular energy than is cell *B*.

WRONG CHOICES EXPLAINED

(1) It is *not* true that one possible conclusion that can be drawn about the activity of these two cells is that *more active transport occurs in cell B than in cell* A. This choice is diametrically opposed to the correct answer described above. More energy is used by cell *A* than by cell *B*.

(3) It is *not* true that one possible conclusion that can be drawn about the activity of these two cells is that *cell B uses some of the extra mitochondria to make food.* Mitochondria are specialized for releasing and transferring energy in the cell, not for manufacturing food.

(4) It is *not* true that one possible conclusion that can be drawn about the activity of these two cells is that *cell A is a plant cell since it has a cell wall.* The diagram shows a thin, irregular line bounding cell *A*, which is a convention used to illustrate a cell membrane, not a cell wall.

43. **3** Row *3* is the correct sequence for processes 1, 2, and 3 represented in the diagram. Process 1—In humans, the testes and ovaries (reproductive organs) carry out meiotic cell divisions that result in the formation of sperm and egg (gametes). Process 2—Sperm and egg fuse to form the zygote in the process of fertilization. Process 3—The zygote undergoes repeated mitotic cell divisions that result in a mass of cells known as the embryo.

WRONG CHOICES EXPLAINED

(1), (2), (4) Rows *1, 2,* and *4* are *not* the correct sequences for processes 1, 2, and 3 represented in the diagram. Each of these rows contains one or more events out of correct order as described above.

PART B–2

44. One credit is allowed for correctly marking an appropriate scale, without any breaks in the data, on each labeled axis. [1]

45. One credit is allowed for correctly plotting the data for the percent of males on the grid, connecting the points, and surrounding each point with a small circle. [1]

46. One credit is allowed for correctly plotting the data for the percent of females on the grid, connecting the points, and surrounding each point with a small triangle. [1]

Sex of Painted Turtles at Various Nest Temperatures

47. **4** The fact that the sex of the painted turtle offspring is controlled by the temperature of the nest is an example of *environment modifying gene expression*. In this case, the temperature of the turtle's environment is an abiotic condition that influences the expression of gender in developing turtle eggs. Low (19°C–20°C) and high (23°C–25°C) temperatures favor the expression of femaleness, while moderate (21°C–22°C) temperatures favor the expression of maleness.

WRONG CHOICES EXPLAINED

(1) The fact that the sex of the painted turtle offspring is controlled by the temperature of the nest is *not* an example of *natural selection causing a new species to form*. No information is provided in the question concerning the evolution of a new turtle species.

(2) The fact that the sex of the painted turtle offspring is controlled by the temperature of the nest is *not* an example of *a predator-prey interaction*. No information is provided in the question concerning nutritional relationships between the turtles and another species.

(3) The fact that the sex of the painted turtle offspring is controlled by the temperature of the nest is *not* an example of *habitat destruction decreasing biodiversity*. No information is provided in the question concerning human activities that would negatively impact the turtles' environment.

48. One credit is allowed for correctly stating the relationship between increasing light intensity and the percentage of disks floating at the conclusion of Investigation 1. Acceptable responses include but are not limited to: [1]

- *With more intense light, a higher percentage of the disks floated.*
- *More light intensity results in more disks floating.*
- *More disks float when the light is brighter.*
- *As light intensity increases, the percentage of floating disks increases.*

49. **2** The substance produced inside the leaf disks that caused them to float to the surface of the solution is *oxygen*. Photosynthesis is a biochemical process carried on by green plants in which the atoms that make up carbon dioxide and water are rearranged to produce molecules of glucose and oxygen gas. At room temperature, oxygen gas is less dense than liquid water. As a result, tiny bubbles of oxygen gas that form inside the leaf disks cause the disks to float to the surface of the water.

WRONG CHOICES EXPLAINED

(1), (4) It is *not* true that the substance produced inside the leaf disks that caused them to float to the surface of the solution is *ozone* or *nitrogen*. Neither of these gases is produced in or associated with the process of photosynthesis. For this reason, neither ozone nor nitrogen could not have been responsible for the buoyancy of the disks in those containers.

(3) It is *not* true that the substance produced inside the leaf disks that caused them to float to the surface of the solution is *water*. Water in the leaf disks has the same density as the water in the experimental containers. For this reason, water could not have been responsible for the buoyancy of the disks in those containers.

50. **3** *Green* is the color of light that appears to be least effective for photosynthesis in spinach leaves. In the graph, the percentage of disks floating under the green light was only about 30%, while the percentages of floating disks under the other colors were higher (white—90%; blue—55%; red—48%). This is largely due to the fact that chlorophyll, the active catalyst in the photosynthetic reactions, is green. This fact indicates that the wavelengths of light we perceive as green are reflected by chlorophyll, so the energy of green light is not absorbed by the leaf as effectively as are other colors of light.

WRONG CHOICES EXPLAINED

(1), (2), (4) *White*, *blue*, and *red* are *not* the colors of light that appear to be least effective for photosynthesis in spinach leaves. The percentages of floating disks under these colors were significantly higher than the percentage for green. See the correct answer above.

51. One credit is allowed for correctly identifying the organelle present in plants where this conversion takes place. Acceptable responses include: [1]

- *Chloroplast*

52. One credit is allowed for correctly providing *one* advantage and *one* disadvantage of owning a solar car. Acceptable responses include but are not limited to: [1]

Advantages:

- *Solar energy is a renewable resource.*
- *There is no air pollution from a solar vehicle.*
- *Using solar cars will mean burning less gasoline.*
- *The cost of operating a solar car will be low.*
- *Sunlight is free.*

Disadvantages:

- *Charging a car at night will be difficult.*
- *Solar cars may not be feasible in all areas due to limited light in winter.*
- *The range of solar cars may make their use at night impossible.*
- *The cost of purchasing a solar car will be high.*
- *Solar panels/converters are expensive.*

53. One credit is allowed for correctly stating *one* inference that could be made based on the fact that this dinosaur had feathers. Acceptable responses include but are not limited to: [1]

- *The dinosaur could fly.*
- *Dinosaurs are the ancestors of/related to birds.*
- *Some types of dinosaurs evolved into birds.*
- *This dinosaur may be an ancestor of modern birds.*

54. One credit is allowed for correctly stating *one* reason why the offspring of crows skilled at using twigs as tools would have the greatest chance of survival. Acceptable responses include but are not limited to: [1]

- *Well-fed parents will probably have well-fed and stronger offspring.*
- *They will have more food.*
- *They will likely be good twig users.*
- *More food will mean more energy and more offspring for these crows.*

55. One credit is allowed for correctly stating *one* reason why some members of a population of crows equally skilled in the use of twigs have different rates of survival. Acceptable responses include but are not limited to: [1]

- *There is a chance that some other variation will affect the crows' ability to survive.*
- *The tool may not be as useful in their environment.*
- *There are many other factors that influence survival (vision, size, and/ or muscle strength).*
- *There may be variations in the populations of wood-boring beetle larvae that affect the crows.*

PART C

56–57. Two credits are allowed for correctly evaluating the student's experiment. As part of your evaluation, be sure to:

- State *one* possible hypothesis for the experiment proposed by the student. [1]
- Describe the type of data that should be collected to determine if the brushing with lotion was having a *negative* effect on the African violet plant. [1]

Acceptable responses include but are not limited to: [2]

Hypothesis:

- *The presence of lotion on the leaves will slow the growth of the plants.*
- *If the leaves of an African violet plant are rubbed with body lotion, growth will be reduced.*

- *Lotion brushed on the leaves of an African violet will have no effect on the leaves.*
- *Lotion on the leaves has a negative effect on the leaves.*
- *Hand lotion will negatively affect the growth of these African violet plants.*

Data:

- *The number of leaves that wilted during the five-day period*
- *The growth of each plant measured in centimeters*
- *The number of leaves that dropped off/turned yellow*
- *Size of area of damage on leaves*
- *The number of damaged leaves*
- *Change in size of the leaves on each plant*
- *Color/texture of leaves*
- *Survival of leaves/plants*
- *Growth rate of the plants*

58. One credit is allowed for correctly stating the role of mutation *or* recombination in the appearance of the trait for black fur color in the pocket mouse population. Acceptable responses include but are not limited to: [1]

- *A mutation resulted in the initial color change.*
- *Recombination provides genetic variability for fur color in offspring.*
- *Mutations for black fur color are already present in the DNA of the pocket mouse.*
- *Recombination allows gene combinations for black fur to assort randomly and to appear more frequently when it is favored by environmental change.*

59. One credit is allowed for correctly explaining why the percentage of black pocket mice changed so much after the volcanic eruption. Acceptable responses include but are not limited to: [1]

- *The black pocket mice blended in with their surroundings better, and more were able to survive and reproduce.*
- *After the color change in the environment, the black mice were more difficult for predators to see.*
- *The black mice are harder to see against the dark volcanic rock.*
- *The dark volcanic rock increased the visibility of light brown mice and decreased their survival rate but increased the survival rate of black mice.*
- *The environmental change had the effect of reducing the frequency of "light brown" genes and increasing the frequency of "black" genes in the gene pool of pocket mice.*

60. One credit is allowed for correctly calculating the percentage of bases that would be C. Acceptable responses include: [1]

- *38%*

[**Note:** This value is found by means of the following calculation:
12% (A) + 12% (T) = 24% (A/T)
100% (A/T/C/G) – 24% (A/T) = 76% (C/G)
76% (C/G) ÷ 2 = 38% (C)]

61. One credit is allowed for circling "yes" and correctly supporting the answer. Acceptable responses include but are not limited to: [1]

- *Because there are high percentages of C and G bases in the sample*
- *Because the percentage of C and G in this DNA segment is over 50%*
- *Because this segment of DNA is most likely a gene*
- *Because C and G bases together represent 76% of all bases in this segment, so is likely to be a protein template for this reason*

62. One credit is allowed for correctly explaining (using the terms antigens and antibodies) why the tumor cells are ignored by the immune system in Tasmanian devils. Acceptable responses include but are not limited to: [1]

- *The tumor cells might lack antigens that the antibodies can recognize and attack.*
- *The antigens on the surface of the tumor cells could be interpreted by the immune system as being on cells that are not harmful, and therefore the immune system does not form antibodies to attack the tumor cells.*
- *The cancer cells might not have any antigens on their surfaces. Without antigens, antibodies will not attack.*
- *Antigens on the tumor cells might mimic the devil's normal antigens, so the devil's immune system doesn't produce antibodies to attack the tumor cells.*

63. One credit is allowed for correctly explaining how cancer cells differ from normal cells. Acceptable responses include but are not limited to: [1]

- *Cancer cells undergo uncontrolled cell division.*
- *Cancer cells have more mutations compared to normal cells.*
- *Cancer cells are more harmful than normal cells and disrupt homeostasis.*
- *Cancer cells divide abnormally and rapidly compared to normal cells.*
- *Cancer cells crowd out normal tissues and cause organs to fail.*

64. One credit is allowed for correctly describing *one* possible way to maintain a population of healthy, uninfected Tasmanian devils until a treat-

ment or cure can be found. Acceptable responses include but are not limited to: [1]

- *Move some uninfected animals to an area where they will not come into contact with infected animals.*
- *Remove baby devils from the population, and place them into a zoo or wildlife refuge.*
- *Separate the animals while they are feeding.*
- *Provide more food to decrease competition/fighting.*
- *Create a wildlife preserve for healthy animals until they can be vaccinated and returned to the wild.*

65. One credit is allowed for correctly explaining why coyotes were *not* in competition with the lynx prior to the presence of packed snow trails. Acceptable responses include but are not limited to: [1]

- *The lynx are adapted to stay on top of the snow, while coyotes are not.*
- *Coyotes could not prey on animals in the same area as the lynx.*
- *Coyotes lack the foot design of the lynx, so coyotes could not chase hares where the snow is soft and deep.*
- *Coyotes could not run/walk in the deep, soft snow.*

[**Note:** In fact, coyotes and lynx compete for the same limited food source (snowshoe hares) under all conditions in this environment. Under deep snow conditions, however, lynx compete for this resource more successfully than do coyotes.]

66. One credit is allowed for correctly describing the niche that the lynx, coyote, and cougar are competing to fill. Acceptable responses include but are not limited to: [1]

- *The niche is to feed on snowshoe hares in a snowy environment.*
- *The niche includes being a predator in an environment that has snow during the winter.*
- *These animals are all predators/carnivores.*
- *This is a nutritional niche that is characterized by the predation of snowshoe hares.*

67. One credit is allowed for correctly explaining how the carrying capacity affects the number of predators in the area. Acceptable responses include but are not limited to: [1]

- *Carrying capacity limits the number of organisms that can live in an area. There are only so many hares, so there can be only so many predators of hares.*
- *The number of prey animals in a given area is limited. The number of prey animals determines the carrying capacity of the area and can support only a certain number of predators.*

- *The carrying capacity of the predators is directly related to the size of the hare population.*
- *Carrying capacity is the number of individuals of a given species that can be supported in a specific environment. As limiting factors (such as the number of hares) are reduced, the carrying capacity for some species (such as the lynx) may be exceeded, causing some individuals of that species to be eliminated from the environment until balance is restored.*

68. One credit is allowed for correctly stating *one* way that the use of pesticides to kill ticks could lead to the decline of the oxpecker population. Acceptable responses include but are not limited to: [1]

- *The ticks could become toxic, and the oxpeckers that eat them could get sick and die.*
- *Pesticides could kill the birds.*
- *The ticks were a food source of the birds, so the birds could starve.*
- *Without ticks for food, the bird population would die off.*
- *The oxpeckers might survive eating the toxic ticks, but the pesticide could become concentrated in the birds' tissues.*

69. One credit is allowed for correctly stating *one* ecological advantage of using oxpeckers to solve the problem with these ticks. Acceptable responses include but are not limited to: [1]

- *Each oxpecker can eat 13,000 ticks in a day, reducing the population of this parasite.*
- *No harmful chemicals are used that might hurt oxpeckers and other animals.*
- *The birds will have ample food, which will enhance their survival.*
- *It will reduce the need for pesticides that might kill beneficial species such as honeybees.*

70. One credit is allowed for correctly stating *one* possible relationship between CO_2 levels and the change in Antarctic ice mass. Acceptable responses include but are not limited to: [1]

- *An increase in CO_2 levels caused the ice to melt because CO_2 is a greenhouse gas.*
- *An increase in CO_2 brought about global warming, which melted the ice.*
- *As the level of CO_2 increased, ice mass decreased.*
- *As CO_2 increased, the temperature increased and the ice mass decreased.*

71. One credit is allowed for correctly explaining why the change in ice mass is resulting in a decline in the penguin populations. Acceptable responses include but are not limited to: [1]

- *The penguins' food source grows and develops under the ice masses.*
- *As the ice mass decreases, there is less krill for the penguins to eat.*
- *Penguins depend on the ice mass as its natural habitat. As that habitat shrinks, the penguin population decreases.*
- *Reduced ice mass leads to a reduced krill population, which in turn leads to a reduction in the penguin population.*

72. One credit is allowed for correctly stating *one* specific way in which humans might have caused the changes in atmospheric CO_2 levels. Acceptable responses include but are not limited to: [1]

- *The burning of fossil fuels by humans is increasing CO_2 levels.*
- *Increased industrialization is releasing more CO_2 into the atmosphere.*
- *The reduction in plant habitat caused by deforestation reduces nature's ability to remove CO_2 from the air.*
- *Cars and trucks release CO_2 into the air as they are driven.*

PART D

73. **1** *The contents of the dialysis tube would turn blue-black* is the statement that best describes what would most likely be observed after 20 minutes. The student must assume that the starch indicator in the beaker is iodine, a small ion dissolved in the water. Over time, the indicator ions would move through the dialysis membrane micropores from the beaker into the tube interior. Once inside the tube, indicator ions would combine with molecules of starch and cause the starch to turn blue-black. Starch molecules are too large to pass through the dialysis membrane micropores from inside the tube to the beaker water.

WRONG CHOICES EXPLAINED
(2) *The liquid in the beaker would turn blue-black* is *not* the statement that best describes what would most likely be observed after 20 minutes. Starch molecules are too large to pass through the dialysis membrane from inside the tube to the beaker water. For this reason, it would not be possible for the starch-indicator reaction to occur in the beaker water outside the tube.

(3) *The dialysis tube would burst* is *not* the statement that best describes what would most likely be observed after 20 minutes. Due to the tensile strength of dialysis tubing, this phenomenon is not likely to occur under any of the circumstances described in this setup.

(4) *There would be no change visible* is *not* the statement that best describes what would most likely be observed after 20 minutes. This setup would result in the visible change described in the correct answer above.

74. **3** *Diffusion* is the term that correctly identifies the process by which molecules move through the dialysis tube membrane. Diffusion is a process by which materials move from a region of higher relative concentration to a region of lower relative concentration without the expenditure of energy. In this case, indicator ions move from the beaker (higher relative concentration) into the dialysis tube (lower relative concentration) by means of diffusion.

WRONG CHOICES EXPLAINED

(1) *Paper chromatography* is *not* the term that correctly identifies the process by which molecules move through the dialysis tube membrane. Paper chromatography is a laboratory technique that is used to help researchers study mixtures of organic chemicals such as plant pigments. This experiment uses none of the materials (such as filter paper, organic solvents, and plant pigments) used in paper chromatography.

(2) *Active transport* is *not* the term that correctly identifies the process by which molecules move through the dialysis tube membrane. Active transport is a process in which materials are moved through a living cell membrane against the concentration gradient (from a region of lower relative concentration to a region of higher relative concentration). Active transport requires the expenditure of cellular energy in order to occur. Active transport is not occurring in this experiment.

(4) *Digestion* is *not* the term that correctly identifies the process by which molecules move through the dialysis tube membrane. Digestion is a life process in which organisms break down complex nutrient molecules into smaller, soluble units that can pass through cell membranes for metabolic purposes. Digestion is not occurring in this experiment.

75. **4** The dialysis tube was placed into the 99% *water* solution. The increase in the mass of the tube and its contents indicates that water moved into the tube by diffusion. Diffusion is a process by which materials move from a region of higher relative concentration to a region of lower relative concentration. The 99% water solution is the only choice of those offered in which the concentration of water is higher outside the tube than inside the tube (which starts at 97% water).

WRONG CHOICES EXPLAINED

(1), (2), (3) It is *not* true that the dialysis tube was placed into the *0% water, 95% water,* or *97% water* solutions. Each of these solutions is lower (0% water and 95% water) or equal to the concentration of water (97% water) inside the dialysis tube. The net movement of water would either be out of the tube (0% water and 95% water) or no net movement into or out of the tube (97% water).

76. **3** *Obtain molecular evidence from all three species and identify similarities* is the statement that best describes the method used as illustrated in the diagram to determine the evolutionary relationships between these three species of finches. The diagram illustrates the process of gel electrophoresis, in which fragmented samples of molecular DNA from different species (*A*, *B*, and *C*) are injected into a gel and drawn through the gel by an electric current. The banding patterns of the DNA fragments in the gel indicate biochemical similarities and differences among the three species of finch studied. More closely related species display more similar banding patterns.

WRONG CHOICES EXPLAINED

(1), (2), (4) *Examine the structure of the beaks and compare them, observe behavioral and physical characteristics of all the finches and group them by similarities,* and *compare common ancestors of all three of the species to see if they are the same* are *not* the statements that best describe the method used as illustrated in the diagram to determine the evolutionary relationships between these three species of finches. These methods may each help scientists to group these species properly by their relative degree of evolutionary similarity. However, these methods are not the ones illustrated in the diagram.

77. One credit is allowed for correctly labeling the branching tree diagram by writing the letters *A*, *B*, and *C* to represent the possible evolutionary relationships among species *A*, *B*, and *C*. Acceptable responses include: [1]

78. One credit is allowed for correctly identifying the finch species that has a slightly smaller, weaker beak and supporting the answer. Acceptable responses include but are not limited to: [1]

Species: 1

Explanations:

- *They have the weaker beak because the majority of the population is at the low end of the graph.*
- *Species 1 has a weaker beak than species 2, except where the two lines overlap.*
- *More of species 1 finches are at the low-strength end of the graph.*
- *The graph shows species 1 (solid line) with lower average beak strength than species 2 (dotted line).*

79. One credit is allowed for correctly selecting the point on the graph where beak strength of the two bird species is equal and supporting the answer. Acceptable responses include but are not limited to: [1]

Point: C

Explanations:

- *At point C, the beak strength has the same value for both species.*
- *At this point, the same relative number of individuals of the two species have the same beak strength.*
- *They are the same at this point because this is where the graph lines intersect.*

80. One credit is allowed for correctly describing what the graph might look like after many years had passed since the environment on the island changed and the seeds of more of the plants became harder to crack open. Acceptable responses include but are not limited to: [1]

- *The curve for species 1 may be lower and shorter.*
- *The curve for species 1 may be eliminated.*
- *Both curves might show an increase in beak strength.*
- *There would be more individuals in species 1 than in species 2.*
- *The curve for species 1 would shift/skew to the right.*

81. **3** The process used to remove excess water from the paramecium by the contractive vacuole is *active transport*. Active transport is a process in which materials are moved through a living cell membrane against the concentration gradient (from a region of lower relative concentration to a region of higher relative concentration). Active transport requires the expenditure of cellular energy in order to operate. The contractile vacuole of the para-

mecium uses energy to concentrate excess water within its membrane and to expel this water to the surrounding environment.

WRONG CHOICES EXPLAINED

(1) It is *not* true that the process used to remove excess water from the paramecium by the contractive vacuole is *synthesis*. Synthesis is a term used to describe metabolic pathways by which complex molecules (such as proteins) are manufactured from simpler subunits (such as amino acids). Synthesis is not illustrated in this diagram.

(2) It is *not* true that the process used to remove excess water from the paramecium by the contractive vacuole is *digestion*. Digestion is a life process in which organisms break down complex nutrient molecules (such as proteins) into smaller, soluble units (such as amino acids) that can pass through cell membranes for metabolic purposes. Digestion is not illustrated in this diagram.

(4) It is *not* true that the process used to remove excess water from the paramecium by the contractive vacuole is *passive transport*. Passive transport is a process in which materials move through a living cell membrane with the concentration gradient (from a region of higher relative concentration to a region of lower relative concentration). Passive transport does not require the expenditure of cellular energy in order to operate. Passive transport is not illustrated in this diagram.

82. **4** This prediction is most likely *correct, because water would be moving out of the cell into the salt solution*. When salt is added to the pond water, the salt goes into solution and reduces the percentage of water in the sample. This allows water to leave the paramecium cell by simple diffusion, making it less necessary for the paramecium to expend cell energy to operate its contractile vacuole to expel excess water.

WRONG CHOICES EXPLAINED

(1) This prediction is *not* most likely *correct, because a high concentration of salt in the environment will force water into the cell, causing the contractile vacuole to pump*. The concentration of salt in the solution will reduce the concentration of water in the solution, which will result in less need for the contractive vacuole to pump.

(2) This prediction is *not* most likely *incorrect, because salt would be entering the cell, and the contractile vacuole would have to pump it out*. Salt does not enter the paramecium under normal circumstances. The contractile vacuole is specialized to pump excess water, not salt.

(3) This prediction is *not* most likely *incorrect, because the concentration of salt in the experiment should not affect the cell*. Single-celled organisms such as paramecia are affected by almost any condition that exists in its environment, including the concentrations of salt and water.

83. One credit is allowed for correctly stating *one* reason why an individual's pulse rate increased during exercise. Acceptable responses include but are not limited to: [1]

- *More nutrients are used when muscle cells are more active.*
- *Muscle cells use more oxygen during exercise.*
- *A higher pulse rate indicates that blood is flowing faster, carrying more nutrients and oxygen to the muscles.*
- *A faster pulse rate will remove wastes from the muscles more rapidly.*
- *Homeostasis is maintained when active cells have enough food and oxygen.*

84. One credit is allowed for correctly calculating the average pulse rate *before* exercise for this group, to the *nearest tenth*. Acceptable responses include: [1]

- *66.2 bpm*

[**Note:** This value is found by means of the following calculation:
68 + 70 + 54 + 64 + 75 = 331
331 ÷ 5 = 66.2]

85. One credit is allowed for correctly stating why the individuals in this group have different pulse rates before exercise. Acceptable responses include but are not limited to: [1]

- *Some individuals are in better physical shape than others.*
- *The individuals may have been involved in different activities before their pulse rates were taken.*
- *Some people may have taken medications that affected their pulse rates.*
- *People have individual variations in their metabolic rates.*
- *Some individuals may have been less active than others before they exercised.*

STANDARDS/KEY IDEAS	JUNE 2016 QUESTION NUMBERS	NUMBER OF CORRECT RESPONSES
STANDARD 1		
Key Idea 1: The central purpose of scientific inquiry is to develop explanations of natural phenomena in a continuing and creative process.		
Key Idea 2: Beyond the use of reasoning and consensus, scientific inquiry involves the testing of proposed explanations involving the use of conventional techniques and procedures and usually requiring considerable ingenuity.	33, 56, 57	
Key Idea 3: The observations made while testing proposed explanations, when analyzed using conventional and invented methods, provide new insights into natural phenomena.	44, 45, 46, 48, 50, 53	
Laboratory Checklist	33, 56, 57	
STANDARD 4		
Key Idea 1: Living things are both similar to and different from each other and from nonliving things.	1, 3, 5, 6, 8, 11, 18, 21, 25, 29, 31, 32, 36, 42	
Key Idea 2: Organisms inherit genetic information in a variety of ways that result in continuity of structure and function between parents and offspring.	10, 19, 40, 41, 47, 60, 61	
Key Idea 3: Individual organisms and species change over time.	9, 14, 16, 17, 54, 55, 58, 59	
Key Idea 4: The continuity of life is sustained through reproduction and development.	4, 12, 15, 26, 43	
Key Idea 5: Organisms maintain a dynamic equilibrium that sustains life.	7, 13, 20, 23, 37, 38, 49, 51, 62, 63	
Key Idea 6: Plants and animals depend on each other and their physical environment.	22, 28, 34, 35, 65, 66, 67	
Key Idea 7: Human decisions and activities have a profound impact on the physical and living environment.	2, 24, 27, 30, 39, 52, 64, 68, 69, 70, 71, 72	

STANDARDS/KEY IDEAS	JUNE 2016 QUESTION NUMBERS	NUMBER OF CORRECT RESPONSES
REQUIRED LABORATORIES		
Lab 1: "Relationships and Biodiversity"	76, 77	
Lab 2: "Making Connections"	83, 84, 85	
Lab 3: "The Beaks of Finches"	78, 79, 80, 81, 82	
Lab 5: "Diffusion Through a Membrane"	73, 74, 75	

Examination August 2016
Living Environment

PART A

Answer all questions in this part. [30]

Directions (1–30): For *each* statement or question, record in the space provided the *number* of the word or expression that, of those given, best completes the statement or answers the question.

1 Which sequence represents structures organized from most complex to least complex?

1 chloroplast → guard cell → leaf → oak tree
2 guard cell → chloroplast → leaf → oak tree
3 oak tree → guard cell → leaf → chloroplast
4 oak tree → leaf → guard cell → chloroplast

1 _____

2 Autotrophs differ from heterotrophs in that only autotrophs

1 require carbon dioxide for cellular respiration
2 release oxygen as a product of cellular respiration
3 synthesize nutrients using carbon dioxide and water
4 break down sugars to assemble other molecules

2 _____

3 Burmese pythons are large snakes that have been introduced into the Florida Everglades ecosystem. Burmese pythons and alligators hunt the same prey. One likely effect of the introduction of the pythons is that

1 alligators will have more prey available
2 pythons will become native to the Everglades
3 alligator populations will decline
4 pythons will become an endangered species   3 _____

4 Which activity enables humans to produce new genetic combinations in other organisms?

1 selecting and breeding the organisms for specific traits
2 increasing the number of enzymes available to the organisms
3 growing organisms that reproduce asexually
4 decreasing the amount of DNA in the diet of the organisms   4 _____

5 Before a cell divides, an exact copy of each chromosome is made by the process of

1 genetic engineering
2 replication
3 mutation
4 recombination   5 _____

6 Some time ago, there were thousands of California condors in North America. Large numbers were poisoned from lead in bullets that were used to kill the animals the condors fed on. An effort was made to help save this large scavenger. There are now more than 350 California condors in North America. The condors most likely increased in number because humans decided to

1 produce lead-resistant condors through asexual reproduction
2 pass laws against using lead bullets to kill animals used by condors for food
3 introduce plants that didn't absorb the lead from discharged bullets
4 produce lead-resistant prey for the condors through genetic engineering   6 _____

7 In humans, the maintenance of a stable internal temperature is a direct result of

1 detection of and reaction to stimuli in the environment
2 digestion of starches and absorption of protein from the internal environment
3 diffusion of water and excretion of glucose to the external environment
4 transport of ATP and locomotion through the environment   7 _____

8 Which molecules are needed to cut and copy segments of DNA?

   1 reproductive hormones
   2 carbohydrates
   3 antibodies
   4 biological catalysts   8 _____

9 Evolution can occur at different rates; however, for evolution to occur, there must be

   1 variations within a species
   2 extinction of the species
   3 asexual reproduction
   4 no change in the genes of an organism   9 _____

10 The two reactions illustrated in the diagrams below often occur when a foreign substance enters the body.

Foreign substance   Specific proteins

Cell A   **Cell B**   produces

The cells labeled A and B are examples of cells known as

   1 guard cells   3 white blood cells
   2 reproductive cells   4 specialized skin cells   10 _____

11 The diagram below represents a protein molecule present in some living things.

This type of molecule is composed of a sequence of

1 amino acids arranged in a specific order
2 simple sugars alternating with starches arranged in a folded pattern
3 large inorganic subunits that form chains that interlock with each other
4 four bases that make up the folded structure   11 _____

12 Three human actions that have been made possible in recent times are:

- Doctors are able to diagnose and treat some fetal problems prior to the birth of a child.
- Cloning can produce large numbers of plants that are resistant to drought.
- Male insects can be sterilized with radiation to prevent them from mating successfully.

Which statement summarizes these three actions?

1 Reproductive technology has medical, agricultural, and ecological applications.
2 Development is a highly regulated process involving mitosis and differentiation.
3 Reproduction and development are subject to environmental effects.
4 Human development, birth, and aging should be viewed as a predictable pattern of events.   12 _____

13 Natural selection is best described as

1 a change in an organism in response to a need of that organism
2 a process of nearly constant improvement that leads to an organism that is nearly perfect
3 differences in survival rates as a result of different inherited characteristics
4 inheritance of characteristics acquired during the life of an organism   13 _____

14 Which statement best describes a situation where competition occurs in an ecosystem?

   1 A deer outruns an attacking wolf.
   2 A deer, during the winter, consumes tree bark.
   3 A deer and a rabbit consume grass in a field.
   4 A deer and a rabbit are both startled by a hawk flying overhead.   14 _____

15 A woman changes her hair color to red; however, her children will not inherit this red hair color because the woman does not have

   1 genes for red hair in her skin
   2 genes for red hair in her sex cells
   3 proteins for red hair in the placenta
   4 proteins for red hair in her egg cells   15 _____

16 Fossils provide evidence that

   1 life on Earth millions of years ago was more complex than life is today
   2 the changes that will occur in species in the future are easy to predict
   3 many species of organisms that lived long ago are now extinct
   4 most species of organisms that lived long ago are exactly the same today   16 _____

17 A male frigatebird displays to the female by inflating its large red throat sac, throwing its head back, vibrating its wings, and producing a "drumming" sound with its throat sac. For the frigatebird, this behavior has most likely resulted in

   1 hiding from predators
   2 greater reproductive success
   3 locating new sources of food
   4 reduced population growth   17 _____

18 Plant species *X* lives in a hot, dry environment. Slowly, over hundreds of years, the climate becomes wetter. Fungi attack species *X* and cause the population of species *X* to decrease. However, plant species *X* could survive if the plants

1 try to mutate quickly and synthesize new proteins
2 are watered often and fertilized with extra nutrients
3 can adapt to the new conditions by mating with the fungus
4 have a few members of the population that are fungus-resistant　　　　　18 ____

19 The diagram below represents reproduction in a yeast cell. The genes in the bud are identical to the genes in the parent.

This type of production of offspring is a form of

1 sexual reproduction
2 asexual reproduction
3 gene manipulation
4 genetic engineering　　　　　19 ____

20 A human cell that contains all of the information
    necessary for the growth and development of a com-
    plete organism is

    1 a sperm cell   3 a zygote
    2 a gamete   4 an egg cell   20 _____

21 When would exposure to a potentially harmful sub-
    stance be most likely to damage many organs in a
    developing embryo?

    1 during the last three months of pregnancy
    2 during the early stages of pregnancy
    3 during the formation of the zygote
    4 during meiosis in both males and females   21 _____

22 The human female reproductive system is repre-
    sented below.

    Within which structure does the placenta normally
    develop?

    1 A   3 C
    2 B   4 D   22 _____

23 An energy-rich organic compound needed by organisms is

    1 water   3 oxygen
    2 salt   4 glucose   23 _____

24 SCIDS (Severe Combined Immunodeficiency Syndrome) is a disorder where a genetic mutation inhibits the production and functioning of T-cells. T-cells are special types of white blood cells that play a role in the body's immune response. A possible symptom of SCIDS would be an increase in the

    1 number of antigens produced
    2 red blood cell count
    3 number of infections by pathogens
    4 ability to maintain homeostasis   24 _____

25 An organ, such as a kidney, used for transplant needs to be tested for compatibility with the person who is to receive the organ. If this is not done, the

    1 donated organ might attack the body
    2 donated organ might attack the immune system
    3 immune system might attack its own body cells
    4 immune system might attack the donated organ 25 _____

26 A researcher concludes from a 10-year study that the biodiversity of an ecosystem had increased. Which set of observations represents evidence for this claim?

    1 There were more niches and greater stability in the ecosystem.
    2 There were more niches and less energy lost as heat in the ecosystem.
    3 There were fewer niches for decomposers and less stability.
    4 There were fewer niches for consumers and greater cycling of materials.   26 _____

27 The final consumers in many food webs are

1 autotrophs   3 herbivores
2 hosts   4 carnivores   27 _____

28 Abandoned farmland that once grew corn is now covered with bushes and small trees. These observed changes resulted directly from

1 evolutionary change
2 ecological succession
3 loss of biodiversity
4 selective breeding   28 _____

29 The overuse of chemical fertilizers has resulted in the growth of some lawns in which decomposers cannot live. This would interfere most directly with the ability of the lawn ecosystem to

1 recycle energy
2 recycle nutrients
3 maintain atmospheric pH
4 reduce biodiversity   29 _____

30 On Long Island, several businesses use geothermal technology. This involves taking heat from within Earth and using it to heat buildings. One benefit of this technology is that it

1 contributes to global warming
2 reduces the ozone shield
3 reduces dependence on fossil fuels
4 decreases resources for many species   30 _____

PART B-1

Answer all questions in this part.   [13]

Directions (31–43): For *each* statement or question, record in the space provided the *number* of the word or expression that, of those given, best completes the statement or answers the question.

31 Recently, researchers from Stanford University have changed mouse skin cells into mouse nerve cells. This was accomplished by inserting genes that control the synthesis of certain proteins into the skin cells. This type of research is often successful in advancing knowledge regarding the functioning of human cells because

  1 cells present in humans often function in similar ways to cells present in other organisms

  2 cells from different types of organisms function differently when transplanted into humans

  3 the cells in all complex organisms contain the same genes and function in similar ways

  4 cellular research using mice can always be applied to human cells since all complex organisms produce the same proteins   31 _____

32 In the experimental setup below, which substance would be used to prove that the gas produced by the yeast in the vacuum bottle could change the pH of the liquid in the flask?

1 an indicator
2 a chemical messenger
3 an enzyme
4 a salt solution   32 _____

33 Which statement best expresses a basic scientific assumption?

1 Interpretation of experimental results has provided explanations for all natural phenomena.
2 If a conclusion is valid, similar investigations by other scientists should result in the same conclusion.
3 For any conclusion to be valid, the design of the experiment requires that only two groups be compared.
4 After a scientist formulates a conclusion based on an experiment, no further investigation is necessary.   33 _____

34 The diagram below represents a cycle that occurs in nature.

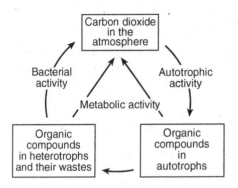

Which phrase describes a human activity that could have a *negative* effect on this cycle?

1 a decrease in the amount of sulfates given off by motor vehicles

2 an increase in recycling programs for plastics and metals

3 the continued deforestation and removal of forest resources

4 development of programs to conserve wildlife   34 _____

Base your answers to questions 35 through 37 on the information and graph below and on your knowledge of biology.

A farmer growing potatoes notices aphids, a type of insect, feeding on the plants. An insecticide was sprayed on the plants several times over a two-year period. The graph represents samples of three different generations of insecticide-resistant and non-resistant aphids over this time period.

Generations of Aphids

35 The resistance gene was present in the aphid population as a result of

1 the need of the potatoes to become resistant to the insecticide

2 changes in the aphids' local habitat by the insecticide

3 a recombination of the proteins in the potato cells

4 a random change in the aphids' DNA sequence   35 _____

36 In year three, the farmer discontinued the use of the insecticide. Which statement would best predict the population in generation 4?

1 The nonresistant aphid would become extinct.

2 The nonresistant aphid population would likely increase.

3 The resistant aphid would mutate to a nonresistant aphid.

4 The plants would be free of insect populations.   36 _____

37 One *negative* consequence of using an insecticide is that it

1 selects for insecticide-resistant organisms

2 keeps a balance of organic compounds

3 encourages biodiversity in plants

4 gives the nonresistant aphids a survival advantage   37 _____

38 The activity of a single-celled organism is represented in the diagram below.

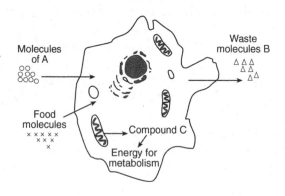

Which concept is best illustrated by this diagram?

1 The life functions performed by single-celled organisms are different from the life functions performed by complex organisms.
2 Single-celled organisms carry out life functions that are essential for survival.
3 Since single-celled organisms lack organs, they can survive only in moist environments.
4 Single-celled organisms contain one organelle that performs all the life functions.

38 _____

39 The chart below provides information about two scientific discoveries in the field of biology.

Early Discovery	Later Discovery
People living near swamps are more likely to get malaria than people who do not live near swamps. Burning swamps early in the summer reduces the amount of malaria.	Mosquitoes breed and lay their eggs in swamps and other pools of still water. Mosquitoes are the carriers of the organisms that cause malaria.
Dark-staining bodies called chromosomes can be seen only in dividing cells. The number of chromosomes doubles during cell division.	Chromosomes contain DNA, which is able to copy itself. DNA carries the genetic code, which is passed from a parent cell to two or more daughter cells.

Which statement is the best interpretation of the material presented in the chart?

1 Scientific explanations are built by combining evidence that can be observed with what people already know.

2 Inquiry involves making judgments about the reliability of the source and relevance of the information.

3 Science provides information, but values are also essential to making ethical decisions.

4 Hypotheses are valuable even if they turn out not to be true, because they may lead to further investigation.   39 _____

Base your answers to questions 40 through 43 on the diagram below and on your knowledge of biology. This diagram represents the roles of different parts of the human body in keeping blood sugar at a balanced, normal level over time.

Homeostasis of Blood Sugar Level

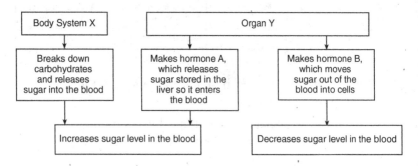

40 The diagram shows human body structures that are coordinated to maintain homeostasis. Which row correctly identifies the functions of these structures?

Row	Body System X	Organ Y
(1)	Digestion	Regulation
(2)	Circulation	Synthesis
(3)	Excretion	Transport
(4)	Locomotion	Nutrition

41 When body system X releases too much sugar into the blood, the body can maintain homeostasis by making

  1 more hormone A, only
  2 more hormone B, only
  3 more hormone A and more hormone B
  4 no hormone A and no hormone B 41 _____

42 If organ Y becomes unable to produce enough hormone B, then homeostasis would be disrupted. To restore homeostasis and compensate for the lack of hormone B, one useful action would be to

  1 increase the production of hormone A
  2 remove organ Y from the body surgically
  3 reduce the carbohydrates in body system X
  4 reduce the synthesis of enzymes in organ Y 42 _____

43 If body system X temporarily stops releasing sugar into the blood, a likely response of the body would be to

  1 stop using enzymes in body system X
  2 stop organ Y from producing hormone A
  3 start to increase synthesis of hormone B
  4 start to increase synthesis of hormone A 43 _____

PART B–2

Answer all questions in this part.   [12]

Directions **(44–55): For those questions that are multiple choice, record in the space provided the *number* of the choice that, of those given, best completes each statement or answers each question. For all other questions in this part, follow the directions given and record your answers in the spaces provided.**

Base your answers to questions 44 through 47 on the information, diagram, and table below and on your knowledge of biology.

A concentrated starch solution was placed in a thistle tube with a semi-permeable membrane covering the wide opening. It was then placed in a beaker of water. The height of the solution in the tube was measured every 5 minutes for 25 minutes. The setup and the data collected are shown below.

Height of Liquid in Thistle Tube

Time (min)	Height (cm)
0	2
5	3
10	6
15	8
20	10
25	11

Directions **(44–45): Using the information in the data table, construct a line graph on the grid below, following the directions below.**

44 Mark an appropriate scale, without any breaks in the data, on each labeled axis. [1]

45 Plot the data for height on the grid. Connect the points and surround each point with a small circle. [1]

Example:

Height of Liquid in Thistle Tube

Height (cm)

Time (min)

46 Explain why the height of the solution in the thistle
tube increased during the 25-minute period. [1]

_____

_____

**Note: The answer to question 47 should be
recorded in the space provided.**

47 The experiment was repeated, and an amber-
colored solution was added to the water in the
beaker. After 10 minutes, the water in the beaker
remained amber-colored and the starch solution
had turned blue-black. The most likely reason for
this observation is that

  1 starch molecules moved out of the thistle tube
  2 water molecules moved into the thistle tube
  3 amber-colored solution moved into the thistle
    tube
  4 water molecules moved out of the thistle tube   47 _____

Base your answer to question 48 on the information and diagrams below and on your knowledge of biology.

When fish of certain species are injured, a chemical substance stored in skin cells of the fish is released into the water. This chemical causes an alarm response among other fish of the same species in the area. Nearby fish of this species become more alert and group together near the bottom.

No Alarm Response   **Alarm Response**

48 Explain why the chemical released from the injured fish may *not* cause an alarm response in other fish species. [1]

_____

_____

Base your answers to questions 49 through 51 on the diagram below and on your knowledge of biology.

Note: The answer to question 49 should be recorded in the space provided.

49 The process represented in the diagram is

  1 DNA replication   3 gel electrophoresis

  2 natural selection   4 genetic engineering   49 _____

Note: The answer to question 50 should be recorded in the space provided.

50 The original gene for the production of a human hormone was most likely removed from a

  1 chromosome   3 mitochondrion

  2 ribosome   4 cell membrane   50 _____

51 State *one* possible reason why a gene for the production of a human hormone would be placed in bacterial DNA. [1]

_____

_____

Base your answers to questions 52 through 55 on the passage below and on your knowledge of biology.

The lake sturgeon is a fish that often grows over six feet long and can weigh close to two hundred pounds. It is currently an endangered species in the Great Lakes area, although the species has lived in those lakes and rivers for millions of years. Now, there is a program to increase the sturgeon population by reintroducing lake sturgeon to areas where they have disappeared.

Like the lake sturgeon, bloater fish are also found in the Great Lakes. Both find their food on or near the bottoms of lakes. They eat a variety of small organisms, including insect larvae, worms, and clams. These small organisms feed on algae.

52 Identify *one* population that will *decrease* in size after the lake sturgeon are added to the new ecosystems. Support your answer. [1]

_____

_____

53 Part of the food web of a lake ecosystem is represented in the diagram below. Indicate the position in the food web where each organism listed below would be placed, by writing the name of each in the appropriate box. [1]

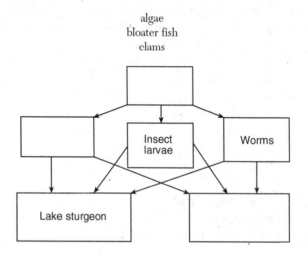

algae
bloater fish
clams

54 Identify which population, other than lake sturgeon, will increase in size after the lake sturgeon are added to the new ecosystems. Support your answer. [1]

_____

_____

55 State what the arrows in the food web represent. [1]

_____

_____

PART C

Answer all questions in this part.   [17]

Directions (56–72): **Record your answers in the spaces provided.**

Base your answers to questions 56 and 57 on the information and drawing below and on your knowledge of biology. The drawing represents a salamander. Salamanders are small amphibians that live in a variety of environments.

Two species of salamander inhabit an island. The habitat on each side of the island is different. One side tends to be wet; the other side tends to be dry. Researchers want to know if the salamanders will survive equally well on either side of the island. Species *A* lives on the wet side of the island, while Species *B* lives on the dry side of the island. Researchers develop two artificial habitats, one that simulates conditions on the wet side and one that simulates conditions on the dry side.

56 Explain why researchers would put the salamanders in an artificial environment, as opposed to conducting the experiment in their natural habitat. [1]

_____

_____

57 Researchers put three salamanders of each species in each of the two different artificial environments. Why would other scientists question the validity of the conclusions based on this setup? [1]

_____

_____

Base your answers to questions 58 through 61 on the information and diagram below and on your knowledge of biology.

Green Roofs

People in Albany and New York City are using "green roofs" to improve the environment. A green roof can be added to many buildings that have large, flat roofs. Green roofs have three parts: a protective layer to separate plant roots from the roof of the building, an absorptive layer to catch and hold rainwater, and a layer of plants. Often, green roofs use _Sedum_, a short, desert plant, because it is efficient at storing water in its leaves and can withstand the colder climate.

A green roof saves energy, reduces carbon dioxide in the atmosphere, and prevents rainwater and melting snow from overloading sewer systems. It can also protect the roof of a building from damage. However, green roofs can be expensive to install, and require care and maintenance.

58 Most varieties of *Sedum* are not native plants in Albany or New York City. State *one* reason why it may be dangerous to introduce a new species to an established ecosystem. [1]

_____

_____

59 State *one* reason why a green roof reduces the amount of carbon dioxide in the atmosphere. [1]

_____

_____

60 State *one* reason why it is important to reduce the amount of carbon dioxide in the atmosphere. [1]

_____

_____

61 State *one* disadvantage of a green roof. [1]

_____

_____

_____

62 A scientist took samples from a culture of *E. coli* bacteria and placed them in each of 100 petri dishes. Once the bacteria began to grow in the dishes, she exposed 50 of the dishes to x-ray radiation and 50 to natural light. After five days, she examined samples of DNA from the bacteria and recorded any differences she found between the DNA of the two groups. State *one* hypothesis the experiment would test. [1]

_____

_____

Base your answers to questions 63 and 64 on the passage below and on your knowledge of biology.

Super Vaccine Could Eliminate Flu

Every flu season, vaccine makers must bet on which strain of influenza A will pose the greatest threat to the public, and millions of Americans must decide whether to get a shot. In August, virologist Gary Nabel at the National Institutes of Health (NIH) announced progress toward a universal flu vaccine: two shots of it could provide years of protection from every known influenza A virus.

"We use a prime-boost strategy, meaning that we immunize with two vehicles that deliver the vaccine in different ways," Nabel says. In their experimental treatment, he and his colleagues injected mice, ferrets, and monkeys with viral DNA, causing their muscle cells to produce hemagglutinin, a protein found on the surface of all flu viruses. The animals' immune systems then began making antibodies that latch onto the protein and disable the virus. The researchers followed the DNA injection with a traditional seasonal flu shot, which contains dead viruses. This one-two punch protected the test subjects against influenza A viruses that had emerged in 1934 and 2007, and other experiments showed that the antibodies it generated successfully neutralized a wide variety of flu strains. Nabel's colleagues at the NIH are already testing similar approaches in humans.

Source: Rowe, A. "Super Vaccine Could Eliminate Flu." Discover, Jan./Feb. 2011, p. 37.

63 Identify *one* specific difference, other than it is a two-step vaccination, between Nabel's vaccination and a traditional flu vaccine. [1]

_____

_____

64 Explain how injecting dead or weakened viruses into a person can help to fight against future infections from that virus. [1]

_____

_____

_____

Base your answers to questions 65 through 68 on the information below and on your knowledge of biology.

Bald Eagle Facts

- Bald eagles eat primarily fish, carrion (dead animals), smaller birds, and rodents. Their most important non-carrion food is fish, which they catch by swooping down and grabbing fish that are near the surface of the water.
- The number of nesting pairs in the lower 48 United States increased from fewer than 450 in the early 1960s to more than 4,500 adult bald eagle nesting pairs in the 1990s. Today, there are an estimated 9,789 nesting pairs of bald eagles.

- Bald eagles are found in large numbers in certain areas during the winter (known as roosts). These winter roosts are located in areas where prey are plentiful. Winter roosts are protected under federal law, and managed with a buffer zone to reduce human disturbance. As winter ends, the eagles return to their summer nesting/hunting areas.

Bald Eagle Research

In the winter of 2009, volunteers from an Audubon group conducted a survey of roosting bald eagles at four locations in an area in the lower Hudson River Valley. The data below show the average number of eagles sighted and the number of visits made by the volunteers each month. Among the other data collected were percent ice cover and percent cloud cover on the surface of the water. The eagles fly freely between these four sites, depending on a variety of conditions.

Some of the data are shown in the table below.

Bald Eagles Sighted at Four Hudson Valley Locations in 2009

Location	January		February		March	
	Average Number of Eagles	Number of Visits	Average Number of Eagles	Number of Visits	Average Number of Eagles	Number of Visits
Croton Reservoir	22.86	7	47.88	8	9.17	6
George's Island Park	27.00	7	18.38	8	5.00	4
George's Island North	12.29	7	4.43	7	2.20	5
Stony Point	3.57	7	3.63	8	0.00	5

65 State *one* reason why the percent ice cover is important to the ability of eagles to obtain food. [1]

_____

_____

66. What inference can be made about the percent ice cover at Croton Reservoir between January and February 2009? Support your answer. [1]

_____

_____

67 State *one* reason why the number of eagles sighted showed a change at all four sites between February and March. [1]

_____

_____

68 State *one* possible reason why a popular hiking trail in this area is closed during eagle roosting seasons. [1]

_____

_____

Base your answers to questions 69 through 72 on the information below and on your knowledge of biology.

Coral Reef Ecosystems

There are many ecological interactions that maintain the biodiversity present in coral reefs. In addition to coral, microscopic algae, seaweed, sea grasses, sponges and worms, and a variety of fish are among the organisms that live in reef ecosystems. Ocean currents often link different reef systems and move organisms from one reef area to another. This movement is a factor in repopulating a reef that has been damaged by environmental changes.

One environmental change involves an increased growth of seaweed. When the population of seaweed increases, the reef shifts from a coral-dominated ecosystem to a seaweed-dominated ecosystem. This change disrupts the relationships between the organisms that live there.

Studies have shown that, as the density of seaweed in a reef area increases, the number of fish that eat the seaweed in that area decreases. This may be due to the presence of more predators, or the taste of the more mature plants. The fish move to areas where there is less seaweed growth. As this trend continues, the reef areas are taken over by the seaweed. Once this happens, it is very hard to remove the seaweed and restore the reef to a healthy ecosystem.

In addition to this problem, temperature changes are threatening the ocean currents that connect the reef systems. A change in the currents would reduce the movement of fish larvae from one area to another. This contributes to the seaweed problem.

69 State the role of the sea grasses in the reef ecosystem.
    [1]

    _____

    _____

70 Identify *one* abiotic factor that is affecting the sta-
    bility of the coral reef ecosystems and state how the
    factor identified is important to the coral reef eco-
    systems. [1]

    Abiotic factor:_____

    Effect:_____

    _____

71 State *one* reason why it is important to maintain the
    stability of the coral reefs. [1]

    _____

    _____

72 State *one* advantage of the fish larvae moving by
    ocean currents into a damaged reef ecosystem. [1]

    _____

    _____

PART D

Answer all questions in this part.   [13]

Directions **(73–85): For those questions that are multiple choice, record in the space provided the *number* of the choice that, of those given, best completes the statement or answers the question. For all other questions in this part, follow the directions given and record your answers in the spaces provided.**

Base your answers to questions 73 and 74 on the diagram below, which represents a process that occurs in living cells, and on your knowledge of biology.

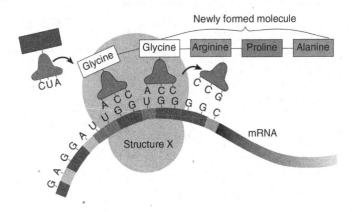

Note: The answer to question 73 should be recorded in the space provided.

73 The process shown in the diagram is

    1 cellular respiration
    2 cellular reorganization
    3 gene recombination
    4 protein synthesis

73 _____

Note: The answer to question 74 should be recorded in the space provided.

74 Structure *X* is a

    1 mitochondrion   3 nucleus

    2 vacuole   4 ribosome   74 _____

Base your answers to questions 75 through 77 on the diagram below and on your knowledge of biology.

Variations in Beaks of Galapagos Islands Finches

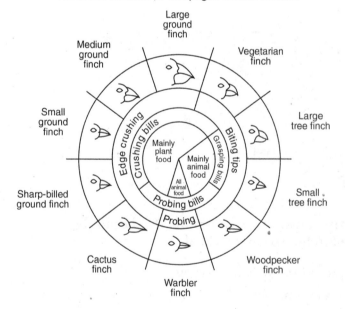

Note: The answer to question 75 should be recorded in the space provided.

75 A finch that picks small insects out from cracks in the bark of trees would most likely have a beak that is

1 sharp and thin   3 rounded and thin
2 sharp and thick   4 rounded and thick   75 _____

Note: The answer to question 76 should be recorded in the space provided.

76 Which statement is a basic assumption from *The Beaks of Finches* lab?

1 The type of beak indicates the type of food the finch eats.
2 Different birds have different songs.
3 Birds with larger beaks can find mates more easily.
4 Nesting behavior of finches is an inherited trait.   76 _____

77 State *two* reasons why the large ground finch and sharp-billed ground finch could live on the same island but not compete for food, even though they both eat mainly plant food. [1]

Reason 1: _____

Reason 2: _____

Base your answer to question 78 on the table below and on your knowledge of biology. The table shows which of four enzymes are present in three related plant species.

Comparison of Four Enzymes

Plant Species	Enzyme W	Enzyme X	Enzyme Y	Enzyme Z
Species A	present	present	absent	present
Species B	absent	absent	present	absent
Species C	present	present	absent	present

The tree diagrams below show two possible evolutionary relationships between the three species.

78 In the space below, write the number of the tree diagram that shows the most probable evolutionary relationship between the three species. Support your answer. [1]

Tree diagram: _____

_____

_____

Base your answers to questions 79 and 80 on the diagram below, which represents the shrinking of a cell in response to an increase in the concentration of a substance outside of the cell.

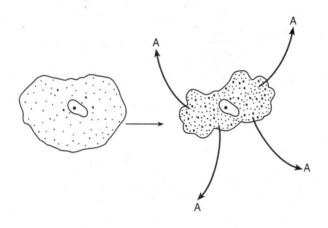

79 Identify substance A. [1]

_____

80 Identify *one* likely substance in the environment of the cell that caused this response. [1]

_____

Note: The answer to question 81 should be recorded in the space provided.

81 A student lifted weights after school and found that his muscles started to burn. He couldn't continue to lift the weights after prolonged exercising. This muscle fatigue is most likely due to

1 the heart beating too fast and tiring out
2 the lungs accumulating oxygen
3 lack of oxygen and build up of waste in the muscles
4 lack of carbon dioxide in the muscles   81____

Base your answers to questions 82 and 83 on the diagram below and on your knowledge of biology. The diagram represents three sections of a cell membrane showing three different methods involved in the transport of various molecules across the membrane.

Method A	Method B	Method C

Note: The answer to question 82 should be recorded in the space provided.

82 Methods *A* and *B* are classified as methods of passive transport because they do not require

1 ATP   3 light
2 carbon dioxide   4 DNA   82 _____

83 Using information from the diagram, state *one* reason why the movement of molecules in method *C* represents active transport. [1]

_____

_____

_____

84 State *one* reason why some species might have similar body structures even if they are *not* closely related. [1]

_____

_____

85 A student went out to the school track and walked two laps, ran two laps, and then walked two more laps. On the grid below, draw a line that shows what most likely happened to the pulse rate of the student during these activities. [1]

Effect of Activity on Pulse Rate

Answers
August 2016

Living Environment

Answer Key

PART A

1. 4	**6.** 2	**11.** 1	**16.** 3	**21.** 2	**26.** 1
2. 3	**7.** 1	**12.** 1	**17.** 2	**22.** 4	**27.** 4
3. 3	**8.** 4	**13.** 3	**18.** 4	**23.** 4	**28.** 2
4. 1	**9.** 1	**14.** 3	**19.** 2	**24.** 3	**29.** 2
5. 2	**10.** 3	**15.** 2	**20.** 3	**25.** 4	**30.** 3

PART B–1

31. 1	**34.** 3	**36.** 2	**38.** 2	**40.** 1	**42.** 3
32. 1	**35.** 4	**37.** 1	**39.** 1	**41.** 2	**43.** 4
33. 2					

PART B–2

44. *See* Answers Explained.
45. *See* Answers Explained.
46. *See* Answers Explained.
47. 3
48. *See* Answers Explained.
49. 4

50. 1
51. *See* Answers Explained.
52. *See* Answers Explained.
53. *See* Answers Explained.
54. *See* Answers Explained.
55. *See* Answers Explained.

PART C. *See* **Answers Explained**.

PART D

73. 4
74. 4
75. 1
76. 1
77. *See* Answers Explained.
78. *See* Answers Explained.
79. *See* Answers Explained.

80. *See* Answers Explained.
81. 3
82. 1
83. *See* Answers Explained.
84. *See* Answers Explained.
85. *See* Answers Explained.

Answers Explained

PART A

1. **4** The sequence *oak tree → leaf → guard cell → chloroplast* is the one that represents structures organized from most complex to least complex. An oak tree is a living organism composed of millions of cells. A leaf is an organ of the tree composed of significantly fewer cells than the tree itself. A guard cell is one type of cell in a leaf and is specialized to regulate gas flow into and out of the leaf. A chloroplast is a cell organelle located within the guard cell that is specialized to carry out the photosynthetic reactions. Each of the structures in this sequence is less complex than the one that precedes it.

WRONG CHOICES EXPLAINED

(1), (2), (3) The sequences *chloroplast → guard cell → leaf → oak tree*, *guard cell → chloroplast → leaf → oak tree*, and *oak tree → guard cell → leaf → chloroplast* are *not* the sequences that represent structures organized from most complex to least complex. Each of these sequences has at least one structure out of order compared to the correct answer above.

2. **3** Autotrophs differ from heterotrophs in that only autotrophs *synthesize nutrients using carbon dioxide and water*. Autotrophs (also known as self-feeders) are organisms that carry on biochemical reactions (photosynthesis) that reorganize inorganic compounds (carbon dioxide and water) into organic nutrient molecules (glucose) using the energy of light to power the reactions. Heterotrophs (also known as other-feeders) are incapable of carrying on photosynthesis. Instead, they must obtain nutrients from the bodies of other living things.

WRONG CHOICES EXPLAINED

(1) It is *not* true that autotrophs differ from heterotrophs in that only autotrophs *require carbon dioxide for cellular respiration*. Carbon dioxide is a by-product, not a requirement, of cellular respiration.

(2) It is *not* true that autotrophs differ from heterotrophs in that only autotrophs *release oxygen as a product of cellular respiration*. Oxygen gas is a requirement, not a by-product, of cellular respiration.

(4) It is *not* true that autotrophs differ from heterotrophs in that only autotrophs *break down sugars to assemble other molecules*. Both autotrophs and heterotrophs reorganize the atoms of sugar molecules to synthesize other molecules.

3. **3** One likely effect of the introduction of the pythons is that *alligator populations will decline.* Pythons and alligators are predators that hunt the same prey. So pythons and alligators are in competition for that same food source. As long as that prey is abundant, completion will be minimal and there will be no immediate effect on either pythons or alligators. If the prey becomes scarce, however, pythons and alligators will compete more directly and intensely. The more successful predator organism will gradually reduce the population of the less successful predator organism. It is unclear whether alligators or pythons will prove to be the more successful predator.

[**Note:** There is not enough information provided in the question to reach any conclusive answer as to the most likely outcome of this scenario.]

WRONG CHOICES EXPLAINED

(1) It is *not* true that one likely effect of the introduction of the pythons is that *alligators will have more prey available.* As a result of interspecies competition, alligators (and pythons) will likely have less, not more, prey available.

(2) It is *not* true that one likely effect of the introduction of the pythons is that *pythons will become native to the Everglades.* The python is an invasive species native to Southeast Asia. So by definition, it cannot be native to North America.

(4) It is *not* true that one likely effect of the introduction of the pythons is that *pythons will become an endangered species.* The python appears to be an exceptionally successful invasive species in the Everglades. It is unlikely that it will become an endangered species unless drastic environmental change occurs in the Everglades that limits the population of pythons there.

4. **1** *Selecting and breeding the organisms for specific traits* is the activity that enables humans to produce new genetic combinations in other organisms. This process, known as selective breeding, is a technique used by animal and plant breeders in which organisms with desirable traits are cross-bred with the hope of producing offspring that also display those traits.

WRONG CHOICES EXPLAINED

(2) *Increasing the number of enzymes available to the organisms* is *not* the activity that enables humans to produce new genetic combinations in other organisms. Enzymes are produced in the cells of living organisms as directed by each organism's specific genetic blueprint, not by human activity.

(3) *Growing organisms that reproduce asexually* is *not* the activity that enables humans to produce new genetic combinations in other organisms. Asexual reproduction limits the production of new genetic combinations in organisms that carry it out.

(4) *Decreasing the amount of DNA in the diet of the organisms* is *not* the activity that enables humans to produce new genetic combinations in other organisms. DNA is an organic molecule that is present in all natural

organic foods. DNA in the diet does not affect the production of new genetic combinations.

5. **2** Before the cell divides, an exact copy of each chromosome is made by the process of *replication*. Replication is the exact self-duplication of genetic material (DNA) in the cell prior to cell division. The process of replication ensures that the same number of genes and the same amount of DNA are present in the daughter cells as were present in the parent cell.

WRONG CHOICES EXPLAINED
(1) It is *not* true that before the cell divides, an exact copy of each chromosome is made by the process of *genetic engineering*. Genetic engineering is a laboratory technique in which a gene for a desired trait is snipped from the DNA of a donor cell and inserted into the genome of a recipient cell. Genetic engineering is not the mechanism responsible for copying genetic material in the cell.
(3) It is *not* true that before the cell divides, an exact copy of each chromosome is made by the process of *mutation*. Mutation is a random process by which genetic material is altered by the effects of mutagenic agents. Mutation is not the mechanism responsible for copying genetic material in the cell.
(4) It is *not* true that before the cell divides, an exact copy of each chromosome is made by the process of *recombination*. Sexual reproduction involves the recombination of genes in offspring through the processes of meiosis and fertilization. Recombination is not the mechanism responsible for copying genetic material in the cell.

6. **2** The condors most likely increased in number because humans decided to *pass laws against using lead bullets to kill animals used by condors for food*. Condors are scavengers that feed on the carcasses of deer and other game animals, some of which are killed by lead bullets commonly used by human hunters. Condors have been known to accidentally ingest spent lead bullets along with the meat of these carcasses. Scientific studies of free-flying condors in California have found some birds with spent lead bullets in their digestive tracts and high levels of lead in their blood. The lead in the spent bullets was thought to have been absorbed into the blood of the condors, causing lead poisoning and eventual death. Passing laws requiring the use of lead-free ammunition is one way to help resolve this problem. This is an example of a trade-off between protecting a wild species (condors) and regulating a human activity (hunting).

WRONG CHOICES EXPLAINED
(1) It is *not* true that the condors most likely increased in number because humans decided to *produce lead-resistant condors through asexual reproduction*. Lead is a naturally occurring metal that can produce adverse effects in all living things exposed to it. Other than limiting exposure to the metal, there is

no practical laboratory procedure to produce lead resistance in condors or in any other species. Condors are birds and as such carry on sexual, not asexual, reproduction.

(3) It is *not* true that the condors most likely increased in number because humans decided to *introduce plants that didn't absorb the lead from discharged bullets*. Condors are scavengers that consume the bodies of dead animals, not plants.

(4) It is *not* true that the condors most likely increased in number because humans decided to *produce lead-resistant prey for the condors through genetic engineering*. Lead is a naturally occurring metal that can produce adverse effects in all living things exposed to it. Other than limiting exposure to the metal, there is no practical laboratory procedure to produce lead resistance in condors or in any other species. The laboratory technique of genetic engineering would have no practical effect on lead toxicity in prey animals and therefore would not reduce lead's toxicity in condors.

7. **1** In humans, the maintenance of a stable internal temperature is a direct result of *detection of and reaction to stimuli in the environment*. The nervous system detects the temperature of the blood (internal environment) and sends command signals to effectors, including the sweat glands and muscles. The effectors act to increase or decrease the temperature of the blood as required to maintain a body temperature of 98.6°F (37°C). This mechanism is known as a feedback loop.

WRONG CHOICES EXPLAINED

(2), (3), (4) In humans, the maintenance of a stable internal temperature is *not* a direct result of *digestion of starches and absorption of protein from the internal environment, diffusion of water and excretion of glucose to the external environment,* or *transport of ATP and locomotion through the environment*. None of these activities would have any direct effect on the maintenance of body temperature in humans. They are all nonsense distracters.

8. **4** *Biological catalysts* are the molecules needed to cut and copy segments of DNA. These catalysts, also known as restriction enzymes, are used by scientists to snip selected genes from donor DNA and insert the genes into recipient DNA. This process transfers genetic traits from one organism to another organism.

WRONG CHOICES EXPLAINED

(1) *Reproductive hormones* are *not* the molecules needed to cut and copy segments of DNA. Reproductive hormones regulate the reproductive processes of humans and other animals. Reproductive hormones are not used in the cutting or copying of DNA.

(2) *Carbohydrates* are *not* the molecules needed to cut and copy segments of DNA. Carbohydrates are a type of organic compound that includes sugars and starches. Carbohydrates are not used in the cutting or copying of DNA.

(3) *Antibodies* are *not* the molecules needed to cut and copy segments of DNA. Antibodies are proteins produced by the body's immune system in response to a specific antigen. Antibodies are not used in the cutting or copying of DNA.

9. **1** For evolution to occur, there must be *variations within a species*. Favorable variations give some individuals an adaptive advantage over others in a changing environment. Individuals with favorable variations will survive to reproduce offspring carrying genes for the same favorable variation. As a result, the frequency of the genes controlling those favorable variations will tend to increase in the gene pool. This action is the basis for biological evolution.

WRONG CHOICES EXPLAINED

(2) It is *not* true that for evolution to occur, there must be *extinction of the species*. Extinction is the complete and permanent elimination of a species from Earth. Extinct species cannot evolve.

(3) It is *not* true that for evolution to occur, there must be *asexual reproduction*. Asexual reproduction is a type of reproduction that results in stable, slowly changing gene pools. Asexually reproducing species can evolve. However, the degree of genetic variation of such species is less than in sexually reproducing species. As a result, evolution of asexually reproducing species is usually slow by comparison to sexually reproducing species.

(4) It is *not* true that for evolution to occur, there must be *no change in the genes of an organism*. If genes are not altered (mutated), there can be no variation within a species. Species whose genes do not change and for which there is no variation cannot evolve.

10. **3** The cells labeled *A* and *B* are examples of cells known as *white blood cells*. White blood cells are components of the blood tissue of humans and other animals. These cells are specialized to protect the body from foreign invaders, such as pathogenic organisms. Cell *A*, known as a phagocyte, is capable of seeking out and engulfing foreign substances in the blood fluid. Cell *B*, known as a lymphocyte, is capable of detecting the antigens on the surfaces of pathogenic organisms and producing specific antibodies to neutralize those pathogens.

WRONG CHOICES EXPLAINED

(1) The cells labeled *A* and *B* are *not* examples of cells known as *guard cells*. A guard cell is one type of cell in a leaf. It is specialized to regulate gas flow into and out of the leaf. Guard cells neither engulf foreign substances nor produce antibodies.

(2) The cells labeled *A* and *B* are *not* examples of cells known as *reproductive cells*. Reproductive cells, such as sperm cells and egg cells, are specialized to carry out fertilization in sexually reproducing species. Reproductive cells neither engulf foreign substances nor produce antibodies.

(4) The cells labeled *A* and *B* are *not* examples of cells known as *specialized skin cells*. Specialized skin cells, such as hair follicles and sweat glands, are specialized to function in the protection and regulation of the body's internal environment. Skin cells neither engulf foreign substances nor produce antibodies.

11. **1** This type of molecule is composed of a sequence of *amino acids arranged in a specific order*. Proteins are complex organic molecules composed of amino acid molecules linked in a particular sequence as determined by the DNA template in the nucleus of a cell.

WRONG CHOICES EXPLAINED
(2), (3), (4) This type of molecule is *not* composed of a sequence of *simple sugars alternating with starches arranged in a folded pattern*, *large inorganic subunits that form chains that interlock with each other*, or *four bases that make up the folded structure*. No such biochemical structures exist in nature. These are nonsense distracters.

12. **1** *Reproductive technology has medical, agricultural, and ecological applications* is the statement that summarizes these three actions. All three of these actions relate to reproductive processes in humans or other organisms. These actions fall into medical ("diagnose and treat fetal problems" or "insects can be sterilized to prevent mating"), agricultural ("produce plants resistant to drought" or "insects can be sterilized to prevent mating"), and ecological (no specific example provided) applications.

WRONG CHOICES EXPLAINED
(2) *Development is a highly regulated process involving mitosis and differentiation* is *not* the statement that summarizes these three actions. No mention is made in the list concerning cell division, development, or differentiation.

(3) *Reproduction and development are subject to environmental effects* is *not* the statement that summarizes these three actions. No mention is made in the list concerning environmental effects on the reproductive process.

(4) *Human development, birth, and aging should be viewed as a predictable pattern of events* is *not* the statement that summarizes these three actions. No mention is made in the list concerning the predictability of human development, birth, or aging.

13. **3** Natural selection is best described as *differences in survival rates as a result of different inherited characteristics*. Natural selection, sometimes referred to as "survival of the fittest," postulates that certain members of a species are better adapted to their environment than others. These better-

adapted members are able to survive adverse conditions at higher rates. They are therefore more likely to pass on their favorable adaptations to succeeding generations than are less-well-adapted members. These factors act together to increase the proportion of individuals displaying highly adaptive traits.

WRONG CHOICES EXPLAINED
(1), (4) Natural selection is *not* best described as *a change in an organism in response to a need of that organism* or *inheritance of characteristics acquired during the life of an organism*. These concepts closely mirror the discredited theory of "acquired characteristics" forwarded by Jean Lamarck in the early 19th century (1801). Natural selection does not occur as a function of need or acquisition of characteristics.

(2) Natural selection is *not* best described as *a process of nearly constant improvement that leads to an organism that is nearly perfect*. Even highly adapted organisms retain characteristics that make them susceptible to drastic changes in environmental conditions. Natural selection does not occur as a function of perfection.

14. **3** *A deer and a rabbit consume grass in a field* is the statement that best describes a situation where competition occurs in an ecosystem. Interspecies competition occurs when members of two different species (deer and rabbits) utilize the same limited resource (grass) in the same environment (field). As long as the resource is abundant, competition is minimized. When the resource becomes scarce, competition intensifies until one of the two species is eliminated from the environment.

WRONG CHOICES EXPLAINED
(1) *A deer outruns an attacking wolf* is *not* the statement that best describes a situation where competition occurs in an ecosystem. This situation is an example of a predator-prey relationship in a natural environment, not interspecies competition.

(2) *A deer, during the winter, consumes tree bark* is *not* the statement that best describes a situation where competition occurs in an ecosystem. This situation is an example of coping behaviors in response to environmental stress, not interspecies competition.

(4) *A deer and a rabbit are both startled by a hawk flying overhead* is *not* the statement that best describes a situation where competition occurs in an ecosystem. This situation is an example of similar stimulus-response behaviors in the deer and rabbit, not interspecies competition.

15. **2** Her children will not inherit this red hair color because the woman does not have *genes for red hair in her sex cells*. Dyeing one's hair a different color is an example of an acquired characteristic. Acquired characteristics cannot be passed on to future generations because these characteristics do not

usually have a genetic basis in the person acquiring them. Only characteristics controlled by genes that occur in sex cells can be passed on to future generations.

WRONG CHOICES EXPLAINED

(1) It is *not* true that her children will not inherit this red hair color because the woman does not have *genes for red hair in her skin*. It is possible that a mutation for red hair color may be present in some of this person's skin cells. However, genes in skin cells cannot be passed on to future generations.

(3), (4) It is *not* true that her children will not inherit this red hair color because the woman does not have *proteins for red hair in the placenta* or *proteins for red hair in her egg cells*. Proteins for red hair color would be present in the skin cells and hair of a person only if the DNA in those cells contains genes for red hair color and if the genes are switched on. These proteins would not be produced in placental cells or in egg cells because the genes for red hair color would be switched off in those cells.

16. **3** Fossils provide evidence that *many species of organisms that lived long ago are now extinct*. Scientific study of fossils reveals many thousands of species that existed in the past that can no longer be found on Earth. It is inferred that these species succumbed to environmental changes that made their survival impossible and, once extinct, can never be recreated.

WRONG CHOICES EXPLAINED

(1) It is *not* true that fossils provide evidence that *life on Earth millions of years ago was more complex than life is today*. Scientific study of fossils reveals a general pattern of increasing complexity from life's origins on Earth 3.7 billion years ago to the present day.

(2) It is *not* true that fossils provide evidence that *the changes that will occur in species in the future are easy to predict*. Because environmental and genetic changes cannot be predicted, making a sound scientific prediction of how evolutionary change will progress in the future is impossible.

(4) It is *not* true that fossils provide evidence that *most species of organisms that lived long ago are exactly the same today*. Scientific study of fossils reveals that modern-day species of organisms are generally very different compared to species that lived millions of years ago.

17. **2** For the frigatebird, this behavior has most likely resulted in *greater reproductive success*. We can infer from the inclusion of both male and female frigatebirds in this scenario that the behaviors described represent a genetically programmed mating display. Such displays are common in males of many bird species. These displays are known to stimulate nesting and mating responses in the female birds. Such mating displays and the resulting mating/ nesting responses help to ensure successful reproduction in, and continuation of, these species.

WRONG CHOICES EXPLAINED

(1), (3) For the frigatebird, this behavior has *not* most likely resulted in *hiding from predators* or *locating new sources of food*. No information is presented in this scenario that would lead a reasonable observer to draw these inferences.

(4) For the frigatebird, this behavior has *not* most likely resulted in *reduced population growth*. These behaviors are likely to result in increased, not decreased, populations of frigatebirds.

18. **4** Plant species *X* could survive if the plants *have a few members of the population that are fungus-resistant*. If fungus-resistant plants mate, their offspring have a high probability of inheriting and displaying the same fungus resistance. Assuming other environmental factors do not intervene, such resistant plants could form a colony that would perpetuate the species in the changing environment.

WRONG CHOICES EXPLAINED

(1) It is *not* true that plant species *X* could survive if the plants *try to mutate quickly and synthesize new proteins*. Mutation is a random event that cannot be controlled at will by species for particular purposes.

(2) It is *not* true that plant species *X* could survive if the plants *are watered often and fertilized with extra nutrients*. Species *X* is adapted to a hot, dry environment. So the species would be unlikely to respond well to being watered and fertilized extensively.

(3) It is *not* true that plant species *X* could survive if the plants *can adapt to the new conditions by mating with the fungus*. It is impossible for members of different species to mate and produce fertile offspring. In addition, fungi reproduce by asexual, not sexual, means.

19. **2** This type of production of offspring is a form of *asexual reproduction*. Asexual reproduction carried on by yeast is a form of mitotic cell division known as budding. Mitotic cell division is characterized by replication (exact self-duplication) of genetic material (DNA) followed by separation of the replicated material into two daughter nuclei/cells. The resulting daughter cells are exact genetic duplicates of each other and of the parent cell. This process helps ensure genetic continuity in the yeast from one generation to the next.

WRONG CHOICES EXPLAINED

(1) This type of production of offspring is *not* a form of *sexual reproduction*. Sexual reproduction involves the fusion of a haploid egg cell and a haploid sperm cell to produce a diploid zygote. This process does not describe budding in yeast cells.

(3), (4) This type of production of offspring is *not* a form of *gene manipulation* or *genetic engineering*. These terms are essentially synonymous for laboratory techniques that involve the use of restriction enzymes to snip genes or gene fragments from the genomes of donor cells and insert the genes into the genomes of recipient cells. These processes do not describe budding in yeast cells.

20. **3** A human cell that contains all of the information necessary for the growth and development of a complete organism is *a zygote*. A zygote, or fertilized egg, is a diploid $(2n)$ cell that results from the fusion of a haploid (n) sperm cell and a haploid (n) egg cell (both gametes). The zygote contains the full set of chromosomes (46) and 100% of the genetic information that describes the entire range of characteristics for a unique human being.

WRONG CHOICES EXPLAINED

(1), (2), (4) It is *not* true that a human cell that contains all of the information necessary for the growth and development of a complete organism is *a sperm cell*, *a gamete*, or *an egg cell*. These cells contain the haploid (n) number of chromosomes (23) and only 50% of the genetic information needed to describe the range of characteristics for a unique human being.

21. **2** *During the early stages of pregnancy* is the time when exposure to a potentially harmful substance would be most likely to damage many organs in a developing embryo. Scientific study of human development seems to demonstrate that the embryo is most at risk from exposure to harmful substances (e.g., alcohol, tobacco, and drugs) in the first trimester (months 1–3) of pregnancy. However, these substances are known to cause potential damage to the reproductive process at other stages, including during the formation of gametes, at fertilization, and at the later stages of pregnancy. For that reason, these substances should be avoided.

WRONG CHOICES EXPLAINED

(1), (3), (4) *During the last three months of pregnancy*, *during the formation of the zygote*, and *during meiosis in both males and females* are *not* the times when exposure to a potentially harmful substance would be most likely to damage many organs in a developing embryo. The potential for embryonic damage from exposure to potentially harmful substances is marginally less likely at these stages of the reproductive process, although damage may still occur.

[**Note:** In the opinion of the author, this question goes well beyond the scope of the *Living Environment* curriculum. Although students should know that a growing body of scientific data indicates that exposure to certain harmful substances can be detrimental to embryonic development at almost any stage of the reproductive process, asking them to differentiate the relative danger of exposure during these various stages is not supported by the *Living Environment* Standards, Key Ideas, and Performance Indicators.]

22. **4** Structure *D* is the structure within which the placenta develops. Structure *D* represents the uterus, where the fertilized egg becomes embedded and where embryonic development occurs. The placenta is a structure composed of both maternal and embryonic tissues that supplies the embryo with dissolved food and oxygen and that removes metabolic wastes during development.

WRONG CHOICES EXPLAINED

(1) Structure *A* is *not* the structure within which the placenta develops. Structure *A* represents the oviduct/Fallopian tube, through which the egg travels after ovulation and where fertilization and early development occur.

(2) Structure *B* is *not* the structure within which the placenta develops. Structure *B* represents the ovary, where egg cells mature before being released into the oviduct.

(3) Structure *C* is *not* the structure within which the placenta develops. Structure *C* represents the vagina/birth canal, through which sperm enter the female reproductive tract and through which the baby passes during birth.

23. **4** An energy-rich organic compound needed by organisms is *glucose*. Glucose is an organic molecule (formula $C_6H_{12}O_6$) that is utilized universally by living things on Earth to supply the energy needed for metabolic activities. This energy is held primarily in the carbon-carbon bonds of the molecule and is released through the life process of respiration.

WRONG CHOICES EXPLAINED

(1), (2), (3) It is *not* true that an energy-rich organic compound needed by organisms is *water, salt,* or *oxygen*. None of these substances is organic, and none is energy rich in terms of normal metabolic activity.

24. **3** A possible symptom of SCIDS would be an increase in the *number of infections by pathogens*. Because SCIDS is known to attack T-cells and because T-cells are known to function in the immune response in humans, it logically follows that SCIDS would adversely affect the body's response to infectious agents (pathogens) such as viruses and bacteria.

WRONG CHOICES EXPLAINED

(1), (2), (4) It is *not* true that a possible symptom of SCIDS would be an increase in the *number of antigens produced, red blood cell count,* or *ability to maintain homeostasis*. No information is provided in the passage that indicates that SCIDS results in an increase in antigen production, red blood cell count, or ability to maintain homeostasis.

25. **4** If this is not done, the *immune system might attack the donated organ*. The human immune system is capable of detecting foreign antigens present in the body and producing specific antibodies designed to neutralize

and destroy those antigens and the cells that carry them. Except in cases of identical twins, donated organs/tissues always contain some foreign antigens. So these organs/tissues must be tested to determine the degree of compatibility in order to minimize the risk of rejection by the recipient's immune system.

WRONG CHOICES EXPLAINED
(1), (2) It is *not* true that if this is not done, the *donated organ might attack the body* or the *donated organ might attack the immune system*. The donated organ lacks an immune system. So it is not capable of detecting foreign antigens or of producing antibodies.

(3) It is *not* true that if this is not done, the *immune system might attack its own body cells*. Under normal circumstances, the human immune system does not attack its own body cells because the antigens present in the body are not foreign antigens.

26. **1** *There were more niches and greater stability in the ecosystem* is the set of observations that represents evidence for this claim. Biodiversity is a measure of the number of different species that inhabit an ecosystem. Biodiverse ecosystems are characterized by a large number of niches (roles) filled by many different species. Biodiverse ecosystems tend to be more stable under environmental stresses than ecosystems that lack biodiversity. More niches and greater stability both indicate that biodiversity increased in that ecosystem.

WRONG CHOICES EXPLAINED
(2) *There were more niches and less energy lost as heat in the ecosystem* is *not* the set of observations that represents evidence for this claim. More niches indicate increasing biodiversity in an ecosystem. Less energy lost as heat is an indication of less metabolic activity (fewer species) and indicates that biodiversity decreased, not increased, in that ecosystem.

(3) *There were fewer niches for decomposers and less stability* is *not* the set of observations that represents evidence for this claim. Fewer niches and less stability in an ecosystem both indicate that biodiversity decreased, not increased, in that ecosystem.

(4) *There were fewer niches for consumers and greater cycling of materials* is *not* the set of observations that represents evidence for this claim. Greater recycling of materials such as carbon and nitrogen is an indication of increasing biodiversity in an ecosystem. Fewer niches indicate that biodiversity decreased, not increased, in that ecosystem.

27. **4** The final consumers in many food webs are *carnivores*. Carnivores (meat eaters) are animals that consume the bodies of other animals for food. Carnivores are considered to be top-order consumers because they are not generally taken as food by any other animal.

WRONG CHOICES EXPLAINED

(1) It is *not* true that the final consumers in many food webs are *autotrophs*. Autotrophs (self-feeders) are producer, not consumer, organisms.

(2) It is *not* true that the final consumers in many food webs are *hosts*. Hosts are organisms that are infected by parasites. Hosts may be plants or animals. Some hosts may also be top-order consumers.

(3) It is *not* true that the final consumers in many food webs are *herbivores*. Herbivores (plant eaters) are animals that consume plants for food. Herbivores are considered to be first-order, not top-order, consumers.

28. **2** These observed changes resulted directly from *ecological succession*. Ecological succession refers to the series of changes that occur to the dominant plant community of an area that has been significantly affected by an environmental change (in this case, farm activity). Over time, several different plant communities inhabit the area, each replacing the last until a stable climax forest community is established.

WRONG CHOICES EXPLAINED

(1) These observed changes did *not* result directly from *evolutionary change*. Biological evolution is a process by which members of a species displaying favorable characteristics survive harsh environmental conditions to reproduce and pass on their favorable characteristics to their offspring. Evolutionary change is not the process described in the question.

(3) These observed changes did *not* result directly from *loss of biodiversity*. Biodiversity is a measure of the number of different species that inhabit an ecosystem. The question describes an ecosystem in which biodiversity is being gained, not lost.

(4) These observed changes did *not* result directly from *selective breeding*. Selective breeding is a technique used by animal and plant breeders in which organisms with desirable traits are cross-bred with the hope of producing offspring that also display those traits. Selective breeding is not the process described in the question.

29. **2** This would interfere most directly with the ability of the lawn ecosystem to *recycle nutrients*. Decomposers, including bacteria and fungi, occupy an ecological niche in which they break down complex organic molecules (such as cellulose) into simple sugars and inorganic compounds. In this role, decomposers play an important role in the recycling of carbon, hydrogen, oxygen, and other materials in the environment. If decomposers cannot live in an ecosystem, such recycling cannot occur effectively, if at all.

WRONG CHOICES EXPLAINED

(1) It is *not* true that this would interfere most directly with the ability of the lawn ecosystem to *recycle energy*. Energy does not recycle in a natural environment. Instead, energy in the natural environment flows in one direction

from light energy from the Sun, to chemical bond energy organic molecules that make up living things, to heat energy that is dissipated into the environment. A lawn is made up of green plants that absorb light and convert it to chemical bond energy via photosynthesis.

(3) It is *not* true that this would interfere most directly with the ability of the lawn ecosystem to *maintain atmospheric pH*. The pH (acidity) of the atmosphere is a function of sulfur and nitrogen oxides dissolved in atmospheric moisture (humidity, fog, rain, and snow). A lawn does not regulate any aspect of atmospheric pH.

(4) It is *not* true that this would interfere most directly with the ability of the lawn ecosystem to *reduce biodiversity*. Biodiversity is a measure of the number of different species that inhabit an ecosystem. By definition, a lawn is a monocrop displaying little or no biodiversity. A lawn does not regulate any aspect of biodiversity.

30. **3** One benefit of this technology is that it *reduces dependence on fossil fuels*. Geothermal technology is one of several alternate energy sources that utilize nonpolluting and renewable sources to produce heat or electrical energy for human uses. The more that these alternate energy sources can be tapped for human use, the less that fossil fuels (coal, gas, and oil) will have to be used to produce the same amount of energy.

WRONG CHOICES EXPLAINED

(1), (2), (4) It is *not* true that one benefit of this technology is that it *contributes to global warming*, *reduces the ozone shield*, or *decreases resources for many species*. Each of these answers represents a negative effect, not a benefit, of any technology on the environment. In fact, competent scientists hold the use of fossil fuels at least partially responsible for each of these negative outcomes.

PART B–1

31. **1** This type of research is often successful in advancing knowledge regarding the functioning of human cells because *cells present in humans often function in similar ways to cells present in other organisms*. Genetic research has shown that complex living things share a large majority of genes in common. As mammals, humans share an even greater percentage of genes in common with other mammals, including mice. It follows, then, that cells that derive their functions from the genes they carry function similarly in humans and other mammals. Because of this fact, research performed using mice and other mammals can provide data useful for human health.

WRONG CHOICES EXPLAINED

(2) It is *not* true that this type of research is often successful in advancing knowledge regarding the functioning of human cells because *cells from different types of organisms function differently when transplanted into humans*. Cells of different organisms cannot be transplanted into humans without triggering the human immune system, which would produce antibodies to counteract the foreign antigens present on those foreign cells.

(3) It is *not* true that this type of research is often successful in advancing knowledge regarding the functioning of human cells because *the cells in all complex organisms contain the same genes and function in similar ways*. Although a majority of genes are held in common among all complex living things, it is not true that these genes function in similar ways in all organisms.

(4) It is *not* true that this type of research is often successful in advancing knowledge regarding the functioning of human cells because *cellular research using mice can always be applied to human cells since all complex organisms produce the same proteins*. Although complex organisms and humans share many genes in common, they do not all produce the same proteins. So research involving mice cannot always be applied to humans.

32. **1** *An indicator* is the substance that would be used to prove that the gas produced by the yeast in the vacuum bottle could change the pH of the liquid in the flask. An acid-base indicator, such as phenolphthalein or bromothymol blue, in the flask would indicate a change of pH (acidity), if it occurred, by means of a visible color change.

WRONG CHOICES EXPLAINED

(2) *A chemical messenger* is *not* the substance that would be used to prove that the gas produced by the yeast in the vacuum bottle could change the pH of the liquid in the flask. Chemical messengers might include neurotransmitters such as acetylcholine. Chemical messengers cannot be used to determine the pH of a mixture.

(3) *An enzyme* is *not* the substance that would be used to prove that the gas produced by the yeast in the vacuum bottle could change the pH of the liquid in the flask. Enzymes, such as pepsin, are organic catalysts that regulate the speed of chemical reactions in the body. Enzymes cannot be used to determine the pH of a mixture.

(4) *A salt solution* is *not* the substance that would be used to prove that the gas produced by the yeast in the vacuum bottle could change the pH of the liquid in the flask. A mixture of water and a salt, such as sodium chloride, constitutes a salt solution. Salt solutions cannot be used to determine the pH of a mixture.

33. **2** *If a conclusion is valid, similar investigations by other scientists should result in the same conclusion* is the statement that best expresses a basic

scientific assumption. The repeatability of experimental results through peer review is a fundamental principle of all scientific investigations.

WRONG CHOICES EXPLAINED
(1) *Interpretation of experimental results has provided explanations for all natural phenomena* is *not* the statement that best expresses a basic scientific assumption. In fact, scientific investigations have provided explanations for only a small fraction of all natural phenomena. Much more remains to be discovered compared to what has already been learned.

(3) *For any conclusion to be valid, the design of the experiment requires that only two groups be compared* is *not* the statement that best expresses a basic scientific assumption. As long as proper scientific methodology is followed, any number of experimental groups may be compared.

(4) *After a scientist formulates a conclusion based on an experiment, no further investigation is necessary* is *not* the statement that best expresses a basic scientific assumption. The repeatability of experimental results though peer review is a fundamental principle of all scientific investigations.

34. **3** *The continued deforestation and removal of forest resources* is the phrase that describes a human activity that could have a negative effect on this cycle. The diagram describes essential aspects of the carbon cycle that occur in nature and that depend on robust interactions among plants, animals, and decomposers. A critical part of this cycle is the role of green plants (such as forest trees) in which carbon dioxide is removed from the atmosphere and incorporated (fixed) into complex carbon compounds such as cellulose. Deforestation and removal of timber resources are human activities that can seriously interfere with this process. They reduce carbon-fixation capacity and remove complex carbon compounds from the environment.

WRONG CHOICES EXPLAINED
(1), (2), (4) *A decrease in the amount of sulfates given off by motor vehicles, an increase in recycling programs for plastics and metals,* and *development of programs to conserve wildlife* are *not* the phrases that describe human activities that could have a negative effect on this cycle. Each of these phrases describes a human activity that could have a positive, not a negative, effect on this cycle and/or on environmental quality in general.

35. **4** The resistance gene was present in the aphid population as a result of *a random change in the aphids' DNA sequence.* Mutations are random events in which genetic material is altered by mutagenic agents. Mutations can be passed on to future generations only if they occur in gametic tissues that actually participate in the formation of offspring. If the phenotype of a mutation provides the individual with some adaptive advantage in a changing environment, the mutation may be passed on to future generations through the reproductive process. In this case, it is probable that a random mutation

provided an individual with a phenotype of resistance to insecticide. This gene then probably increased in frequency in the population when most nonresistant aphids were killed by the insecticide but the resistant aphids were not. The resistant aphids survived and mated with other resistant aphids to produce an insecticide-resistant population of aphids.

WRONG CHOICES EXPLAINED

(1) It is *not* true that the resistance gene was present in the aphid population as a result of *the need of the potatoes to become resistant to the insecticide*. Mutation does not occur as a function of need in any species. Insecticides are toxic to insects such as the aphid; they are not toxic to potatoes.

(2) It is *not* true that the resistance gene was present in the aphid population as a result of *changes in the aphids' local habitat by the insecticide*. Although the presence of the insecticide did change the aphids' local habitat by introducing selection pressure, this change did not create the resistance gene in the aphids.

(3) It is *not* true that the resistance gene was present in the aphid population as a result of *a recombination of the proteins in the potato cells*. Changes in the potatoes' biochemistry did not create the resistance gene in the aphids.

36. **2** *The nonresistant aphid population would likely increase* is the statement that would best predict the population in generation 4. Without the selection pressure of the insecticide, more nonresistant aphids would likely survive to mate and produce offspring that were also nonresistant. This would have the effect of increasing the proportion of nonresistant aphids in the population.

WRONG CHOICES EXPLAINED

(1) *The nonresistant aphid would become extinct* is *not* the statement that would best predict the population in generation 4. It is very likely that unless the resistant gene provides an adaptive disadvantage to some other environmental condition, it will remain in the aphid's gene pool and that insecticide resistance will continue to be a part of the phenotype of the aphid population.

(3) *The resistant aphid would mutate to a nonresistant aphid* is *not* the statement that would best predict the population in generation 4. Mutation is a random event in which a gene is altered due to the effect of a mutagenic agent. So it is extremely unlikely that the mutated gene will receive an alteration that would reverse the insecticide-resistant phenotype.

(4) *The plants would be free of insect populations* is *not* the statement that would best predict the population in generation 4. It is likely that in the absence of the insecticide, the aphids would continue to form populations on the plants.

37. **1** One *negative* consequence of using an insecticide is that it *selects for insecticide-resistant organisms*. By selecting for those individuals that carry the gene for insecticide resistance, the insecticide helps to ensure that

insecticide-resistant aphids will survive at a higher rate than nonresistant aphids. This action has the consequence that the gene for insecticide resistance will increase in frequency in the aphid gene pool, making future generations more and more resistant to the insecticide and compounding the problem of aphid infestations.

WRONG CHOICES EXPLAINED

(2) It is *not* true that one *negative* consequence of using an insecticide is that it *keeps a balance of organic compounds*. This answer has no meaning in modern science. This is a nonsense distracter.

(3) It is *not* true that one *negative* consequence of using an insecticide is that it *encourages biodiversity in plants*. The insecticide would have no known effect on the biodiversity of plants in this scenario.

(4) It is *not* true that one *negative* consequence of using an insecticide is that it *gives the nonresistant aphids a survival advantage*. In fact, the nonresistant aphids will be killed by the insecticide, providing them with a distinct disadvantage compared to resistant aphids.

38. **2** *Single-celled organisms carry out life functions that are essential for survival* is the concept that is best illustrated by the diagram. The diagram depicts a single-celled organism as it simultaneously carries out multiple life functions utilizing its specialized organelles and enzymes for this purpose. Organelles shown include the cell membrane, nucleus, endoplasmic reticulum, vacuole, and mitochondrion. Life functions illustrated include respiration, absorption, and excretion.

WRONG CHOICES EXPLAINED

(1) *The life functions performed by single-celled organisms are different from the life functions performed by complex organisms* is *not* the concept that is best illustrated by the diagram. In fact, the life functions carried out by single-celled organisms are nearly exactly the same as those carried out by complex multicellular organisms, including humans.

(3) *Since single-celled organisms lack organs, they can survive only in moist environments* is *not* the concept that is best illustrated by the diagram. Most single-celled organisms live in moist environments because they lack a means to prevent water loss in dry environments. In complex organisms, the organ known as the skin functions to reduce water loss.

[**Note:** This is a possible correct answer to the question as written, but it is not the best answer.]

(4) *Single-celled organisms contain one organelle that performs all the life functions* is *not* the concept that is best illustrated by the diagram. In fact, single-celled organisms contain several types of organelles, each type specialized to perform specific life functions on behalf of the cell.

39. **1** *Scientific explanations are built by combining evidence that can be observed with what people already know* is the statement that is the best interpretation of the material presented in the chart. The chart presents information about "early discoveries" made by scientists who had incomplete information about their observations of the natural world. Subsequently, "later discoveries" were made by other scientists that shed light on the earlier discoveries and provided more complete explanations of the observed phenomena. This is a central tenet of modern science that drives scientific knowledge forward from one generation to the next.

WRONG CHOICES EXPLAINED

(2), (4) *Inquiry involves making judgments about the reliability of the source and relevance of the information* and *hypotheses are valuable even if they turn out to not be true, because they may lead to further investigation* are *not* the statements that are the best interpretation of the material presented in the chart. These are important tasks that should be part of any scientific investigation in order to ensure freedom from bias. However, these concepts are not reflected in the information provided in the chart.

(3) *Science provides information, but values are also essential to making ethical decisions* is *not* the statement that is the best interpretation of the material presented in the chart. This is an important consideration that should be made before scientific and technological discoveries are implemented. However, this concept is not reflected in the information provided in the chart.

40. **1** Row *1* in the table correctly identifies the functions of these structures. Body system *X*, which is described as functioning to break down carbohydrates and release sugar into the blood, is the human digestive system. Organ *Y*, which is described in the table as functioning to make and release hormones *A* and *B* that regulate sugar concentration in the blood, is the pancreas, which is part of the human regulatory (endocrine) system.

WRONG CHOICES EXPLAINED

(2) Rows 2, 3, and 4 in the table do *not* correctly identify the functions of these structures. Each of these rows contains information contrary to the correct answer as explained above.

41. **2** When body system *X* releases too much sugar into the blood, the body can maintain homeostasis by making *more hormone B, only*. Hormone *B* is insulin, a hormone produced by the pancreas (organ *Y*) that regulates blood sugar concentrations by removing excess sugar from the blood. This excess sugar is stored in the liver and muscles as the complex carbohydrate glycogen.

WRONG CHOICES EXPLAINED

(1) It is *not* true that when body system *X* releases too much sugar into the blood, the body can maintain homeostasis by making *more hormone A,*

only. Hormone *A* is glucagon, a hormone produced by the pancreas (organ *Y*) that regulates blood sugar concentrations by converting glycogen to sugar and releasing it into the blood. This is the opposite reaction called for in the question.

(3) It is *not* true that when body system *X* releases too much sugar into the blood, the body can maintain homeostasis by making *more hormone A and more hormone B*. Releasing more of both hormones would result in a counteractive reaction, with glucagon (hormone *A*) increasing and insulin (hormone *B*) decreasing blood sugar concentrations simultaneously.

(4) It is *not* true that when body system *X* releases too much sugar into the blood, the body can maintain homeostasis by making *no hormone A and no hormone B*. Releasing neither of these hormones would result in an unregulated situation in which blood sugar concentrations would rise to dangerously high levels whenever the digestive process released sugar into the bloodstream.

42. **3** To restore homeostasis and compensate for the lack of hormone *B*, one useful action would be to *reduce the carbohydrates in body system X*. By limiting carbohydrate consumption, fewer complex carbohydrates would be digested into sugar and less sugar would enter the bloodstream. If properly planned and carried out, this method could control the concentration of sugar in the blood at homeostatic levels.

WRONG CHOICES EXPLAINED

(1) To restore homeostasis and compensate for the lack of hormone *B*, one useful action would *not* be to *increase the production of hormone A*. Increasing the production of hormone *A* (glucagon) would have the effect of releasing even more sugar into the blood from the liver and muscles. This would have the opposite effect of the one needed in this case and would further disrupt homeostasis.

(2) To restore homeostasis and compensate for the lack of hormone *B*, one useful action would *not* be to *remove organ Y from the body surgically*. Removing organ *Y* (pancreas) from the body would stop the production of both hormone *A* (glucagon) and hormone *B* (insulin). This would result in an unregulated situation in which blood sugar concentrations would rise and fall to dangerously high and low levels depending on the digestive process and further disrupt homeostasis.

(4) To restore homeostasis and compensate for the lack of hormone *B*, one useful action would *not* be to *reduce the synthesis of enzymes in organ Y*. Enzymes in organ *Y* (pancreas) are responsible for the regulation of all metabolic activities in this organ. Reducing enzyme production would slow or stop these metabolic activities and further disrupt homeostasis.

43. **4** If body system *X* temporarily stops releasing sugar into the blood, a likely response of the body would be to *start to increase synthesis of hormone A*. Hormone *A* (glucagon) is produced by the pancreas (organ *Y*). Glucagon

regulates blood sugar concentrations by converting glycogen to sugar and releasing it into the blood. Increased levels of hormone *A* would compensate for the lack of sugar being released from body system *X* (digestive system).

WRONG CHOICES EXPLAINED

(1) If body system *X* temporarily stops releasing sugar into the blood, a likely response of the body would *not* be to *stop using enzymes in body system X*. Body system *X* (digestive system) depends on the production of enzymes to carry out its digestive function. Stopping enzyme production in this system would completely block the digestive process.

(2) If body system *X* temporarily stops releasing sugar into the blood, a likely response of the body would *not* be to *stop organ Y from producing hormone A*. Stopping organ *Y* (pancreas) from producing hormone *A* (glucagon) would cause sugar to be retained in the liver instead of being released into the blood. This would have the opposite effect of the one needed in this case.

(3) If body system *X* temporarily stops releasing sugar into the blood, a likely response of the body would *not* be to *start to increase synthesis of hormone B*. Increasing production of hormone *B* (insulin) would have the effect of storing even more sugar as glycogen and would further reduce blood sugar levels. This would have the opposite effect of the one needed in this case.

PART B–2

44. One credit is allowed for correctly marking an appropriate scale, without any breaks in the data, on each labeled axis. [1]

45. One credit is allowed for correctly plotting the data on the grid, connecting the points, and surrounding each point with a small circle. [1]

Height of Liquid in Thistle Tube

46. One credit is allowed for correctly explaining why the height of the solution in the thistle tube increased during the 25-minute period. Acceptable responses include but are not limited to: [1]

- *Water diffused into the thistle tube.*
- *Water moved from an area of high concentration to an area of low concentration.*
- *Osmosis took place.*
- *Water diffused through the membrane.*
- *Water molecules migrated through the membrane with the concentration gradient.*

47. **3** The most likely reason for this observation is that the *amber-colored solution moved into the thistle tube.* First, we must assume that the amber-colored liquid is iodine (Lugol's) solution, an indicator that is known to react with starch molecules to produce a distinct blue-black coloration. The molecules that make up Lugol's solution are small and can readily pass through the pores of a semipermeable membrane from the beaker to the thistle tube. Starch molecules are too large to pass through this same semipermeable membrane and so remain inside the thistle tube. These facts explain why the contents of the beaker remain amber-colored while the contents of the thistle tube turn blue-black.

WRONG CHOICES EXPLAINED

(1) The most likely reason for this observation is *not* that *starch molecules moved out of the thistle tube.* Starch molecules are too large to pass through this same semipermeable membrane and so remain inside the thistle tube.

(2), (4) The most likely reason for this observation is *not* that *water molecules moved into the thistle tube* or that *water molecules moved out of the thistle tube.* Water molecules can readily pass in either direction through a semipermeable membrane, but its net movement is always with the concentration gradient. In this experiment, water serves only as a suspension medium and solvent, not as a reactant.

48. One credit is allowed for correctly explaining why the chemical released from the injured fish may *not* cause an alarm response in other fish species. Acceptable responses include but are not limited to: [1]

- *Other species lack the receptors for the chemical.*
- *The alarm chemicals are specific to the species.*
- *Other species lack the ability to sense/recognize the chemical.*
- *Animal behaviors are genetic traits that are unique to the species displaying them.*

49. **4** The process represented in the diagram is *genetic engineering.* Genetic engineering is a laboratory technique in which a gene for a desired trait is snipped from the DNA of a donor cell and inserted into the genome of a recipient cell. This is the process illustrated in the diagram.

WRONG CHOICES EXPLAINED

(1) It is *not* true that the process represented in the diagram is *DNA replication.* Replication is the exact self-duplication of genetic material (DNA) in the cell prior to cell division. DNA replication is not illustrated in the diagram.

(2) It is *not* true that the process represented in the diagram is *natural selection.* Natural selection, sometimes referred to as survival of the fittest, postulates that certain members of a species are better adapted to their environment than others. Natural selection is not illustrated in the diagram.

(3) It is *not* true that the process represented in the diagram is *gel electrophoresis.* In this laboratory technique, electrically charged fragments of DNA are placed into a gel and then drawn through the gel by an electric current until they separate based on size. Gel electrophoresis is not illustrated in the diagram.

50. **1** The original gene for the production of a human hormone was most likely removed from a *chromosome.* Chromosomes are cell organelles that contain DNA molecules that carry the genetic codes needed for the production of specific proteins in the cell. The snipped gene for a human hormone was taken from a human chromosome.

WRONG CHOICES EXPLAINED

(2), (3), (4) The original gene for the production of a human hormone was *not* most likely removed from a *ribosome,* a *mitochondrion,* or a *cell membrane.* None of these organelles contains a gene for the production of a human hormone. The mitochondrion contains DNA that codes for the production of enzymes involved in respiration.

51. One credit is allowed for correctly stating *one* possible reason why a gene for the production of a human hormone would be placed in bacterial DNA. Acceptable responses include but are not limited to: [1]

- *The bacteria will produce the human hormone quickly.*
- *This is done in order to increase the production of insulin or some other hormone.*
- *The human hormone produced can be given to people who need it, and the number of allergic reactions will be reduced.*
- *By inserting the gene, the bacteria can be made to produce the human hormone.*

52. One credit is allowed for correctly identifying *one* population that will *decrease* in size after the lake sturgeon are added to the new ecosystem and for supporting the answer. Acceptable responses include but are not limited to: [1]

- *Populations of insect larvae/clams/worms will decrease in size because they will be eaten by both lake sturgeon and bloater fish.*
- *The bloater fish may temporarily decrease in number due to competition with the newly introduced lake sturgeon.*
- *If the lake sturgeon are more successful competitors, then they will gradually eliminate bloater fish from their habitat.*
- *If the bloater fish are more successful in the environment, then at some time after its reintroduction, the lake sturgeon may decrease in number due to competition with the bloater fish.*

53. One credit is allowed for correctly indicating the position in the food web where each organism listed would be placed and for writing the name of each in the appropriate box. Acceptable responses include: [1]

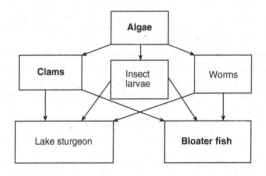

54. One credit is allowed for correctly identifying *one* population, other than lake sturgeon, that will increase in size after the lake sturgeon are added to the new ecosystems and for supporting the answer. Acceptable responses include but are not limited to: [1]

- *Algae, because there will be fewer organisms that feed on them.*
- *More algae will survive and reproduce more because populations of insect larvae, clams, and worms that eat algae will decrease in size because they will be eaten by both lake sturgeon and bloater fish.*

55. One credit is allowed for correctly stating what the arrows in the food web represent. Acceptable responses include but are not limited to: [1]

- *The arrows show the direction of energy movement from one organism to another.*
- *The arrows show the flow of food energy through the ecosystem.*
- *The arrows indicate the movement of nutrient energy through the food web.*
- *Each arrow represents what each organism is eaten by or what each organism it eats.*

PART C

56. One credit is allowed for correctly explaining why researchers would put the salamander in an artificial environment, as opposed to conducting the experiment in their natural habitat. Acceptable responses include but are not limited to: [1]

- *It would be too difficult to track the movements of individual salamanders in the natural habitat.*
- *The salamanders are easier to watch in a laboratory setting.*
- *The conditions the salamanders are kept in can be better controlled.*
- *There are fewer uncontrolled variables in a laboratory setup.*
- *Conducting the research in the laboratory protects the natural habitat from disruption.*
- *In the lab, scientists can exclude predators and other adverse conditions from affecting the experiment.*

57. One credit is allowed for correctly explaining why other scientists would question the validity of the conclusions based on this setup. Acceptable responses include but are not limited to: [1]

- *The sample size is too small to provide adequate data.*
- *In such a small experimental group, it would be easy to have some abnormal salamanders.*
- *No hypothesis is stated and no experimental method is outlined.*
- *The artificial environment lacks many factors present in the natural habitat.*
- *The lab environments might not accurately duplicate the natural habitats.*
- *No statement is made concerning what is being measured and how data will be collected and analyzed.*
- *It is not clear on what basis experimental inferences/conclusions will be made.*

58. One credit is allowed for correctly stating *one* reason why it may be dangerous to introduce a new species to an established ecosystem. Acceptable responses include but are not limited to: [1]

- *New species might compete with native species for limited resources.*
- *Native species can have their space, light, and water taken away by invasive plants.*
- *New plant species could have nonnative insects/pathogens on them.*
- *Invasive plants can crowd out native plants, which also eliminate animals that use the native plants for food/habitat.*

59. One credit is allowed for correctly stating *one* reason why a green roof reduces the amount of carbon dioxide in the atmosphere. Acceptable responses include but are not limited to: [1]

- *Green plants take in carbon dioxide to perform photosynthesis.*
- *Plants use CO_2 to produce food.*
- *Plants take in carbon dioxide.*
- *Green roofs increase the area of photosynthesizing plants in an urban area that doesn't have other green spaces.*

60. One credit is allowed for correctly stating *one* reason why it is important to reduce the amount of carbon dioxide in the atmosphere. Acceptable responses include but are not limited to: [1]

- *Carbon dioxide in the atmosphere adds to the greenhouse effect.*
- *Excessive CO_2 in the air traps solar heat and raises the temperature of Earth.*
- *Carbon dioxide has been associated with global warming/global climate change.*
- *High concentrations of CO_2 in the atmosphere have been linked to melting of the polar ice caps and rising sea levels.*

61. One credit is allowed for correctly stating *one disadvantage* of a green roof. Acceptable responses include but are not limited to: [1]

- *Green roofs are expensive to install.*
- *Green roofs need more maintenance than traditional roofs.*
- *Such roofs might be more susceptible to leakage and rot.*
- *Green roofs might be too heavy for the building structure to support.*
- *The damp absorptive layer might attract burrowing rodents or insects that could damage the roof.*

62. One credit is allowed for correctly stating *one* hypothesis the experiment would test. Acceptable responses include but are not limited to: [1]

- *Bacteria exposed to x-ray radiation will show a greater number of mutations than bacteria exposed to natural light.*
- *Bacteria exposed to x-rays will show less DNA damage than bacteria exposed to natural light.*
- *If bacteria are exposed to x-rays, they will show more mutations than bacteria not exposed to x-rays.*
- *The DNA of bacteria exposed to x-rays will change more than that of bacteria not exposed to x-rays.*
- *X-ray exposure will affect bacterial DNA by increasing the frequency of mutation.*

63. One credit is allowed for correctly identifying *one* specific difference, other than it is a two-step vaccination, between Nabel's vaccination and a traditional flu vaccine. Acceptable responses include but are not limited to: [1]

- *Nabel's vaccination protects from every known influenza A virus.*
- *The new NIH vaccine directly injects viral DNA.*
- *This new approach may add years of protection compared to traditional vaccines.*
- *It makes muscle cells produce hemagglutinin, which triggers the immune response against flu.*
- *It causes the body to make a protein found on the surface of all flu viruses.*

64. One credit is allowed for correctly explaining how injecting dead or weakened viruses into a person can help fight against future infections from that virus. Acceptable responses include but are not limited to: [1]

- *The body responds by making antibodies that protect against future infections.*
- *There are antigens on the surface of the dead virus, which then trigger the body to produce antibodies specific to that antigen.*
- *The body makes memory cells that can fight the virus in the future.*

65. One credit is allowed for correctly stating *one* reason why the percent ice cover is important to the ability of eagles to obtain food. Acceptable responses include but are not limited to: [1]

- *The access to fish would be limited by the amount of ice cover.*
- *Eagles would have to search for food other than fish.*
- *The ice cover might limit the number of organisms that are available for the eagles to eat.*
- *During periods of ice cover, fish tend to stay near the bottom of the river where the temperature is stable at 4°C, so eagles must find other food sources.*

[**Note:** Because no data are provided concerning percent of ice cover and other environmental factors affecting eagle survival, this question can be answered only in the form of a hypothesis.]

66. One credit is allowed for correctly stating what inference can be made about the percent ice cover at Croton Reservoir between January and February 2009 and for supporting the answer. Acceptable responses include but are not limited to: [1]

- *The ice cover was probably less in February because there were more eagles sighted.*
- *Since fewer eagles were reported in January, there might have been more ice on the reservoir.*

[**Note:** Because no data are provided concerning percent of ice cover and other environmental factors affecting eagle survival, this question can be answered only in the form of a hypothesis.]

67. One credit is allowed for correctly stating *one* reason why the number of eagles sighted showed a change at all four sites between February and March. Acceptable responses include but are not limited to: [1]

- *Winter was ending, so the eagles returned to their summer nesting areas.*
- *As winter ended, the eagles migrated away.*
- *The warming weather and longer days stimulated the eagles to begin their annual migration north.*
- *The eagles are free flying and so have left their winter roosts for their summer nesting areas.*

[**Note:** No data are provided concerning percent of ice cover, late-winter storms, and other environmental factors affecting eagle migration. So responses including these factors are speculative and should not be acceptable. In addition, the data on number of eagles in all months at all sites are averaged. So the smaller number of visits in March becomes less important as a factor affecting the data.]

68. One credit is allowed for correctly stating *one* possible reason why a popular hiking trail in this area is closed during eagle roosting seasons. Acceptable responses include but are not limited to: [1]

- *Winter roosts are protected under federal law and managed with a buffer zone to reduce human interference.*
- *Humans might disturb the eagles.*
- *People using the hiking trails in these areas might make noise, leave trash, or otherwise disrupt the roosting behaviors of the eagles.*

69. One credit is allowed for correctly stating the role of the sea grasses in the reef ecosystem. Acceptable responses include but are not limited to: [1]

- *Their role is to convert inorganic materials into organic matter/food.*
- *They capture radiant energy.*
- *They serve as the producer in this food web/ecosystem.*
- *They serve as food for fish.*
- *They produce oxygen/absorb carbon dioxide.*
- *They disrupt the coral-dominated ecosystem.*
- *They change the coral reef food web by eliminating seaweed-eating fish.*
- *They reduce the amount of sunlight reaching the sea floor near the reefs.*

70. One credit is allowed for correctly identifying *one* abiotic factor that is affecting the stability of the coral reef ecosystems and for stating how the factor identified is important to the coral reef ecosystems. Acceptable responses include but are not limited to: [1]

- *Temperature—causes change in ocean currents*
- *Water temperature—can affect the survival of coral polyps*
- *Light—used by plants for photosynthesis*
- *Ocean currents—carry organisms from one reef to another*
- *Carbon dioxide—used by plants for photosynthesis*
- *Dissolved oxygen—used by all organisms for respiration*
- *Salinity—provides proper salt/water concentration for the survival of marine life*
- *Acidity—can affect the efficiency of enzyme-catalyzed reactions/survival of reef organisms*

71. One credit is allowed for correctly stating *one* reason why it is important to maintain the stability of coral reefs. Acceptable responses include but are not limited to: [1]

- *Reefs should be protected in order to maintain the food webs that exist there.*
- *This is important because many organisms depend on them for food/ shelter/nesting sites.*
- *By keeping coral reefs healthy, we contribute to biodiversity in the marine environment.*
- *Healthy reefs contribute to healthy oceans and to the fish we depend on for food.*
- *Maintaining healthy coral reefs helps to prevent the extinction of reef species.*
- *Reef ecosystems are indicators of the overall health of the global ecosystem that we depend on for human survival.*

72. One credit is allowed for correctly stating *one* advantage of the fish larvae moving by ocean currents into a damaged reef ecosystem. Acceptable responses include but are not limited to: [1]

- *The fish larvae could repair/repopulate the damaged reefs.*
- *The reef might become more stable.*
- *They might keep the seaweed under control/eat the seaweed.*
- *The reef would provide the fish larvae with food/shelter.*
- *The fish larvae would have less competition/more resources.*

PART D

73. **4** The process shown in the diagram is *protein synthesis*. The diagram illustrates a molecule of messenger RNA (mRNA) attached to Structure *X* (a ribosome). Smaller bell-shaped molecules of transfer RNA (tRNA) carrying specific amino acids (glycine, arginine, proline, alanine) are shown linking to the mRNA codon in a predetermined sequence to form a chain of amino acids ("newly formed molecule"), also known as a protein.

WRONG CHOICES EXPLAINED

(1) It is *not* true that the process shown in the diagram is *cellular respiration*. An illustration of cellular respiration would show molecules of glucose and oxygen reacting to produce water, carbon dioxide, and cellular energy as ATP.

(2) It is *not* true that the process shown in the diagram is *cellular reorganization*. An illustration of cellular reorganization would show a cell about to divide as it gathers its chromosomes and morphs its organelles prior to spindle formation and mitosis.

(3) It is *not* true that the process shown in the diagram is *gene recombination*. An illustration of gene recombination would show the reuniting of segregated chromosome pairs during fertilization, a process that brings alleles together in a new genetic combination.

74. **4** Structure *X* is a *ribosome*. The ribosome is a cell organelle that functions as the site for protein synthesis in the cell.

WRONG CHOICES EXPLAINED

(1) Structure *X* is *not* a *mitochondrion*. The mitochondrion is a cell organelle that functions as the site for aerobic respiration in the cell.

(2) Structure *X* is *not* a *vacuole*. The vacuole is a cell organelle that functions as the site for food and/or waste storage in the cell.

(3) Structure *X* is *not* a *nucleus*. The nucleus is a cell organelle that functions as the site for genetic regulation in the cell.

75. **1** A finch that picks small insects out from cracks in the bark of trees would most likely have a beak that is *sharp and thin*. An examination of the diagram reveals that finches eating all or mainly animal food (including insects) tend to have more pointed bills than finches that eat mainly plant food. In addition, insect-eating finches whose bills are described as "probing" (warbler finch and woodpecker finch) have the sharpest and thinnest bills of those illustrated.

WRONG CHOICES EXPLAINED
(2), (3), (4) A finch that picks small insects out from cracks in the bark of trees would *not* most likely have a beak that is *sharp and thick*, *rounded and thin*, or *rounded and thick*. Each of these combinations of characteristics contains at least one characteristic (i.e., thick, rounded) that would not be efficient at probing for insects in the cracks in the bark of trees.

76. **1** *The type of beak indicates the type of food the finch eats* is the statement that represents a basic assumption from *The Beaks of Finches* lab. In this lab exercise, students are asked to pick up small objects using different tools with different shapes and gripping surfaces and then record data concerning their observations. The lab is designed to help students understand the mechanisms (natural selection) that favored some variations (beak shapes) over others in the evolution of the Galapagos finches.

WRONG CHOICES EXPLAINED
(2), (3), (4) *Different birds have different songs*, *birds with larger beaks can find mates more easily*, and *nesting behavior of finches is an inherited trait* are *not* the statements that represent a basic assumption from *The Beaks of Finches* lab. None of these statements relates its concept to the actual activities of this lab. Exploring these concepts would require laboratory investigations using different tools and techniques than those used in *The Beaks of Finches* lab.

77. One credit is allowed for correctly stating *two* reasons why the large ground finch and sharp-billed ground finch could live on the same island but *not* compete for food, even though they both eat mainly plant food. Acceptable responses include but are not limited to: [1]

- *They have different beak structures.*
- *They might eat different types of plants.*
- *They might eat at different times of day or night.*
- *They might live in different areas of the island.*
- *They might feed at different elevations on the same island.*
- *They are adapted to eat different kinds of seeds.*
- *They would compete only if they both required the same limited food resources.*

78. One credit is allowed for correctly writing the number "2" to designate the tree diagram that shows the most probable evolutionary relationships among the three species and for supporting the answer. Acceptable responses include but are not limited to: [1]

- *Tree 2—Species B has enzyme Y, which is not present in either species A or C.*
- *2—Enzymes W and X are present in species A and C but not in B.*
- *2—A and C have enzymes W, X, and Z, but B doesn't.*
- *Tree 2—A and C have three enzymes in common: W, X, and Z.*
- *2—A and C have more enzymes in common.*

79. One credit is allowed for correctly identifying substance A. Acceptable responses include: [1]

- *Water*
- *H_2O*
- *H-O-H*

80. One credit is allowed for correctly identifying *one* likely substance in the environment of the cell that caused this response. Acceptable responses include but are not limited to: [1]

- *Salt*
- *Sugar*
- *Seawater*
- *Sodium chloride (or any other soluble salt)*

81. **3** This muscle fatigue is most likely due to *lack of oxygen and build up of waste in the muscles.* During periods of strenuous exercise, muscle cells quickly use up available dissolved oxygen required for aerobic respiration. As a result, the cells resort instead to a less-efficient form of anaerobic respiration known as lactic acid fermentation or glycolysis. This process results in the production of a small amount of ATP energy and the metabolic waste lactic acid. Lactic acid stimulates a pain response in the muscles where it forms. This combination of factors is what we describe as muscle fatigue.

WRONG CHOICES EXPLAINED

(1) This muscle fatigue is *not* most likely due to *the heart beating too fast and tiring out.* Although an increased heart rate is typical during exercise, it is not the cause of muscle fatigue.

(2) This muscle fatigue is *not* most likely due to *the lungs accumulating oxygen.* The lungs function to absorb oxygen from the air and transfer the oxygen to the blood. This activity is not the cause of muscle fatigue.

(4) This muscle fatigue is *not* most likely due to *lack of carbon dioxide in the muscles*. Carbon dioxide is a waste product of aerobic respiration but not of lactic acid fermentation. Carbon dioxide is not the cause of muscle fatigue.

82. **1** Methods *A* and *B* are classified as methods of passive transport because they do not require *ATP*. In the illustrations of both of these methods, a higher concentration of molecules is shown outside the cell compared to inside the cell. In these cases, the molecules move across the membrane by the process of diffusion. Diffusion does not require the expenditure of ATP energy in order to occur.

WRONG CHOICES EXPLAINED

(2), (3), (4) It is *not* true that methods *A* and *B* are classified as methods of passive transport because they do not require *carbon dioxide, light,* or *DNA*. None of these factors has any direct effect on the diffusion of molecules from a region of higher relative concentration to an area of lower relative concentration of those molecules.

83. One credit is allowed for correctly stating *one* reason why the movement of molecules in method *C* represents active transport. Acceptable responses include but are not limited to: [1]

- *Method C shows the transport of molecules from an area of lower concentration to an area of higher concentration.*
- *The molecules are moving against the concentration gradient.*

84. One credit is allowed for correctly stating *one* reason why some species might have similar body structures even if they are *not* closely related. Acceptable responses include but are not limited to: [1]

- *They may have evolved in the same or similar environments, and these structures and traits were advantageous adaptations.*
- *They have two separate mutations that produce the same appearance.*
- *These structures might help them use similar food/adapt to their environment.*
- *The species inhabit similar environments.*
- *They occupy similar niches.*
- *Natural selection may favor the body structures for survival in a particular environment.*

85. One credit is allowed for correctly drawing a line that shows what most likely happened to the pulse rate of the student during these activities. Acceptable responses include but are not limited to: [1]

Effect of Activity on Pulse Rate

Note: Allow credit for a graph that shows a gradual increase followed by a gradual decrease.

STANDARDS/KEY IDEAS	AUGUST 2016 QUESTION NUMBERS	NUMBER OF CORRECT RESPONSES
STANDARD 1		
Key Idea 1: The central purpose of scientific inquiry is to develop explanations of natural phenomena in a continuing and creative process.	39, 47	
Key Idea 2: Beyond the use of reasoning and consensus, scientific inquiry involves the testing of proposed explanations involving the use of conventional techniques and procedures and usually requiring considerable ingenuity.	56, 57, 62	
Key Idea 3: The observations made while testing proposed explanations, when analyzed using conventional and invented methods, provide new insights into natural phenomena.	33, 36, 38, 44, 45	
Laboratory Checklist	32	
STANDARD 4		
Key Idea 1: Living things are both similar to and different from each other and from nonliving things.	1, 14, 31, 40, 46, 48, 52, 53, 54, 55, 69	
Key Idea 2: Organisms inherit genetic information in a variety of ways that result in continuity of structure and function between parents and offspring.	4, 5, 8, 11, 35, 49, 50, 51	
Key Idea 3: Individual organisms and species change over time.	9, 13, 15, 16, 17, 18, 37	
Key Idea 4: The continuity of life is sustained through reproduction and development.	12, 19, 20, 21, 22	
Key Idea 5: Organisms maintain a dynamic equilibrium that sustains life.	2, 7, 10, 23, 24, 25, 41, 42, 43, 63, 64	
Key Idea 6: Plants and animals depend on each other and their physical environment.	26, 27, 28, 65, 66, 67, 70, 71, 72	
Key Idea 7: Human decisions and activities have a profound impact on the physical and living environment.	3, 6, 29, 30, 34, 58, 59, 60, 61, 68	

STANDARDS/KEY IDEAS	AUGUST 2016 QUESTION NUMBERS	NUMBER OF CORRECT RESPONSES
REQUIRED LABORATORIES		
Lab 1: "Relationships and Biodiversity"	73, 74, 78, 84	
Lab 2: "Making Connections"	81, 85	
Lab 3: "The Beaks of Finches"	75, 76, 77	
Lab 5: "Diffusion Through a Membrane"	79, 80, 82, 83	